FUNDAMENTALS OF BUILDING CONSTRUCTION

FUNDAMENTALS OF BUILDING CONSTRUCTION
MATERIALS AND METHODS

—

EDWARD ALLEN

DRAWINGS BY
JOSEPH IANO

—

JOHN WILEY & SONS

NEW YORK   CHICHESTER   BRISBANE   TORONTO   SINGAPORE

Cover and Text Design: Karin Gerdes Kincheloe

Cover Photograph: Mona Zamdmer

Cover: One Dallas Centre, a 30-floor office
building by I. M. Pei & Partners; Henry Cobb,
FAIA partner in charge of design

Drawings: Joseph Iano

Production Supervisor: Janice Weisner

Copy Editor: Tamara Lee

Library of Congress Cataloging in Publication Data:

Allen, Edward, 1938-
    Fundamentals of building construction.

    Includes bibliographies and index.
    1. Building.   2. Building materials.
I. Iano, Joseph. II. Title.

TH145.A417   1985   690   85-12221
ISBN 0-471-79976-9

Printed in the United States of America

10

PREFACE

This book is an introduction to the art of building. It begins by describing the simple but wondrously rich materials of the ancients—wood, stone, and brick—and the techniques by which they are made into buildings today, before proceeding to structural steel, reinforced and prestressed concrete, float glass, extruded aluminum, advanced gypsum products, synthetic rubber compounds, and plastics, the miraculous materials of construction developed over the past two centuries. Insofar as possible it deals with whole systems of building. The structural systems are those widely used today—heavy timber framing, wood platform framing, masonry loadbearing wall, structural steel framing, and concrete framing systems. The nonstructural systems are those that provide enclosure and interior finishes. Electrical and mechanical systems, which are treated in detail in other books, are considered here only to the extent that they interact with the fabric of a building.

The discussion of each building system begins with a brief summary of its historical development and continues with a review of how the major material is obtained and processed, an outline of the people and organizations who work with the material, a description of their tools and working methods, and a discussion of the role of the system in relation to alternative systems. The intent of this organization is to help the reader understand the evolutionary development of the system, the properties of its major materials, the possibilities and limitations of the building method, and the basis for choosing among systems.

At the end of each chapter are a group of features designed to assist the student in reviewing, remembering, and learning to use the essential material. A listing of the key terms and concepts introduced in the chapter will be helpful in mastering the extensive technical terminology of the building field. The review questions relate to the most important ideas contained in the chapter, and the exercises encourage the application of the lessons of the chapter to real-life situations. Several carefully selected references point the way toward further reading for those who wish to delve more deeply into the topic.

The book contains three hundred drawings and nearly seven hundred photographs, a thousand illustrations in all. Each chapter includes three types of illustrations: line drawings to show in detail the major features of building assemblies, photographs of manufacturing and assembly operations to relate the drawings to everyday building practice, and photographs of finished buildings to exemplify the use of the materials and techniques described in the chapter. Key progressions of construction operations are illustrated with sequential drawings or photographs wherever possible.

At the back of the book there is a brief appendix containing tables of physical quantities that the student will find useful while studying any of the chapters: densities and thermal expansion coefficients of common construction materials, and metric and English units of conversion. This is followed by a glossary that defines the approximately nine hundred technical terms introduced in this book.

Several aids are available from the publisher to the teacher who uses this as a textbook, including a set of black-and-white 35-mm slides excerpted from the illustrations in the book, and a teacher's manual that incorporates a script to accompany the slides. The teacher's manual also includes extensive suggestions on lecture topics and schedules, and further questions and exercises for use in homework problems, examinations, and the laboratory.

The student will soon recognize that full expertise in the materials and methods of construction can best be acquired through practice—in homework exercises, in the design studio, in the professional office, in the field. One must use the information in order to learn it, extend it, and shape it into a working methodology. This is all the more important because there are qualities of construction materials that cannot be conveyed adequately in graphic form: their colors, their fragrances, the sounds they make when struck or rubbed, the way they feel beneath the hand, the ways light plays off them, the ways they interact with tools, the manner in which they change with age. And there is a potential in construction methods that can only be fully realized, whether aesthetically or technically, by one who has observed firsthand, or better yet, labored beside, skilled construction artisans over many a season.

ACKNOWLEDGMENTS

This book has many coauthors. Joseph Iano, while preparing his admirable drawings, shaped the book in important ways through his ideas, criticism, and enthusiastic sharing of his extensive experience in construction. Clifford Boehmer, architect and craftsman, lent an expert hand with the inking of drawings and furnished valuable advice on many construction details. Carol Davidson, with efficiency, skill, and good humor, accomplished much of the heavy task of obtaining and organizing the photographs, edited the manuscript in a way that greatly enhanced its clarity and readability, and prepared the index. Mary M. Allen, as she has done so many times before, willingly and unstintingly gave every kind of support required to get the job done.

Merrill Smith, Margaret DePopolo, and Stewart Roberts gave special assistance in obtaining particular illustrations from outside sources. For the remainder of the illustrations I am indebted to several hundred representatives of the building industry in North America and Europe, the names of whose firms appear in the credit lines that accompany the figures.

A well-coordinated team of exceptionally patient, skilled, and creative professionals at John Wiley & Sons undertook the daunting task of converting a mountain of rough manuscript and pictures into a finished book. Judith R. Joseph, the acquiring editor, deftly guided the preparation of the book from earliest idea to bound volumes, with assistance from Cindy Zigmund. Ishaya Monokoff helped develop the graphic guidelines for the drawings. Elizabeth Doble ably coordinated the editorial and production functions. Tamara Lee and Richard Christopher edited the manuscript, and Lilly Kaufman and Janice Weisner managed the production process. Karin Kincheloe applied her exceptional talent to the design of the book and worked tirelessly to lay out its pages one by one. To all these people, and to those who worked still further in the background in the editing and production of the book, I express my gratitude for making a protracted and potentially difficult process into a delightful and uniformly pleasurable collaboration.

A number of teachers at colleges and universities read portions of the manuscript and offered advice that was instrumental in sharpening its focus. These include Charles C. Benton, Georgia Institute of Technology; Robert J. Bradley, Delaware Technical and Community College; Stanley W. Crawley, the University of Utah; R. George Cunningham, P.E., Rice University; Jonathan Faulkner, Columbus Technical Institute; Alan Levy, University of Pennsylvania; Enrique Limosner, San Francisco City College; Fred Ludwig, San Antonio College; William J. Morrow, Hillsborough Community College; D. F. Regan, State University of New York Agricultural & Technical College at Delhi; John S. Reynolds, University of Oregon; G. Fred Sheets, Jr., California State Polytechnic University; J. Mark Taylor, Auburn University; and James E. White, A.I.A., Texas Tech University.

I am especially indebted to the specialists who reviewed the various chapters for technical accuracy. Chapter 2 could not have been written without the expert assistance of Dr. Francisco Silva-Tulla. Chapter 8 was considerably improved in response to specific suggestions by Robert Beiner of the International Masonry Institute. David Ricker of the Berlin Steel Construction Company and Emile Troup, Regional Engineer of the American Institute of Steel Construction, contributed in many valuable ways to the preparation and revision of Chapter 9. Chapters 10 and 11 were reviewed by David Gustafson and Anthony Feldner of the Concrete Reinforcing Steel Institute. Alvin Ericson of the Prestressed Concrete Institute gave graciously of his time and talent to help assemble the material for Chapter 12, and also acted as its reviewer. Robert A. LaCosse of the National Roofing Contractors Association read Chapter 13 and offered many helpful suggestions. Chapter 14 was reviewed by Arnold J. Thimons of PPG Industries, and Chapter 15 by John Gurniak, on behalf of the Architectural Aluminum Manufacturers Association; both made numerous recommendations that were incorporated into the manuscript. Martin Cook of the United States Gypsum Company contributed his expertise in reviewing Chapters 16, 17, and 18. Any errors and misconceptions that have found their way into the book did so in spite of the conscientious and highly professional work of these people, and remain my responsibility alone.

Finally I wish to thank Professor Albert G. H. Dietz, scientist, teacher, author, expert carpenter, longtime colleague, and friend, for his invaluable work in reviewing Chapters 3 through 7. It is to Al Dietz and his remarkable wife Ruth that I dedicate this volume.

E.A.

South Natick, Massachusetts
May 1985

DISCLAIMER

The drawings, tables, descriptions, and photographs in this book have been obtained from many sources, including trade associations, suppliers of building materials, governmental organizations, and architectural firms. They are presented in good faith, but the author, illustrator, and publisher, while they have made every reasonable effort to make this book accurate and authoritative, do not warrant, and assume no liability for, its accuracy or completeness or its fitness for any particular purpose. It is the responsibility of users to apply their professional knowledge in the use of information contained in this book, to consult the original sources for additional information when appropriate, and to seek expert advice when appropriate.

CONTENTS

—

FUNDAMENTALS OF BUILDING CONSTRUCTION

MAKING BUILDINGS

An ironworker connects a steel wide-flange beam to a column. *(Courtesy of Bethlehem Steel Company)*

We build because little that we do can take place outdoors. We need shelter from sun, wind, rain, and snow. We need dry, level platforms for our activities. Often we need to stack these platforms to multiply available groundspace. On these platforms, and within our shelter, we need air that is warmer or cooler, more or less humid, than outdoors. We need less light by day, and more by night, than is offered by the natural world. We need services that provide energy, communications, and water, and dispose of wastes. So we gather materials and assemble them into the constructions we call buildings in an attempt to satisfy these needs.

DESIGNING BUILDINGS

A building begins as an idea in someone's mind, a desire for new and ample accommodations for a family, many families, an organization, or an enterprise. For any but the smallest of buildings, the next step for the owner of the prospective building is to engage, either directly or through a hired construction manager, the services of building design professionals. An architect helps to consolidate the owner's ideas about the new building and assembles a group of engineering specialists to help work out concepts and details of foundations, structural support, and mechanical, electrical, and communications services. This team of designers, working with the owner, then develops a scheme for the building in progressively finer degrees of detail. Drawings and written specifications are produced by the architect-engineer design team to document how the building is to be made and of what. A general contractor is selected, either by negotiation or by competitive bidding, and it is the general contractor who hires subcontractors to carry out many specialized portions of the work. The drawings and specifications are submitted to the municipal inspector of buildings, who checks them for conformance with zoning ordinances and building codes before issuing a permit to build. Construction may then begin, with the building inspector, the architect, and the engineering consultants inspecting the work at frequent intervals to be sure it is carried out according to plan.

CHOOSING BUILDING SYSTEMS: CONSTRAINTS

Although a building begins as an abstraction, it is built in a world of material realities. The designers of a building—the architects and engineers—work constantly from a knowledge of what is possible and what is not. They are able, on the one hand, to employ any of a limitless palette of building materials and any of a number of structural systems to produce a building of almost any desired form and texture. On the other hand, they are inescapably bound by certain physical limitations: how much land there is with which to work; how heavy a building the soil can support; how long a structural span is feasible; what sorts of materials will perform well in the given environment; and so on. They are also constrained by a construction budget and by a complex web of legal restrictions. Those who work in the building professions need a broad understanding of many things, including people, climate, the physical principles by which buildings work, the technologies available for utilization in buildings, the legal restrictions on building, and the contractual arrangements under which buildings are built.

This book is concerned primarily with the technologies of construction materials—what the materials are, how they are produced, what their properties are, and how they are crafted into buildings. But these must be studied with reference to many other factors that bear on the design of buildings, some of which require explanation here.

Zoning Ordinances

The legal restrictions on building begin with local *zoning ordinances*, which govern such matters as what types of activities may take place on a given piece of land, how much of the land may be covered by the building or buildings, how far buildings must be set back from each of the property lines, how many parking spaces must be provided, how large a total floor area may be constructed, and how tall the building may be. Copies of the zoning ordinance for a municipality are available for purchase or reference at the office of the building inspector or the planning department, or they may consulted at public libraries.

Building Codes

In addition to its zoning ordinance, each local government also regulates building activity by means of a *building code*. The intent of a building code is to pro-

Table 501
HEIGHT AND AREA LIMITATIONS OF BUILDINGS
Height limitations of buildings (shown in upper figure as stories and feet above grade), and area limitations of one or two story buildings facing on one street or public space not less than 30 feet wide (shown in lower figure as area in square feet per floor). See Note a.

N.P. — Not permitted
N.L. — Not limited

Use Group	Note a	Noncombustible Type 1 — 1A Protected Note b	Type 1 — 1B	Type 2 — 2A Protected	Type 2 — 2B Protected	Type 2 — 2C Unprotected	Noncombustible/Combustible Type 3 — 3A Protected	Type 3 — 3B Unprotected	Type 4 — 4 Heavy timber	Combustible Type 5 — 5A Protected	Type 5 — 5B Unprotected
A-1-A Assembly, theaters	With stage and scenery	N.L.	6 St. 75' 14,400	4 St. 50' 11,400	2 St. 30' 7,500	1 St. 20' 4,800	2 St. 30' 6,600	1 St. 20' 4,800	2 St. 30' 7,200	1 St. 20' 5,100	N.P.
A-1-B Assembly, theaters	Without stage (motion picture theaters)	N.L.	N.L.	5 St. 65' 19,950	3 St. 40' 13,125	2 St. 30' 8,400	3 St. 40' 11,550	2 St. 30' 8,400	3 St. 40' 12,600	1 St. 20' 8,925	1 St. 20' 4,200
A-2 Assembly, night clubs and similar uses		N.L.	4 St. 50' 7,200	3 St. 40' 5,700	2 St. 30' 3,750	1 St. 20' 2,400	2 St. 30' 3,300	1 St. 20' 2,400	2 St. 30' 3,600	1 St. 20' 2,550	1 St. 20' 1,200
A-3 Assembly	Lecture halls, recreation centers, terminals, restaurants other than night clubs	N.L.	N.L.	5 St. 65' 19,950	3 St. 40' 13,125	2 St. 30' 8,400	3 St. 40' 11,550	2 St. 30' 8,400	3 St. 40' 12,600	1 St. 20' 8,925	1 St. 20' 4,200
A-4 Assembly, churches	Note d	N.L.	N.L.	5 St. 65' 34,200	3 St. 40' 22,500	2 St. 30' 14,400	3 St. 40' 19,800	2 St. 30' 14,400	3 St. 40' 21,600	1 St. 20' 15,300	1 St. 20' 7,200
B Business		N.L.	N.L.	7 St. 85' 34,200	5 St. 65' 22,500	3 St. 40' 14,400	4 St. 50' 19,800	3 St. 40' 14,400	5 St. 65' 21,600	3 St. 40' 15,300	2 St. 30' 7,200
E Educational	Note c,d	N.L.	N.L.	5 St. 65' 34,200	3 St. 40' 22,500	2 St. 30' 14,400	3 St. 40' 19,800	2 St. 30' 14,400	3 St. 40' 21,600	1 St. 20' 15,300	1 St. 20' 7,200
F Factory and industrial		N.L.	N.L.	6 St. 75' 22,800	4 St. 50' 15,000	2 St. 30' 9,600	3 St. 40' 13,200	2 St. 30' 9,600	4 St. 50' 14,400	2 St. 30' 10,200	1 St. 20' 4,800
H High hazard	Note e	5 St. 65' 16,800	3 St. 40' 14,400	3 St. 40' 11,400	2 St. 30' 7,500	1 St. 20' 4,800	2 St. 30' 6,600	1 St. 20' 4,800	2 St. 30' 7,200	1 St. 20' 5,100	N.P.
I-1 Institutional, residential care		N.L.	N.L.	9 St. 100' 19,950	4 St. 50' 13,125	3 St. 40' 8,400	4 St. 50' 11,550	3 St. 40' 8,400	4 St. 50' 12,600	3 St. 40' 8,925	2 St. 35' 4,200
I-2 Institutional, incapacitated		N.L.	8 St. 90' 21,600	4 St. 50' 17,100	2 St. 30' 11,250	1 St. 20' 7,200	2 St. 30' 9,900	1 St. 20' 7,200	2 St. 30' 10,800	1 St. 20' 7,650	N.P.
I-3 Institutional, restrained		N.L.	6 St. 75' 18,000	4 St. 50' 14,250	2 St. 30' 9,375	1 St. 20' 6,000	2 St. 30' 8,250	1 St. 20' 6,000	2 St. 30' 9,000	1 St. 20' 6,375	N.P.
M Mercantile		N.L.	N.L.	6 St. 75' 22,800	4 St. 50' 15,000	2 St. 30' 9,600	3 St. 40' 13,200	2 St. 30' 9,600	4 St. 50' 14,400	2 St. 30' 10,200	1 St. 20' 4,800
R-1 Residential, hotels		N.L.	N.L.	9 St. 100' 22,800	4 St. 50' 15,000	3 St. 40' 9,600	4 St. 50' 13,200	3 St. 40' 9,600	4 St. 50' 14,400	3 St. 40' 10,200	2 St. 35' 4,800
R-2 Residential, multi-family		N.L.	N.L.	9 St. 100' 22,800	4 St. 50' 15,000 Note f	3 St. 40' 9,600	4 St. 50' 13,200 Note f	3 St. 40' 9,600	4 St. 50' 14,400	3 St. 40' 10,200	2 St. 35' 4,800
R-3 Residential, one and two family		N.L.	N.L.	4 St. 50' 22,800	4 St. 50' 15,000	3 St. 40' 9,600	4 St. 50' 13,200	3 St. 40' 9,600	4 St. 50' 14,400	3 St. 40' 10,200	2 St. 35' 4,800
S-1 Storage, moderate		N.L.	N.L.	5 St. 65' 19,950	4 St. 50' 13,125	2 St. 30' 8,400	3 St. 40' 11,550	2 St. 30' 8,400	4 St. 50' 12,600	2 St. 30' 8,925	1 St. 20' 4,200
S-2 Storage, low	Note g	N.L.	N.L.	7 St. 85' 34,200	5 St. 65' 22,500	3 St. 40' 14,400	4 St. 50' 19,800	3 St. 40' 14,400	5 St. 65' 21,600	3 St. 40' 15,300	2 St. 30' 7,200
U Utility, miscellaneous		N.L.	N.L.								

Notes applicable to Table 501

Note a. See the following sections for general exceptions to Table 501:
Section 501.4 Allowable area reduction for multi-story buildings.
Section 502.2 Allowable area increase due to street frontage.
Section 502.3 Allowable area increase due to automatic fire suppression system installation.
Section 503.1 Allowable height increase due to automatic fire suppression system installation.
Section 504.0 Unlimited area one story buildings.

Note b. Buildings of Type 1 construction permitted to be of unlimited tabular heights and areas are not subject to special requirements that allow increased heights and areas for other types of construction (see Section 502.5).

Note c. The tabular area of one story school buildings of Use Group E may be increased 200 percent provided every classroom has at least one door opening directly to the exterior of the building. Not less than one-half of the required exits from any assembly room included in such buildings shall also open directly to the exterior of the building (see Section 502.4).

Note d. Auditoriums in buildings of Use Groups A-4 and E of Type 2B, 3A, 4 or 5A construction may be erected to 65 feet in height, and of Type 2C, 3B or 5B construction to 45 feet in height (see Section 503.2).

Note e. For exceptions to height and area limitations of buildings of Use Group H, see Article 6 governing the specific use. For other special fireresistive requirements governing specific uses, see Section 1405.0.

Note f. For exceptions to height of buildings of Use Group R-2 of Types 2B and 3A construction, see Section 1405.5.

Note g. For height and area exceptions for open parking structures, see Section 617.0.

Note h. 1 foot = 304.8 mm; 1 square foot = 0.093 m².

FIGURE 1.1

Height and area limitations of buildings of various types of construction, as defined in the BOCA Basic/National Building Code/1984.

Table 401
FIRERESISTANCE RATINGS OF STRUCTURE ELEMENTS (IN HOURS)

Structure element Note a		Noncombustible					Noncombustible/Combustible			Combustible	
		Type 1 Section 402.0		Type 2 Section 403.0			Type 3 Section 404.0		Type 4 Section 405.0	Type 5 Section 406.0	
		Protected		Protected		Unprotected	Protected	Unprotected	Heavy timber	Protected	Unprotected
		1A	1B	2A	2B	2C	3A	3B	4	5A	5B
Exterior walls (Section 1406.0 and Note b)											
1 Fire separation of 30' or more	Bearing	4	3	2	1	0	2	2	2	1	0
	Nonbearing	0	0	0	0	0	0	0	0	0	0
Fire separation of less than 6'	Bearing	4	3	2	1½	1	2	2	2	1	1
	Nonbearing	2	2	1½	1	1	2	2	2	1	1
Fire separation of 6' or more but less than 11'	Bearing	4	3	2	1	0	2	2	2	1	1 Note i
	Nonbearing	2	2	1½	1	0	2	2	2	1	1 Note i
Fire separation of 11' or more but less than 30'	Bearing	4	3	2	1	0	2	2	2	1	1 Note i
	Nonbearing	1½	1½	1	1	0	1½	1½	see Sec. 405.0	1	1 Note i
2 Fire walls and party walls (Section 1407.0)		4	3	2	2	2	2	2	2	2	2
		← Not less than fire grading of use group—(see Table 1402) →									
3 Fire separation assemblies (Sections 313.0, 1409.0 and 1412.0)		← Fireresistance rating corresponding to fire grading of use group—(see Table 1402) →									
4 Fire enclosures of exits, exit hallways and stairways (Section 816.9.2, 1409.0 and Note c)		2	2	2	2	2	2	2	2	2	2
5 Shafts (other than exits) and elevator hoistways (Section 1410.0 and Note c)		2	2	2	2	2	2	2	2	1	1
		← Noncombustible →									
6 Exit access corridors (Note g)		1	1	1	1	1	1	1	1	1	1
		← Note e →									
Separation of tenant spaces		1	1	1	1	0	1	0	1	1	0
		← Note e →									
7 Dwelling unit separations		1	1	1	1	1	1	1	1	1	1
		← Note e →									
Other nonbearing partitions		0	0	0	0	0	0	0	0	0	0
		← Note e →									
8 Interior bearing walls, bearing partitions, columns, girders, trusses (other than roof trusses) and framing (Section 1411.0)	Supporting more than one floor	4	3	2	1	0	1	0	see Sec. 405.0	1	0
	Supporting one floor only	3	2	1½	1	0	1	0	see Sec. 405.0	1	0
	Supporting a roof only	3	2	1½	1	0	1	0	see Sec. 405.0	1	0
9 Structural members supporting wall (Section 1411.0 and Note h)		3	2	1½	1	0	1	0	1	1	0
		← Not less than fireresistance rating of wall supported →									
10 Floor construction including beams (Section 1412.0)		3	2	1½	1	0	1	0	see Sec. 405.0 Note d	1	0
11 Roof construction, including beams, trusses and framing, arches and roof deck (Section 1412.0 and Note f)	15' or less in height to lowest member	2	1½	1	1	0	1	0	see Sec. 405.0 Note d	1	0
		← Note e →									
	More than 15 but less than 20' in height to lowest lowest member	1	1	1	0	0	0	0	see Sec. 405.0 Note d	1	0
		← Note e →									
	20' or more in height to lowest member	0	0	0	0	0	0	0	see Sec. 405.0 Note d	0	0
		← Note e →									

Note a. For special high hazard uses involving a higher degree of fire severity and higher concentration of combustible contents, the fireresistance rating requirements for structural elements shall be increased accordingly (see Section 600.2).

Note b. The fire separation or fire exposure in feet as herein limited applies to the distance measured from the building face to the closest interior lot line, the center line of a street or public space or an imaginary line between two buildings on the same property (see definition of fire separation, exterior fire exposure in Section 201.0).

Note c. Exit and shaft enclosures connecting three floor levels or less shall have a fireresistance rating of not less than 1 hour (see Sections 1409.1.3 and 1410.3).

Note d. In Type 4 construction, members which are of material other than heavy timber shall have a fireresistance rating of not less than 1 hour (see Section 1224.2).

Note e. Fire-retardant treated wood, complying with Section 1403.5.1 may be used as provided in Section 1403.5.2 (see Section 1405.7).

Note f. Where the omission of fire protection from roof trusses, roof framing and decking is permitted, roofs in buildings of Type 1 and Type 2 construction shall be constructed of noncombustible materials without a specified fireresistance rating, or of Type 4 construction in buildings not over five stories or 65 feet in height (see Section 1413.4).

Note g. In all occupancies except Use Groups R-1 and R-2, exit access corridors serving 30 or fewer occupants may have a zero fireresistance rating. Exit access corridors contained within a dwelling unit shall not be required to have a fireresistance rating (see Section 810.4).

Note h. Structural members supporting fireresistance rated exit access corridor walls in buildings of Types 2C, 3B or 5B construction shall not be required to be fireresistance rated unless required by other provisions of this code (see Section 1409.4).

Note i. The exterior walls of buildings or structures of Type 5B construction with a fire separation of less than 15 feet shall have a fireresistance rating of not less than 1 hour except buildings of Use Group R-2, R-3 or U (see Section 1406.2).

Note j. 1 foot = 304.8 mm.

tect public health and safety by setting a minimum standard of construction quality. A typical code begins by defining *use groups* for buildings: buildings of assembly, business, industry, high-hazard industry (such as plants working with highly flammable or explosive substances), institutional buildings, mercantile buildings, residential buildings, and simple buildings used only for storage or agricultural purposes.

These definitions are followed by a set of definitions of *construction types*. At the head of this list are highly fireresistive types of construction such as reinforced concrete and fire-protected steel. At the foot of it are types of construction that are relatively combustible because they are framed with small wood members. In between are a range of construction types with varying levels of resistance to fire.

With use groups and construction types carefully defined, the code proceeds to match the two, setting forth which use groups may be housed in which types of construction, and under what limitations. Figure 1.1 is reproduced from a building code prepared by the Building Officials and Code Administrators International, Incorporated (BOCA). This table concentrates a great deal of useful information into a very small space. A designer may enter it with a particular use group in mind—an electronics plant, for example—and find out very quickly what types of construction will be permitted and what shape the plant may take. An electronics plant obviously fits under Use Group F, Factory and Industrial. Reading across the chart, one finds immediately that this factory may be built to any desired size, without limit, using Type 1 construction.

Type 1 construction is defined in nearby table in the BOCA Code, reproduced here as Figure 1.2. Looking down this table under Type 1 construction, one finds a rather detailed listing of the required *fireresistance ratings*, measured in hours, of the various parts of either a Type 1A or a Type 1B building. Fireresistance ratings of actual construction components are not found in the BOCA Code. Instead, they are tabulated in a variety of catalogs and handbooks issued by building material manufacturers, construction trade associations, and organizations concerned with fire protection of buildings. In each case the ratings are derived from full-scale laboratory fire tests of building components carried out in accordance with standards set by the American Society for Testing and Materials, to assure uniformity of results. Figures 1.3 through 1.5 reproduce small sections of tables from catalogs and handbooks to illustrate how this type of information is presented.

It is not possible in this book to reproduce a comprehensive listing of fireresistance ratings for every type of building component, but what can be said in a very general way (and with many exceptions) is that the higher the degree of fire resistance, the higher the cost. In general, therefore, buildings are built with the least level of fire resistance that is permitted by the applicable building code. The hypothetical electronics plant could be built using Type 1 construction, but does it really need to be built to this standard?

Let us suppose that the owners want the electronics plant to be a two-story building with 10,000 square feet on each floor. The table in Figure 1.1 makes it clear that it cannot be built of unprotected metal (Type 2C), unprotected joisted construction (Type 3B), or unprotected wood frame (Type 5B), because none of these types will allow construction of floors as large as 10,000 square feet. But it can be built of steel with a relatively small amount of applied fire protective material (Type 2B), of protected wood joists with exterior walls either of masonry or of wood (Type 3A), or of Heavy Timber construction (Type 4). (The names associated with the various construction types are defined in later chapters of this book.)

In reality the situation can be a little more complicated than this; under

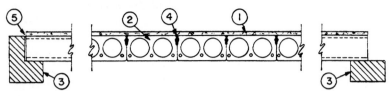

Design No. J921
Restrained Assembly Ratings—2 and 3 Hr. (See Item 1)
Unrestrained Assembly Rating—2 Hr.

Restrained
End Detail

Unrestrained
End Detail

1. **Concrete Topping**—3000 psi compressive strength. Normal weight 150 ± 3 pcf unit weight or lightweight, 112 ± 3 pcf unit weight.

Rating	Thickness-In.
2	0
3	1¼

2. **Precast Concrete Units***—Light-weight aggregate. Cross-section similar to the above illustration. Units 8, 10 or 12 in. thick, 16, 20 or 24 in. wide with 2 or 3 core holes.

3. **End Details**—Restrained and unrestrained. Min bearing 3 in.
4. **Joint**—Clearance between slabs at bottom, full length, min ¹/₁₆ in., max ⁵/₁₆ in. grouted full length with sand-cement grout, 3500 psi min.
   Note—A ³/₄-in. lateral expansion joint to be provided the full length and depth of the slabs every 14 ft. Expansion should be obtained with noncombustible, compressible material, for example; 12 sheets of 1/16 in. thick asbestos paper.
5. **End Clearance**—Clearance for expansion at each end of slabs shall be equal to L/17 (¼ ± 1/16) in., where "L" equal to length of span in feet.
*Bearing the UL Classification Marking

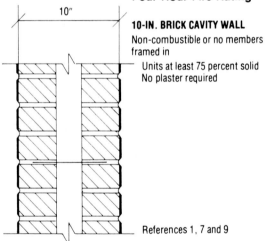

Four-Hour Fire Rating

10"

10-IN. BRICK CAVITY WALL
Non-combustible or no members framed in

Units at least 75 percent solid
No plaster required

References 1, 7 and 9

FIGURE 1.3
Examples of fireresistance ratings for concrete and masonry structure elements. The upper detail, taken from the Underwriters Laboratory Fire Resistance Directory, *is for a precast concrete hollow-core plank floor with poured concrete topping. "Restrained" and "unrestrained" refer to whether or not the floor is prevented from expanding longitudinally when exposed to the heat of a fire. The lower detail is from literature published by the Brick Institute of America. (Reprinted by permission of Underwriters Laboratory, Inc., and Brick Institute of America, respectively)*

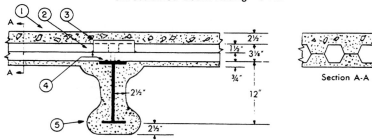

Design No. A814
Restrained Assembly Rating—3 Hr.
Unrestrained Assembly Rating—3 Hr.
Unrestrained Beam Rating—3 Hr.

Section A-A

Beam—W12×27, min size.
1. **Sand-Gravel Concrete**—150 pcf unit weight 4000 pcf compressive strength.
2. **Steel Floor and Form Units***—Non-composite 3 in. deep galv units. All 24 in. wide, 18/18 MSG min cellular units. Welded to supports 12 in. O.C. Adjacent units button-punched or welded 36 in. O.C. at joints.

3. **Cover Plate**—No. 16 MSG galv steel.
4. **Welds**—12 in. O.C.
5. **Fiber Sprayed***—Applied to wetted steel surfaces which are free of dirt, oil or loose scale by spraying with water to the final thickness shown above. The use of adhesive and sealer and the tamping of fiber are optional. The min ind density of the finished fiber should be 11 pcf and the specified fiber thicknesses require a min fiber density of 11 pcf. For areas where the fiber density is between 8 and 11 pcf, the fiber thickness shall be increased in accordance with the following formula:

$$\text{Thickness, in.} = \frac{(11)\ (\text{Design Thickness, in.})}{\text{Actual Fiber Density, pcf.}},$$

Fiber density shall not be less than 8 pcf. For method of density determination refer to General Information Section.

*Bearing the UL Classification Marking.

Design No. X511
Rating—3 Hr.

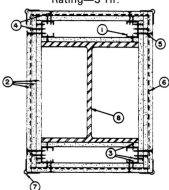

1. **Steel Studs**—1⅝ in. wide with leg dimensions of 1-5/16 and 1-7/16 in. with a ¼-in. folded flange in legs, fabricated from 25 MSG galv steel, ¾- by 1¾-in. rectangular cutouts punched 8 and 16 in. from the ends. Steel stud cut ½ in. less in length than assembly height.
2. **Wallboard, Gypsum***—½ in. thick, three layers.

3. **Screws**—1 in. long, self-drilling, self-tapping steel screws, spaced vertically 24 in. O.C.
4. **Screws**—1⅝ in. long, self-drilling, self-tapping steel screws, spaced vertically 24 in. O.C., except on the outer layer of wallboard on the flange, which are spaced 12 in. O.C.
5. **Screws**—2¼ in. long, self-drilling, self-tapping steel screws, spaced vertically 12 in. O.C.
6. **Tie Wire**—One strand of 18 SWG soft steel wire placed at the upper one-third point, used to secure the second layers of wallboard only.
7. **Corner Beads**—No. 28 MSG galv steel, 1¼ in. legs or 27 MSG uncoated steel, 1⅜ in. legs,

FIGURE 1.4
Fireresistance ratings for a steel floor structure and column, respectively, taken from the Underwriters Laboratory Fire Resistance Directory. (Reprinted by permission of Underwriters Laboratory, Inc.)

Fire Rating	Sound Rating STC	GA File No.	DETAILED DESCRIPTION	SKETCH AND DESIGN DATA	
				Fire	Sound
1 HR	**30 to 34**	WP 3620	**Construction Type: Gypsum-Veneer Base, Veneer Plaster, Wood Studs** One layer ½″ type X gypsum veneer base applied at right angles to each side of 2 x 4 wood studs 16″ o.c. with 5d etched nails, 1¾″ long, 0.099″ shank, ¼″ heads, 8″ o.c. Minimum 1/16″ gypsum-veneer plaster over each face. Stagger vertical joints 16″ and horizontal joints each side 12″. Sound tested without veneer plaster. (LB)	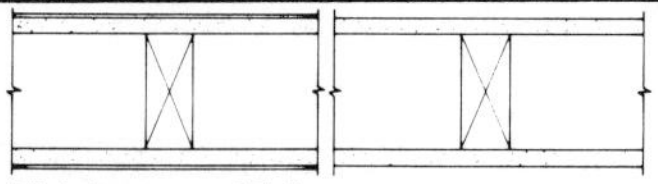 Thickness: 4⅞″ Approx. Weight: 7 psf Fire Test: UC, 1-12-66 Sound Test: G & H IBI-35FT. 5-26-64	

FIGURE 1.5
A sample of fireresistance ratings published by the Gypsum Association, in this case for an interior partition. (Courtesy of Gypsum Association)

certain conditions, many codes permit the allowable heights and areas to be increased if an automatic fire extinguishing system is installed, and in some urban situations, zoning ordinances may further restrict construction types, but this example serves to illustrate in a general way how construction types are selected. Again and again in succeeding chapters of this book reference will be made to these tables to see how they relate to particular building materials and techniques.

A building code goes far beyond what is illustrated here. A typical code establishes standards for minimum room dimensions, natural light, ventilation, means of emergency egress, structural design, floor, wall, ceiling, and roof construction, chimney construction, and energy efficiency. Codes are generally established at the local or state level and vary considerably from one area to another, but organizations such as BOCA, the International Conference of Building Officials (ICBO), and the Southern Building Code Congress (SBCC), have been instrumental in helping states and localities to adopt standardized or *model* codes, both to raise the level of quality of codes in general, and to make life easier for building professionals whose buildings are constructed in many different geographic areas.

The building code is not the only code with which a new building must

> ...the architect should have construction at least as much at his fingers' ends as a thinker his grammar.
>
> Le Corbusier, *Towards a New Architecture,* 1927.

comply. There are also health codes, fire codes, plumbing codes, and electrical codes in force in most communities. Some of these are locally written, but many are model national codes that have been adopted locally.

Other Legal Constraints

Other types of legal restrictions must also be observed in the design and construction of buildings. *Access standards* regulate the design of entrances, stairs, doorways, elevators, and toilet facilities to assure that they are usable by physically handicapped members of the population. The U.S. *Occupational Safety and Health Act (OSHA)* controls the design of workplaces to minimize threats to the health and safety of workers. OSHA sets safety standards under which a building may be constructed and also has an important effect on the design of industrial and commercial buildings. Many localities have established standards of

energy efficiency for buildings, and a national standard seems likely to be instituted in the near future. Fire insurance companies exert a major influence on construction standards through their testing organizations (Underwriters Laboratories and Factory Mutual, for example), and through their rate structures for building insurance coverage, which offer strong financial incentives for more fire-resistant construction. Building contractors and construction labor unions also have standards, both formal and informal, that affect the ways in which buildings are built. Unions have work rules and safety rules that must be observed; contractors have particular types of equipment, certain kinds of skills, and customary ways of going about things. All of these vary significantly from one place to another.

CHOOSING BUILDING SYSTEMS: INFORMATION RESOURCES

The tasks of the architect and the engineer would be impossible without the support of dozens of organizations that produce and disseminate information on materials and methods of construction. Several organizations and types of organizations whose work has the most direct impact on day-to-day building design and construction operations are discussed in the sections that follow.

ASTM and CSA

The *American Society for Testing and Materials (ASTM)* establishes standard specifications for commonly used materials of construction. These specifications are accepted throughout the United States, and are generally referred to by number (ASTM C150, for example, is a specification for portland cement). These numbers are frequently used in construction specifications as precise shorthand designations for the quality of material that is required. Throughout this book ASTM specification numbers will be provided for the major building materials. Should you wish to examine the contents of any ASTM specification, they can be found in the ASTM publication listed at the end of this chapter. In Canada, corresponding standards are set by the *Canadian Standards Association (CSA)*.

Construction Trade and Professional Associations

Building professionals, building materials manufacturers, and building contractors have formed a large number of organizations that work toward the development of technical standards and the dissemination of information with relation to their respective fields of interest. The *Construction Specifications Institute (CSI)*, whose *Masterformat* standard is described in the following section, is but one example. It is composed both of independent building professionals, such as architects and engineers, and of members from industry. The *American Plywood Association*, to choose an example from among a hundred or more *trade associations*, is made up of producers of plywood and wood panel products. It carries out extensive programs of research on wood panel products, establishes uniform standards of product quality, certifies mills and products that conform to its standards, and publishes authoritative technical literature concerning the use of plywood and related products. Associations with a similar range of activities exist for virtually every material and product used in building. All publish technical data relating to their fields of interest, and many of these publications are indispensable references for the architect or engineer. A considerable number are incorporated by reference into various building codes and standards. Selected publications from professional and trade associations are identified in the reference lists at the ends of the chapters in this book. The reader is encouraged to obtain these publications and to request complete publication lists from the various organizations.

Masterformat

The Construction Specifications Institute of the United States and *Construction Specifications Canada* have evolved over a period of many years a standard outline called *Masterformat* for organizing information about construction materials and components. Masterformat is used as the outline for construction specifications for nearly all large construction projects in the two countries and forms the basis on which trade associations' and manufacturers' technical literature is catalogued and filed. Its sixteen primary divisions are as follows:

Division 1	General Requirements
Division 2	Sitework
Division 3	Concrete
Division 4	Masonry
Division 5	Metals
Division 6	Wood and Plastics
Division 7	Thermal and Moisture Protection
Division 8	Doors and Windows
Division 9	Finishes
Division 10	Specialties
Division 11	Equipment
Division 12	Furnishings
Division 13	Special Construction
Division 14	Conveying Systems
Division 15	Mechanical
Division 16	Electrical

Within these broad divisions, several levels of subdivision are established to allow the user to reach any desired degree of detail. A five-digit code, in which the first two digits correspond to the division numbers above, gives the exact reference to any category of information. Within Division 5—Metals, for example, some standard reference codes are

05120	Structural Steel
05210	Steel Joists
05310	Steel Deck
05400	Cold-Formed Metal Framing
05725	Ornamental Metal Castings

Each chapter of this book gives the major Masterformat designations for the information presented, to help the reader know where to look in construction specifications and in the technical literature for further information. The full Masterformat system is contained in the volume referenced at the end of this chapter.

CHOOSING BUILDING SYSTEMS: THE WORK OF THE DESIGN PROFESSIONAL

The designers of a building must make many choices before its design is complete and ready for construction. In making these choices they confront four basic questions:

1. What will give the required functional performance?

2. What will give the desired aesthetic result?

3. What is possible legally?

4. What is most economical?

This book examines primarily the first of these questions, the functional performance of the materials and methods of building construction most commonly employed in North America. But reference is made constantly to the other three questions, and the reader should always have them uppermost in mind.

The chapters that follow describe many alternative ways of doing things: different structural systems, different systems of enclosure, different systems of interior finish. Each system has characteristics that distinguish it from the alternatives. Sometimes a system is distinguished chiefly by its visual qualities, as one might acknowledge in choosing one type of granite over another, one color of paint over another, or one type of tile pattern over another. However, visual distinctions can extend beyond surface qualities; a designer may prefer the massive appearance of a masonry bearing wall building to the slender look of an exposed steel frame on one project, yet would choose the steel for another building whose situation is different. Again, one may choose for purely functional reasons, as in selecting terrazzo flooring instead of carpet or wood in a restaurant kitchen. One could choose on purely technical grounds, as, for example, in electing to posttension a long concrete beam rather than merely to reinforce it. A designer is often forced into a particular choice by some of the legal constraints identified earlier in this chapter. And frequently the selection is made on purely economic grounds. The economic criterion can mean any of several things: Sometimes one system is chosen over another because its first cost is less; sometimes the entire *life-cycle costs* of competing systems are compared by means of formulas that include first cost, maintenance cost, energy consumption cost (if any), the useful lifetime and replacement cost of the system, and interest rates on invested money; and, finally, a system

may be chosen because there is keen competition among local suppliers and/or installers that keeps the price at the lowest possible level. This is often a reason to specify a very standard type of roofing material, for example, that can be furnished and installed by any of a number of companies, instead of a new system that is theoretically better from a functional standpoint, but can only be furnished by a single company that has the new equipment and skills required to install it.

One cannot gain all the knowledge needed to make such decisions from a textbook. It is incumbent upon the reader to go far beyond what can be presented here—to other books, to catalogs, to trade publications, to professional periodicals, and especially to the design office, the workshop, and the building site. There is no other way to gain much of the required information than to get involved in the art and business of building. One must learn how materials feel in the hand; how they look in a building; how they are manufactured, worked, and put in place; how they perform in service; how they deteriorate with time. One must become familiar with the people and organizations that produce buildings—the architects, engineers, materials suppliers, contractors, subcontractors, workers, inspectors, managers, and building owners—and learn to understand their respective methods, problems, and points of view. In the meantime, this long and hopefully enjoyable process of education in

G̲o into the field where you can see the machines and methods at work that make the modern buildings, or stay in construction direct and simple until you can work naturally into building-design from the nature of construction.

Frank Lloyd Wright, *To the Young Man in Architecture,* 1931.

the materials and methods of building construction can begin with the information presented in the chapters that follow.

Recurring Concerns

Certain themes are woven throughout this book, and surface again and again in different, often widely varying, contexts. These represent a set of concerns that fall into two broad categories: one that has to do with building *performance*, and one that has to do with building *construction*.

The performance concerns relate to the inescapable problems that must be confronted in every building: fire; building movement of every kind, including foundation settlement, structural deflections, and expansion and contraction due to changes in temperature and humidity; water vapor migration and condensation; water leakage; acoustical privacy; deterioration, cleanliness, and building maintenance.

The construction concerns are associated with the everyday problems of getting a building built safely, on time, within budget, and to the required standard of quality: division of work between the shop and the field; optimum use of the various building trades; sequencing of construction operations for maximum productivity; convenient worker access to construction operations; dealing with inclement weather; making building components fit together; and quality assurance in construction materials and components, through grading, testing, and inspection.

To the novice, these matters may seem of minor consequence when compared to the larger and often more interesting themes of building form and function. To the experienced building professional, who has seen buildings fail both aesthetically and physically for want of attention to one or more of these concerns, they are issues that must each be resolved as a matter of course before the work of a building project can be allowed to proceed.

S E L E C T E D R E F E R E N C E S

1. Allen, Edward. *How Buildings Work: The Natural Order of Architecture*. New York, Oxford University Press, 1980.

What do buildings do, and how do they do it? This book sets forth in easily understandable terms the physical principles by which buildings stand up, enclose a piece of the world, and modify it for human use.

2. American Society for Testing and Materials. *ASTM Standards in Building Codes*, 20th Edition. Philadelphia, 1982.

In this volume are found the ASTM standards for all the materials used in the construction of buildings. (Address for ordering: 1916 Race Street, Philadelphia, PA 19103.)

3. The Construction Specifications Institute and Construction Specifications Canada. *Masterformat—Master List of Section Titles and Numbers*. Alexandria, Virginia and Toronto, Ontario, 1983.

This book is the key to the Masterformat organization for construction information, and gives the full set of numbers and titles under which such information is filed and utilized. (Address for ordering: 601 Madison Street, Alexandria, VA 22314, or One St. Clair Avenue West, Suite 1206, Toronto, Ontario M4V 1K6.)

4. Building Officials & Code Administrators International, Inc. *The BOCA Basic Building Code/1984*, 9th Edition. Country Club Hills, Illinois, 1984.

The BOCA code is a so-called "model" code, one carefully developed over a period of many years by an organization of building experts and intended for adoption by local governments. The BOCA code has been adopted by many municipalities, especially in the eastern United States, and is used in this book. In the West, the *Uniform Building Code*, published by the International Conference of Building Officials in Whittier, California, predominates. The *Standard Building Code* of the Southern Building Code Congress, Birmingham, Alabama, has been widely adopted throughout the South. (Address for ordering BOCA Code: 4051 West Flossmoor Road, Country Club Hills, IL 60477.)

K E Y T E R M S A N D C O N C E P T S

zoning ordinance
building code
BOCA
ICBO
SBCC
use group

construction type
fireresistance rating
emergency egress
OSHA
ASTM
CSA

access standards
CSI
CSC
Masterformat
trade association

R E V I E W Q U E S T I O N S

1. Who are the members of the typical team that designs a major building? What are their respective roles?

2. What are the major constraints under which the designers of a building must work?

3. What types of subjects are covered by zoning ordinances? By building codes?

E X E R C I S E S

1. Have each class member write to two or three trade associations at the beginning of the term to request their lists of publications, and then have each send for some of the publications. Display and discuss the publications.

2. Repeat the above exercise for manufacturers' catalogs of building materials and components.

3. Obtain copies of your local zoning ordinance and building code. Look up the applicable provisions of these documents for a specific site and building. What setbacks are required? How large a building is permitted? What construction types may be employed? What types of roofing materials are permitted? What are the restrictions on interior finish materials? Outline in complete detail the requirements for emergency egress from the building.

FOUNDATIONS

---

*F*oundation work in progress for a major new hotel in Boston. The earth surrounding the excavation is retained with steel sheet piling supported by steel walers and tiebacks. Equipment enters and leaves the site via the earth ramp at the bottom of the picture. Although a large backhoe continues excavation work at the right, digging around piles from a previous building on the site, the installation of pressure-injected concrete footings is already well underway, with two pile drivers at work in the near and far corners, and clusters of completed piles visible in the left foreground. Concrete pile caps and column reinforcing are under construction in the center of the excavation. *(Courtesy of Franki Foundation Company)*

The function of a foundation is to transfer the structural loads from a building safely into the ground. Every building needs a foundation of some kind: A backyard toolshed will not be damaged by slight movement of its foundations and may need only wooden skids to spread its load across a sufficient area of the surface of the ground. A house needs greater stability than a toolshed, so its foundation reaches through the unstable surface to underlying soil that is free of organic matter and unreachable by winter's frost. A larger building of masonry, steel, or concrete weighs many times more than a house, and its foundations probe into the earth until they reach soil that is competent to carry its massive loads; on some sites this means going a hundred feet or more below the surface. Because of the variety of soil and water conditions that are encountered below the surface of the ground and the unique demands that many buildings make upon their foundations, foundation design is a highly specialized field of engineering that can be sketched here only in its broad outlines.

FOUNDATION LOADS

A foundation supports a number of different kinds of loads:

- The *dead load* of the building, which is the sum of the weights of the frame, the floors, roofs, and walls, the electrical and mechanical equipment, and the foundation itself.
- The *live load*, which is the sum of the weights of the people in the building, the furnishings and equipment they use, and snow, ice, and water on the roof.
- *Wind loads*, which can apply lateral, downward, and uplift loads to a foundation.
- Horizontal pressures of earth and water against basement walls.
- In some buildings, horizontal *thrusts* from arches, rigid frames, domes, vaults, or tensile structures.
- On some building sites, buoyant uplift forces from underground water, identical to the forces that cause a boat to float.
- During earthquakes, horizontal and vertical forces caused by the motion of the ground relative to the building.

FOUNDATION SETTLEMENT

All foundations *settle* to some extent as the soil around and beneath them adjusts itself to these loads. Foundations on bedrock settle a negligible amount. Foundations on certain types of clay, such as that found in Mexico City, settle to an alarming degree, allowing buildings to subside by amounts that are measured in feet or meters. Foundation settlement in most buildings is measured in millimeters or fractions of inches. If *total settlement* occurs at roughly the same rate from one side of the building to the other, no harm is likely to be done to the building, but if large amounts of *differential settlement* occur, in which the various columns and loadbearing walls of the building settle by substantially different amounts, the frame of the building may become distorted, floors may slope, walls and glass may crack, and doors and windows may refuse to work properly (Figure 2.1). Accordingly, a primary objective in foundation design is to minimize differential settlement by loading the soil in such a way that equal settlement occurs under the various parts of the building. This is not difficult when all parts of the building rest on the same kind of soil, but can become a problem when a building occupies a piece of ground that is underlain by two or more areas of dif-

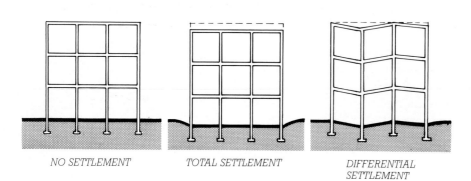

NO SETTLEMENT   TOTAL SETTLEMENT   DIFFERENTIAL SETTLEMENT

FIGURE 2.1
Total settlement is usually of little consequence in a building, but differential settlement can cause severe structural damage.

ferent types of soil with very different loadbearing capacities. Most foundation failures are attributable to excessive differential settlement; gross failure of a foundation, in which the soil fails completely to support the building, is extremely rare.

SOILS

Types of Soils

The following distinctions may be helpful in acquiring an initial understanding of how soils are classified for engineering purposes:

- *Rock* is a continuous mass of solid mineral material, such as granite or limestone, that can only be removed by drilling and blasting. Rock is not completely monolithic, but is crossed by a system of joints that divide it into irregular blocks. Despite these joints, rock is generally the strongest and most stable material on which a building can be founded.

- *Soil* is a general term referring to earth material that is particulate.

- If an individual particle of soil is too large to lift by hand, or requires two hands to lift, it is known as a *boulder*.

- If it takes the whole hand to lift a particle, it is a *cobble*.

- If a particle can be lifted easily with thumb and forefinger, the soil is *gravel*. In the Unified Soil Classification System (Figure 2.2), gravels are classified visually as having more than half their particles larger than 0.25 inch (6.5 mm) in diameter.

- If the individual particles can be seen but are too small to be picked up individually, the soil is *sand*. Sand particles range in size from about 0.25 to 0.002 inch (6.5 to 0.06 mm). Sand and gravel are considered to be *coarse-grained soils*.

- If the particles are too small to be seen individually, the soil is classified as a *fine-grained soil* and is either *silt* or *clay*.

FIGURE 2.2

A soil classification chart based on the Unified Soil Classification System. The group symbols are a universal set of abbreviations for soil types, as used in Figure 2.7.

			Group Symbols	Typical Names
Coarse-grained Soils	Gravels	Clean Gravels	GW	Well-graded gravels, gravel-sand mixtures, little or no fines
			GP	Poorly graded gravels, gravel-sand mixtures, little or no fines
		Gravels with Fines	GM	Silty gravels, poorly graded gravel-sand-silt mixtures
			GC	Clayey gravels, poorly graded gravel-sand-clay mixtures
	Sands	Clean Sands	SW	Well-graded sands, gravelly sands, little or no fines
			SP	Poorly graded sands, gravelly sands, little or no fines
		Sands with Fines	SM	Silty sands, poorly graded sand-silt mixtures
			SC	Clayey sands, poorly graded sand-clay mixtures
Fine-grained Soils	Silts and Clays	(Liquid limit greater than 50)	ML	Inorganic silts and very fine sands, rock flour, silty or clayey fine sands with plasticity
			CL	Inorganic clays of low to medium plasticity, gravelly clays, sandy clays, silty clays, lean clays
			OL	Organic silts and organic silt-clays of low plasticity
		(Liquid limit less than 50)	MH	Inorganic silts, micaceous or diatomaceous fine sandy or silty soils, elastic silts
			CH	Inorganic clays of high plasticity, fat clays
			OH	Organic clays of medium to high plasticity
	Highly Organic Soils		Pt	Peat and other highly organic soils

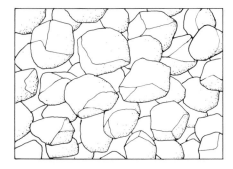

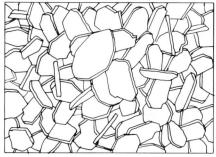

FIGURE 2.3

Silt particles (top) are approximately equidimensional granules, while clay particles (bottom) are platelike and generally much smaller than silt. The clay structure shown here is flocculated. In some other types of clays the particles lie in parallel relationships, either more closely packed, or dispersed by electrostatic forces.

- *Silt* particles are approximately equidimensional, and range in size from 0.002 to 0.00008 inch (0.06 to 0.002 mm). Because of their low surface-area-to-volume ratio, which approximates that of sand and gravel, the behavior of silt is controlled by the same mass forces that control the behavior of the coarse-grained soils.

- *Clay* particles are plate-shaped rather than equidimensional (Figure 2.3) and smaller than silt particles, less than 0.00008 inch (0.002 mm). Clay particles, because of their smaller size and flatter shape, have a surface-area-to-volume ratio hundreds or thousands of times greater than that of silt. As particle size decreases, the size of the pores between the particles also decreases, and soil behavior increasingly depends on surface forces (osmotic repulsion, van der Waals forces, electrostatic attraction, and repulsion due to water adsorption). As particle shape becomes more platelike, the *fabric* (arrangement of particles and pores) becomes more important: The volume of the pores and the amount of water in the pores greatly influences the properties of a clay soil.

Clay soils are generally referred to as being *cohesive*, which means that they retain a measurable shear resistance in the absence of confining forces. This cohesive behavior can result from true cohesion (due to cementation, electrostatic and electromagnetic attraction, and primary valence bonding) or from apparent cohesion (due to capillary stresses or apparent mechanical forces). In contrast, in a *frictional* or *cohesionless* soil such as sand, the shear resistance is directly proportional to the confining force pushing the particles together. In the absence of a confining force, the shear resistance disappears and the soil cannot stand with a vertical face (Figure 2.4).

A building site is usually underlain by a number of superimposed layers

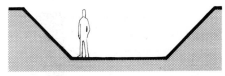

EXCAVATION IN FRICTIONAL SOIL

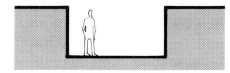

EXCAVATION IN HIGHLY COHESIVE SOIL

FIGURE 2.4

Excavations in highly cohesive and frictional soils.

PRESUMPTIVE SURFACE BEARING VALUES OF FOUNDATION MATERIALS

Class of material	Tons per square foot
1. Massive crystalline bedrock including granite, diorite, gneiss, trap rock, hard limestone and dolomite	100
2. Foliated rock including bedded limestone, schist and slate in sound condition	40
3. Sedimentary rock including hard shales, sandstones, and thoroughly cemented conglomerates	25
4. Soft or broken bedrock (excluding shale), and soft limestone	10
5. Compacted, partially cemented gravels, and sand and hardpan overlying rock	10
6. Gravel and sand gravel mixtures	6
7. Loose gravel, hard dry clay, compact coarse sand, and soft shales	4
8. Loose, coarse sand and sand-gravel mixtures and compact fine sand (confined)	3
9. Loose medium sand (confined), stiff clay	2
10. Soft broken shale, soft clay	1.5

Note 1 ton per square foot = 9765 kg/m².

FIGURE 2.5

Presumptive surface bearing values of various soils, as given in the BOCA Basic/National Building Code, 1984, reproduced by permission.

(*strata*) of different soils. These strata were deposited one after another in very ancient times by volcanic action and by the action of water, wind, and ice. Most of the soils in these strata are mixtures of several different sizes and types of particles and bear such names as silty gravels, gravelly sands, clayey sands, silty clays, and so on (Figure 2.2). The distribution of particle sizes and types in a soil is important to know when designing a foundation because it is helpful in predicting the loadbearing capacity of the soil, its stability, and its drainage characteristics.

Figure 2.5 gives some typical ranges of loadbearing capacities for various types of soils. These values give only a general idea of the relative strengths of different soils; the strength of an actual soil is also dependent on factors such as the presence or absence of water, the depth of the soil beneath the surface, and the size of the foundation applying the load to the soil. The designer may also arbitrarily choose to reduce the pressure of the foundations on the soil to well below these values in order to reduce the anticipated amount of settlement of the building. Rock is generally the best material on which to found a building; the proliferation of tall office buildings in Manhattan is attributable in part to the bedrock that lies a short distance

beneath the surface. But rock is too deep on most sites to be reached by foundations, so the designer must choose from the strata of different soils that lie closer to the surface, and design a foundation to perform satisfactorily in the selected soil.

The stability of a soil is its ability to retain its engineering properties under the varying conditions that may occur during the lifetime of the building. Rock, gravels, sands, and many silts tend to be stable soils. Organic soils are usually considered unstable because their properties may change through bacterial decomposition or through a change in water content. And many clays are dimensionally unstable under changing subsurface moisture conditions. Clay swells considerably as it absorbs water and shrinks as it dries. When wet clay is put under pressure, water can be squeezed out of it, with a corresponding reduction in volume. Taken together, these properties make clays that are subject to changes in water content the least stable and least predictable soils for supporting buildings.

The drainage characteristics of a soil are important in predicting how water will flow on and under building sites and around building substructures. Water passes readily through clean gravels and sands, slowly through very fine silts and sands, and almost not at

all through many clays. A building site with clayey or silty soils near the surface drains poorly, and is likely to be muddy and covered with puddles during rainy periods, while a gravelly site is likely to remain virtually dry. Underground, water passes quickly through strata of gravel and sand, but accumulates above strata of clay and fine silt. An excellent way to keep a basement dry is to surround it with a thick layer of clean gravel or crushed stone. Water passing through the soil toward the building cannot reach the basement without first falling to the bottom of the gravel layer, from where it can be drawn off in perforated pipes before it accumulates (Figures 2.54, 2.55). It does little good to place perforated drainage pipes directly in clay or silt because water cannot flow easily toward the pipes.

Subsurface Exploration and Soil Testing

Prior to designing a foundation for any building larger than a single-family house, it is necessary to determine the soil and water conditions beneath the site. This can be done by digging test pits, driving sounding rods, or drilling. Test pits are useful when the foundation is not expected to extend deeper than about 8 feet (2.5 m), which is the

FIGURE 2.6

Examples of soil exploration drilling equipment. (a) A portable cathead drilling rig. A small gasoline engine on the leg of the tripod next to the operator spins a metal drum called a "cathead."

By alternately tightening and loosening the rope around the cathead, the operator lifts and drops the cylindrical weight to drive a casing or a sampling tube into the earth. (b) A trailer-

mounted hydraulic feed core drill. (c) A truck-mounted hydraulic drill rig with core augers. (Courtesy of Acker Drill Company, Inc.)

maximum practical reach of small excavating machines. The strata of the soil can be observed and measured in the pit, and soil samples can be taken for laboratory testing. The level of the *water table* (the elevation at which the pressure of the ground water is atmospheric) will be readily apparent if it falls within the depth of the pit because water will seep through the walls of the pit up to the level of the water table. If desired, a load test can be performed on the soil in the bottom of the pit to determine the stress the soil can safely carry and the amount of settlement that should be anticipated. If a pit is not dug, a standard penetration test using driven rods can give an indication of the bearing capacity of the soil by the number of blows of a standard driving hammer required to advance the rod into the soil by a fixed amount. The rods can also help locate bedrock. Drilling (Figure 2.6) extends the possible range of exploration much deeper into the earth than test pits or sounding, and returns information on the dimensions and locations of the soil strata and the depth of the water table,

as well as laboratory-quality soil samples for testing. A number of holes are drilled across the site; the information from the holes is coordinated and interpolated in the preparation of drawings that document the subsurface conditions for the use of the engineer

who will design the foundation (Figure 2.7).

Laboratory testing of soil samples is an important preliminary to foundation design. By passing a dried sample of soil through a set of sieves with graduated mesh sizes, the particle size

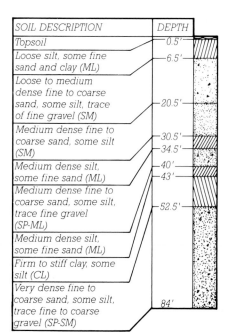

SOIL DESCRIPTION	DEPTH
Topsoil	0.5'
Loose silt, some fine sand and clay (ML)	6.5'
Loose to medium dense fine to coarse sand, some silt, trace of fine gravel (SM)	20.5'
Medium dense fine to coarse sand, some silt (SM)	30.5'
	34.5'
Medium dense silt, some fine sand (ML)	40'
	43'
Medium dense fine to coarse sand, some silt, trace fine gravel (SP-ML)	52.5'
Medium dense silt, some fine sand (ML)	
Firm to stiff clay, some silt (CL)	
Very dense fine to coarse sand, some silt, trace fine to coarse gravel (SP-SM)	84'

FIGURE 2.7

A typical log from a soil test boring, indicating the type of soil in each stratum and the depth in feet at which it was found. The abbreviations in parentheses refer to the Unified Soil Classification System, and are explained in Figure 2.2.

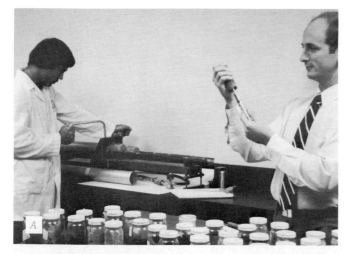

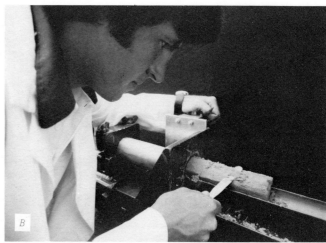

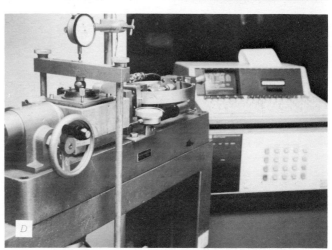

distribution in the soil can be determined. Further tests on fine-grained soils assist in their identification and provide information on their engineering properties. Important among these are tests that establish the *liquid limit*, the water content at which the soil passes from a plastic state to a liquid state, and the *plastic limit*, the water content at which the soil loses its plasticity and begins to behave as a solid. Additional tests can determine the water content of the soil, its permeability to water, its shrinkage when dried, its shear and compressive strengths, the amount by which the soil can be expected to consolidate under load, and the rate at which this consolidation will take place (Figure 2.8). These latter two qualities are helpful in predicting the rate and magnitude of foundation settlement in a building.

FIGURE 2.8

Some laboratory soil testing procedures. (a) To the right, the hardness of split spoon samples is checked with a penetrometer to be sure they come from the same stratum of soil. To the left, a soil sample from a Shelby sampling tube is cut into sections for testing. (b) A section of undisturbed soil from a Shelby tube is trimmed to a smaller diameter to become a sample for a triaxial load test, permability test, or consolidation test. (c) A cylindrical sample of soil is set up for a triaxial load test, the principal method for determining the shear strength of soil. The sample will be loaded axially by the piston in the top of the apparatus, and also circumferentially by water pressure in the transparent cylinder. (d) A direct shear test, used to measure the shear strength of cohesionless soils. A rectangular prism of soil is placed in a split box and sheared by applying pressure in opposite directions to the two halves of the box. (e) One-dimensional consolidation tests in

(continued)

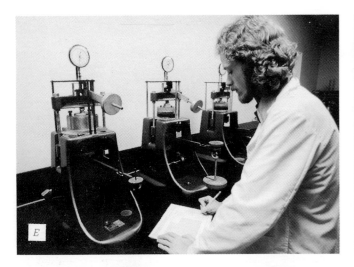

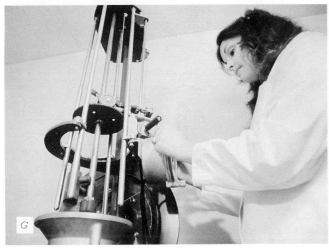

FIGURE 2.8 *(continued)*
progress on fine-grained soils, to determine their compressibility and expected rate of settlement. Each sample is compressed over an extended period of time by a porous piston that allows water to flow out of the sample. (f) A panel for running 30 simultaneous constant-head permeability tests, to determine the rate at which a fluid, usually water, moves through a soil. (g) A Proctor compaction test, in which successive layers of soil are compacted with a specified tamping force. The test is repeated for the same soil with varying moisture content, and a curve is plotted of density achieved versus moisture content of the soil, to identify an optimum moisture content for compacting the soil in the field. Not shown here are some common testing procedures for grain size analysis, liquid limit, plastic limit, specific gravity, and unconfined compression. (Courtesy of Ardaman and Associates, Inc., Orlando, Florida)

EXCAVATION

At least some excavation is required for every building. Foundations generally should not be placed on organic soils, which are subject to decomposition and to shrinking and swelling with changing moisture content. Organic topsoil, which is excellent for growing lawns and landscape plants, is scraped away from the building area and stockpiled for redistribution over the site after construction of the building is complete. When the topsoil has been removed, further digging is necessary to place the footings out of reach of water and wind erosion and, in colder climates, to place them below the level to which the ground freezes in winter. Soil expands slightly as it freezes, with a force that can lift and break buildings. Under certain soil and temperature conditions, upward migration of water vapor from the pores in the soil can result in the formation of *ice lenses*, thick layers of frozen crystals that can lift buildings by even larger amounts, unless the foundations are placed below the frost line.

Excavation is required on many sites to place the footings at a depth where soil of the appropriate bearing capacity is available. Excavation is frequently undertaken so one or more levels of basement space can be added to a building, whether for additional habitable rooms or for mechanical equipment and storage. Where footings must be taken deep to get below the frost line or reach competent soil, a basement is often bargain-rate space, adding little to the overall cost of the building.

In particulate soils, a variety of machines can be used to loosen and lift

*I*f...solid ground cannot be found, but the place proves to be nothing but a heap of loose earth to the very bottom, or a marsh, then it must be dug up and cleared out and set with piles made of charred alder or olive wood or oak, and these must be driven down by machinery, very closely together...

Marcus Vitruvius Pollio,
first century B.C.

the soil from the ground: bulldozers, shovel dozers, backhoes, bucket loaders, scrapers, trenching machines, and power shovels of every type. If the soil must be moved more than a short distance, dumptrucks come into use.

In rock, excavation is slower and many times more costly. Weak or decomposing rock can sometimes be fractured and loosened with power shovels, tractor-mounted rippers, pneumatic hammers, or drop balls such as those used in building demolition. Blasting, in which explosives are placed and detonated in lines of closely spaced holes drilled deep into the rock, is often necessary. Once the rock has been reduced to fragments, it can be scooped up and transported as a particulate soil.

Slope Support

If the site is sufficiently larger than the area to be covered by the building, the edges of the excavation can be sloped back or *benched* at an angle such that

the soil will not slide back into the hole. This angle can be steep for cohesive soils such as clays, but must be shallow for frictional soils such as sand and gravel. On constricted sites, the soil around an excavation must be held back by temporary walls (called *sheeting*). The sheeting, in turn, must be supported against the pressure of the earth and water it retains (Figure 2.9). If the sheeting is less than a story tall, this can often be done by driving it deeply enough into the soil that it can act as a vertical cantilever beam. For taller walls, and in less stable soils, some form of *bracing* is required.

Sheeting

Sheeting can take any of several forms, depending on the qualities of the soil, and circumstances such as the equipment and preferences of the contractor, the proximity of surrounding buildings, the level of the water table, the desire to retain the sheeting as a permanent part of the substructure of the building, and so on. Among the most common forms of sheeting are *soldier beams and lagging*, *sheet piling*, and *slurry walls*. *Soldier beams* are steel wide-flange sections driven vertically into the earth at close intervals around an excavation site before excavation begins. As earth is removed, *lagging*, usually consisting of heavy wood planks, is placed against the flanges of the soldier beams to retain the soil outside the excavation (Figures 2.10, 2.11). The soldier beams and lagging are removed after construction of the substructure is complete.

Sheet piling consists of vertical planks of wood, steel, or precast concrete that

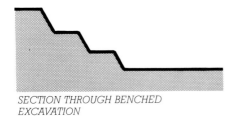

SECTION THROUGH BENCHED
EXCAVATION

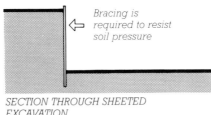

Bracing is required to resist soil pressure

SECTION THROUGH SHEETED
EXCAVATION

FIGURE 2.9
Sheeting and benching an excavation, viewed in cross section.

are placed tightly against one another and driven into the earth to form a solid wall before excavation begins (Figures 2.12, 2.13). Steel and precast concrete sheet piling are often left in place as part of the substructure of the building, or like wood sheet piling, they may be pulled from the soil when their work is finished.

A *slurry wall* is a more complicated and expensive type of sheeting that is usually economical only if it becomes the permanent foundation wall of the building. The first step in creating a slurry wall is to lay out the wall location on the surface of the ground with surveying instruments and to define the location and thickness of the wall

with shallow poured concrete *guide walls* (Figures 2.14, 2.15). When the formwork has been removed from the guide walls, a special narrow clamshell bucket, mounted on a crane, is used to excavate the soil from between the guide walls. As the narrow trench deepens, the tendency of its earth walls to collapse is counteracted by filling

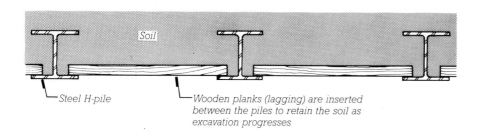

— Steel H-pile   — Wooden planks (lagging) are inserted
between the piles to retain the soil as
excavation progresses

FIGURE 2.10

Soldier beams and lagging, seen in horizontal section.

FIGURE 2.11

Soldier beams and lagging. Lagging planks are added at the bottom as excavation proceeds. The drill rig is boring a hole for a tieback to support the soldier beams. (Courtesy of Franki Foundation Company)

TIMBER SHEET PILING

FIGURE 2.12

Horizontal sections through three types of sheet piling. The shading represents the retained earth.

STEEL SHEET PILING

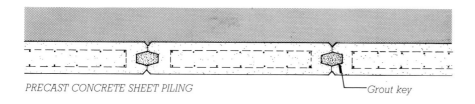

PRECAST CONCRETE SHEET PILING   — Grout key

the trench with a slurry of bentonite clay and water, which exerts a pressure back against the earth. The clamshell bucket is lowered and raised through the slurry to continue excavating the soil from the bottom of the trench until the desired depth, often a number of stories below ground, has been reached.

Meanwhile, workers have welded together cages of steel bars designed to reinforce the concrete wall that will replace the slurry in the trench. Steel tubes with a diameter corresponding to the width of the trench are driven vertically into the trench at predetermined intervals to divide it into sections of a size that can be reinforced and concreted conveniently. The concreting of each section begins with the lowering of a cage of reinforcing bars into the slurry. Then concrete is poured into the trench, filling it from the bottom up, using a funnel-and-tube arrangement called a *tremie*. As the concrete rises in the trench, the displaced slurry is pumped out and into holding

FIGURE 2.13

Drilling tieback holes for a wall of steel sheet piling. Notice the completed tieback connections to the waler in the foreground and background. The hole in the top of each piece of sheet piling allows it to be pulled from the ground by a crane. (Courtesy of Franki Foundation Company)

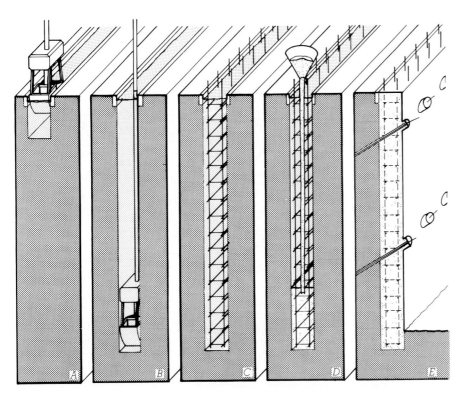

FIGURE 2.14

Steps in constructing a slurry wall. A. The concrete guide walls have been installed, and the clamshell bucket begins excavating the trench through a Bentonite clay slurry. B. The trench is dug to the desired depth, with the slurry serving to prevent collapse of the walls of the trench. C. A welded cage of steel reinforcing bars is lowered into the slurry. D. The trench is concreted from the bottom up with the aid of a tremie. The displaced slurry is pumped from the trench, filtered, and stored for reuse. E. The reinforced concrete wall is tied back as excavation progresses.

tanks, where it is stored for reuse. This process is repeated for each section of the wall. When the concrete in the trenches is sufficiently strong, earth removal begins inside the wall, with the reinforced concrete serving as sheeting for the excavation.

In addition to the *sitecast* slurry wall described in the preceding paragraphs, *precast* concrete slurry walls are also commonly built. The slurry for precast walls is a mixture of water, bentonite clay, and portland cement. The wall is cast in sections and pre-tensioned (see Chapter 12) in a precasting plant, then trucked to the construction site. Before each section is lowered by a crane into the slurry, its face is coated with a compound that prevents the clay—cement slurry from adhering to it (Figure 2.16). The sections are installed side by side in the trench, joined by tongue-and-groove edges or synthetic rubber gaskets. After the cement has caused the slurry to harden, excavation can begin, with the slurry on the inside face of the wall dropping away from the coated surface as soil is removed. The primary advantages of a precast slurry wall over a sitecast one are better surface quality, more accurate wall alignment, a thinner wall (due to the structural efficiency of prestressing), and improved watertightness of the wall because of the continuous layer of solidified clay outside.

Bracing

Soldier beams, sheet piling, and slurry walls all need to be braced against soil

FIGURE 2.15
Constructing a slurry wall. (a) *The guide walls are formed and poured in a shallow trench.* (b) *The narrow clamshell bucket discharges a load of soil into a waiting dump truck. Most of the trench is covered with wood pallets for safety.* (c) *A detail of the narrow clamshell bucket used for slurry wall excavation.* (d) *Hoisting a reinforcing cage from the area where it was assembled, getting ready to lower it into* the trench. The depth of the trench and height of the slurry wall are equal to the height of the cage, which is about four stories for this project. (e) Concreting the slurry wall with a tremie. The pump just to the left of the tremie removes slurry from the trench as concrete is added. (Photos b, c, and d courtesy of Franki Foundation Company. Photos a and e courtesy of Soletanche)

FIGURE 2.16
Workers apply a nonstick coating to a section of precast concrete slurry wall as it is lowered into the trench.
(Courtesy of Soletanche)

and water pressures as the excavation deepens (Figure 2.17). In *crosslot* bracing, temporary steel wide-flange columns are driven into the earth at points where braces will cross. As the earth is excavated down around the sheeting and the columns, tiers of horizontal bracing are added to support *walers* that span across the face of the sheeting. Where the excavation is too wide for crosslot bracing, *rakers* are used instead, bearing against *heel* *blocks* or other temporary footings.

Both rakers and crosslot bracing, especially the latter, are a hindrance to the excavation process. A clamshell bucket on a crane must be used to remove the earth between the braces, which is much less efficient and much more costly than removing soil with a shovel dozer or backhoe in an open excavation. Where subsoil conditions permit, *tiebacks* can be used instead of braces to support the sheeting while maintaining an open excavation. At each level of walers, holes are drilled at intervals through the sheeting and the surrounding soil into rock or a stratum of stable soil. Steel cables or tendons are then inserted into the holes, grouted to anchor them to the rock or soil, and stretched tight with hydraulic jacks (*post-tensioned*) before they are fastened to the walers (Figures 2.18, 2.19).

Excavations in fractured rock can often avoid sheeting altogether, either

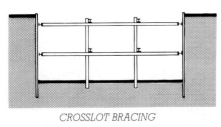

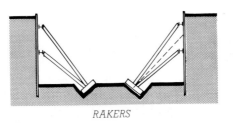

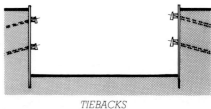

CROSSLOT BRACING

RAKERS

TIEBACKS

FIGURE 2.17

Three methods of bracing sheeting, drawn in cross section. The connection between the brace, raker, or tieback and the waler needs careful structural design. The broken line between rakers

indicates the mode of excavation: The center of the hole is excavated first, with sloping sides as indicated by the broken line. The footblocks and uppermost tier of rakers are installed.

As the sloping sides are excavated deeper, more tiers of rakers are installed.

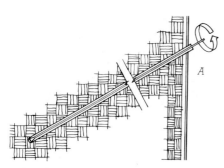

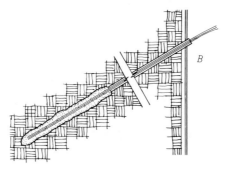

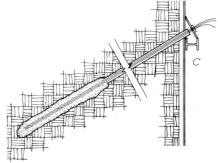

FIGURE 2.18

*Three steps in the installation of a tieback to a soil anchor: **A.** A rotary drill bores a hole through the sheeting and into stable soil or rock. A steel pipe*

*casing keeps the hole from caving in. **B.** Steel prestressing tendons are inserted into the hole and grouted under pressure to anchor them to the soil.*

***C.** After the grout has set, the tendons are tensioned with a hydraulic jack and anchored to a waler.*

A

FIGURE 2.19

Installing tiebacks. (a) Drilling through a slurry wall for a tieback. The ends of hundreds of completed tiebacks protrude from the wall. (b) Inserting prestressing tendons into the steel casing for a tieback. The apparatus in the center of the picture is for pressure-injecting grout around the tendons. (c) After the tendons have been tensioned, they are anchored to a steel chuck that holds them under stress, and the cylindrical hydraulic jack is moved to the next tieback. (d) Slurry walls and tiebacks used to support historic buildings around deep excavations for a station of the Paris Metro. (photos a, b, and c courtesy of Franki Foundation Company. Photo d courtesy of Soletanche)

(continued)

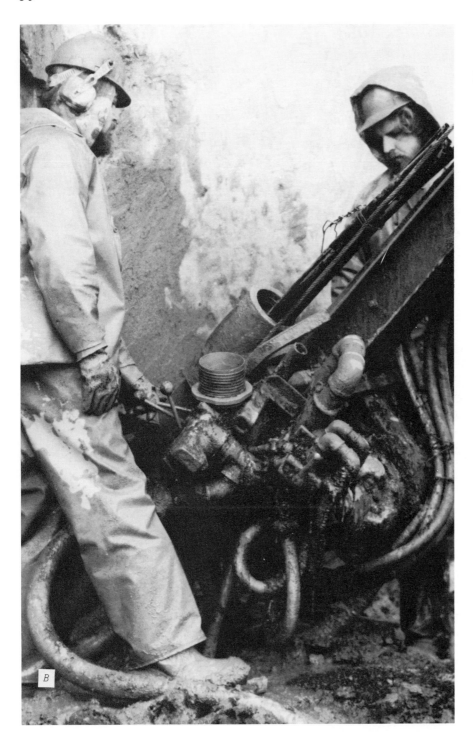

FIGURE 2.19b

FIGURE 2.19c

FIGURE 2.19d

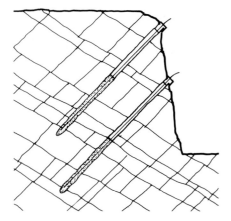

FIGURE 2.20
Rock anchors are similar to tiebacks, but are used to hold jointed rock formations in place around an excavation.

by injecting grout into the joints of the rock to stabilize it or by drilling into the rock and inserting *rock anchors* that fasten the blocks together (Figure 2.20).

Bracing and tiebacks in excavations are usually temporary. Their function is taken over by the floor structure of the basement levels of the building, which is designed specifically to resist lateral loads from the surrounding earth as well as ordinary floor loads.

Dewatering

When excavation work is carried lower than the water table in the surrounding soil, water begins to flow from the soil into the excavation and makes further work difficult or impossible. To continue excavating, it is necessary to keep ground water from entering the excavation, either by pumping water from the surrounding soil to depress the water table below the level of the

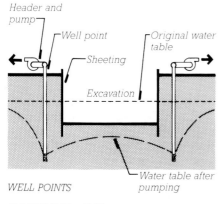

WELL POINTS

FIGURE 2.21
Two methods of keeping an excavation dry, viewed in cross section. The water sucked from well points draws the water table in the immediate vicinity below the level of the excavation.

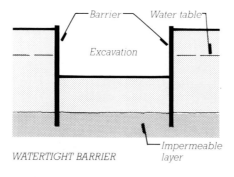

WATERTIGHT BARRIER

Watertight barrier walls work only if their bottom edges are inserted into an impermeable stratum that prevents water from working its way under the walls.

bottom of the excavation, or by erecting a watertight barrier around the excavation (Figure 2.21).

Well points are commonly used to depress the water table. These are vertical pieces of pipe with screened openings at the foot that keep out soil particles while allowing water to enter. Closely spaced well points are driven into the soil around the entire circumference of the excavation, and are connected to horizontal header pipes leading to pumps that continually suck water from the system and discharge it away from the building site. Once pumping has drawn down the water table in the area of the excavation, work can continue "in the dry" (Figure 2.22). For very deep excavations, two rings of well points may be required, the inner ring being driven to a deeper level than the outer ring, or a single ring of deep wells with deep-well pumps may have to be installed.

In some cases, the lowering of the water table by well points or deep wells can have serious adverse effects on neighboring buildings, by causing consolidation and settling of soil under their foundations, or by exposing untreated wood foundation piles, previously protected by total immersion in water, to decay. In these cases a *watertight barrier wall* is used as an alternative to well points. A slurry wall makes an excellent watertight barrier. Sheet piling can also work, but tends to leak. A watertight barrier must resist the hydrostatic pressure of the surrounding water, which grows greater as the foundation grows deeper, so a strong system of bracing or tiebacks is required. A watertight barrier only works if it reaches into an impermeable stratum that lies beneath the water table; otherwise, water will flow beneath the barrier and up into the excavation.

FOUNDATIONS

It is convenient to think of a building as consisting of three major parts: the *superstructure*, which is the above-

FIGURE 2.22

The pump in the foreground, drawing water through a header that connects to well points spaced every meter or so *around the perimeter of this excavation, allows work to continue in the dry despite a high water table maintained* *by the nearby lake.* (Courtesy of Griffin Dewatering Corporation)

ground portion of the building; the *substructure*, which is the habitable below-ground portion; and the *foundations*, which are the components of the building that transfer its loads into the soil (Figure 2.23).

There are two basic types of foundations: *shallow* and *deep*. Shallow foundations are those that transfer the load to the earth at the base of the column or wall of the substructure; deep foundations transfer the load at a point far below the substructure. Shallow foundations are generally less expensive than deep ones and can be used where suitable soil is found at the level at which the substructure of the building ends, whether this be a few feet or a few stories below the surface. Deep foundations, either piles or caissons, penetrate through upper layers of incompetent soil in order to reach competent bearing soil deeper within the earth.

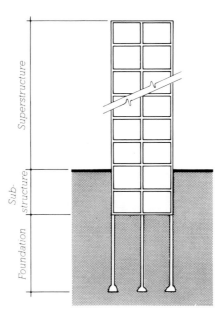

FIGURE 2.23

Superstructure, substructure, and foundation. The substructure in this example contains two levels of basements, and the foundation consists of bell caissons.

Shallow Foundations

Most shallow foundations are simple concrete *footings*. A *column footing* is a square pad of concrete, with or without steel reinforcing, that accepts the concentrated load placed on it from above by a building column and spreads this load across an area of soil large enough that the allowable stress of the soil is not exceeded. A *wall footing* or *strip footing* is a continuous strip of concrete that serves the same function for a loadbearing wall (Figures 2.24, 2.25).

To minimize settlement, footings are placed on undisturbed soil. (The only exception to this rule is that some buildings are built on *engineered fill*, which is earth that has been deposited and compacted in accordance with detailed procedures that assure a known degree of long-term stability.) The last few inches of soil in the excavation for a footing are removed with hand tools to avoid the loosening of the soil caused by digging machinery.

If the projection of the footing beyond the face of the wall or column is less than half the depth of the footing, steel reinforcing is not required. If the footing needs to project farther than this to achieve the necessary bearing

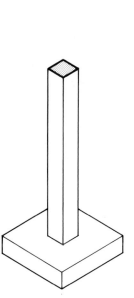
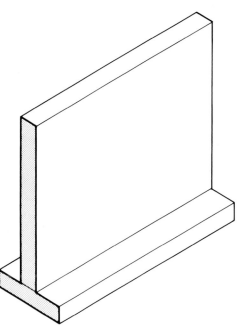

COLUMN FOOTING

WALL FOOTING

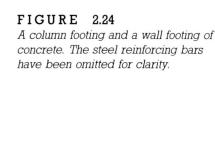

FIGURE 2.24
A column footing and a wall footing of concrete. The steel reinforcing bars have been omitted for clarity.

FIGURE 2.25
These concrete foundation walls for an apartment building, with their steel formwork not yet stripped, rest on wall footings. For more extensive illustrations of wall and column footings, see Figures 11.4, 11.9, and 11.10. (Courtesy of Portland Cement Association, Skokie, Illinois)

area, reinforcing bars are placed near the bottom of the footing to accept the tensile stresses that occur in this area, as discussed in Chapter 11.

Footings appear in many forms in different foundation systems. In climates with little or no ground frost, a concrete *slab on grade* with thickened edges is the least expensive foundation and floor system one can use and is applicable to one- and two-story buildings of any type of construction. For floors raised above the ground, either over a *crawlspace* or a *basement*, support is provided by concrete or masonry foundation walls supported on concrete strip footings (Figure 2.26). When building on slopes, it is often necessary to step the footings to maintain the required depth of footing at all points around the building (Figure 2.27). If soil conditions or earthquake precautions require it, column footings on steep slopes are linked together with reinforced concrete *tie beams* to avoid possible differential slippage between footings.

Footings cannot legally extend beyond a property line, even for a building built tightly against it. If the outer

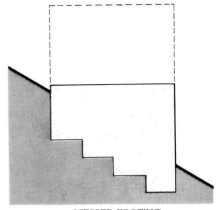

STEPPED FOOTING

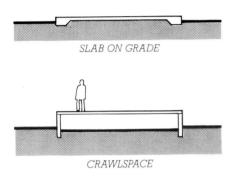

SLAB ON GRADE

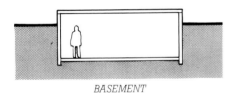

CRAWLSPACE

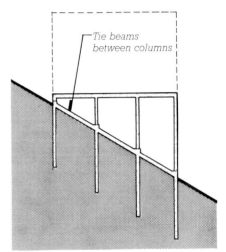

Tie beams between columns

TIE BEAMS

BASEMENT

FIGURE 2.26
Three types of substructures using simple wall footings. The slab on grade is most economical under many circumstances, especially where the water table lies near the surface of the ground. A crawlspace is often used under a floor structure of wood or steel, and gives much better access to underfloor piping and wiring than a slab on grade.

FIGURE 2.27
Foundations on sloping sites, viewed in a cross section through the building. The broken line indicates the outline of the superstructure. Wall footings are stepped to maintain the necessary distance between the bottom of the footing and the surface of the ground. Separate column foundations, whether caissons, as shown here, or column footings are often connected with reinforced concrete tie beams to reduce differential movement between the columns. A grade beam differs from a tie beam by being reinforced to distribute the continuous load from a bearing wall to separate foundations.

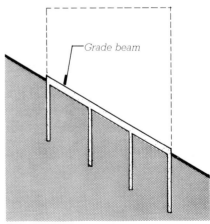

Grade beam

GRADE BEAM

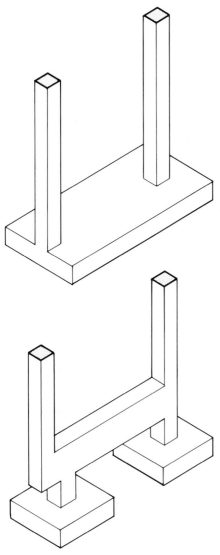

FIGURE 2.28

Either the combined footing (above) or the cantilevered footing (below) is used when column footings must abut a property line. By combining the foundation for the column against the property line, at the left, with the foundation for the next interior column, to the right, in a single structural unit, a balanced footing design can be achieved. The concrete reinforcing has been omitted from these drawings for the sake of clarity.

FIGURE 2.29

Pouring a large foundation mat. Six truck-mounted pumps receive concrete from a continuous procession of transit-mix concrete trucks and deliver this concrete to the heavily reinforced mat. Concrete placement continues nonstop around the clock until the mat is finished to avoid "cold joints" between hardened concrete and fresh concrete. The soil around this excavation is supported with a sitecast concrete slurry wall. Most of the slurry wall is tied back, but a set of rakers is visible at the lower right. (Courtesy of Schwing America, Inc.)

FIGURE 2.30

A cross section through a building with a floating foundation. The six stories of superstructure weigh approximately the same as the soil excavated for the substructure, so the stress in the soil beneath the building has not changed.

toe of the footing were simply cut off, the footing would be eccentrically loaded by the column or wall and would tend to rotate and fail. *Combined footings* and *cantilever footings* solve this problem by tying the footings for the outside row of columns to those of the first inside row in such a way that any rotational tendency is neutralized (Figure 2.28).

In situations where the allowable bearing capacity of the soil is low in relation to the weight of the building, column footings may become large enough that it is more economical to merge them into a single *mat* or *raft* foundation that supports the entire building. Mats for very tall buildings are often 6 feet (1800 mm) or more thick and are heavily reinforced (Figure 2.29).

Where the bearing capacity of the soil is low and settlement must be carefully controlled, a *floating* foundation is sometimes used. A floating foundation is essentially the same as a mat, but is placed beneath a substructure of a depth such that the weight of the soil removed from the excavation is equal to the weight of the building. Roughly speaking, one story of excavated soil weighs about the same as five to eight stories of superstructure, depending on the density of the

soil and the construction of the building. A floating foundation under two stories of basement space, for example, can support a ten- to sixteen-story building while imparting little new load to the soil beneath (Figure 2.30).

Deep Foundations

Caissons

A *caisson* (Figure 2.31) is similar to a column footing in that it spreads the load from a column over a large enough area of soil that the allowable stress in the soil is not exceeded. It differs from a column footing in that it reaches through strata of unsatisfactory soil beneath the substructure of a building until it reaches a satisfactory bearing stratum, such as rock, dense sands and gravels, or firm clay. A caisson is constructed by drilling or hand-digging a hole, belling out the bottom as necessary to achieve the required bearing area, and filling the hole with concrete. Large auger drills (Figures 2.32, 2.33) are used for drilling caissons; hand excavation is used only if the soil is too bouldery for the drill. A temporary cylindrical steel casing is usually lowered around the drill as it progresses, to support the soil around the hole. When firm bearing is reached, the bell, if required, is created at the bottom of the shaft either by hand excavation or by a special *belling bucket* on the drill

We must never trust too hastily to any ground... I have seen a tower at Mestre, a place belonging to the Venetians, which, in a few years after it was built, made its way through the ground it stood upon...and buried itself in earth up to the very battlements.

Leon Battista Alberti, 1404–1472

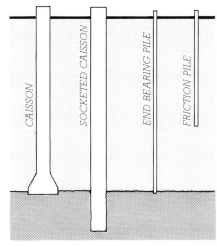

FIGURE 2.31

Deep foundations. The caissons are concrete cylinders poured into drilled holes. They reach through weaker soil (light shading) to bear on competent soil beneath. The end bearing caisson at the left is belled as shown when additional bearing capacity is required. The socketed caisson is drilled into a hard stratum and transfers its load primarily by friction between the soil or rock and the sides of the caisson. Piles are driven into the earth. The end bearing pile acts in the same way as a caisson. The friction pile derives its load-carrying capacity from friction between the soil and the sides of the pile.

(Figure 2.34). The bearing surface of the soil at the bottom of the hole is then inspected to be sure it is of the anticipated quality, and the hole is filled with concrete, with the casings withdrawn as the concrete rises. Reinforcing is seldom used in the concrete except near the top of the caisson, where it joins the column of the superstructure.

Caissons are large, heavy-duty units. Their shaft diameters range up to 6 feet (1800 mm) and more, from a minimum of 18 inches (460 mm) for a mechanical drill with a belling bucket, or 32 inches (800 mm) for hand excavation. Belled caissons are practical only where the bell can be excavated from a cohesive soil (such as clay) that can retain its shape until concrete is poured, and where the bearing stratum is im-

FIGURE 2.32
A six-foot (1828-mm) diameter auger on a telescoping 70-foot (21-m) bar brings up a load of soil from a caisson hole. The auger will be rotated rapidly to spin off the soil before being reinserted in the hole. (Courtesy of Calweld, Inc.)

FIGURE 2.33
For cutting through hard material, the caisson drill is equipped with a carbide-toothed coring bucket. (Courtesy of Calweld, Inc.)

FIGURE 2.34
The bell is formed at the bottom of the caisson shaft by a bell bucket with retractable cutters. The example shown here is for an 8-foot (2.44-m) shaft, and makes a bell 21 feet (6.40 m) in diameter. (Courtesy of Calweld, Inc.)

FIGURE 2.35

Installing a rock caisson foundation. (a) The shaft of the caisson has been drilled through softer soil to the rock beneath, and cased with a steel pipe. A churn drill is being lowered into the casing to begin advancing the hole into the rock. (b) When the hole has penetrated the required distance into the rock stratum, a heavy steel H section is lowered into the hole and suspended on steel channels across the mouth of the casing. The space between the casing and the H section is then filled with concrete, producing a caisson with a very high load-carrying capacity. (Courtesy of Franki Foundation Company)

pervious to the passage of water, to prevent flooding of the hole.

A *socketed caisson* (Figure 2.31) is one that is drilled into rock at the bottom, rather than belled. Its bearing capacity comes not only from its end bearing, but from the frictional forces between the sides of the caisson and the rock as well. Figure 2.35 shows the installation of a *rock caisson*, a special type of socketed caisson with a steel H-section core.

The term "caisson" originally referred to a watertight chamber within which foundation work could be carried out underwater. In common usage, it has also come to mean the type of deep foundation unit described in the preceding paragraphs. To avoid confusion, the term *pier* is sometimes used in the literature to describe the foundation unit.

Piles

A *pile* is distinguished from a caisson by being driven into place rather than drilled and poured. The simplest kind of pile is a tree trunk with its bark removed; it is supported small end down in a *piledriver* and beaten into the earth with repeated blows of a very heavy mechanical hammer. If this *timber pile* is driven until its tip encounters firm resistance from rock or dense sands and gravels it is an *end bearing pile*. If it is driven only into softer material, without encountering a firm bearing layer, it will still develop a considerable load-carrying capacity through the frictional resistance between the sides of the pile and the soil through which it is driven; in this case it is known as a *friction pile* (Figure 2.31). Piles are generally driven closely

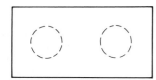

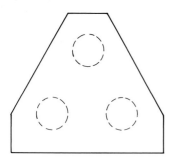

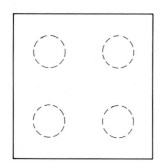

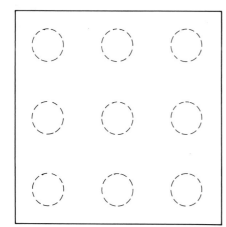

FIGURE 2.36

Clusters of 2, 3, 4, and 9 piles with their concrete caps, viewed from above. The caps are reinforced to transmit column loads equally into all the piles in the cluster, but the reinforcing has been omitted here for the sake of clarity.

together in *clusters* containing from two to twenty-five piles. Each cluster is later joined at the top by a reinforced concrete *pile cap*, which distributes the force of the column or wall equally among the piles (Figures 2.36, 2.37).

End bearing piles are used on sites where a firm bearing stratum exists at a depth that can be reached by piles. End bearing piles work exactly the same as caissons, but can be used when noncohesive soils or subsurface water conditions preclude the installation of caissons. Each pile is driven "to refusal," the point at which little additional penetration is made with continuing blows of the hammer, indicating that the pile has penetrated the bearing layer and is firmly embedded in it.

Friction piles are used when no firm bearing stratum exists at a reasonable depth, or when a high water table would make footings or caissons prohibitively expensive. Friction piles work best in silty, clayey, and sandy soils, and are driven to a predetermined depth or resistance rather than to refusal. Clusters of friction piles have the effect of distributing a column load to a large volume of soil around and below the cluster, at stresses that lie safely within the capability of the soil (Figure 2.38). The loadbearing capacities of piles are calculated in advance based on soil test results and the properties of the piles and piledriver. To verify the correctness of the calculation, test piles are often driven and loaded on the building site before foundation work begins.

Where piles are used to support loadbearing walls, reinforced concrete *grade beams* are constructed between the pile caps to transmit the wall loads to the piles (Figure 2.39). Grade beams are also used with caisson foundations for the same purpose.

Pile Driving Pile hammers are massive weights lifted by the energy of steam, compressed air, compressed hydraulic fluid, or a diesel explosion, then dropped against a block that is in firm contact with the top of the pile. Single-acting hammers fall by gravity

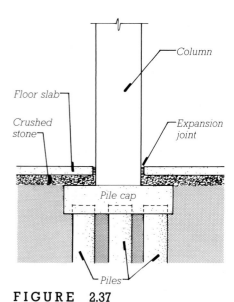

FIGURE 2.37

An elevation view of a pile cap, column, and floor slab. Where possible, the cap is poured against earth sides, as shown, for economy and for better resistance to lateral loads.

FIGURE 2.38

A single friction pile (left) transmits its load into the earth as an equal pressure along the bulb profile indicated by the dotted line. As the size of the pile cluster increases, the piles act together *to create a single, larger bulb that reaches deeper into the ground. A building with many closely spaced clusters of piles (right) creates a very large, deep bulb. Care must be taken* *on some sites to be sure that pressure bulbs do not become so large as to extend completely through a stratum of competent soil into softer or less stable material beneath.*

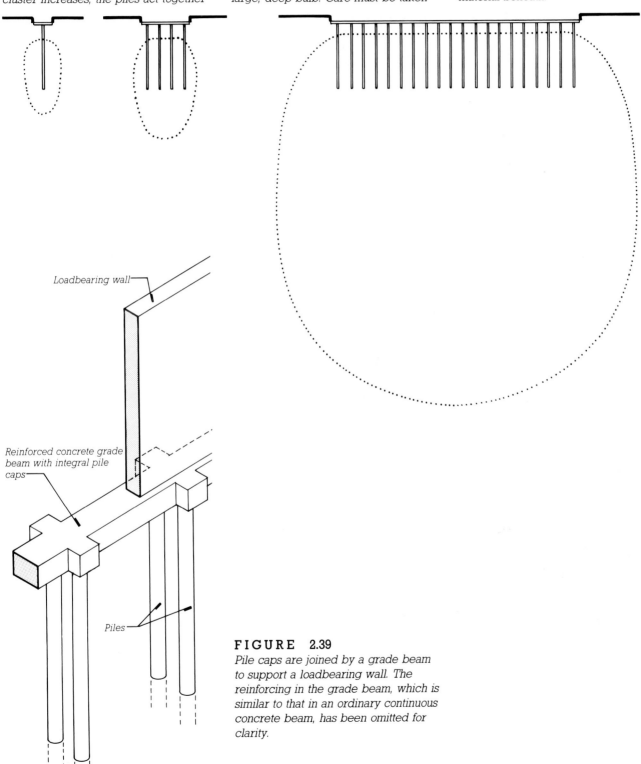

Loadbearing wall

Reinforced concrete grade beam with integral pile caps

Piles

FIGURE 2.39

Pile caps are joined by a grade beam to support a loadbearing wall. The reinforcing in the grade beam, which is similar to that in an ordinary continuous concrete beam, has been omitted for clarity.

FIGURE 2.40

A steam piledriver hammers a precast concrete pile into the ground. The pile is supported by the vertical structure (leads) of the piledriver, and driven by a heavy piston mechanism that follows it down the leads as it penetrates deeper into the soil. (Courtesy of Lone Star/San-Vel Concrete)

alone, while double-acting hammers are forced downward by reverse application of the energy source that lifts the hammer. The hammer travels on tall vertical rails called *leads* (pronounced "leeds") at the front of a piledriver (Figure 2.40). It is first hoisted up the leads to the top of each pile as driving commences, then follows the pile down as it penetrates the earth. The piledriver mechanism includes lifting machinery to raise each pile into position before driving.

In certain types of soil, piles can be driven more efficiently by vibration than by a heavy hammer, using a special vibratory hammer mechanism. When piles must be driven beneath an existing building to increase the capacity of its foundations, as is often necessary when increasing the height of a building, they are driven by hydraulic jacks acting against the weight of the building.

Pile Materials Piles may be made of timber, steel, sitecast or precast concrete, and various combinations of these materials (Figure 2.41). Timber piles have been in use since Roman times, at which time they were driven by large mechanical hammers hoisted by muscle power. Their main advantage is that they are economical for lightly loaded foundations. On the minus side, they cannot be spliced during driving, and are therefore limited in length to the length of available tree trunks, approximately 60 feet (18 m) maximum. Unless pressure treated, or completely submerged below the water table, they will decay. Relatively small hammers must be used in driving timber piles to avoid splitting them. Capacities of timber piles lie in the range of 8 to 30 tons each (7000 to 27,000 kg).

Two forms of steel piles are used, *H-piles* and *pipe piles*. H-piles are special hot-rolled, wide-flange sections, 8 to 14 inches deep (200 to 360 mm), that are approximately square in cross section. They are used mostly in end bearing applications. H-piles displace relatively little soil during driving, minimizing upward heaving of adjacent structures, which is often a problem on urban sites with large numbers of piles. They can be brought to the site in any convenient lengths, welded together as driving progresses to form any necessary length of pile, and cut off with an oxyacetylene torch when the required depth is reached. Cut-off ends can then be spliced onto other piles to avoid waste. Corrosion can be a problem in some soils, however, and unlike closed pipe piles and hollow precast concrete piles, H-piles cannot be inspected after driving to be sure they are straight and undamaged. Allowable loads on H-piles run from 30 to 100 tons (27,000 to 91,000 kg).

Steel pipe piles, with diameters of 8 to 15 inches (200 to 380 mm), may be driven with the lower end either open, or closed with a heavy steel plate. An open pile is easier to drive than a closed one, but its interior must be cleaned of soil and inspected before being filled with concrete, while a closed pile can be inspected and concreted immediately after driving. Pipe piles are stiff and can carry heavy loads (50 to 80 tons, or 45,000 to 73,000 kg). They displace relatively large amounts of soil during driving, which can lead to heaving of nearby soil and buildings. The larger sizes of pipe piles require a very heavy hammer for driving.

Precast concrete piles are square, octagonal, or round in section, and in the larger sizes often have open cores to allow inspection (Figures 2.41 through 2.43). Most are prestressed, but some for smaller buildings are merely reinforced. Sizes range from 10 to 54 inches (250 to 1370 mm), and bearing capacities from 60 to 940 tons (55,000 to 850,000 kg). Advantages of precast piles include high load capacities, an absence of corrosion or decay problems, and in most situations, a relative economy of cost. Precast piles must be handled carefully to avoid bending and cracking before installation. Splices between lengths of precast piling can be made effectively with mechanical fastening devices cast into the ends of the sections.

A sitecast concrete pile is made by driving a hollow steel shell into the ground and filling it with concrete. The shell is sometimes corrugated to in-

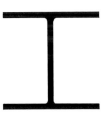
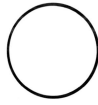

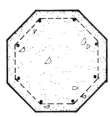

STEEL H-PILE STEEL PIPE PILE PRECAST CONCRETE PILE WOOD PILE

FIGURE 2.41
Common types of piles. Precast concrete piles may be square or round instead of the octagonal cross section shown, and may be hollow in the larger sizes.

FIGURE 2.42
Precast, prestressed concrete piles. The ends of the prestressing tendons can be seen in the piles on the ground. Lifting loops are cast into the sides of the piles as crane attachments for lifting them into a vertical position. (Courtesy of Lone Star/San-Vel Concrete)

crease its stiffness; if the corrugations are circumferential, a heavy steel *mandrel* (a tight-fitting liner) is inserted in the shell during driving to protect the shell from collapse, then withdrawn before concreting. Some shells with longitudinal corrugations are stiff enough that they do not require mandrels. Some types of man-drel-driven piles are limited in length, and the larger diameters of sitecast piles (up to 16 inches, or 400 mm) can cause ground heaving. Load capacities range up to 80 tons (73,000 kg) for sitecast piles with shells, or 60 tons (55,000 kg) without shells. The primary reason to use sitecast concrete piles is their economy.

There are many proprietary systems of sitecast concrete piles, each with various advantages and disadvantages (Figures 2.44, 2.45). Pressure-injected footings (Figure 2.45) are interesting because they share characteristics of piles, piers, and footings and because they are highly resistant to uplift forces.

FIGURE 2.43
A driven cluster of six precast concrete piles, ready for cutting off and capping. (Photo by Alvin Ericson)

FIGURE 2.44
Some proprietary types of sitecast concrete piles. All are cast into steel casings that have been driven into the ground; the uncased piles are made by withdrawing the casing as the concrete is poured, and saving it for subsequent reuse. The numbers refer to the methods of driving that may be used with each: **1.** *Mandrel driven.* **2.** *Driven from the top of the tube.* **3.** *Driven from* the bottom of the tube to avoid buckling it. **4.** *Jetted. Jetting is accomplished by advancing a high-pressure water nozzle ahead of the pile to wash the soil back alongside the pile to the surface. Jetting has a tendency to disrupt the soil around the pile, so is not a favored method of driving under most circumstances.*

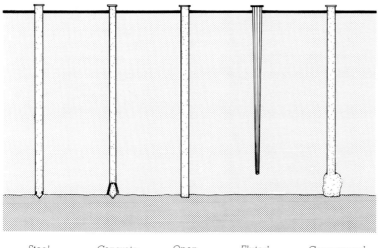

Steel Point 1,2,3,4	Concrete Plug 3	Open Ended 2	Fluted Tapered 2	Compressed Base 1

CASED PILES

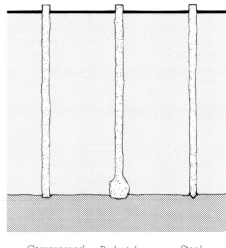

Compressed Concrete 1	Pedestal Pile 1	Steel Point 2

UNCASED PILES

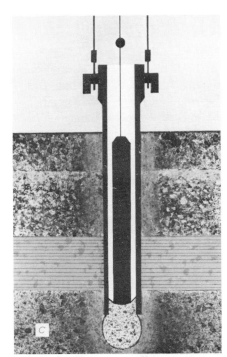

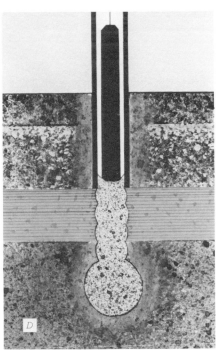

FIGURE 2.45

Steps in the construction of a proprietary pressure-injected, bottom-driven concrete footing. (a) A charge of a very dry concrete mix is inserted into the bottom of the steel drive tube at the surface of the ground and compacted into a sealing plug with repeated blows of a drop hammer. (b) As the drop hammer drives the sealing plug into the ground, the drive tube is pulled along by the friction between the plug and the tube. (c) When the desired depth is reached, the tube is held and a bulb of concrete is formed by adding small charges of concrete and driving the concrete out into the soil with the drop hammer. The bulb provides an increased bearing area for the pile and strengthens the bearing stratum by compaction. (d, e) The shaft is formed of additional compacted concrete as the tube is withdrawn. (f) Charges of concrete are dropped into the tube from a special bucket supported on the leads of the driving equipment. (Courtesy of Franki Foundation Company)

UNDERPINNING

Underpinning is the process of strengthening and stabilizing the foundations of an existing building. It may be required for any of a number of reasons: The existing foundations may never have been large enough to carry their loads, leading to excessive settlement of the building. A change in building use or additions to the building may overload the existing foundations. New construction near a building may disturb the soil around its foundations or require that the foundations be carried deeper. Whatever the cause, underpinning is a slow, expensive, highly specialized task that is seldom the same for any two buildings. Three different paths are available when foundation capacity needs to be increased: The foundations may be enlarged; new, deep foundations can be inserted under shallow ones to carry the load to a deeper, stronger stratum of soil; or the soil itself can be strengthened by grouting or by chemical treatment. Figures 2.46 and 2.47 illustrate in diagrammatic form some concepts of underpinning.

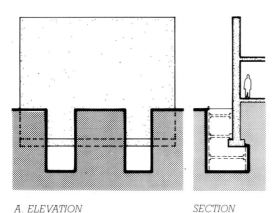

A. ELEVATION   SECTION

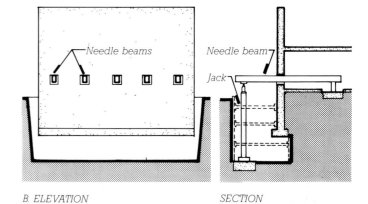

B. ELEVATION   SECTION

FIGURE 2.46

Two methods of supporting a building while carrying out underpinning work beneath its foundation, each shown in elevation and section. **A.** *Trenches are dug beneath the existing foundation at intervals, leaving the majority of the foundation supported by the soil. When* underpinning has been completed in the trenches, another set of trenches may be dug to expose more of the foundation, and so on. **B.** *The foundations of an entire wall can be exposed at once by* needling, *in which the wall is supported temporarily on* needle beams *threaded through holes cut in the wall. After underpinning has been accomplished, the jacks and needle beams are removed and the trench is backfilled.*

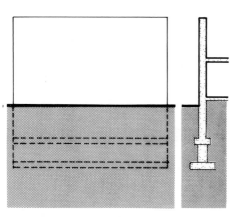

A. ELEVATION   SECTION

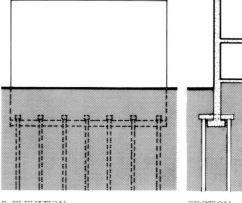

B. ELEVATION   SECTION

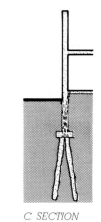

C. SECTION

FIGURE 2.47

Three types of underpinning. **A.** *A new foundation wall and footing are constructed beneath the existing foundation.* **B.** *New piles or caissons* are constructed on either side of the existing foundation. **C.** *Sitecast concrete mini-piles are cast into holes drilled diagonally through the existing* foundation. *Mini-piles do not generally require excavation or temporary support of the building.*

RETAINING WALLS

A *retaining wall* is a wall that holds soil back to create an abrupt change in the elevation of the ground. A retaining wall must resist the pressure of the earth that bears against it on the uphill side. Retaining walls may be made of masonry, preservative-treated wood, coated or galvanized steel, precast concrete, or, most commonly, sitecast concrete.

The structural design of retaining walls is complicated by such factors as the height of the wall, the character of the soil behind the wall, the presence or absence of ground water behind the wall, any structures whose foundations apply pressure to the soil behind the wall, and the character of the soil beneath the base of the wall, which must support the footing that keeps the wall in place. The rate of structural failure in retaining walls is high relative to the rate of failure in other types of structures, and may occur through fracture of the wall, overturning of the wall due to soil failure, lateral sliding of the wall, or undermining of the wall by flowing ground water (Figure 2.48). Careful engineering design and site supervision are crucial to the success of a retaining wall.

There are many ways of building retaining walls. For walls less than 3 feet (900 mm) in height, simple, unreinforced walls of various types are often appropriate (Figure 2.49). For taller walls, and ones subjected to unusual loadings or ground water, the type most frequently employed today is the *cantilevered concrete retaining wall*, two examples of which are shown in Figure 2.50. The shape and reinforcing of a cantilevered wall can be custom designed to suit almost any situation.

Earth reinforcing (Figure 2.51) is an economical alternative to conventional retaining walls in many situations. Soil is compacted in thin layers, alternating with strips or meshes of galvanized steel, polymer fibers, or glass fibers, which stabilize the soil in much the same manner as the roots of plants.

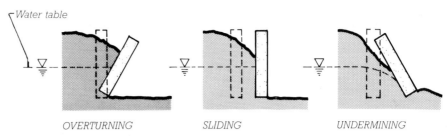

OVERTURNING   SLIDING   UNDERMINING

FIGURE 2.48
Three failure mechanisms in retaining walls.

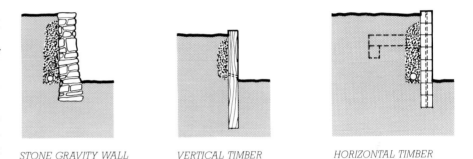

STONE GRAVITY WALL

VERTICAL TIMBER CANTILEVERED WALL

HORIZONTAL TIMBER WALL WITH DEADMEN

FIGURE 2.49
Three types of simple retaining walls, usually used for heights not exceeding 3 feet (900 mm). The deadmen in the horizontal timber wall are timbers embedded in the soil behind the wall and connected to it with timbers inserted into the wall at right angles. The timbers, which should be pressure treated, are held together with very large spikes, or with steel reinforcing bars driven into drilled holes. The crushed stone drainage trench behind each wall is important as a means of relieving water pressure against the wall to prevent wall failure. With proper engineering design, any of these types of construction can also be used for taller retaining walls.

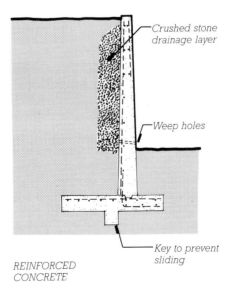

REINFORCED CONCRETE

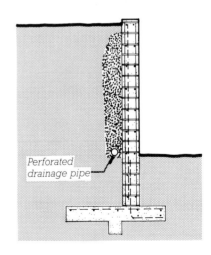

Crushed stone drainage layer

Weep holes

Key to prevent sliding

Perforated drainage pipe

REINFORCED CONCRETE MASONRY

FIGURE 2.50
Cantilevered retaining walls of concrete and concrete masonry. The footing is shaped to resist sliding and overturning, and drainage behind the wall reduces the likelihood of undermining. The pattern of steel reinforcing (broken lines) is designed to resist the tensile forces in the wall.

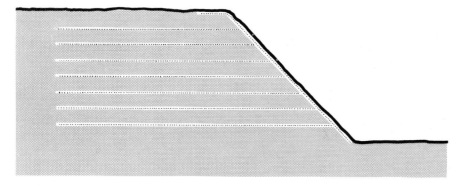

FIGURE 2.51
Earth reinforcing. The embankment in the top section is placed in layers of earth alternated with layers of synthetic mesh fabric. The retaining wall in the lower section is made of precast concrete panels fastened to long galvanized steel straps running back into the soil.

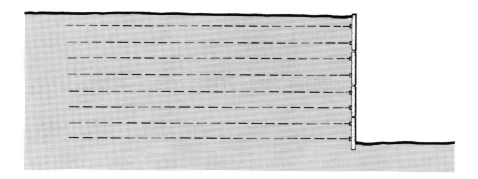

Waterproofing and Drainage

The substructure of a building is subject to penetration of ground water, especially if it lies below the water table. Even if the concrete walls of a basement are perfectly constructed, water can work its way to the inside through the microscopic pores in the concrete, and basement walls are usually far from perfect, being riddled with shrinkage cracks, form tie holes, and joints between pours of concrete.

There are two fundamental approaches to waterproofing: *waterproof membranes* and *drainage*. In theory either will work independently of the other, but in practice they are usually combined. Waterproof membranes may be made of plastic, asphaltic, or synthetic rubber sheets; of asphaltic mastics or coatings; of special cementitious plasters; or of bentonite clay.

The cementitious plasters can generally be applied either inside or outside the foundation, with some preference for the inside because in this location the membrane may be inspected and repaired as needed. The major problem with plasters is that they are brittle and will crack and leak if the foundation cracks. Sheets, mastics, coatings, and clay membranes must generally be applied to the outside of the wall where they can be supported against water pressures by the wall. Membranes outside the wall cannot be reached for repair, so the installation must be inspected with extreme care before the excavation around the foundation wall is *backfilled* with soil. Once successfully installed and backfilled, such membranes are shielded from weather and will last for a very long time, provided they are not subsequently torn by foundation cracking. The safest of all membranes is perhaps bentonite clay, which, when wetted,

swells to several times its dry volume and becomes practically impervious to the passage of water. Moist bentonite clay is sufficiently flexible to adjust even to fairly large movements in the wall it protects. The clay is either sprayed onto the wall as a slurry, or applied in the form of corrugated cardboard sheets whose cells are filled with dry clay. When the panels are wetted, the clay expands to form a continuous membrane, and the cardboard disintegrates.

At expansion joints and joints between pours of concrete in basement walls, structural movement is likely to cause cracking and leakage in most types of waterproof membranes. Preformed synthetic rubber *waterstops* are cast into the mating concrete edges at these joints to maintain a seal despite subsequent movement (Figures 2.52, 2.53).

Drainage is more secure than membranes in keeping a basement dry and

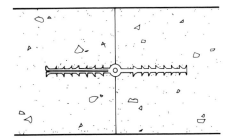

SECOND POUR   FIRST POUR

FIGURE 2.52

A synthetic rubber waterstop is used to seal against water penetration at expansion joints and at joints between pours of concrete in a foundation. The type shown here is split on one side so its halves can be placed flat against the end of the formwork. After the formwork is removed from the first pour, the split halves are reunited before the next pour.

FIGURE 2.53

Installing waterstops in a building foundation. (a) Workers unroll the waterstop around a concrete form, where it will be inserted in the top of the footing. (b) The top half of the waterstop is left projecting above the concrete of the footing and will seal the joint between the footing and the concrete wall that will be poured atop the footing. (c) A waterstop ready for the next pour of a concrete wall, as diagrammed in Figure 2.52. (Courtesy of Vulcan Metal Products, Inc., Birmingham, Alabama)

has the advantage of preventing the buildup of potentially destructive water pressure against basement walls and slabs. A slab, in particular, is many times less expensive if it does not have to be reinforced against hydrostatic pressure. Several schemes for basement drainage are shown in Figures 2.54 and 2.55. The perforated pipes are usually 4 inches (100 mm) in diameter, and serve to maintain an open channel in the rock bed through which water can flow by gravity either "to daylight" below the building on a sloping site, or to a sump that is periodically pumped dry. With a properly installed drainage system, the only function of membrane waterproofing is to keep soil humidity from permeating the basement walls.

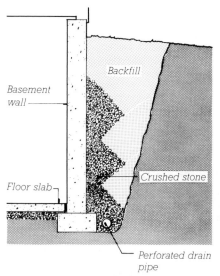

FIGURE 2.54

Two methods of relieving water pressure around a building substructure. The gravel drain (left) is the standard method, but it is hard to do well because of the difficulty of depositing the crushed stone and backfill soil in neatly separated, alternating layers. The drainage blanket or drainage panel (right) is much easier and often more economical to install. It is a manufactured component that may

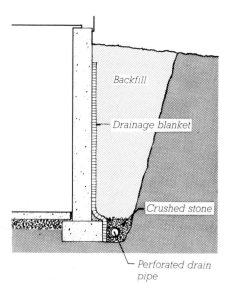

be made of a loose mat of inert fibers, a plastic eggcrate structure, or some other very open, porous material. It is faced on the outside with a filter fabric that prevents fine soil particles from entering and clogging the drainage passages in the blanket or panel. Water approaching the wall falls through the porous material to the drain pipe at the footing.

DESIGNING FOUNDATIONS

It is a good idea to begin the design of the foundations of a building at the same time as architectural design work commences. Subsurface conditions beneath a site can strongly influence several fundamental decisions about a building—its location, its size and shape, its weight, and the required degree of flexibility of its construction. On a large building project at least three designers are involved in these decisions: the architect, who has primary responsibility for the location and form of the building; the structural engineer, who has primary responsibility for its physical integrity; and the foundation engineer, who must decide on the basis of site exploration and laboratory reports how best to support it in the earth. More often than not it is possible for the foundation engineer to design foundations for a building design dictated entirely by the architect, but in some cases the cost of the foundations may consume a much larger share of the construction budget than the architect has anticipated, unless certain compromises can be reached on the form and location of the building. It is safer and more productive for the architect to work with the foun-

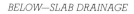
BELOW—SLAB DRAINAGE

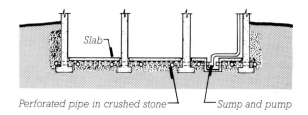

Slab

Perforated pipe in crushed stone — Sump and pump

ABOVE—SLAB DRAINAGE

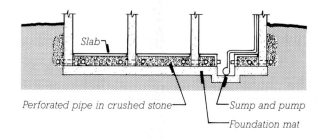

Slab

Perforated pipe in crushed stone — Sump and pump — Foundation mat

FIGURE 2.55

For a high degree of security against substructure flooding, drainage both around and under the basement is required. Above-slab drainage is used in buildings with mat foundations.

Foundations
ought to be twice as thick as
the wall to be built on them;
and regard in this should be
had to the quality of the
ground, and the largeness of
the edifice; making them
greater in soft soils, and very
solid where they are to
sustain a considerable weight.

The bottom of the trench
must be level, that the weight
may press equally, and not
sink more on one side than
on the other...

Andrea Palladio, in *The Four Books
of Architecture,* first published
in 1570.

dation engineer from the outset, seeking alternative site locations and building configurations that will result in the fewest foundation problems and the lowest foundation cost.

In designing a foundation, a number of different design thresholds need to be kept in mind. If the designer crosses any of these thresholds, foundation costs take a sudden jump. Some of these thresholds are

- *Building below the water table.* If the substructure and foundations of a building are above the water table, little or no effort will be required to keep the excavation dry during construction. If the water table is penetrated, whether by an inch or a hundred feet, expensive steps will have to be taken to dewater the site. For an extra inch or foot of depth, the expense would probably not be justified; for another story or two of useful building space, it might be.

- *Building close to an existing structure.* If the excavation can be kept well away from adjacent structures, the foundations of these structures can remain undisturbed and no effort and expense are required to

protect them. When digging close to an existing structure, and especially when digging deeper than its foundations, the structure will probably have to be braced temporarily against structural collapse, and may require permanent underpinning with new foundations. Furthermore, an excavation at a distance from an existing structure may not require sheeting, while one immediately adjacent almost certainly will.

- *Increasing the column or wall load from a building beyond what can be supported by a shallow foundation.* Shallow foundations are far less expensive than piles or caissons under most conditions. If the building grows too tall, however, a shallow foundation may no longer be able to carry the load, and a threshold must be crossed into the realm of deep foundations. If this has happened for the sake of an extra story or two of height, the designer should consider reducing the height by broadening the building. If individual column loadings are too high for shallow foundations, perhaps they can be re-

duced by increasing the number of columns in the building and decreasing their spacing.

For very small buildings, at the scale of one- and two-family dwellings, foundation design is usually much simpler than for large buildings because foundation loadings are low, and the uncertainties in foundation design can be reduced with reasonable economy by adopting a large factor of safety in calculating the bearing capacity of the soil. Unless the designer has reason to suspect poor soil conditions, the footings are usually designed using rule-of-thumb allowable soil stresses and standardized footing proportions. The designer then examines the actual soil when the excavations have been made. If it is not of the quality that was expected, the footings can be hastily redesigned using a revised estimate of soil-bearing capacity before construction continues. If unexpected ground water is encountered, better drainage provisions may have to be provided around the foundation, or the depth of the basement decreased.

C.S.I./C.S.C. Masterformat Section Numbers for Foundations	
02010	SUBSURFACE INVESTIGATION
02050	DEMOLITION
02100	SITE PREPARATION
02110	Site Clearing
02110	Structure Moving
02140	DEWATERING
02150	SHORING AND UNDERPINNING
02160	EXCAVATION SUPPORT SYSTEMS
02162	Cribbing and Walers
02164	Soil and Rock Anchors
02168	Slurry Wall Construction
02200	EARTHWORK
02210	Grading
02220	Excavation, Backfilling, and Compacting
02350	PILES AND CAISSONS
02355	Pile Driving
02360	Driven Piles
02380	Caissons
02500	PAVING AND SURFACING

SELECTED REFERENCES

1. Ambrose, James E. *Simplified Design of Building Foundations*. New York, John Wiley and Sons, Inc., 1981.

After an initial summary of soil properties, this small book covers simplified foundation computation procedures for both shallow and deep foundations.

2. Liu, Cheng, and Jack B. Evett.

Soils and Foundations. Englewood Cliffs, New Jersey, Prentice-Hall, Inc., 1981.

This book is a fairly detailed discussion of the engineering properties of soils, subsurface exploration techniques, soil mechanics, and shallow and deep foundations, but is well suited to the beginner.

3. Schroeder, W. L. *Soils in Construction*. New York, John Wiley and Sons, Inc., 1980.

A well-illustrated, clearly written, moderately detailed survey of soils, soil testing, subsurface construction, and foundations.

KEY TERMS AND CONCEPTS

foundation
dead load
live load
wind load
thrust
foundation settlement
total settlement
differential settlement
rock
soil
boulder
cobble
gravel
sand
fine-grained soil
silt
clay
strata
excavation
sheeting
soldier beams
lagging
sheet piling
tremie
slurry wall

precast slurry wall
bracing
crosslot bracing
walers
rakers
tiebacks
rock anchors
dewatering
well points
watertight barrier
superstructure
substructure
shallow foundations
deep foundations
footing
column footing
wall or strip footing
engineered fill
slab on grade
crawlspace
basement
combined footing
cantilever footing
mat or raft foundation
floating foundation

caisson
belled caisson
socketed caisson
pier
pile
end bearing pile
friction pile
piledriver
pile hammer
grade beam
H-pile
pipe pile
precast concrete pile
sitecast concrete pile
mandrel
pressure-injected footing
underpinning
retaining wall
waterproof membrane
waterstop
drainage
backfill
bentonite clay

REVIEW QUESTIONS

1. What is the nature of the most common type of failure in foundations?

2. Explain in detail the differences among fine sand, silt, and clay, especially as they relate to foundations for buildings.

3. List three different ways of sheeting an excavation.

4. Under what conditions would you probably use a watertight barrier instead of well points when digging below the water table?

5. If shallow foundations are substantially less costly than deep foun-

dations, why do we use deep foundations?

6. What particular soil conditions are favorable to the use of belled caissons?

7. Which types of friction piles can carry the heaviest loads per pile?

8. List some cost thresholds frequently encountered in foundation design, and explain each of them.

9. The Bible advises us not to build a house upon sand. Under what conditions can we build safely on sand, and under what conditions can we not?

E X E R C I S E S

1. Obtain the foundation drawings for a nearby building. Look first at the log of test borings. What sorts of soils are found beneath the site? How deep is the water table? What type of foundation do you think should be used in this situation (keeping in mind the relative weight of the building)? Now look at the foundation drawings. What type is actually used? Why?

2. What type of foundation is normally used for houses in your area? Why?

3. Look at several excavations for major buildings under construction. Note carefully the arrangements made for slope support and dewatering. How is the soil being loosened and carried away? What is being done with the excavated soil? What type of foundation is being installed? What provisions are being made for keeping the substructure permanently dry?

WOOD

(*Weyerhaeuser Company Photo*)

Wood is probably the best loved of all the materials we use for building. It delights the eye with its endlessly varied colors and grain patterns. It invites the hand to feel its subtle warmth and varied textures. When it is fresh from the saw, its fragrance enchants. We treasure its natural, organic qualities and take pleasure in its genuineness. Even as it ages, bleached by the sun, eroded by rain, worn by the passage of feet and the rubbing of hands, we find beauty in its transformations of color and texture.

Wood earns our respect as well as our love. It is strong and stiff, yet by far the least dense of the materials used for the beams and columns of buildings. It is worked and fastened easily with small, simple, relatively inexpensive tools. It is readily re-cycled from demolished buildings for use in new ones, and when finally discarded, it biodegrades rapidly to become natural soil. It is our only renewable building material, one that will be available to us for as long as we manage our forests with an eye to the perpetual production of wood.

But wood, like a valued friend, has its idiosyncrasies. A piece of lumber is never perfectly straight or true, and its size and shape can change significantly with changes in the weather. Wood is peppered with defects that are relics of its growth and processing. Wood can split; wood can warp; wood can give splinters. If ignited, wood burns. If left in a damp location, it decays and harbors destructive insects. But the skillful designer and the seasoned carpenter know all these things and understand how to build with wood to bring out its best qualities, while neutralizing or minimizing its problems.

TREES

Since wood comes from trees, an understanding of tree physiology is essential to knowing how to build with wood.

Tree Growth

The trunk of a tree is covered with a protective layer of dead *bark* (Figure 3.1). Inside the dead bark is a layer of living bark composed of hollow longitudinal cells that conduct nutrients down the trunk from the leaves to the roots. Inside this layer of living bark lies a very thin layer, the *cambium*, which creates new bark cells toward the outside of the trunk and new wood cells toward the inside. The thick layer of living wood cells inside the cambium is the *sapwood*. In this zone of the tree, nutrients are stored and sap is pumped upward from the roots to the leaves and distributed laterally in the trunk. At the inner edge of this zone, sapwood dies progressively and becomes *heartwood*. In many species of trees, heartwood is easily distinguished from sapwood by its darker

FIGURE 3.1

Summerwood rings are prominent and a few rays are faintly visible in this cross section of an evergreen tree, but the cambium, which lies just beneath the thick layer of bark, is too thin to be seen, and heartwood cannot be distinguished visually from sapwood. (Courtesy of Forest Products Laboratory, Forest Service, USDA)

color. Heartwood no longer participates in the life processes of the tree but continues to contribute to its structural strength. At the very center of the trunk, surrounded by heartwood, is the *pith* of the tree, a small zone of weak wood cells that were the first year's growth.

An examination of a small section of wood under a low-powered microscope shows that it consists primarily of hollow, cylindrical cells whose long axes are parallel to the long axis of the trunk. The cells are structured of tough *cellulose* and are bound together by a softer cementing substance called *lignin*. The direction of the long axes of the cells is referred to as the *direction of grain* of the wood. Grain direction is important to the designer of wooden buildings because the properties of wood parallel to grain and perpendicular to grain are very different.

In temperate climates, the cambium begins to manufacture new sapwood cells in the spring, when the air is cool and ground water is plentiful, conditions that favor rapid growth. Growth is slower during the heat of the summer when water is scarce. *Springwood* or *earlywood* cells are therefore larger and less dense in substance than the *summerwood* or *latewood* cells. Concentric bands of springwood and summerwood make up the annual growth rings in a trunk that can be counted to determine the age of a tree. The relative proportions of springwood and summerwood also have a direct bearing on the structural value of the wood a given tree will yield because summerwood is stronger and stiffer than springwood. A tree grown under continuously moist, cool conditions grows faster than another tree of the same species grown under warmer, drier conditions, but its wood is not as dense or as strong.

Softwoods and Hardwoods

Softwoods come from coniferous trees and *hardwoods* from broad-leafed trees. The names can be deceptive because many coniferous trees produce harder

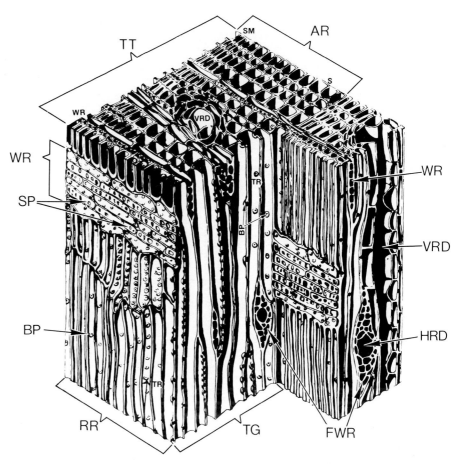

Cell structure of a softwood

FIGURE 3.2

Vertical cells (tracheids, labeled TR) dominate the structure of a softwood, seen here greatly enlarged, but rays (WR), which are cells that run radially from the center of the tree to the outside, are clearly in evidence. An annual ring (labeled AR) consists of a layer of smaller summerwood cells (SM) and a layer of larger springwood cells (S). Simple pits (SP) allow sap to pass from ray cells to longitudinal cells and vice versa. Resin is stored in vertical and horizontal resin ducts (VRD and HRD), with the horizontal ducts centered in fusiform wood rays (FWR). Border pits (BP) allow the transfer of sap between longitudinal cells. The face of the sample labeled RR represents a radial cut through the tree, and TG, a tangential cut. (Courtesy of Forest Products Laboratory, Forest Service, USDA)

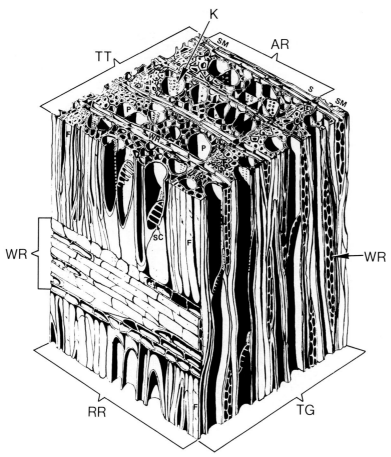

Cell structure of a hardwood

FIGURE 3.3

Rays (WR) constitute a large percentage of the mass of a hardwood, as seen in this sample, and are largely responsible for the beautiful grain figures associated with many species. The vertical cell structure is more complex than that of a softwood, with large pores (P) to transport the sap, and smaller wood fibers (F) to give the tree structural strength. Pore cells in some hardwood species end with crossbars (SC), while those of other species are entirely open. Pits (K) pass sap from one cavity to another. (Courtesy of Forest Products Laboratory, Forest Service, USDA)

woods than many broad-leafed trees, but the distinction is nevertheless a useful one. Softwood trees have a relatively simple microstructure, consisting mainly of large longitudinal cells *(tracheids)* together with a small percentage of radial cells *(rays)*, whose function is the storage and radial transfer of nutrients (Figure 3.2). Hardwood trees are more complex in structure, with a much larger percentage of rays and two different types of longitudinal cells: small-diameter *fibers*, and larger-diameter *vessels* or *pores*, which transport the sap of the tree (Figure 3.3).

When cut into lumber, softwoods generally have a coarse and relatively uninteresting grain structure, while many hardwoods show beautiful patterns of rays and vessels. Most of the lumber used today for building framing comes from softwoods, which are comparatively plentiful and inexpensive. For fine furniture and interior finish details, hardwoods are often chosen. A few softwood and hardwood species widely used in North America are listed in Figure 3.4, along with the principal uses of each; however, it should be borne in mind that literally thousands of species of wood are used in construction around the world and that the available species vary considerably with geographic location. The major lumber-producing forests in North America are in the western and eastern mountains of both the United States and Canada, and the southeastern United States, but other regions also produce significant quantities.

LUMBER

Sawing

The production of *lumber*, lengths of squared wood for use in construction, begins with the felling of trees and the transportation of the logs to a sawmill (Figure 3.5). Sawmills range in size

SOFTWOODS	HARDWOODS

Used for Framing, Sheathing, Paneling

Alpine fir
Balsam fir
Douglas fir
Eastern hemlock
Eastern spruce
Eastern white pine
Englemann spruce
Idaho white pine
Larch
Loblolly pine
Lodgepole pine
Longleaf pine
Mountain hemlock
Ponderosa pine
Red spruce
Shortleaf pine
Sitka spruce
Western hemlock
White spruce

Used for Moldings, Window and Door Frames

Ponderosa pine
Sugar pine
White pine

Used for Finish Flooring

Douglas fir
Longleaf pine

Decay-Resistant Woods, Used for Shingles, Siding, Outdoor Structures

California redwood
Southern cypress
Western red cedar
White cedar

Used for Moldings, Paneling, Furniture

Ash
Beech
Birch
Black walnut
Butternut
Cherry
Lauan
Mahogany
Pecan
Red oak
Rosewood
Teak
Tupelo gum
White oak
Yellow poplar

Used for Finish Flooring

Pecan
Red oak
Sugar maple
Walnut
White oak

FIGURE 3.4

Some species of woods commonly used in construction in North America, listed alphabetically in groups according to end use. All are domestic except Lauan (Asia), Mahogany (Central America), Rosewood (South America and Africa), and Teak (Asia).

FIGURE 3.5
Loading Southern pine logs onto a truck for their trip to the sawmill. (Courtesy of Southern Forest Products Association)

from tiny family operations to giant, semiautomated factories, but the process of lumber production is much the same regardless of scale. Each log is first passed repeatedly through a large *headsaw*, which may be either a circular saw or a bandsaw, to reduce it to untrimmed slabs of lumber (Figures 3.6, 3.7). The *sawyer* (with the aid of a computer in the larger mills) judges how to obtain the maximum marketable wood from each log and uses hydraulic machinery to rotate and advance the log between passes through the saw in order to achieve the required succession of cuts. As the slabs fall from the log at each pass, they are carried away on a conveyor belt to smaller saws that reduce them to square-edged pieces of the desired widths (Figure 3.8). The sawn pieces at this stage of production have rough-textured surfaces and may vary slightly in dimension from one end to the other.

Most lumber intended for use in the framing of buildings is *plainsawed*, a method of dividing the log that

FIGURE 3.6
In this small sawmill, a spruce log is passed repeatedly through a circular headsaw to cut it into rough lumber, which is carried away by the conveyor belt in the foreground. (Photo by the author)

FIGURE 3.7
In a large, mechanized mill the operator controls the high speed bandsaw from an overhead booth. (Weyerhaeuser Company Photo)

FIGURE 3.8
Separating the sawn boards from the bark slabs, which are used for fuel or as feedstock for wood products of various kinds. (Courtesy of Southern Forest Products Association)

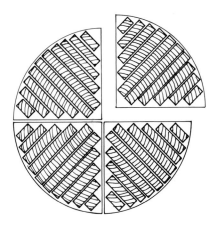 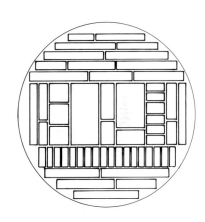

PLAINSAWING

QUARTERSAWING

TYPICAL SAWING OF A LARGE LOG

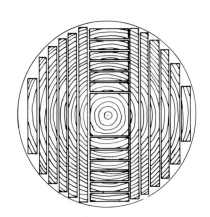 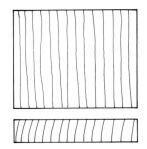

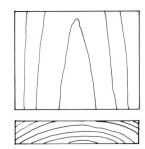

FIGURE 3.9
Plainsawing produces boards with a broad grain figure, as seen in end and top views below the plainsawed log. Quartersawing produces a vertical grain structure which is seen on the face of

the board as tightly spaced parallel summerwood lines. A large log of softwood is typically sawn to produce some large timbers, some plainsawed dimension lumber, and, in the horizontal

row of small pieces seen just below the heavy timbers, some pieces of vertical-grain decking.

produces the maximum yield and therefore the greatest economy (Figure 3.9). In plainsawed lumber, some pieces have the annual rings running virtually perpendicular to the faces of the piece, some have the rings on various diagonals, and some have the rings running almost parallel to the faces. These varying grain orientations cause the pieces to distort differently during seasoning, to have very different surface appearances from one another, and to erode at different rates when used in applications such as flooring and exterior siding. For uses where any of these variations will cause a problem, especially for finish flooring, interior trim, and furniture, hardwoods are often *quartersawn* to produce pieces of lumber that have the annual rings running more or less perpendicular to the face of the piece (Figure 3.9). Boards with this *vertical grain* orientation tend to remain flat despite changes in moisture content. The visible grain on the surface of a quartersawn piece makes a tighter and more pleasing *figure*; the wearing qualities

> **N**or do I ever come to a lumber yard with its citylike, graduated masses of fresh shingles, boards and timbers, without taking a deep breath of its fragrance, seeing the forest laid low in it by processes that cut and shaped it to the architect's scale of feet and inches...
>
> Frank Lloyd Wright

of the piece are improved because there are no broad areas of soft springwood exposed in the face, as there are when the annual rings run more nearly parallel to the face.

Seasoning

Growing wood contains a quantity of water that can vary from about 30 per-

cent to as much as 300 percent of the oven-dry weight of the wood. After the wood is cut, this water slowly starts to evaporate. First to leave the wood is the *free water* that is held in the cavities of the cells. When the free water is gone, the wood still contains about 26 to 32 percent moisture, and the *bound water* held by the cell walls themselves begins to evaporate. As the first of the bound water disappears, the wood starts to shrink, and the strength of the wood begins to increase. The shrinkage and strength increase steadily as the moisture content decreases. Wood can be dried to any desired moisture content, but framing lumber is considered to be *seasoned* when its moisture content is 19 percent or less. It is of little use to season ordinary framing lumber to a moisture content below about 13 percent because wood is hygroscopic and will take on or give off moisture, swelling or shrinking as it does so, in order to stay in equilibrium with the moisture in the air.

Most lumber is seasoned at the saw-

FIGURE 3.10
For proper air drying, lumber is supported well off the ground. The stickers, which keep the boards separated for ventilation, are carefully placed above one another to avoid bending the lumber, and a watertight roof protects each stack from rain and snow. (Courtesy of Forest Products Laboratory, Forest Service, USDA)

FIGURE 3.11
Drying kilns season huge stacks of lumber in the span of a few days. (Photo by the author)

mill, either by being air-dried in loose stacks for a period of months, or more commonly, by being dried in a kiln under closely controlled conditions of temperature and humidity for a period of days (Figures 3.10, 3.11). Seasoned lumber is stronger and stiffer than unseasoned *(green)* lumber, more dimensionally stable, and lighter and more economical to ship. For most purposes, kiln drying is preferred to air drying because it is much faster and produces better quality lumber.

Wood does not shrink and swell uniformly with changes in moisture content. Moisture shrinkage along the length of the log *(longitudinal shrinkage)* is essentially negligible in building practice. Shrinkage in the radial direction is very large by comparison, and shrinkage around the circumference of the log *(tangential shrinkage)* is about half again greater than radial shrinkage (Figure 3.12). If an entire log is seasoned before cutting, it will shrink very little along its length, but it will grow noticeably smaller in diameter, and the difference between

the tangential and radial shrinkage will cause it to *check*, that is, to split open all along one face (Figure 3.13).

These differences in shrinkage rates are so large that they cannot be ignored in building design. In constructing frames of plainsawed lumber, a simple distinction is made between parallel-to-grain shrinkage, which is negligible, and perpendicular-to-grain shrinkage, which is considerable. The difference between radial and tangential shrinkage is usually ignored because the orientation of the annual rings in plainsawed lumber is random and unpredictable. Wood building frames are carefully detailed to equalize the amount of wood loaded perpendicular to grain from one side of the structure to the other in order to avoid the noticeable tilting of floors and tearing of wall finish materials that would otherwise occur.

The position in a log from which a piece of lumber is sawn determines in large part how it will distort as it dries. Figure 3.14 shows how the differences between tangential and radial shrink-

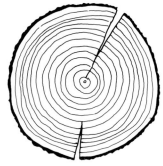

FIGURE 3.13
Because tangential shrinkage is so much greater than radial shrinkage, high internal stresses are created in a log as it dries, resulting inevitably in the formation of radial cracks called checks.

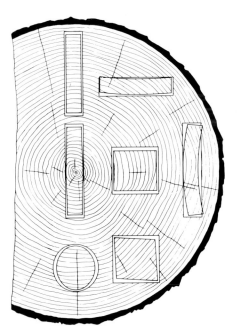

FIGURE 3.14
The difference between tangential and radial shrinkage also produces seasoning distortions in lumber. The nature of the distortion depends on the position the piece of lumber occupied in the tree. The distortions are the most pronounced in plainsawed lumber (upper right, extreme right, lower right. Courtesy of Forest Products Laboratory, Forest Service, USDA).

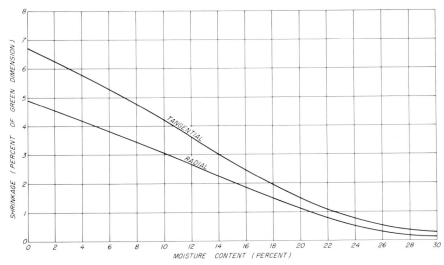

FIGURE 3.12
Shrinkage of a typical softwood with decreasing moisture content. Longitudinal shrinkage, not shown on this graph, is so small by comparison to tangential and radial shrinkage that it is of no consequence in wood buildings. (Courtesy of Forest Products Laboratory, Forest Service, USDA)

FIGURE 3.15
Automatic planing mills transport each piece of lumber at high speed past rotating cutters that smooth its four faces and reduce it to its finished dimension. (Photo by the author)

age cause this to happen. These effects are pronounced and are readily predicted and observed in everyday practice.

Surfacing

Lumber is *surfaced* to make it smooth and more dimensionally precise. Rough (unsurfaced) lumber is often available and is used for many purposes, but surfaced lumber is easier to work with because it is more square and uniform in dimension and less damaging to the hands of the carpenter. Surfacing is done by high-speed automatic machines whose rotating blades plane the surfaces of the piece and round the edges slightly (Figure 3.15). Most lumber is surfaced on all four sides (*S4S*), but hardwoods are often surfaced on only two sides (*S2S*), leaving the two edges to be finished by the craftsman.

Lumber is usually seasoned before it is planed, which allows the planing process to remove some of the distortions that occur during seasoning, but for some framing lumber this order of operations is reversed. The designation *S-DRY* in a lumber gradestamp indicates that the piece was surfaced (planed) when in a seasoned (dry) condition, and *S-GRN* indicates that it was planed when green.

Lumber Defects

Almost every piece of lumber contains one or more discontinuities in its structure caused by *growth characteristics* of the tree from which it came, or *manufacturing characteristics* that were created at the mill (Figures 3.16, 3.17). Among the most common growth characteristics are *knots*, which are places where branches joined the trunk of the tree; *knotholes*, which are holes left by loose knots dropping out of the wood; *decay*; and *insect damage*. Knots and knotholes reduce the structural strength of a piece of lumber, make it more difficult to cut and shape, and are often considered to be detrimental to its appearance. Decay and insect damage that occurred during the life of the tree may or may not affect the useful properties of the piece of lumber, depending on whether the organisms are still alive in the wood and the extent of the damage they have done.

Manufacturing characteristics arise largely from changes that take place during the seasoning process because of the differences in rates of shrinkage with varying orientations to the grain. Splits and checks are usually caused by shrinkage stresses. *Crooking, bowing, twisting,* and *cupping* all occur because of nonuniform shrinkage. *Wane* is caused by sawing pieces too close to

FIGURE 3.16
Surface features often observed in lumber include, in the lefthand column from top to bottom, a knot cut crosswise, a knot cut longitudinally, and a bark pocket. To the right are a gradestamp, wane on two edges of the same piece, and a small check. The gradestamp indicates that the piece was graded according to the rules of the American Forest Products Association, that it is #2 grade Spruce-Pine-Fir, and that it was surfaced after drying. The 27 is a code number for the mill that produced the lumber. (Photos by the author)

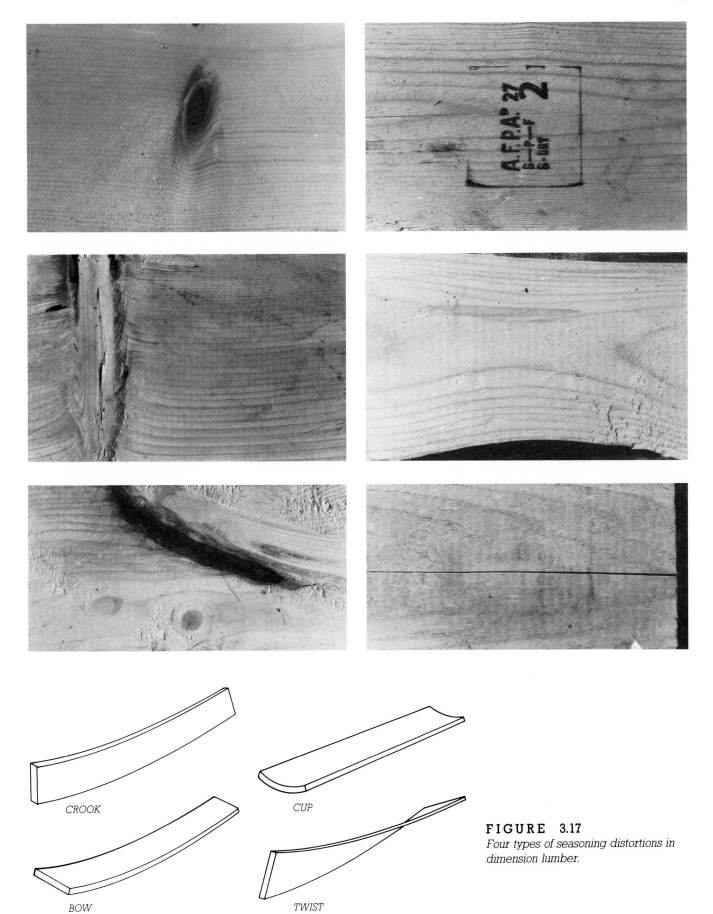

FIGURE 3.17
Four types of seasoning distortions in dimension lumber.

CROOK

CUP

BOW

TWIST

the perimeter of the log. Experienced carpenters judge the extent of these defects and distortions in each piece of lumber and make decisions accordingly on where and how to use the piece. Checks and shakes are of little consequence in framing lumber, but a joist or rafter with a crook in it is usually placed with the convex edge (the "crown") facing up, to allow the floor loads to straighten the piece. Badly bowed pieces are straightened as subflooring or sheathing is applied over them, and badly twisted pieces are put aside to be cut up for blocking. Cupping can be a major problem in flooring and interior baseboards and trim, but this is usually minimized by using quartersawn stock and by shaping the pieces so as reduce the likelihood of distortion (Figure 3.18).

Lumber Grading

Lumber is graded either for appearance or for structural strength and stiffness, depending on its intended use, before it leaves the mill. Lumber is sold by species and grade; the higher the grade, the higher the price. Grading offers the architect and the engineer the opportunity to build as economically as possible by using only as high a grade as is required for a par-

ticular use. In a specific building, a few main beams or columns may require a very high structural grade of lumber, while the remainder of the framing members will perform adequately in an intermediate, less expensive grade. For blocking, the lowest grade is perfectly adequate. For finish trim that will be coated with a clear varnish, a high appearance grade is desirable; for painted trim, a lower grade will suffice.

Stress grading of structural lumber may be done either visually or by machine. In visual grading, trained inspectors examine each piece for ring density and for growth and manufacturing characteristics, then judge it and stamp it with a grade in accordance with industry-wide grading rules. Grading machines flex each piece between rollers and measure its resistance to bending; from this measurement its stiffness is calculated and a grade stamped automatically on the piece. Figure 3.19 outlines a typical grading scheme for framing lumber, and Figure 3.20 outlines the appearance grades for nonstructural lumber. Light framing lumber for houses and other small buildings is usually ordered as "#2 and better" (a mixture of #1 and #2 grades) for floor joists and roof rafters, and as "Stud" grade for wall framing.

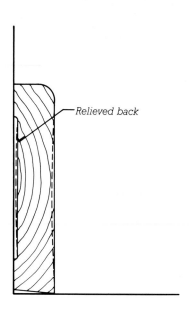

— Relieved back

FIGURE 3.18

The effects of seasoning distortions can often be minimized through knowledgeable detailing practices. As an example, this wood baseboard, seen in cross section, has been formed with a relieved back, a shallow groove that allows the piece to lie flat against the wall even if it cups. The sloping bottom on the baseboard assures that it can be installed tightly against the floor despite the cup. The grain orientation in this piece is the worst possible with respect to cupping. If quartersawn lumber were not available, the next best choice would have been to mill the baseboard with the center of the tree toward the room rather than toward the wall.

FIGURE 3.19

Standard grades for softwood framing lumber. For each species of wood, the allowable structural stresses for each of these grades are tabulated in the structural engineering literature on wood. (See reference 2. This table courtesy of Western Wood Products Association)

FIGURE 3.20

Nonstructural boards are graded according to appearance. (Courtesy of Western Wood Products Association)

Dimension/Stress-Rated Framing Lumber 2 x 2 Through 4 x 16

LIGHT FRAMING 2 x 2 Through 4 x 4	CONSTRUCTION (40.11) STANDARD (40.12) UTILITY (40.13)	This category for use where high strength values are not required; such as studs, plates, sills, cripples, blocking, etc. (Tables 1, 1a, and 1b).
STUDS 2 x 2 Through 4 x 6 10' and Shorter	STUD (41.13)	An optional all-purpose grade limited to 10' and shorter. Characteristics affecting strength and stiffness values are limited so that the "Stud" grade is suitable for all stud uses, including load bearing walls. (Table 1 — 2 x 2 through 4 x 4, Table 3 — 2 x 6 through 4 x 6.)
STRUCTURAL LIGHT FRAMING 2 x 2 Through 4 x 4	SELECT STRUCTUAL (42.10) NO. 1 (42.11) NO. 2 (42.12) NO. 3 (42.13)	These grades are designed to fit those engineering applications where higher bending strength ratios are needed in light framing sizes. Typical uses would be for trusses, concrete pier wall forms, etc. (Table 2.)
STRUCTURAL JOISTS & PLANKS 2 x 5 Through 4 x 16	SELECT STRUCTURAL (62.10) NO. 1 (62.11) NO. 2 (62.12) NO. 3 (62.13)	These grades are designed especially to fit in engineering applications for lumber 5" and wider, such as joists, rafters and general framing uses. (Table 3.)

Timbers 5" and thicker (For design values, see Tables 4 and 5.)

BEAMS & STRINGERS 5" and thicker Width more than 2" greater than thickness (i.e. 6 x 10)	SELECT STRUCTURAL (70.10) NO. 1 (70.11) NO. 2 (70.12) NO. 3[3] (70.13)		**POST & TIMBERS** 5" and larger Width not more than 2" greater than thickness (i.e. 6 x 6)	SELECT STRUCTURAL (80.10) NO. 1 (80.11) NO. 2 (80.12) NO. 3[3] (80.13)

3. Design values are not assigned.

Decking/Stress-Rated 2x4 Through 4x12 (see "Standard Patterns" G-16 for shapes and sizes, see Table 7 for design values.)

SELECTED DECKING (55.11)	COMMERCIAL DECKING (55.12)

Grade Selector Charts/All Species
Boards 1" (4/4) and Thicker. Non Stress-Rated (Numbers in Parentheses refer to WWPA Grading Rules Section Numbers.)

APPEARANCE GRADES	**SELECTS**	B & BETTER (IWP—SUPREME)[1] (10.11) C SELECT (IWP—CHOICE)[1] (10.12) D SELECT (IWP—QUALITY)[1] (10.13)	
	FINISH	SUPERIOR (10.51) PRIME (10.52) E (10.53)	
	PANELING	ANY SELECT OR FINISH GRADE OR SELECTED 2 COMMON FOR KNOTTY PANELING (30.22) SELECTED 3 COMMON FOR KNOTTY PANELING (30.23)	
	BEVEL OR BUNGALOW SIDING	SUPERIOR (16.11) PRIME (16.12) (Refer to WWPA "Wood Siding" Catalog for other siding grades)	
GENERAL PURPOSE BOARDS	**COMMON[2] BOARDS (WWPA)**	1 COMMON (IWP—COLONIAL)[1] (30.11) 2 COMMON (IWP—STERLING)[1] (30.12) 3 COMMON (IWP—STANDARD)[1] (30.13) 4 COMMON (IWP—UTILITY)[1] (30.14) 5 COMMON (IWP—INDUSTRIAL)[1] (30.15)	
	ALTERNATE[2] BOARDS (WCLIB)	SELECT MERCHANTABLE (118-a) CONSTRUCTION (118-b) STANDARD (118-c) UTILITY (118-d) ECONOMY (118-e)	

SPECIFICATION CHECK LIST
- ☐ Grades listed in order of quality.
- ☐ Include all species suited to project.
- ☐ Specify lowest grade that will satisfy job requirement.
- ☐ Specify surface texture desired.
- ☐ Specify moisture content suited to project.
- ☐ For appearance grades, specify the lumber to be stamped on back or ends.
- ☐ See publication A-2 "Lumber Specification Information."
- ☐ Vertical, flat or mixed grain.
- ☐ Pattern number when applicable.

Western Red Cedar

FINISH PANELING AND CEILING	CLEAR HEART (20.11) A (20.12) B (20.13)
BEVEL SIDING	CLEAR — V.G. HEART (21.11) A — BEVEL SIDING (21.12) B — BEVEL SIDING (21.13) C — BEVEL SIDING (21.14)

1. Idaho White Pine carries its own comparable grade designations.   2. Structural applications (sheathing) are regulated by model codes.

The Structural Strength of Wood

The strength under loading of a piece of wood depends chiefly on its species, its grade, and the direction in which the load is acting with respect to the direction of grain of the piece. Wood is several times stronger parallel to grain than perpendicular to grain. With defects, it is stronger in compression than in tension. *Allowable strengths* (structural stresses that include factors of safety) vary tremendously with species and grade. Allowable compressive strength parallel to grain, for example, varies from 325 to 1850 pounds per square inch (2.24 to 12.76 MN/m²) for commercially available grades and species of framing lumber, a difference of nearly six times. Figure 3.21 compares the structural properties of an "average" framing lumber to those of the other common structural materials—brick masonry, steel, and concrete. Of the four materials, only wood and steel have useful tensile strength. Defect-free wood is comparable to steel on a strength-per-unit-weight basis, but with the ordinary run of defects, wood is somewhat inferior to steel by this yardstick.

When designing a wooden structure, the architect or engineer determines the stresses that are expected in each of the structural members and selects an appropriate species and grade of lumber for each. In a given locale, a very limited number of species are usually available in the lumberyards, and it is from these that the selection is made. It is not uncommon to use a stronger but more expensive species (Douglas fir or Southern pine, for example) for certain highly stressed members, and to use a weaker, less expensive species (such as Eastern hemlock) or species group (Hem-Fir, Spruce-Pine-Fir) for the remainder of the structure. Within each species, grades are selected based on published tables of allowable stresses. The higher the grade, the higher the allowable stress, but the lower the grade, the less costly the lumber.

There are many factors other than species and grade that influence the useful strength of wood: These include the length of time the wood will be subjected to its maximum load, the temperature and moisture conditions under which it serves, and the size and shape of the piece. Certain fire retardant treatments also reduce the strength of wood slightly. All these factors are taken into account when engineering a building structure of wood.

Lumber Dimensions

Lumber sizes in the United States are given as *nominal dimensions* in inches.

A piece nominally 1 by 2 inches in cross section is a 1 × 2 ("one by two"), a piece 2 by 10 inches is a 2 × 10, and so on. At the time a piece of lumber is sawn, it may approach these dimensions. Subsequent to sawing, however, seasoning and surfacing diminish its size substantially. By the time a kiln-dried 2 × 10 reaches the lumberyard its actual dimensions are 1½ by 9¼ inches (38 × 235 mm). The relationship between nominal lumber dimensions (which are always written without inch marks) and actual dimensions (which are written *with* inch marks) is given in simplified form in Figure 3.22 and in complete form in Figure 3.23. Anyone who designs or constructs wooden buildings soon commits the simpler of these relationships to memory. Because of changing moisture content and manufacturing tolerances, however, it is never wise to assume that a piece of lumber will conform precisely to its intended dimension. The experienced detailer of wooden buildings knows not to treat wood as if it had the dimensional accuracy of steel. The designer working with an existing wooden building will find a great deal of variation in lumber sizes; wood members in hot, dry locations such as attics often will have shrunk to dimensions substantially below their original measurements, and members in older buildings may have been manufactured to full dimension, or to ear-

Material	Allowable Tensile Strength	Allowable Compressive Strength	Density	Modulus of Elasticity
Wood (average)	700 psi (4.83 MN/m²)	1,100 psi (7.58 MN/m²)	30 pcf (480 kg/m³)	1,200,000 psi (8,275 MN/m²)
Brick masonry (average)	0	250 psi (1.72 MN/m²)	120 pcf (1,920 kg/m³)	1,200,000 psi (8,275 MN/m²)
Steel (ASTM A36)	22,000 psi (151.69 MN/m²)	22,000 psi (151.69 MN/m²)	490 pcf (7,850 kg/m³)	29,000,000 psi (200,000 MN/m²)
Concrete (average)	0	1,350 psi (9.31 MN/m²)	145 pcf (2,320 kg/m³)	3,150,000 psi (21,720 MN/m²)

FIGURE 3.21
Comparative physical properties of the four common structural materials: wood, masonry, steel, concrete.

FIGURE 3.22

The relationship between nominal and actual dimensions for the most common sizes of kiln dried lumber is given in this simplified chart, which is extracted from the complete chart in Figure 3.23.

Nominal Dimension	Actual Dimension
1″	$\frac{3}{4}$″ (19 mm)
2″	$1\frac{1}{2}$″ (38 mm)
3″	$2\frac{1}{2}$″ (64 mm)
4″	$3\frac{1}{2}$″ (89 mm)
5″	$4\frac{1}{2}$″ (114 mm)
6″	$5\frac{1}{2}$″ (140 mm)
8″	$7\frac{1}{4}$″ (184 mm)
10″	$9\frac{1}{4}$″ (235 mm)
12″	$11\frac{1}{4}$″ (286 mm)
over 12″	$\frac{3}{4}$″ less (19 mm less)

FIGURE 3.23

A complete chart of nominal and actual dimensions for both framing lumber and finish lumber. (Courtesy of Western Wood Products Association)

Standard Lumber Sizes/Nominal and Dressed, Based on WWPA Rules

Product	Description	Nominal Size Thickness In.	Nominal Size Width In.	Dry Dressed Dimensions Thickness In.	Dry Dressed Dimensions Width In.	Dry Dressed Dimensions Lengths Ft.
SELECTS AND COMMONS	S1S, S2S, S4S, S1S1E, S1S2E .	4/4 5/4 6/4 7/4 8/4 9/4 10/4 11/4 12/4 16/4	2 3 4 5 6 7 8 and wider	3/4 1 5/32 1 13/32 1 19/32 1 13/16 2 3/32 2 3/8 2 9/16 2 3/4 3 3/4	1 1/2 2 1/2 3 1/2 4 1/2 5 1/2 6 1/2 3/4 Off nominal	6' and longer in multiples of 1' except Douglas Fir and Larch Selects shall be 4' and longer with 3% of 4' and 5' permitted.
FINISH AND BOARDS	S1S, S2S, S4S, S1S1E, S1S2E . Only these sizes apply to Alternate Board Grades.	3/8 1/2 5/8 3/4 1 1 1/4 1 1/2 1 3/4 2 2 1/2 3 3 1/2 4	2 3 4 5 6 7 8 and wider	5/16 7/16 9/16 5/8 3/4 1 1 1/4 1 3/8 1 1/2 2 2 1/2 3 3 1/2	1 1/2 2 1/2 3 1/2 4 1/2 5 1/2 6 1/2 3/4 Off nominal	3' and longer. In Superior grade, 3% of 3' and 4' and 7% of 5' and 6' are permitted. In Prime Grade, 20% of 3' to 6' is permitted.
RUSTIC AND DROP SIDING	(D&M) . . . . If 3/8″ or 1/2″ T&G specified, same over-all widths apply. (Shiplapped, 3/8″ or 1/2″ lap).	1	6 8 10 12	23/32	5 3/8 7 1/8 9 1/8 11 1/8	4' and longer in multiples of 1'
PANELING AND SIDING	T&G or Shiplap	1	6 8 10 12	23/32	5 7/16 7 1/8 9 1/8 11 1/8	4' and longer in multiples of 1'
CEILING AND PARTITION	T&G	5/8 1	4 6	9/16 23/32	3 3/8 5 3/8	4' and longer in multiples of 1'
BEVEL SIDING	Bevel or Bungalow Siding . . . . . . Western Red Cedar Bevel Siding available in 1/2″, 5/8″, 3/4″ nominal thickness. Corresponding thick edge is 15/32″, 9/16″ and 3/4″. Widths for 8″ and wider, 1/2″ off nominal.	1/2 3/4	4 5 6 8 10 12	15/32 butt, 3/16 tip 3/4 butt, 3/16 tip	3 1/2 4 1/2 5 1/2 7 1/4 9 1/4 11 1/4	3' and longer in multiples of 1' 3' and longer in multiples of 1'

Product	Description	Nominal Size Thickness In.	Nominal Size Width In.	Surfaced Dry	Surfaced Green	Surfaced Dry	Surfaced Green	Lengths Ft.
STRESS-RATED BOARDS	S1S, S2S, S4S, S1S1E, S1S2E .	1 1 1/4 1 1/2	2 3 4 5 6 7 8 and Wider	3/4 1 1 1/4	25/32 1 1/32 1 9/32	1 1/2 2 1/2 3 1/2 4 1/2 5 1/2 6 1/2 Off 3/4	1 9/16 2 9/16 3 9/16 4 5/8 5 5/8 6 5/8 Off 1/2	6' and longer in multiples of 1'

MINIMUM ROUGH SIZES. Thickness and Widths, Dry or Unseasoned, All Lumber
80% of the pieces in a shipment shall be at least 1/8″ thicker than the standard surfaced size, the remaining 20% at least 3/32″ thicker than the surfaced size. Widths shall be at least 1/8″ wider than standard surfaced widths.
When specified to be full sawn, lumber may not be manufactured to a size less than the size specified.

A NATURALLY GROWN BUILDING MATERIAL

by

JOSEPH IANO

Wood is the only major building material that is organic in origin. This accounts for much of its uniqueness as a structural material. Because trees grow naturally, wood is a renewable resource, and most of the work of "manufacturing" wood is done for us by the processes of life and growth within the tree. The consequences of this are both economic and practical. Wood is cheap to produce. Most lumber need only be harvested, cut to size, and dried before it is ready for use. However, because wood is produced by trees, we have little control over the product. Unlike other structural materials such as steel or concrete, we can do little to adjust or refine wood's properties to suit our needs. Rather, we must accept its natural strengths and limitations.

Though we use wood much as we find it in the living tree, it is remarkably well suited to our building needs. This is because a tree is itself a structure made from wood. From a mechanical point of view, a tree is a tower erected for the purpose of displaying leaves to the sun.* Thus a tree is subjected to many of the same forces as the buildings we erect: It supports itself against the pull of gravity. It withstands the forces of wind, and the accumulation of snow and ice on its limbs. And it resists the natural stresses of our environment, including temperature and moisture extremes, attack by other organisms, and physical abuse. Since wood is the material that the tree uses to resist these forces, it is not surprising that we find wood a suitable material for our structures as well.

One of wood's natural strengths is its high resistance to loads or forces that act only for short periods of time. The greatest forces that trees experience in nature are from the wind, particularily in combination with ice or heavy snow. Because of this, wood has evolved to be more capable of withstanding forces such as the wind that are applied over shorter periods of time than forces applied over much longer periods.

When we build with wood, we can take advantage of this unusual short-term strength. In the engineering design of any wooden structure, the maximum forces that the structural members may carry are increased as the length of time these forces are expected to act decreases. This adjustment is called the *duration of load factor*. Increases from 15 to 100 percent in the allowable stresses are applied to wooden structures when considering the effects of wind, snow, and earthquake—the types of forces also resisted best by the tree.

The structural form of the tree is also well suited to its environment. Tree branches are supported at one end only—an arrangement called a *cantilever*. The joint supporting a cantilevered branch is a strong, stiff, efficient connection that utilizes the tree's material resources to the utmost. Despite the stiffness of this joint, however, the branch it supports can still deflect relatively great amounts since it is fastened at one end only. This high deflection can be advantageous for the tree. A branch that droops under a heavy load of snow can drop that snow and relieve its load. Similarly, a tree that sways and bends in the wind can shed much of the force acting upon it. This capacity to deflect helps a tree to survive forces that might otherwise break it.

When we build with wood, we cannot directly exploit the efficient structural form of the tree. The high deflections characteristic of cantilevered beams are unsuitable for our buildings because such large movements would cause discomfort to the occupants and undue distress to other building components. Nor can we take advantage of the naturally strong joints that support these cantilevers, though they could offer benefits in their stiffness and economy of materials. In preparing wood for our uses, we saw the tree into pieces of convenient size and shape. This destroys the structural continuity between the tree's parts. When we reassemble these pieces, we must devise new ways to join the wood. Despite the many methods developed for fabricating wood connections, joints with strength and stiffness comparable to the tree's are impossible to make on the building site. This is why most of our wood buildings rely on the much simpler type of connections that suffice when a beam is supported at both ends simultaneously.

Wood in a tree performs many functions that wood in a building does not, because wood is fundamentally involved with all aspects of the tree's life processes and growth. In order to meet numerous and specialized tasks, wood has evolved a complex internal structure that is highly directional. The fact that wood is not a uniform and "isotropic" material (the same in all directions) places basic limitations on how it can be used. Virtually all of wood's physical properties vary greatly, depending on whether they are measured parallel to the direction of the grain, or across it. This directional quality of wood is often taken for granted, even though no other major building material is like this. In fact, the direction of the grain in a piece of wood affects every aspect of how we can use it, including its production, shaping, and fastening, its structural capacity, its durability, and its beauty.

An example that illustrates this is a comparison of two early American wood building types: the log cabin and the heavy timber frame. The log cabin was a building system appropriate to the most primitive technology. The logs were found on or near the site and prepared for use with a minimum of labor and tools. In their final form they served many functions, including structural support, interior finish, exterior cladding, and insulation. But because the logs were stacked with their grain running horizontally, the wood was being used inefficiently from a structural point of view. Not only did this arrangement of the timbers produce a relatively weak wall, it also maxi-

mized the effects of vertical shrinkage and expansion in the structure, resulting in a high degree of dimensional instability. Where more sophisticated building techniques and a wider range of building materials were available, the log cabin was never used. Rather, buildings were framed with heavy timbers, utilizing a structural system that more closely resembled the structure of the tree. This more natural configuration of timbers in which a wall was framed with vertical posts and spanned with horizontal beams could be stronger, required less wood, was more dimensionally stable, and permitted the incorporation of other materials better suited to the wall's nonstructural requirements. A simple difference in orientation of the timbers produced a radically different building system that was more efficient, durable, and comfortable.

Many of the modern developments in the wood industry reflect a desire to overcome the natural limitations of wood. Plywood sheets are larger than could normally be obtained from trees, are dimensionally more stable than natural wood, and minimize the adverse effects of grain and natural defects in the wood. Newer panel products such as particleboard, waferboard, and hardboard use scraps and waste that would otherwise be discarded. Glue-laminated beams are larger in size and of higher quality than can be obtained from nature, and successfully utilize techniques of joinery that are as strong as those of the original tree. Chemical treatments can increase wood's resistance to fire and decay. Thus the trend is toward using wood less in its natural state, and more as a raw material in sophisticated manufacturing processes. These new products share more of the qualities of being highly refined and carefully controlled that are characteristic of other man-made building materials. However, despite such technological changes the tree remains not only a valuable resource, but also an inspiration in its grace and strength, and in the lessons it offers in understanding our own building materials and methods.

*Brayton F. Wilson and Robert R. Archer, "Tree Design: Some Biological Solutions to Mechanical Problems," *BioScience*, Vol. 29, No. 5, p. 293.

lier standards of actual dimensions such as $1\frac{5}{8}''$ or $1\frac{3}{4}''$ (41 or 44 mm) for a nominal 2-inch member.

Pieces of lumber less than 2 inches in nominal thickness are called *boards*. Pieces ranging from 2 to 4 inches in nominal thickness are referred to collectively as *dimension lumber*. Pieces nominally 5 inches and more in thickness are termed *timbers*.

Dimension lumber is usually supplied in 2-foot (610-mm) increments of length. The most commonly used lengths are 8, 10, 12, 14, and 16 feet (2.44, 3.05, 3.66, 4.27, and 4.88 m), but most yards stock rafter material in lengths to 24 feet (7.32 m). Actual lengths are usually a fraction of an inch longer than nominal lengths.

Lumber in the United States is priced by the *board foot*. Board foot measurement is based on nominal dimensions, not actual dimensions. A board foot of lumber is defined as a solid volume 12 square inches in nominal cross-sectional area and 1 foot long. A 1×12 or 2×6 10 feet long contains 10 board feet. A 2×4 10 feet long contains $[(2 \times 4)/12] \times 10 = 6.67$ board feet, and so on. Prices of dimension lumber and timbers in the United States are usu-

ally quoted on the basis of dollars per thousand board feet. In other parts of the world, lumber is sold by the cubic meter.

The architect and engineer specify lumber for a particular construction use by designating its species, grade, seasoning, surfacing, nominal size, and treatment, if any. When ordering lumber, the contractor must additionally give the required lengths of pieces, and the required number of pieces of each length.

Wood PRODUCTS

Much of the wood used in construction has been processed into laminated wood or wood panel products in order to overcome various of the shortcomings of solid wood structural members.

Laminated Wood

Large structural members are often produced by joining many small strips of wood together with glue to form *laminated wood* (also called "glulam"

for short). There are three major reasons to laminate: size, shape, and quality. Any desired size of structural member can be laminated, up to the capacities of the hoisting and transportation machinery needed to deliver and erect it, without having to search for a tree of sufficient girth and height. Wood can be laminated into shapes that cannot be obtained in nature: curves, angles, varying cross sections (Figure 3.24). Quality can be specified and closely controlled in laminated members because defects can be cut out of the wood before laminating and the direction of grain can be closely controlled in curved members. Seasoning is carried out before the wood is laminated (largely eliminating the checks and distortions that characterize solid timbers), and the strongest, highest-quality wood can be placed in the parts of the member that will be subjected to the highest structural stresses. The fabrication of laminated members obviously adds to their cost per board foot, but this is often overcome by the smaller size of the laminated member that can replace a solid timber of equal load-carrying capacity, and in many cases, solid timbers are

FIGURE 3.24
Glue-laminating a U-shaped timber for a ship. Several smaller members have also been glued and clamped and are drying alongside the larger timber. (Courtesy of Forest Products Laboratory, Forest Service, USDA)

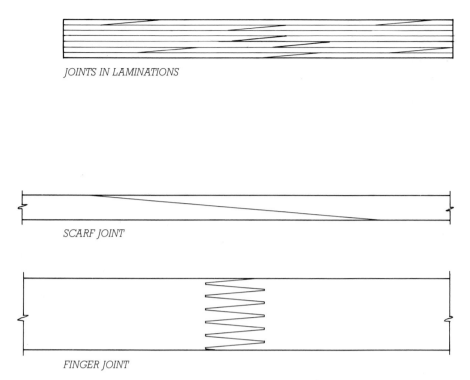

JOINTS IN LAMINATIONS

SCARF JOINT

FINGER JOINT

FIGURE 3.25
Joints within a lamination of a glue-laminated beam, seen in the upper drawing in a small-scale elevation view, must be scarf jointed or finger jointed to transmit the tensile and compressive forces from one piece of wood to the next. The individual pieces of wood are prepared for jointing by high speed machines that mill the scarf or fingers with rotating cutters of the appropriate shape.

simply not available at any price in the required size, shape, or quality.

Individual laminations are $1\frac{1}{2}$ inches (38 mm) thick, except in curved members with small bending radii, where $\frac{3}{4}$-inch (19-mm) stock is used. End joints between individual pieces are either *finger-jointed* or *scarf-jointed*. These types of joints allow the glue to transmit tensile and compressive forces longitudinally from one piece to the next within a lamination (Figure 3.25). Adhesives are chosen according to the moisture conditions under which the member will serve. Any size member can be laminated, but standard depths range from 3 to 75 inches (76 to 1905 mm) in increments of $1\frac{1}{2}$ inches (38 mm). Standard widths are $2\frac{1}{8}$ inches, $3\frac{1}{8}$ inches, $5\frac{1}{8}$ inches, $6\frac{3}{4}$ inches, $8\frac{3}{4}$ inches, and $10\frac{3}{4}$ inches (54, 79, 130, 171, 222, and 273 mm).

Wood Panel Products

Wood in panel form is advantageous for many building applications (Figure 3.26). Panels require less labor for installation than boards because fewer pieces must be handled, and wood panel products are fabricated in such a way that they minimize many of the problems of boards and dimension lumber: Panels are more nearly equal in strength in their two principle directions than solid wood; shrinking, swelling, checking, and splitting are greatly reduced. Additionally, panel products make more efficient use of forest resources than solid wood products through less wasteful ways of reducing logs to building products and through utilization in some types of panels of material that would otherwise be thrown away—branches, undersized trees, and mill wastes.

Structural wood panel products fall into three general categories (Figure 3.27):

1. *Plywood* panels, which are made up of thin wood *veneers* glued together. The grain on the front and back face veneers runs in the same direction, while the grain in one or more interior *crossbands* runs in the perpendicular direction. There is always an odd number of *layers* in plywood, which equalizes the effects of moisture movement, but an interior layer may be made up of a single veneer, or of two veneers with their grain running in the same direction.

2. *Composite* panels, which have two parallel face veneers bonded to a core of reconstituted wood fibers.

3. *Nonveneered* panels, which are of three different types:

   a. *Oriented strand board (OSB)*, which is made of long strand-like wood particles compressed and glued into three to five layers, with the strands oriented in much the same manner in each layer as the grains of the veneer layers in plywood. Because of the length and controlled orientation of the strands, oriented strand board is generally stronger and stiffer than the other two types of nonveneered panels.

   b. *Waferboard*, which is composed of large wafer-like flakes of wood compressed and bonded into panels.

   c. *Particleboard*, which is manufactured in several different classes, all of which are made up

FIGURE 3.26

Plywood is made of veneers selected to give the optimum combination of economy and performance for each application. This sheet of roof sheathing plywood is faced with a D veneer on the underside (seen here pointed toward the camera) and a C veneer on the top side. These veneers, though unattractive, perform well structurally and are much less costly than the higher grades incorporated into plywoods made for uses where appearance is important. (Courtesy of American Plywood Association)

FIGURE 3.27

Five types of wood panel products, from top to bottom: Plywood, composite panel, waferboard, oriented strand board, and particleboard. (Courtesy of American Plywood Association)

of smaller wood particles than oriented strand board or waferboard, compressed and bonded into panels.

The American Plywood Association has established performance standards that allow considerable interchangeability among these various types of panels for many construction uses. The standards are based on structural adequacy for a specified use, dimensional stability under varying moisture conditions, and the durability of the adhesive bond that holds the panel together.

Plywood Production

Veneers for structural panels are *rotary sliced:* Logs are soaked in hot water to soften the wood, then each is rotated in a large lathe against a stationary knife that peels off a continuous strip of veneer, much as paper is unwound from a roll (Figures 3.28, 3.29). The strip of veneer is clipped into sheets that are passed through a drying kiln where in a few minutes they are reduced to a moisture content of roughly 5 percent. The sheets are then assembled into larger sheets, repaired as necessary with patches glued into the sheet to fill open defects, and graded and sorted according to quality (Figure 3.30). A machine spreads glue onto the veneers as they are laid atop one another in the required sequence and grain orientations. The glued veneers are transformed into plywood in presses that apply elevated temperatures and pressures to create dense, flat panels. The panels are trimmed to size, sanded as required, graded and gradestamped before shipping. Grade B veneers and higher are always sanded smooth; panels intended for sheathing are always left unsanded because sanding slightly reduces the thickness, which diminishes the structural strength of the panel. Panels intended for subfloors and floor underlayment are lightly *touch-sanded* to produce a more flat and uniform surface without seriously affecting their structural performance.

Veneers for hardwood plywoods intended for interior paneling and cabinetwork are usually sliced from square blocks of wood called *flitches* in a machine that moves the flitch vertically against a stationary knife (Figure 3.29). Flitch-sliced veneers are analogous to quartersawn lumber: They exhibit a much tighter and more interesting grain figure than rotary-sliced veneers. They can also be arranged on the plywood face in such a way as to produce symmetrical grain patterns.

Standard plywood panels are 4 by 8 feet (1219 × 2438 mm) in surface area and range in thickness from $\frac{1}{4}$ to 1 inch (6.1 to 25.4 mm). Longer sheets are manufactured for siding and industrial use. Actual surface dimensions of the structural grades of plywood are slightly less than nominal to permit the panels

"**Y**our chessboard, sire, is inlaid with two woods: ebony and maple. The square on which your enlightened gaze is fixed was cut from the ring of a trunk that grew in a year of drought: you see how its fibers are arranged? Here a barely hinted knot can be made out: a bud tried to burgeon on a premature spring day, but the night's frost forced it to desist...Here is a thicker pore: perhaps it was a larvum's nest...." The quantity of things that could be read in a little piece of smooth and empty wood overwhelmed Kublai; Polo was already talking about ebony forests, about rafts laden with logs that come down the rivers, of docks, of women at the windows...

Italo Calvino, in *Invisible Cities,* New York, Harcourt Brace Jovanovich, 1978

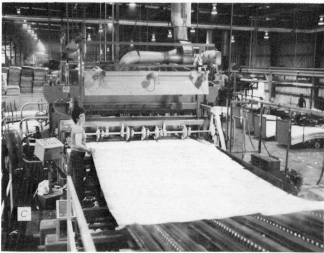

FIGURE 3.28

Plywood manufacture. (a, b) A 250-horsepower lathe spins a softwood log as a knife peels off a continuous sheet of veneer for plywood manufacture. (c) An automatic clipper removes unusable areas of veneer and trims the rest into sheets of the proper size for plywood panels. (d) The clipped sheets are fed into a continuous forced-air dryer, along whose 150-foot (45-m) path they will lose about half their weight in moisture. (e) Leaving the dryer, the sheets have a moisture content of about 5 percent. They are graded and sorted at this point in the process. (f) The higher grades of veneer are patched on this machine which punches out defects and replaces them with tight-fitting wood plugs. (g) In the layup line, automatic machinery applies glue to one side of each sheet of veneer, and alternates the grain direction of the sheets to produce loose plywood panels. (h) After layup, the loose panels are prepressed with a force of 300 tons per panel to consolidate them for easier handling. (i) Following prepressing, panels are squeezed individually between platens heated to 300 degrees Fahrenheit (150° C) to cure the glue. (j) After trimming, sanding, or grooving as specified for each batch, the finished plywood panels are sorted into bins by grade, ready for shipment. (Photos b and i courtesy of Georgia-Pacific; others courtesy of American Plywood Association)

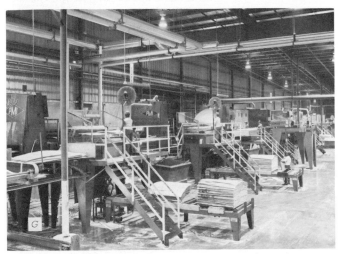

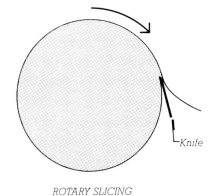

ROTARY SLICING

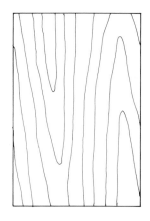

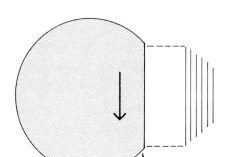

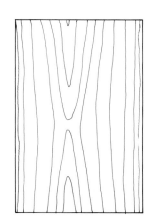

PLAIN SLICING

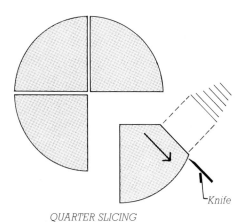

QUARTER SLICING

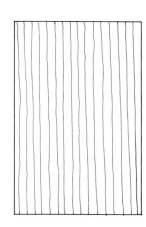

FIGURE 3.29

Veneers for structural plywood are rotary sliced, which is the most economical method. For better control of grain figure in face veneers of hardwood plywood, flitches are plain sliced or quarter sliced. The grain figure produced by rotary slicing, as seen to the right, is extremely broad and uneven. The finest figures are produced by quarter slicing, which results in a very close grain pattern with prominent rays.

to be installed with small spaces between to allow for moisture expansion. Composite panels and nonveneered panels are manufactured by analogous processes to the same set of standard sizes as plywood panels and to some larger sizes as well.

Specifying Structural Wood Panels

For structural uses such as subflooring and sheathing, wood panels may be specified either by thickness or by *span rating*. The span rating is determined by laboratory load testing and is given as a part of the gradestamp on the back of the panel, as shown in Figures 3.31 and 3.32. The purpose of the span rating system is to permit the use of many different species of woods and types of panels to achieve the same structural objectives. A panel with a span rating of 32/16, as an example, may be used as roof sheathing over rafters spaced 32 inches (813 mm) apart, or as subflooring over joists spaced 16 inches (406 mm) apart. The long dimension of the sheet must be placed perpendicular to the length of the supporting members. A 32/16 panel may be plywood, composite, or nonveneered, may be composed of any accepted wood species, and may be any of several thicknesses, so long as it passes the structural tests for a 32/16 rating.

The designer must also select from three *exposure durability classifications* for structural wood panels: *Exterior*, *Exposure 1*, and *Exposure 2*. Panels marked "Exterior" are suitable for use as siding or in other continuously exposed applications. "Exposure 1" panels have fully waterproof glue but do not have veneers of as high a quality as those of "Exterior" panels; they are suitable for structural sheathing and subflooring, which must often endure repeated wetting during construction. "Exposure 2" is suitable for panels that will be fully protected from weather and will be subjected to a minimum of wetting during construction.

Table 1 Veneer Grades

N	Smooth surface "natural finish" veneer. Select, all heartwood or all sapwood. Free of open defects. Allows not more than 6 repairs, wood only, per 4 x 8 panel, made parallel to grain and well matched for grain and color.
A	Smooth, paintable. Not more than 18 neatly made repairs, boat, sled, or router type, and parallel to grain, permitted. May be used for natural finish in less demanding applications.
B	Solid surface. Shims, circular repair plugs and tight knots to 1 inch across grain permitted. Some minor splits permitted.
C Plugged	Improved C veneer with splits limited to 1/8-inch width and knotholes and borer holes limited to 1/4 x 1/2 inch. Admits some broken grain. Synthetic repairs permitted.
C	Tight knots to 1-1/2 inch. Knotholes to 1 inch across grain and some to 1-1/2 inch if total width of knots and knotholes is within specified limits. Synthetic or wood repairs. Discoloration and sanding defects that do not impair strength permitted. Limited splits allowed. Stitching permitted.
D	Knots and knotholes to 2-1/2 inch width across grain and 1/2 inch larger within specified limits. Limited splits are permitted. Stitching permitted. Limited to Interior (Exposure 1 or 2) panels.

FIGURE 3.30

Veneer grades for softwood plywood. (Courtesy of American Plywood Association)

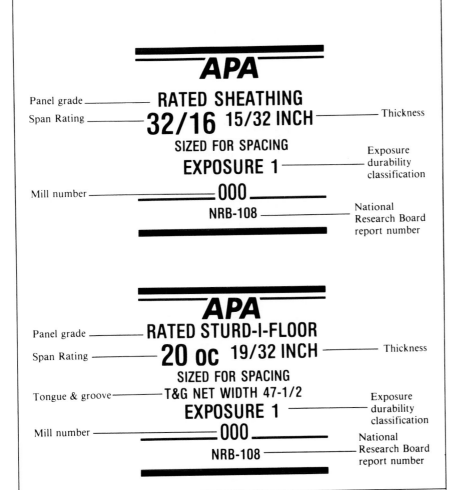

FIGURE 3.31

Typical gradestamps for structural wood panels. Gradestamps are found on the back of each panel. (Courtesy of American Plywood Association)

Guide to APA Performance-Rated Panels[1][2]

	Grade Designation	Description & Common Uses	Typical Trademarks
PROTECTED OR INTERIOR USE	**APA RATED SHEATHING EXP 1 or 2**	Specially designed for subflooring and wall and roof sheathing, but can also be used for a broad range of other construction and industrial applications. Can be manufactured as conventional veneered plywood, as a composite, or as a nonveneered panel. For special engineered applications, including high load requirements and certain industrial uses, veneered panels conforming to PS 1 may be required. Specify Exposure 1 when long construction delays are anticipated. Common thicknesses: 5/16, 3/8, 7/16, 15/32, 1/2, 19/32, 5/8, 23/32, 3/4.	**APA** RATED SHEATHING 32/16 15/32 INCH SIZED FOR SPACING EXPOSURE 1 000 NRB-108
	APA STRUCTURAL I & II RATED SHEATHING EXP 1	Unsanded all-veneer plywood grades for use where strength properties are of maximum importance: structural diaphragms, box beams, gusset plates, stressed-skin panels, containers, pallet bins. Made only with exterior glue (Exposure 1). STRUCTURAL I more commonly available. Common thicknesses: 5/16, 3/8, 15/32, 1/2, 19/32, 5/8, 23/32, 3/4. (3)	**APA** RATED SHEATHING STRUCTURAL I 24/0 3/8 INCH SIZED FOR SPACING EXPOSURE 1 000 PS 1-83 C-D NRB-108
	APA RATED STURD-I-FLOOR EXP 1 or 2	For combination subfloor-underlayment. Provides smooth surface for application of carpet and possesses high concentrated and impact load resistance. Can be manufactured as conventional veneered plywood, as a composite, or as a nonveneered panel. Available square edge or tongue-and-groove. Specify Exposure 1 when long construction delays are anticipated. Common thicknesses: 19/32, 5/8, 23/32, 3/4.	**APA** RATED STURD-I-FLOOR 24 OC 23/32 INCH SIZED FOR SPACING T&G NET WIDTH 47-1/2 EXPOSURE 1 000 NRB-108
	APA RATED STURD-I-FLOOR 48 oc (2-4-1) EXP 1	For combination subfloor-underlayment on 32- and 48-inch spans and for heavy timber roof construction. Provides smooth surface for application of carpet and possesses high concentrated and impact load resistance. Manufactured only as conventional veneered plywood and only with exterior glue (Exposure 1). Available square edge or tongue-and-groove. Thickness: 1-1/8.	**APA** RATED STURD-I-FLOOR 48 OC 1-1/8 INCH 2-4-1 SIZED FOR SPACING EXPOSURE 1 T&G 000 PS 1-83 UNDERLAYMENT NRB-108 FHA-UM-66
EXTERIOR USE	**APA RATED SHEATHING EXT**	Exterior sheathing panel for subflooring and wall and roof sheathing, siding on service and farm buildings, crating, pallets, pallet bins, cable reels, etc. Manufactured as conventional veneered plywood. Common thicknesses: 5/16, 3/8, 15/32, 1/2, 19/32, 5/8, 23/32, 3/4.	**APA** RATED SHEATHING 48/24 23/32 INCH SIZED FOR SPACING EXTERIOR 000 NRB-108
	APA STRUCTURAL I & II RATED SHEATHING EXT	For engineered applications in construction and industry where resistance to permanent exposure to weather or moisture is required. Manufactured only as conventional veneered plywood. Unsanded. STRUCTURAL I more commonly available. Common thicknesses: 5/16, 3/8, 15/32, 1/2, 19/32, 5/8, 23/32, 3/4. (3)	**APA** RATED SHEATHING STRUCTURAL I 32/16 1/2 INCH SIZED FOR SPACING EXTERIOR 000 PS 1-83 C-C NRB-108
	APA RATED STURD-I-FLOOR EXT	For combination subfloor-underlayment under carpet where severe moisture conditions may be present, as in balcony decks. Possesses high concentrated and impact load resistance. Manufactured only as conventional veneered plywood. Available square edge or tongue-and-groove. Common thicknesses: 19/32, 5/8, 23/32, 3/4.	**APA** RATED STURD-I-FLOOR 20 OC 19/32 INCH SIZED FOR SPACING EXTERIOR 000 NRB-108

(1) Specific grades, thicknesses, constructions and exposure durability classifications may be in limited supply in some areas. Check with your supplier before specifying.

(2) Specify Performance-Rated Panels by thickness and Span Rating.

(3) All plies in STRUCTURAL I plywood panels are special improved grades and limited to Group 1 species. All plies in STRUCTURAL II plywood panels are special improved grades and limited to Group 1, 2, or 3 species.

FIGURE 3.32

A guide to specifying structural panels. Plywood panels for use in paneling, furniture, and other uses where appearance is important are graded by the quality of their face veneers. (Courtesy of American Plywood Association)

For panels intended as finish surfaces, the quality of the face veneers is of obvious concern and should be specified by the designer. For some types of work, fine flitch-sliced hardwood face veneers may be selected rather than rotary-sliced softwood veneers, and the matching pattern of the veneers specified.

Other Wood Panel Products

Several types of nonstructural or semistructural panels of wood fiber are often used in construction. *Hardboard* is a thin, dense panel made of highly compressed wood fibers. It is made in several thicknesses and surface finishes, and in some formulations it is durable against weather exposure. Hardboard is produced in configurations for residential siding and roofing as well as in general-purpose panels of standard dimension. *Cane fiber board* is a thick, low-density panel with some thermal insulating value; it is used in wood construction chiefly as a nonstructural or semistructural wall sheathing. Panels made of recycled paper are low in cost and are useful for wall sheathing, carpet underlayment, and incorporation in certain proprietary types of roof decking and insulating assemblies.

Wood TREATMENT

Various treatments have been developed to counteract two major weaknesses of wood: its combustibility, and its susceptibility to attack by decay and insects. Fire-retardant treatment is accomplished by placing lumber in a pressure vessel and impregnating it with certain chemical salts that greatly reduce its flammability. The cost of fire-retardant treated wood is such that it is little used in single-family residential construction; its major use is in nonstructural partitions and other interior components in buildings of fire-resistant construction.

Decay and insect resistance is very

I shall always remember how as a child I played on the wooden floor. The wide boards were warm and friendly, and in their texture I discovered a rich and enchanting world of veins and knots. I also remember the comfort and security experienced when falling asleep next to the round logs of an old timber wall; a wall which was not just a plain surface but had a plastic presence like everything alive. Thus sight, touch, and even smell were satisfied, which is as it should be when a child meets the world.

Christian Norberg-Schulz

important in wood that is used in or very near to the ground, and in wood used for exposed outdoor structures such as marine docks, fences, decks, and porches. Decay-resistant treatment is accomplished by pressure impregnation with any of several types of preservatives. *Creosote* is an oily derivative of coal that is widely used in engineering structures, but because of its odor, toxicity, and unpaintability, it is unsuitable for most purposes in building construction. *Pentachlorophenol* is also impregnated as an oil solution, and as with other oily preservatives, wood treated with it cannot be painted. The most widely used preservatives in building construction are the *waterborne salts*. Most are based on copper salts and impart a greenish color to the wood, but do not prevent subsequent painting or staining. Preservatives of any of these types can be brushed or sprayed onto wood, but long-lasting protection (30 years and more) can only be accomplished by pressure impregnation. Because of the

poisonous nature of wood preservatives, their use is often controlled by environmental regulations.

The heartwood of some species of wood is naturally resistant to decay and insects and can be used instead of preservative-treated wood. The most commonly used decay-resistant species are Redwood, Bald cypress, and various cedars. The sapwood of these species is no more resistant to attack than that of any other tree, so "All-Heart" grades should be specified.

Most wood-attacking organisms need both air and moisture to live. Most can therefore be kept out of wood by constructing and maintaining a building so its wood components are kept dry at all times. This includes keeping all wood well clear of the soil, ventilating attics and crawlspaces to remove moisture, using good construction detailing to keep wood dry, and fixing roof and plumbing leaks as soon as they occur.

Wood FASTENERS

Fasteners have always been the weak link in wood construction. The interlocking timber connections of the past, laboriously mortised and pegged, were weak because much of the wood in a joint had to be removed to make the connection. In today's wood connections it is usually impossible to insert enough nails, screws, or bolts in a connection to develop the full strength of the members being joined. Adhesives and toothed plates are often capable of achieving this strength, but are largely limited to factory installation. Fortunately, most connections in wood structures depend primarily on direct bearing of one member on another for their strength, and a variety of simple fasteners are adequate for the majority of purposes.

Nails

Nails are sharpened metal pins driven into wood with a hammer or a me-

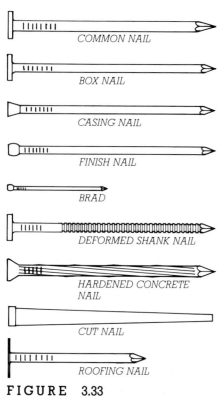

COMMON NAIL

BOX NAIL

CASING NAIL

FINISH NAIL

BRAD

DEFORMED SHANK NAIL

HARDENED CONCRETE NAIL

CUT NAIL

ROOFING NAIL

FIGURE 3.33

All nailed framing connections are made with common nails or their machine-driven equivalent. Box nails, which are made of lighter gauge wire, do not have as much holding power as common nails; they are used in construction for attaching wood shingles. Casing nails, finish nails, and brads are used for attaching finish components of a building. Their heads are set below the surface of the wood with a steel punch, and the holes are filled before painting. Deformed shank nails, which are very resistant to withdrawal from the wood, are used for such applications as attaching gypsum wallboard and floor underlayment, materials that cannot be allowed to work loose in service. Concrete nails can be driven short distances into masonry or concrete for attaching furring strips and sleepers. Cut nails, once used for framing connections, are now used mostly for attaching finish flooring because their square tips punch through the wood rather than wedging through, which tends to minimize splitting of brittle woods. The large head on roofing nails is needed to apply sufficient holding power to the soft material of which asphalt shingles are made.

chanical nail gun. Common nails and finish nails are the two types most frequently used. *Common nails* have flat heads and are used for most structural connections in light frame construction. *Finish nails* are virtually headless and are used to fasten finish woodwork, where they are less obtrusive than common nails (Figure 3.33).

In the United States the size of a nail is measured in *pennies*. This strange unit originated earlier in history as the price of a hundred nails of a given size and persists in use despite the effects of inflation on nail prices. Figure 3.34 shows the dimensions of the various sizes of common nails. Finish nails are the same length as common nails of the corresponding penny designation.

Nails are ordinarily furnished *bright*, meaning that they are made of plain, uncoated steel. Nails that will be exposed to the weather should be of a corrosion-resistant type, either *hot dip galvanized*, *aluminum*, or *stainless steel*. (The zinc coating on *electro-galvanized* nails is very thin, and is easily damaged during driving.) This is particularly important in exterior siding, trim, and decks, which would be stained by rust leaching from bright nails. The natural acids in redwood and cedar are corrosive to zinc under some conditions of exposure, so stainless steel nails are preferred in these woods.

The three methods of fastening with nails are shown in Figure 3.35. Each of these methods has its uses in building construction, as illustrated in Chapters 5 and 6. Nails are a favored means of fastening wood because they require no predrilling of holes under most conditions and they are driven with a hammer, which makes it extremely fast to install them.

Wood Screws and Lag Screws

Screws are inserted into drilled holes and turned into place with a screwdriver or wrench (Figure 3.36.) They are little used in ordinary light framing because they take much longer to in-

FIGURE 3.34

Standard sizes of common nails, reproduced full size. The abbreviation "d" stands for "penny." The length of each nail is given below its size designation. The three sizes of nail used in light frame construction, 16d, 10d, and 8d, are shaded.

FIGURE 3.35

Face nailing is the strongest of the three methods of nailing. End nailing is weak and is useful primarily for holding framing members in alignment until gravity forces and applied sheathing make a stronger connection. Toe nailing is used in situations where access for end nailing is not available. Toe nails are surprisingly strong—load tests show them to carry about two-thirds as much load as face nails of the same size.

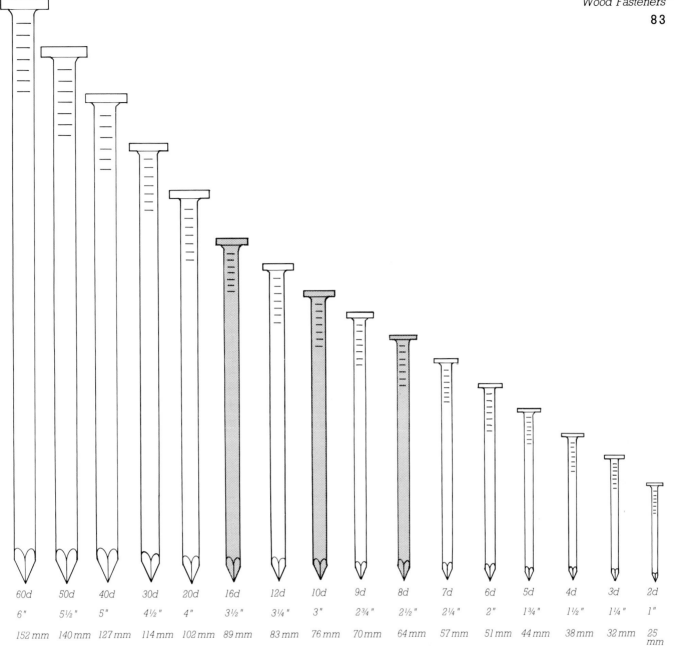

60d	50d	40d	30d	20d	16d	12d	10d	9d	8d	7d	6d	5d	4d	3d	2d
6"	5½"	5"	4½"	4"	3½"	3¼"	3"	2¾"	2½"	2¼"	2"	1¾"	1½"	1¼"	1"
152 mm	140 mm	127 mm	114 mm	102 mm	89 mm	83 mm	76 mm	70 mm	64 mm	57 mm	51 mm	44 mm	38 mm	32 mm	25 mm

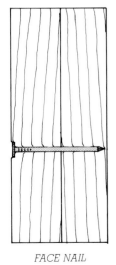

FACE NAIL

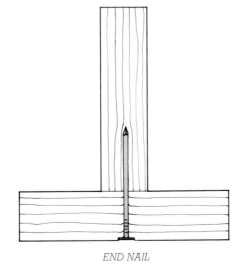

END NAIL

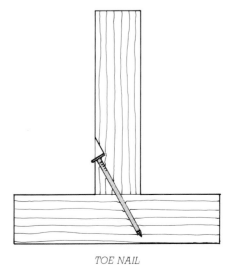

TOE NAIL

stall than nails, but they are often used in cabinetwork and furniture and for mounting hardware. Screws form tighter, stronger connections than nails and can be backed out and reinserted if a component needs to be adjusted or remounted. Very large screws for heavy structural connections are called *lag screws*. They have square or hexagonal heads and are driven with a wrench rather than a screwdriver. Small *drywall screws*, which can be driven by a power screwdriver without first drill-

ing holes in most cases, are made for use in attaching gypsum board to wood framing members and can be at least as fast to install as nails. They are weak and extremely brittle, however, so they cannot be used for structural connections.

Bolts

Bolts are used for major structural connections in heavy timber framing.

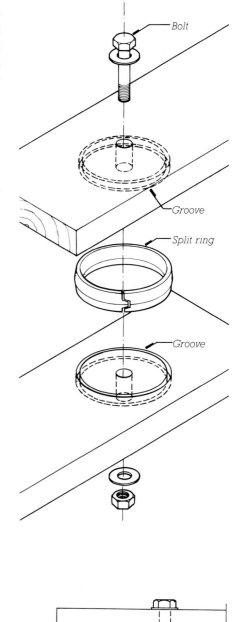

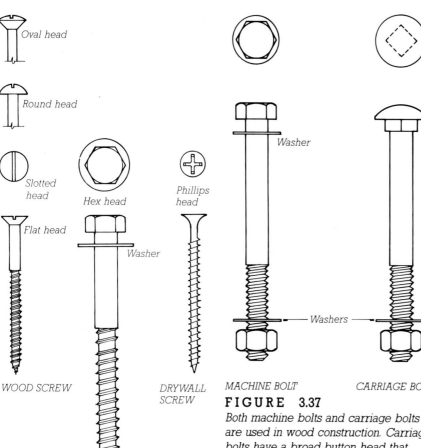

FIGURE 3.36

Flat head screws are used without washers and are driven flush with the surface of the wood. Round head screws are used with flat washers, and oval head with countersunk washers. Drywall screws do not use washers, and are the only screw shown here that does not require a predrilled hole. Slotted, hex, and Phillips heads are all common.

FIGURE 3.37

Both machine bolts and carriage bolts are used in wood construction. Carriage bolts have a broad button head that needs no washer and a square shoulder under the head that is forced into the drilled hole in the wood to hold the bolt against turning as the nut is tightened.

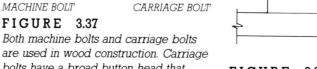

FIGURE 3.38

Split rings are high-capacity connectors used in heavily loaded joints of timber frames and trusses. After the center hole is drilled through the two pieces, they are separated and the matching grooves are cut with a special rotary cutter driven by a power drill. The joint is then reassembled with the ring in place.

Commonly used bolts range in diameter from $\frac{3}{8}$ to 1 inch (9.53 to 25.4 mm), in any desired length. Flat steel disks called *washers* are inserted under heads and nuts of bolts to distribute the compressive force from the bolt across a greater area of wood (Figure 3.37).

Timber Connectors

Various types of timber connectors have been developed to increase the load-carrying capacity of bolts. The most widely used of these is the *split ring connector* (Figure 3.38). The split ring is used in conjunction with a bolt and is inserted in matching circular grooves in the mating pieces of wood. Its func-tion is to spread the load across a much greater area of wood than can be done with a bolt alone. The split permits the ring to adjust to wood shrinkage. Split rings are used almost exclusively in heavy timber construction.

Toothed Plates

Toothed plates (Figure 3.39) are used in factory-produced roof and floor trusses. They are inserted into the wood with hydraulic presses, pneumatic presses, or mechanical rollers and act as metal splice plates, each with a very large number of built-in nails (Figures 3.40 through 3.42). They are extremely effective connectors because their multiple, closely spaced points interlock tightly with the fibers of the wood.

Sheet Metal and Metal Plate Framing Devices

Dozens of ingenious sheet metal and metal plate devices are manufactured for strengthening common connections in wood framing. The most frequently used is the *joist hanger*, but all of the devices shown in Figure 3.43 find extensive use. There are two parallel series of this type of device: one made of sheet metal for use in light framing and one of metal plate for heavy timber and laminated wood framing.

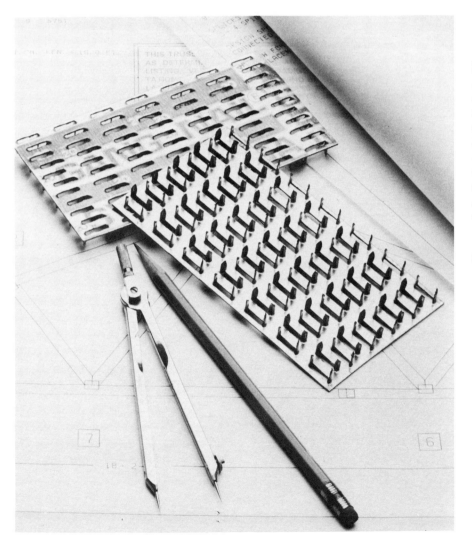

FIGURE 3.39
Manufacturers of toothed plate connectors also manufacture the machinery to install them and provide computer programs to aid local truss fabricators in designing and detailing trusses for specific buildings. The truss drawing in this photograph was generated by a plotter driven by such a program. The small rectangles on the drawing indicate the positions of toothed plate connectors at the joints of the truss. (Courtesy of Gang-Nail Systems, Inc.)

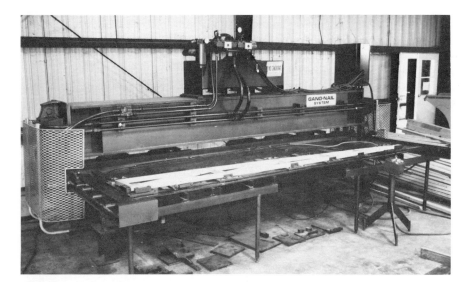

FIGURE 3.40
A hydraulic machine aligns the wood members of a small roof truss for a factory-built sectional house and presses toothed plate connectors into the joints. (Courtesy of Gang-Nail Systems, Inc.)

FIGURE 3.41
The trusses are assembled into a roof panel. Applications of toothed-plate trusses to site built framing are illustrated in Chapter 5. (Courtesy of Gang-Nail Systems, Inc.)

FIGURE 3.42
Toothed plate connectors are inserted into long floor trusses by a press that moves along the truss from one line of connections to the next. (Courtesy of Gang-Nail Systems, Inc.)

FIGURE 3.43

Joist hangers are used to make strong connections in floor framing wherever wood joists bear on one another at right angles. The heavier steel beam hangers are used primarily in laminated wood construction. Post bases serve the twofold function of preventing water from entering the end of the post and anchoring the post to the foundation. The bolts and lag screws used to connect the wood members to the heavier connectors are omitted from this drawing.

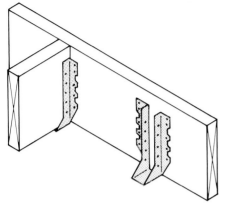

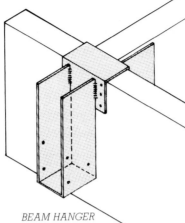

JOIST HANGER

BEAM HANGER

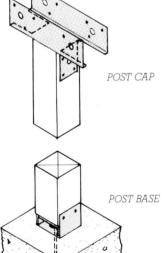

POST CAP

POST CAP

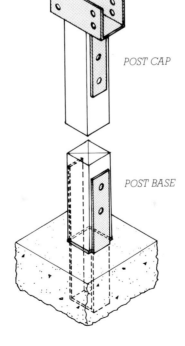

POST BASE

POST BASE

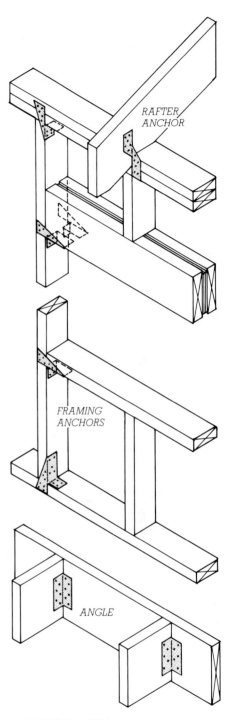

RAFTER ANCHOR

FRAMING ANCHORS

ANGLE

FIGURE 3.44

The sheet metal connectors shown in this diagram are less commonly used than those in Figure 3.43, but are invaluable in solving special framing problems and in reinforcing frames against wind uplift and earthquake forces.

Machine-Driven Staples and Nails

The speed of wood frame construction can be significantly increased through the use of pneumatically operated *nail guns* and *staple guns*. The nails or staples are of special design and are prepackaged by the manufacturer in self-feeding strips, each containing several dozen fasteners. These are loaded into guns powered by air fed through a hose from a small air compressor. They can be driven one by one as fast as the operator can pull the trigger. Several uses of machine-driven nails and staples are illustrated in Chapters 5 and 6.

Adhesives

Adhesives are widely employed in the factory production of plywood and panel products, laminated wood, wood structural components, and cabinetwork, but have only a few uses on the construction job site. This disparity is explained by the need to clamp and hold adhesive joints under controlled environmental conditions until the adhesives have cured, which is easy to do in the shop but much more cumbersome and unreliable in the field. The area in which job site adhesives are most common is in securing subflooring and wall panels to wood framing. Adhesives for these purposes are applied from a sealant gun in mastic form, and are clamped by simply nailing the panels to the framing. The nails serve also as a backup structural connection, and the adhesive enables the panel and the framing members on which it is mounted to act as a composite structure, which increases the overall stiffness of the assembly.

WOOD MANU-FACTURED BUILDING COMPONENTS

Dimension lumber, structural panel products, mechanical fasteners, and adhesives are used in combination to manufacture a number of highly efficient structural components that offer certain advantages to the designer of wooden buildings.

Trusses

Trusses for both roof and floor construction are manufactured in small, highly efficient plants in every part of North America. Most are based on 2×4s and 2×6s joined with toothed plate connectors. The designer or builder needs to specify only the span, the roof pitch, and the desired overhang detail. The truss manufacturer either uses a preengineered design for the specified truss, or employs a sophisticated computer program to engineer the truss and develop the necessary cutting patterns for its constituent parts. The manufacture of trusses is shown in Figures 3.40 and 3.42, and several of their uses are depicted in Chapter 5.

Roof trusses use less wood than a comparable frame of conventional rafters and ceiling joists. They, like floor trusses, span the entire width of the building in most applications, allowing the designer complete freedom to locate interior partitions anywhere they are needed. The chief disadvantages of roof trusses are that they make the attic space unusable, and they generally restrict the designer to the spatial monotony of a flat ceiling throughout the building, though truss shapes can be designed and manufactured to overcome this latter limitation.

Plywood Beams

I-beams and *box beams* manufactured of a combination of plywood and dimension lumber are useful for longer-span framing of both roofs and floors (Figures 3.45 through 3.47). Like trusses, these components use wood more efficiently than conventional rafters and joists, and they free the interior of the building from loadbearing partitions. Plywood box beams can be

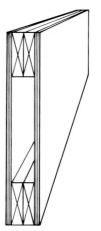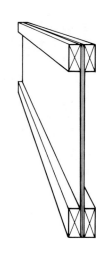

FIGURE 3.45
Box beams (left) and I-beams (right) are shop fabricated from a combination of dimension lumber and plywood, using adhesives to form the primary structural connections and nails or screws to hold the parts together until the adhesive cures.

custom designed and manufactured to carry major loads over fairly long spans, and can be shaped to fit particular structural and architectural requirements.

Panel Components and Box Components

Dimension lumber and wood structural panels lend themselves readily to the prefabrication of many kinds of floor, wall, and roof panels. At their simplest, such panels consist of a section of framing, usually 4 feet (1219 mm) wide, sheathed with a sheet of plywood or waferboard. In light frame construction, these are trucked to the construction site and rapidly nailed together into a complete frame, sheathed and ready for finishing. Panels can also be larger, encompassing whole walls or floors in a single component. They can become more sophisticated, including insulation, wiring, windows, doors, and exterior and interior finishes, if the thorny problems of joinery and code acceptance can be resolved.

For greatest structural efficiency and economy, wall, floor, and roof panels

FIGURE 3.46
A proprietary design of I-beam joist, seen in two different sizes to the right of the picture, is made up entirely of softwood veneers, including the top and bottom flanges, allowing the manufacturer to produce the flange material in any desired length without splicing. The remainder of the structural members in the photograph are beams and headers fabricated entirely from softwood veneers. (Courtesy of Trus Joist Corporation)

FIGURE 3.47
Installing a plywood box beam. (Courtesy of American Plywood Association)

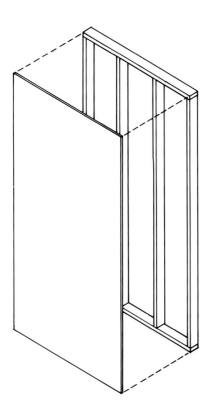

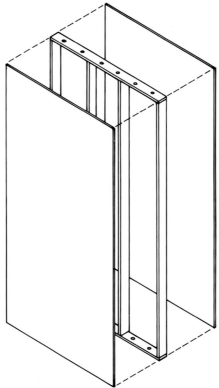

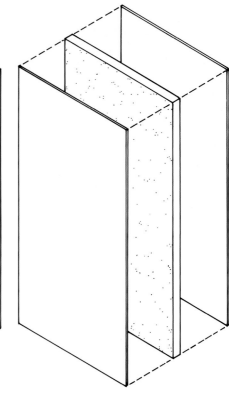

FRAMED PANEL   STRESSED SKIN PANEL   SANDWICH PANEL

FIGURE 3.48

Three types of prefabricated wood panels. The framed panel is identical to a segment of a conventionally framed wall, floor, or roof. The facings on the *stressed skin panel are adhesive bonded to thin wood spacers to form a structural unit in which the facings carry the major stresses. A sandwich panel* *functions structurally in the same way as a stressed skin panel, but its facings are bonded to a core of insulating foam instead of wood spacers.*

FIGURE 3.49

Installing stressed skin roof panels over plywood box beams. (Courtesy of American Plywood Association)

can utilize top and bottom sheets of plywood or wood composition board as the primary load-carrying members of the panel, if they are joined firmly together by a stiff plastic foam core to make a *sandwich panel*, or by dimension lumber spacers to make a *stressed-skin panel* (Figures 3.48 and 3.49). Panels of either of these types can also be joined to make folded plate structures (Figure 3.50) or curved to make vaulted roofs (Figure 3.51).

Dimension lumber and structural wood panels have become the favored materials of the *manufactured housing industry*, which factory-builds entire houses as finished boxes, often complete with furnishings, and trucks them to prepared foundations where they are set in place and made ready for occupancy in a matter of hours. If the house is 14 feet (4.27 m) or less in width, it is constructed on a rubber-tired frame, is completely finished in the factory, and is known as a *mobile home*. If wider than this, or more than one story high, it is built in two or more completed sections that are joined at the site, and is known as a *sectional* or *modular home*. Mobile homes are sold at a fraction of the price of conventionally constructed houses. This is due in part to the economies of factory production and mass marketing, and in part to the use of components that are lighter and less costly, and therefore of shorter life expectancy. But at prices that more closely approach the cost of conventional on-site construction, many companies manufacture modular housing to the same standards as conventional construction.

FIGURE 3.50
Erecting a plywood folded plate roof from prefabricated components. (Courtesy of American Plywood Association)

FIGURE 3.51
Prefabricated plywood barrel vaults. (Courtesy of American Plywood Association)

TYPES OF WOOD CONSTRUCTION

Wood construction has evolved into two major systems of on-site construction: The hand-hewn frames of centuries past have become the heavy timber frames of today, used primarily for structures larger than single-family houses. And from the heavy timber frame has sprung the light frame construction so popular for houses, small apartment buildings, and small commercial structures. These two systems of framing are detailed in chapters 4 and 5.

C.S.I./C.S.C. Masterformat Section Numbers for Wood	
06050	FASTENERS AND ADHESIVES
06100	ROUGH CARPENTRY
06110	Wood Framing
06115	Wood Sheathing
06120	Structural Panels
06130	HEAVY TIMBER CONSTRUCTION
06170	PREFABRICATED STRUCTURAL WOOD
06180	Glued-Laminated Construction
06190	Wood Trusses
06195	Prefabricated Wood Beams and Joists
06300	WOOD TREATMENT
06310	Preservative Treatment
06320	Fire Retardant Treatment
06330	Insect Treatment

SELECTED REFERENCES

1. The American Institute of Timber Construction. *AITC 117-79--Design Standard Specifications for Structural Glued Laminated Timber of Softwood Species*. Englewood, Colorado, 1980.

This booklet sets the standards for the design and fabrication of laminated wood structural members. (Address for ordering: 333 West Hampden Avenue, Englewood, CO 80110.)

2. National Forest Products Association. *National Design Specification for Wood Construction*, 1982 Edition. Washington, D.C., 1982.

This booklet sets the standards for structural design procedures for wood structures and covers the calculation of stresses in beams, columns, and wood fastenings. Its companion volume, *Design Values for Wood Construction*, gives the allowable structural stresses for every species and grade of wood commonly used in North America. (Address for ordering: 1619 Massachusetts Avenue, N.W., Washington, D.C. 20036. Ask also for their catalog of publications.)

3. American Plywood Association. *Performance-Rated Panels*. Tacoma, Washington, 1984.

The full range of structural wood panel products and their grading system are explained in this booklet. (Address for ordering: P.O. Box 11700, Tacoma, WA 98411. Ask also for the catalog of APA publications.)

KEY TERMS AND CONCEPTS

bark
cambium
sapwood
heartwood
cellulose
lignin
grain
springwood (earlywood)
summerwood (latewood)
softwood

hardwood
annual rings
tracheids
rays
fibers
vessels or pores
lumber
headsaw
sawyer
plainsawed

quartersawn
figure
seasoning
free water
bound water
air drying
kiln drying
longitudinal, radial, tangential
    shrinkage
check

parallel-to-grain
perpendicular-to-grain
surfacing
S4S, S2S
S-DRY
S-GRN
growth characteristics
manufacturing characteristics
knot, knothole
decay
insect damage
split
crook
bow
twist
cup
wane
visual grading
machine grading
nominal dimension
actual dimension

boards
dimension lumber
timbers
board foot
laminated wood (glulam)
structural wood panel
plywood
veneer
composite panel
nonveneered panel
oriented strand board (OSB)
waferboard
particleboard
rotary-sliced
flitch-sliced
span rating
hardboard
creosote
pentachlorophenol
waterborne salts
pressure impregnation

nail
common nail
finish nail
penny
bright, hot-dip galvanized nail
wood screw
lag screw
bolt
washer
split ring connector
toothed plate
joist hanger
machine-driven staples or nails
adhesive
I-beam
box beam
stressed-skin panel
sandwich panel
sectional or modular home
mobile home

REVIEW QUESTIONS

1.   Discuss the changes in moisture content of the wood and their effects on a piece of dimension lumber, from the time the tree is cut, through its processing, until it has been in service in a building for an entire year.

2.   Give the actual cross-sectional dimensions of the following pieces of kiln-dried lumber: 1×4, 2×4, 2×6, 2×8, 4×4, 4×12.

3.   Why is wood laminated?

4.   What is meant by a span rating of 32/24? What types of wood products are rated in this way?

5.   For what reasons might you specify pressure-treated wood?

6.   Which common species of wood have decay- and insect-resistant heartwood?

EXERCISES

1.   Visit a nearby lumber yard and list the species, grades, and sizes of lumber carried in stock. For what uses are each of these intended? While at the yard, look also at the available range of fasteners.

2.   Pick up a number of wood scraps from a construction site. Examine each to see where it was located in the log before sawing. Note any drying distortions in each piece: How well do

these correspond to the distortions you would have predicted? Measure accurately the width and thickness of each scrap and compare your measurements to the specified actual dimensions for each.

3.   Assemble samples of as many different species of wood as you can. Learn how to tell the different species apart, by color, odor, grain figure, ray structure, relative hardness, and so on. What

are the most common uses for each species?

4.   Visit a construction site and list the various types of lumber and wood products being used. Look for a grade stamp on each, and determine why the given grade is being used for each use. If possible, look at the architect's written specifications for the project and see how the lumber and wood products were specified.

## CHAPTER FOUR

HEAVY TIMBER FRAME CONSTRUCTION

Heavy Timber (Type 4) Construction

The Fire Resistance of Heavy Timbers

Wood Shrinkage in Heavy Timber Construction

Floor and Roof Decks for Heavy Timber Buildings

Bracing of Heavy Timber Buildings

Combustible (Type 5) Buildings Framed With Heavy Timber

Longer Spans in Heavy Timber

Large Beams

Rigid Frames

Trusses

Arches and Domes

Heavy Timber and the Building Codes

The Uniqueness of Heavy Timber Framing

*T*hree cranes erect curved laminated timbers for a sports arena. The vertical poles are temporary supports and will be removed when the dome is completely framed, as will the small wood cleats nailed to the timbers to aid workers in climbing up and down. Welded steel connectors are used at all the joints. *(Courtesy of American Institute of Timber Construction)*

Wood beams have been used for spanning roofs and floors of buildings since the beginnings of civilization. The first timber-framed buildings were crude pit houses, lean-tos, teepees, and basketlike assemblages of bent saplings. In earliest historic times roof timbers were combined with masonry loadbearing walls to build houses and public buildings. In the Middle Ages braced wall frames of timber were built for the first time (Figures 4.1, 4.2). The British carpenters who emigrated to North America in the seventeenth and eighteenth centuries brought with them a fully developed knowledge of how to build efficiently braced frames, and for two centuries North Americans lived and worked almost exclusively in buildings spanned with hand-hewn wooden timbers joined by interlocking wood-to-wood connections (Figures 4.3, 4.4). Nails were rare and expensive, so they were used only in door and window construction and, sometimes, for fastening siding boards to the frame.

Until two centuries ago, logs could be converted to boards and timbers only by human muscle power. To make timbers, axemen skillfully scored and hewed logs to reduce them to a rectangular profile. Boards were produced slowly and laboriously with a long, two-man pit saw, one man standing in a pit beneath the log pulling the saw down and the other standing above, pulling it back up. But at the beginning of the nineteenth century, water-powered sawmills began to take over the work of transforming tree trunks into lumber, squaring timbers and slicing boards in a fraction of the time it took to do the same work by hand.

Most of the great industrial mills of the nineteenth century, which manufactured textiles, shoes, machinery, and all the goods of civilization, consisted of heavy, sawn timber floors and roofs supported by masonry exterior walls (Figures 4.5, 4.6). The housebuilders and barnraisers of the early nineteenth century switched from hard-hewn to sawn timbers as soon as they became available. Many of these mills, barns, and houses still survive, and with them survives a rich tradition of heavy timber building that continues to the present day.

FIGURE 4.1

Braced wall framing was not developed until the late Middle Ages and early Renaissance, when it was often exposed on the face of the building in the style of construction known as halftimbering. The space between the timbers was filled with brickwork, or with wattle and daub, a crude plaster of sticks and mud, as seen here in Wythenshawe Hall, a sixteenth-century house near Manchester, England. (Photo by James Austin, Cambridge, England)

FIGURE 4.2

*European timber house forms generally
followed a progression of development
from crude pit dwellings, made of earth
and tree trunks, to cruck frames, to
braced frames. The crucks (curved
timbers), hewn by hand from
appropriately shaped trees, were
precursors of the laminated wood
arches and rigid frames that are widely
used today.*

PIT DWELLING   CRUCK FRAME   BRACED FRAME

FIGURE 4.3

*The European tradition of heavy timber
framing was brought to North America
by the earliest settlers and was used for
houses and barns until well into the
nineteenth century. (Drawing by Eric
Sloane, courtesy of the artist)*

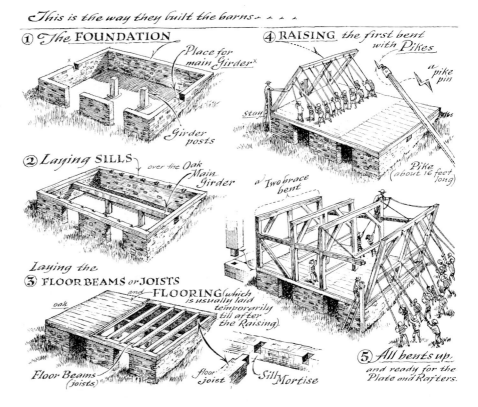

FIGURE 4.4
Traditional timber framing has been revived in recent years by a number of builders who have learned the old methods of joinery and updated them with the use of modern power tools and equipment. (a) Assembling a bent. (b) The completed bents are laid out on the floor, ready for raising. (c) Raising the bents and installing floor framing, using a truck mounted crane. (d) The completed frame. (e) Enclosing the timber-framed house with sandwich panels consisting of waferboard faces bonded to an insulating foam core. (Courtesy of Benson Woodworking, Alstead, New Hampshire)

The old country builder, when he has to get out a cambered beam or a curved brace, goes round his yard and looks out the log that grew in the actual shape, and taking off two outer slabs by handwork in the sawpit, chops it roughly to shape with his side-axe and works it to the finished face with the adze, so that the completed work shall for ever bear the evidence of his skill...

Gertrude Jekyll, English garden designer, writing in 1900

FIGURE 4.5

Most nineteenth-century industrial mills in America were constructed of heavy timber roofs and floors supported at the perimeter on masonry loadbearing walls. This impressive group of textile mills stretches for a distance of 2 miles (3 km) along the Merrimac River in Manchester, New Hampshire. (Photo by Randolph Langenbach)

FIGURE 4.6

The windows in the mills were generous, which provided plenty of daylight to work by. Columns were of wood or cast iron. Most New England mills, like this one, were framed very simply, with decking carried by beams running at right angles to the exterior walls, supported on two lines of columns. Notice that the finish flooring runs at right angles to the structural decking. Overhead sprinklers add a considerable additional measure of fire safety. (Photo by Randolph Langenbach)

HEAVY TIMBER (TYPE 4) CONSTRUCTION

The Fire Resistance of Heavy Timbers

Large timbers, because of their greater capacity to absorb heat, are much slower to catch fire and burn than smaller lumber. In building fires a heavy timber beam, though deeply charred by gradual burning, will continue to support its load long after an unprotected steel beam would have collapsed. If the fire is not prolonged, a heavy timber beam or column can often be sandblasted afterward to re- move the char and can continue in service. For these reasons, building codes recognize heavy timber construction that meets certain specific code requirements as having fire-resistive properties.

Heavy Timber construction with exterior walls of masonry or concrete is listed in the building code table in Fig-

FIGURE 4.7

Minimum member sizes for Heavy Timber construction, as typically specified in a building code. The dimensions given are nominal rather than actual. It is difficult to develop rule-of-thumb values for member sizes in Heavy Timber construction because of the highly variable effects of species, grade, and loading, but for preliminary design purposes, prior to the exact computation of member sizes, timber beams can be assumed to have an actual depth of about one-sixteenth of their span, and a width equal to one-third to one-half of their depth. Girders will be slightly deeper. An initial rule-of-thumb value for the thickness of heavy timber roof decking is one-fortieth of the span, and for floor decking, one-thirty-second.

	Supporting Floor Loads	Supporting Roof and Ceiling Loads Only
Columns	8 × 8 (184 × 184 mm)	6 × 8 (140 × 184 mm)
Beams and Girders	6 × 10 (140 × 235 mm)	4 × 6 (89 × 140 mm)
Trusses	8 × 8 (184 × 184 mm)	4 × 6 (89 × 140 mm)
Decking	3″ + 1″ finish (64 mm + 19 mm finish)	2″, or 1⅛″ plywood (38 mm, or 29 mm plywood)

ure 1.1 as Type 4 construction, Heavy Timber, and is often referred to as *Mill construction*. Minimum sizes for the timbers and decking of Heavy Timber construction are specified in another section of the code; these requirements are summarized in Figure 4.7. Either sawn or laminated timbers are permitted. Traditionally, the edges of the timbers are *chamfered* to eliminate the thin corners of wood that catch fire most easily, but many codes no longer require this.

Wood Shrinkage in Heavy Timber Construction

The perimeter of a Heavy Timber building must be supported on concrete or masonry (unless the building is at least 30 feet (9 m) from any surrounding buildings), while the interior may be supported on heavy wood columns. Wood, unlike masonry, is subject to large amounts of expansion and contraction caused by seasonal changes in moisture content, particularly in the direction perpendicular to its grain. A Heavy Timber building is detailed to minimize the effects of this differential shrinkage by eliminating cross-grain wood from the interior lines of support. In the traditional Mill construction shown in Figures 4.8 through 4.12, iron *pintle caps* carry the column loads past the cross-grain of the beams at each floor, so the beams and girders can shrink without causing the floors and roof to sag. In current practice, a laminated column may be fabricated as a single piece running the entire height of the building, with the beams supported by wood bearing blocks or welded metal connectors; or columns may be butted directly to one another at each floor with the aid of metal connectors (Figures 4.11, 4.13, 4.15, 4.16).

Anchorage of Timber Beams to Masonry Walls

Where heavy timber beams join masonry or concrete walls, three problems must be solved: First, the beam must be protected from possible decay caused by moisture that may seep through the wall. This is done by leaving a ventilating airspace of at least ½ inch (13 mm) between the masonry and all sides of the beam except the bottom, unless the beam is treated against decay. The second and third problems have to be solved together: The beam must be securely anchored to the wall so it cannot pull away under normal service, yet must be able to rotate freely so it does not pry the wall apart if it burns through during a severe fire (Figure 4.10). Two methods of accomplishing these dual needs are shown in Figure 4.9, and another in Figure 4.13.

Floor and Roof Decks for Heavy Timber Buildings

Heavy Timber (Type 4) buildings are required by code to have floors and roofs of solid wood construction, without internal cavities. Figure 4.14 shows several different types of decking used for these purposes. Minimum permissible thicknesses of decking are given in Figure 4.7. Floor decking must be covered with a finish floor consisting of nominal 1-inch (25-mm) tongue-and-groove boards laid at right angles or diagonally to the structural decking, or with ½-inch (13-mm) plywood or particle board.

Bracing of Heavy Timber Buildings

A Heavy Timber building with an exterior masonry or concrete bearing wall is normally braced by the shear resistance of the walls, working together with the diaphragm action of the roof and floor decks. In areas of high seismic risk, the walls must be reinforced both vertically and horizon-

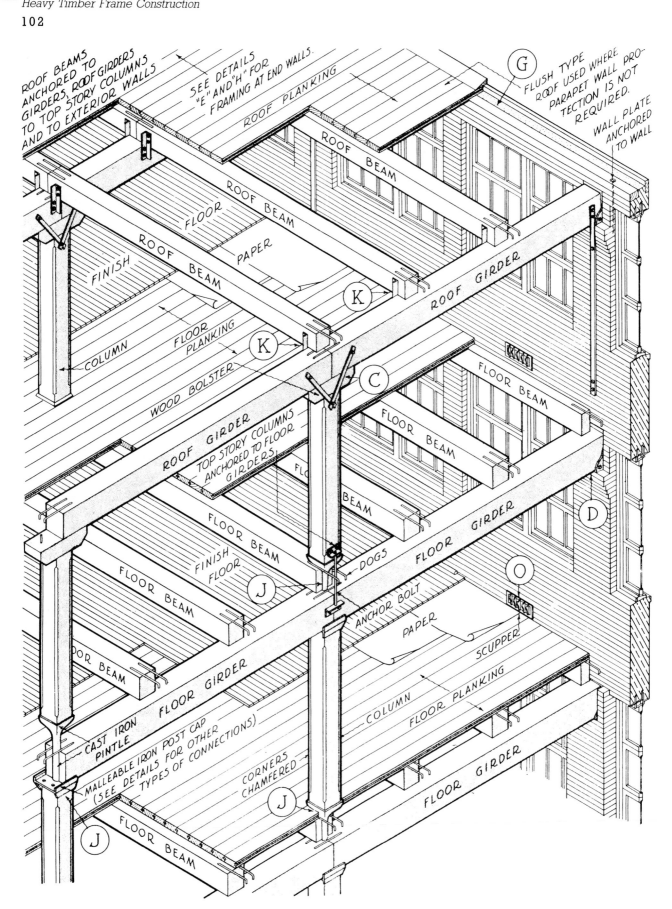

ROOF BEAMS ANCHORED TO GIRDERS, ROOF GIRDERS TO TOP STORY COLUMNS AND TO EXTERIOR WALLS

SEE DETAILS "E" AND "H" FOR FRAMING AT END WALLS.

ROOF PLANKING

FLUSH TYPE ROOF USED WHERE PARAPET WALL PROTECTION IS NOT REQUIRED.

WALL PLATE ANCHORED TO WALL

ROOF BEAM

G

ROOF BEAM

FLOOR

PAPER

ROOF BEAM

FINISH

ROOF BEAM

K

ROOF GIRDER

COLUMN

FLOOR PLANKING

K

WOOD BOLSTER

C

FLOOR BEAM

FLOOR BEAM

ROOF GIRDER

TOP STORY COLUMNS ANCHORED TO FLOOR GIRDERS

FLOOR BEAM

FLOOR

BEAM

FLOOR BEAM

FINISH FLOOR

J

DOGS

FLOOR GIRDER

D

O

FLOOR BEAM

ANCHOR BOLT

PAPER

SCUPPER

FLOOR BEAM

CAST IRON PINTLE

FLOOR GIRDER

COLUMN

FLOOR PLANKING

MALLEABLE IRON POST CAP (SEE DETAILS FOR OTHER TYPES OF CONNECTIONS)

CORNERS CHAMFERED

J

FLOOR GIRDER

J

FLOOR BEAM

J

FLOOR BEAM

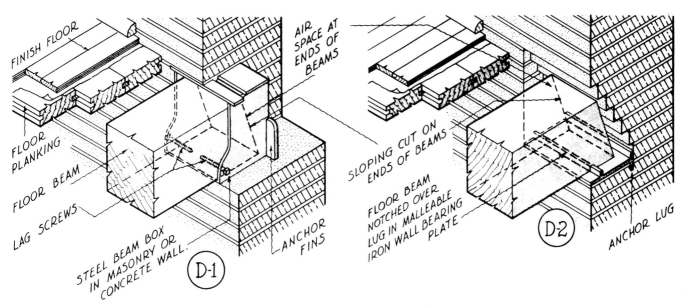

FINISH FLOOR

AIR SPACE AT ENDS OF BEAMS

FLOOR PLANKING

FLOOR BEAM

LAG SCREWS

STEEL BEAM BOX IN MASONRY OR CONCRETE WALL.

ANCHOR FINS

(D-1)

SLOPING CUT ON ENDS OF BEAMS

FLOOR BEAM NOTCHED OVER LUG IN MALLEABLE IRON WALL BEARING PLATE

(D-2)

ANCHOR LUG

FIGURE 4.9

Two alternative details for the bearing of a beam on masonry in tradition Mill construction. In each case, the beam end is firecut to allow it to rotate out of the wall if it burns through (Figure 4.10), but anchored against pulling away from the wall, either with lag screws or a lug on the iron bearing plate. (Courtesy of National Forest Products Association)

FIGURE 4.10

A timber beam that burns through in a prolonged fire is likely to topple its supporting masonry wall (left) unless its end is firecut (right). The beam is anchored to the wall in each case by a steel strap anchor of the type shown in Figure 4.13.

FIGURE 4.8

Traditional Mill construction bypasses problems of wood shrinkage at the the interior lines of support by using cast iron pintles to transmit the column loads through the beams and girders. Iron dogs tie the beams together over the girders. A long steel strap anchors the roof girder to a point sufficiently lower in the outside wall that the weight of the masonry above the anchorage point is enough to resist wind uplift on the roof. (Courtesy of National Forest Products Association)

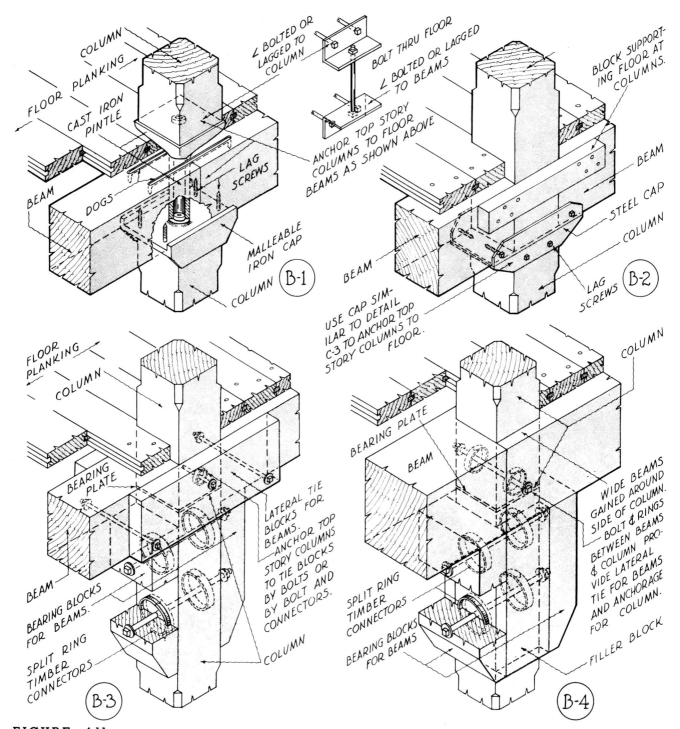

FIGURE 4.11

Four alternative details for interior girder/column intersections. Detail B-1 avoids wood shrinkage problems with an iron pintle, while the other three details bring the columns through the beams with only a steel bearing plate between the ends of the column sections. Split ring connectors are used in the lower two details to form a strong enough connection between the bearing blocks and the columns to support the loads from the beams; it would take a much larger number of bolts to do the same job. (Courtesy of National Forest Products Association)

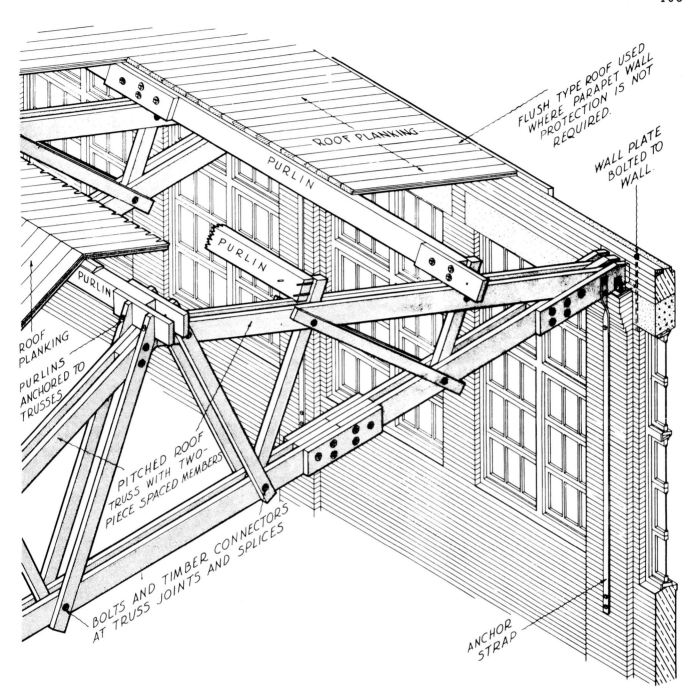

Labels within the figure:

FLUSH TYPE ROOF USED WHERE PARAPET WALL PROTECTION IS NOT REQUIRED.

ROOF PLANKING

PURLIN

PURLIN

WALL PLATE BOLTED TO WALL.

ROOF PLANKING

PURLINS ANCHORED TO TRUSSES

PURLIN

PITCHED ROOF TRUSS WITH TWO-PIECE SPACED MEMBERS

BOLTS AND TIMBER CONNECTORS AT TRUSS JOINTS AND SPLICES

ANCHOR STRAP

FIGURE 4.12

Heavy timber roof trusses for Mill construction. Split ring connectors are used to transmit the heavy loads between the overlapping members of the truss. A long anchor strap is again used at the outside wall, as explained in the legend to Figure 4.8. (Courtesy of National Forest Products Association)

FIGURE 4.13
An example of contemporary Heavy Timber construction, based on steel plate connectors and an insulated cavity wall with brick facings on both sides. As shown here, the masonry is unsuitable for use in zones of high seismic risk because it is unreinforced. Details of reinforced masonry construction are shown in Chapter 8.

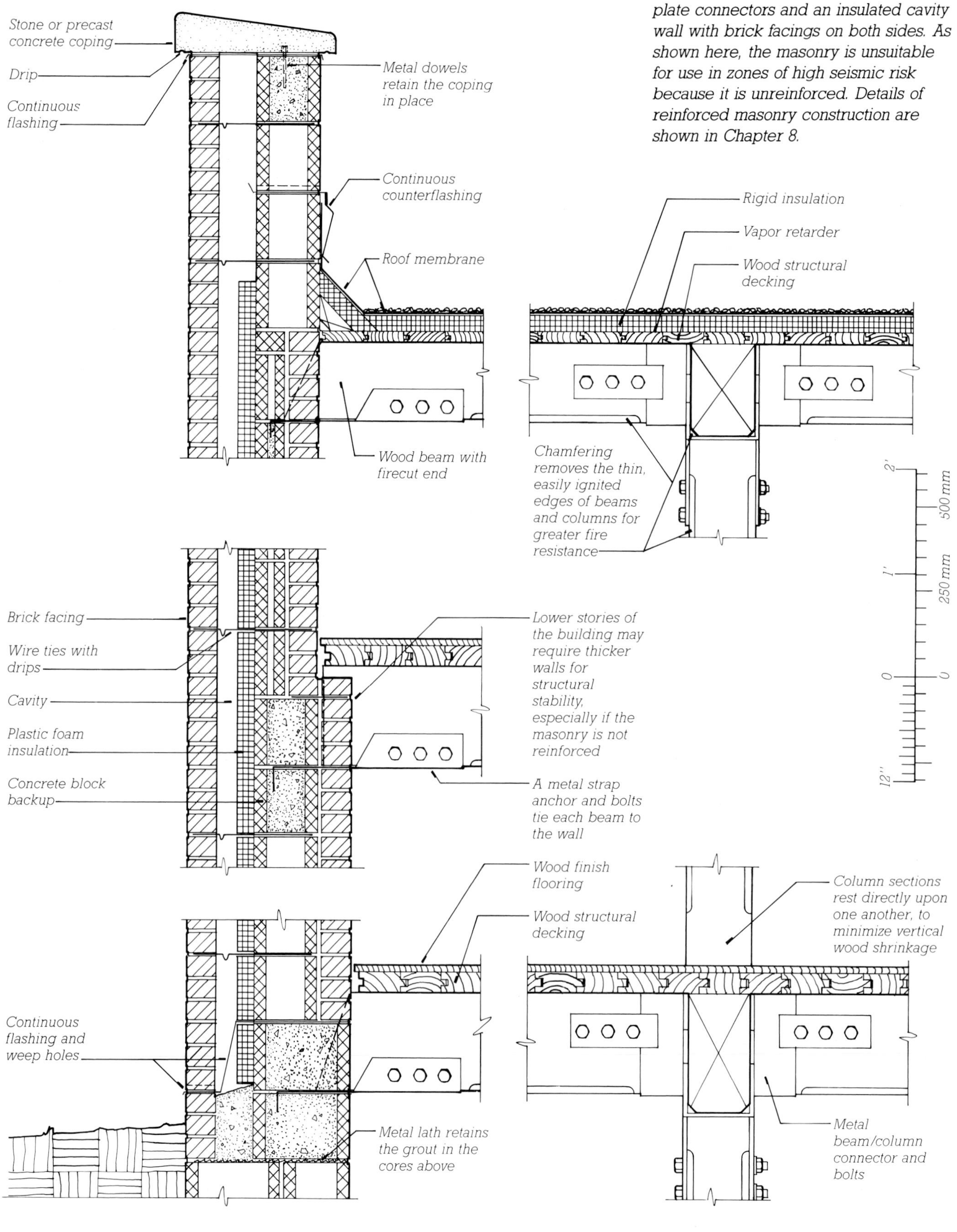

Stone or precast concrete coping

Drip

Continuous flashing

Metal dowels retain the coping in place

Continuous counterflashing

Roof membrane

Wood beam with firecut end

Rigid insulation

Vapor retarder

Wood structural decking

Chamfering removes the thin, easily ignited edges of beams and columns for greater fire resistance

Brick facing

Wire ties with drips

Cavity

Plastic foam insulation

Concrete block backup

Lower stories of the building may require thicker walls for structural stability, especially if the masonry is not reinforced

A metal strap anchor and bolts tie each beam to the wall

2'

500 mm

1'

250 mm

0

0

12"

Wood finish flooring

Wood structural decking

Column sections rest directly upon one another, to minimize vertical wood shrinkage

Continuous flashing and weep holes

Metal lath retains the grout in the cores above

Metal beam/column connector and bolts

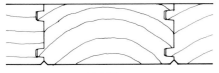

TONGUE AND GROOVE

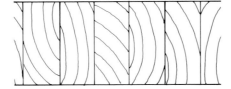

LAMINATED DECK

Spline

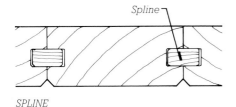

SPLINE

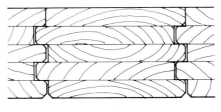

GLUE LAMINATED
DECKING

FIGURE 4.14

Large-scale cross sections of four types of heavy timber decking. Tongue-and-groove is the most common, but the other three types are slightly more economical of lumber because wood is not wasted in the milling of the tongues. Laminated deck is a traditional type for longer spans and heavier loads; it consists of ordinary dimension lumber nailed together. Glue-laminated decking is a modern type. As shown here, five separate boards are glued together to make each piece of decking. Decking of any type is usually furnished and installed in random lengths, with end joints staggered in the roof to avoid zones of structural weakness. The splines, tongues, or nails allow the narrow strips of decking to share concentrated structural loads as if they constituted a continuous sheet of solid wood.

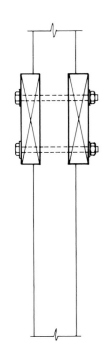

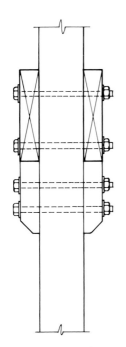

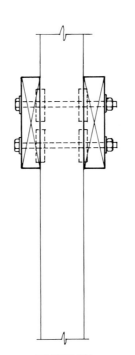

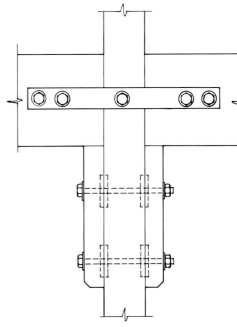

SHOULDER BEARING   BEARING BLOCKS   SPLIT RINGS   BEARING BLOCKS WITH
SPLIT RINGS

FIGURE 4.15

Some typical beam-column connections for Heavy Timber construction. The first three examples are for doubled beams sandwiched on either side of the column, and the fourth shows single beams in the same plane as the column. In the shoulder bearing connection, the beams are recessed into the column by an amount that allows the load to be *safely transferred by wood-to-wood bearing; the bolts serve only to keep the beams in the recesses. Bearing blocks allow more bolts to be inserted in a connection than can fit through the beams, and each bolt in the bearing blocks can hold several times as much load as one through the beam because it acts parallel to the grain of the wood* *rather than perpendicular. It is generally impossible to place enough bolts in a beam-column joint to transfer the load successfully without bearing blocks unless split rings are used, as shown in the third example. The steel straps and bolts in the fourth example hold the beams on the bearing blocks.*

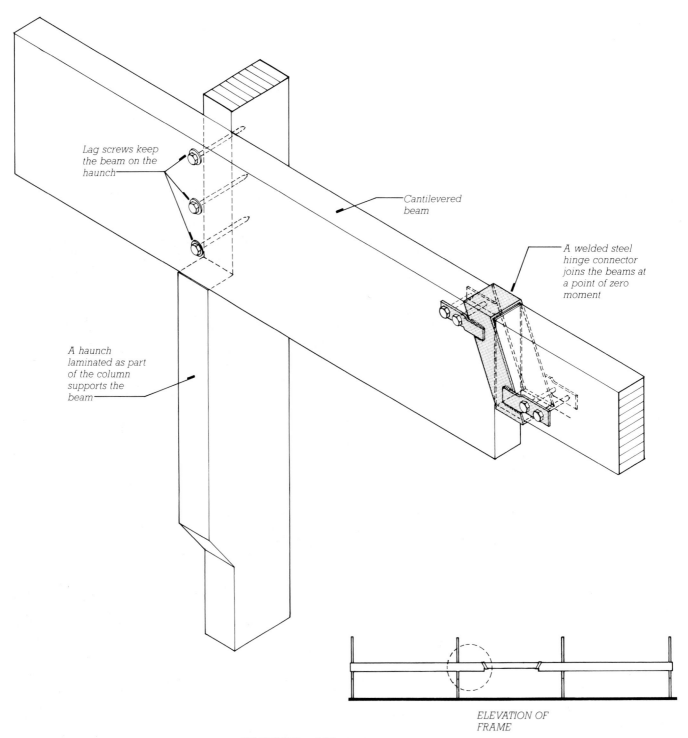

Lag screws keep the beam on the haunch

Cantilevered beam

A welded steel hinge connector joins the beams at a point of zero moment

A haunch laminated as part of the column supports the beam

ELEVATION OF FRAME

FIGURE 4.16

Typical connections for a laminated wood frame. The cantilevered beam and hinge connector save wood by connecting the beams at points of zero moment rather than at the columns to take full advantage of continuous bending action in the beams.

tally, and the decks may have to be specially nailed, or overlaid with plywood, to increase their shear resistance. In buildings with walls of materials other than concrete or masonry, bracing or shear panels must be provided to stabilize the building against wind and earthquake loads.

COMBUSTIBLE (TYPE 5) BUILDINGS FRAMED WITH HEAVY TIMBER

Heavy timbers are often used in combination with smaller wood framing members to construct buildings that do not meet all the fire-resistive requirements for Type 4 construction. These buildings, classified under the BOCA Code as Type 5 (Wood Light Frame) construction, bring the appearance and structural performance of beam-and-decking framing to small freestanding residential, commercial, religious, and institutional buildings. In these applications there are no restrictions on minimum size of timber, minimum thickness of decking, or exterior wall material, and light framing of nominal 2-inch (51-mm) lumber can be incorporated as desired, so long as the code requirements for Type 5 construction are met.

The use of exposed beams and decking in buildings that are normally made of light framing poses some new problems for the designer because the heavy timber structure does not have the concealed cavities that are normally present in the light frame structure. Thermal insulation cannot be simply inserted between ceiling joists but must be placed on top of the roof deck. If the roof is flat, it can be insulated and roofed in the manner shown for flat roofs in Chapter 13, but if it is pitched, a nailing surface for the shingles must be installed on the outside of the insulation. Electrical wiring for lighting fixtures on the ceiling will have to be run either through exposed metal conduit, which may be visually unsatis-

factory, or through the insulation above the deck. If the walls and partitions are made of masonry or stressed skin panels instead of light framing, special arrangements are necessary for installing wiring, plumbing, and heating devices in the walls as well.

LONGER SPANS IN HEAVY TIMBER

For buildings requiring spans longer than the 20 feet (6 m) or so that is the maximum usually associated with

FIGURE 4.17

Installing tongue-and-groove roof decking over laminated beams and girders. Note the hinge connectors in the beams at the extreme lower left corner of the photograph. (Courtesy of American Institute of Timber Construction)

framing of sawn timbers, the designer may select from among several different types of timber structural devices.

Large Beams

With large, virgin trees no longer readily available, very large timber beams are usually laminated rather than sawn. Laminated beams are stronger and more dimensionally stable than sawn wood beams and can be made in the exact size and shape desired (Figures 4.17 through 4.23).

Rigid Frames

The cruck (Figure 4.2), cut from a bent tree, was a form of *rigid frame* or *portal frame*. Today's rigid frames are laminated to shape, and find wide use in longer-span buildings. Standard configurations are readily available (Figures 4.20 through 4.23), or the designer may order a custom shape. Rigid frames exert a horizontal thrust, so must be tied together at the base. In laminated wood construction, rigid frames are often called *arches*, acknowledging that the two structural forms act in very nearly the same manner.

FIGURE 4.18
This panelized roof uses proprietary longspan trusses made of wood with steel tube diagonals. (Courtesy of Truss Joist Corporation)

FIGURE 4.19
Installing a roof deck of prefabricated panels, each consisting of a beam, joists, and plywood, with hangers attached to the beam. Plywood sheets are installed in the gaps over the girders after the panels are in place to create a fully continuous plywood diaphragm to stiffen the building against wind loads. Note the hinge connectors in the girders. (Courtesy of American Institute of Timber Construction)

FIGURE 4.20
Three-hinged arches of laminated wood carry laminated wood roof purlins. The short crosspieces of wood between the purlins are temporary ladders for workers. (Courtesy of American Institute of Timber Construction)

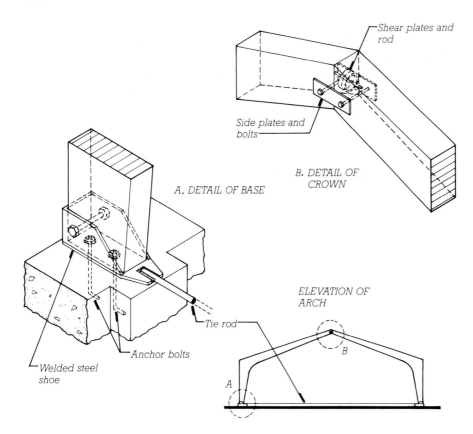

A. DETAIL OF BASE

B. DETAIL OF CROWN

—Shear plates and rod

Side plates and bolts

—Tie rod

—Anchor bolts

—Welded steel shoe

ELEVATION OF ARCH

A

B

FIGURE 4.21

Typical details for three-hinged arches of laminated wood. The tie rod is later covered by the floor slab.

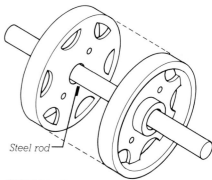

Steel rod—

SHEAR PLATES

FIGURE 4.22

The shear plates in the crown connection of the arch, shown here in a larger scale detail, are recessed into grooves in the wood and serve to spread any load from the steel rod across a much wider surface area of wood to avoid crushing and splitting.

FIGURE 4.23

Curved three-hinged arches in a church. (Courtesy of American Institute of Timber Construction)

Trusses

The largest number of wood *trusses* built each year are light roof trusses of nominal 2-inch (51-mm) lumber joined by toothed plates (see Figures 3.39 through 3.41, 5.55, and 5.56), but heavy timber trusses are often used in larger buildings. Their joints are made with steel bolts and welded steel plate connectors, or with split ring connectors. Both sawn and laminated timbers are used, sometimes in combination with steel rod tension members. Many shapes of heavy timber truss are possible, and spans of over 100 feet (30 m) are common (Figures 4.12, 4.24, 4.25).

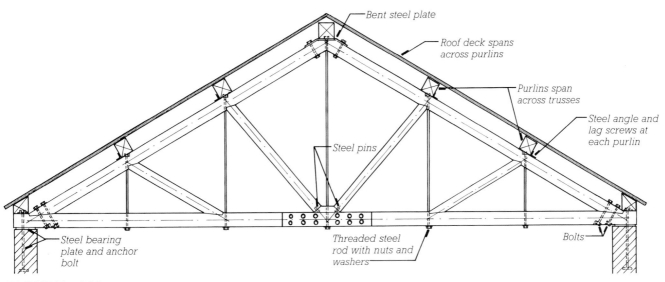

Bent steel plate
Roof deck spans across purlins
Purlins span across trusses
Steel angle and lag screws at each purlin
Steel pins
Steel bearing plate and anchor bolt
Threaded steel rod with nuts and washers
Bolts

FIGURE 4.24

A heavy timber roof truss with steel rod tension chords. This type of truss is easy to construct but cannot be used where wind uplift can cause the forces in the tension members to be reversed.

The center splice in the lower chord of the truss is required only if it is impossible to obtain lumber long enough to reach in one piece from one end of the truss to the other. Compare this mode of truss construction with that shown in Figure 4.12; still another common mode is to form the truss of a single layer of heavy members connected by steel side plates and bolts.

FIGURE 4.25

Laminated wood roof trusses with steel connector plates and steel rods for lateral stability. (Architects: Woo and Williams. Photo by Richard Bonarrigo. Courtesy of the architects)

FIGURE 4.26
A typical foundation connection for a two-hinged arch or dome, made of welded steel plates. The hinge pin allows for rotation between the arch and the foundations.

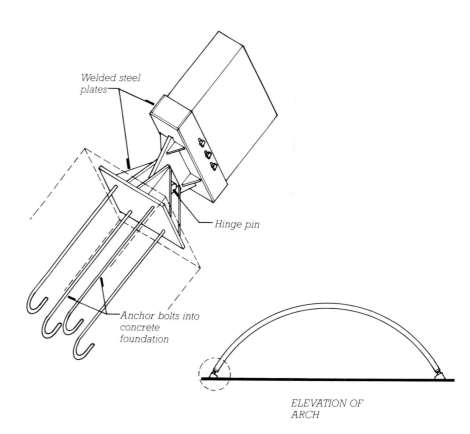

Welded steel plates

Hinge pin

Anchor bolts into concrete foundation

ELEVATION OF ARCH

FIGURE 4.27
This laminated wood dome spans 530 feet (161.5 m) to cover a 25,000-seat stadium and convention center in Tacoma, Washington. (Architects: McGranahan, Messenger Associates. Structural engineers: Chalker Engineers, Inc. Photo by Gary Vannest. Courtesy of American Wood Council)

Arches and Domes

Long curved timbers for making *vaults* and *domes* are easily fabricated in laminated wood and are widely used in athletic arenas, auditoriums, suburban retail stores, warehouses, and factories (Figures 4.26, 4.27). Arched structures, like rigid frames, exert lateral thrusts that must be countered by tie rods or suitably designed foundations.

Heavy timber and the building codes

The table in Figure 1.1 shows the range of building types that may be built of Heavy Timber construction (Type 4), and of Type 5 construction utilizing heavy timbers. Notice that the allowable heights and areas for Heavy Timber construction are superior to those for unprotected steel (Type 2C construction), and are comparable to those for lightly protected noncombustible construction (Type 2B). The references in the table to "Section 405.0" are to several paragraphs of this particular code (the *BOCA Basic/National Building Code*), which are summarized here in Figure 4.7. The allowable floor areas can be increased by installing an automatic fire suppression sprinkler system in the building: Sprinklers triple the allowable area under this code for a one- or two-story building and double it for taller buildings.

The uniqueness of heavy timber framing

Heavy Timber buildings cannot span as far or with such lightness of structure as steel, and they cannot mimic the structural continuity or smooth shell forms of concrete, yet many people respond more positively to the idea of a timber building than they do to one

FIGURE 4.28
Architects Greene and Greene of Pasadena, California were known for their carefully wrought timber frame *houses such as this one, built for David B. Gamble in 1909.* (Photo by Wayne Andrews)

FIGURE 4.29
Each of the attached dwellings in Sea Ranch Condominium #1 in northern California, built in 1965, is framed with a simple cage of unplaned timbers sawn from trees taken from another portion of *the site. The diagonal members are wind braces.* (Architects: Moore, Lyndon, Turnbull, and Whitaker. Photo by the author)

FIGURE 4.30

(a) *The Sea Ranch Condominium #1 is sheathed with vertical 2 inch (51 mm) unplaned tongue-and-groove decking, and clad in ¾ inch (19 mm) redwood* *tongue-and-groove siding. (b, c) Inside the dwelling units, the massive timbers and their connectors are on display.* (Photos by Morley Baer)

of steel or concrete. To some degree this response may stem from the color, grain figure, and warmer feel of wood. In larger part, it probably comes from the pleasant associations people have with the sturdy, satisfying houses our ancestors erected from hand-hewn timbers only a few generations ago. Most people today live in dwellings where none of the framing is exposed. A beamed ceiling in one's house or apartment has become a much-desired amenity, and a restaurant dinner or shopping in a converted mill building of heavy timber and masonry construction is generally thought to be a pleasant experience. There is something in all of us that derives satisfaction from seeing wood beams at work.

In economic reality, heavy timber must compete successfully on the ba-sis of price with other materials of con-struction, and as our forests have di-minished in size and shipping costs have risen, timber is no longer an automatic choice for building a mill or any other type of structure. But for many build-ings heavy timber is an economic al-ternative to steel and concrete, par-ticularly in situations where the appearance and feel of a heavy timber structure will be highly valued by those who use it, or where code provisions or fire insurance premiums create a financial incentive.

C.S.I./C.S.C. Masterformat Section Numbers for Heavy Timber	
06130	HEAVY TIMBER CONSTRUCTION
06132	Mill-framed Structures
06133	Pole Construction
06135	Timber Trusses
06140	Timber Decking
06170	PREFABRICATED STRUCTURAL WOOD
06180	Glued-Laminated Construction

SELECTED REFERENCES

1. The American Institute of Timber Construction. *Glulam Systems*. Englewood, Colorado, 1980.

A practical how-to manual for the design of glue-laminated timber structures. (Address for ordering: 333 West Hampden Avenue, Englewood, CO 80110.)

2. National Forest Products Association. *Heavy Timber Construction Details*. Washington, D.C., 1983.

In drawings, photographs, and a brief text, this booklet summarizes construction requirements and major details for heavy timber buildings. (Address for ordering: 1619 Massachusetts Avenue N.W., Washington, D.C. 20036.)

KEY TERMS AND CONCEPTS

Heavy Timber construction or
    Mill construction
chamfer
pintle cap
firecut

decking
tongue-and-groove
spline
bearing block
rigid frame

portal frame
truss
arch
vault
dome

REVIEW QUESTIONS

1. Why does heavy timber framing receive a relatively favorable fire rating from building codes and insurance companies?

2. What are the important factors in detailing the junction of a wood beam with a masonry loadbearing wall? Sketch several ways of making this joint.

3. Sketch from memory one or two typical details for the intersection of a wood column with a floor of a building of Type 4 construction.

E X E R C I S E S

1. Determine from the code table in Figure 1.1 whether a building you are currently designing could be built of Type 4 construction, or what modifications you might make in your design to make it conform to the requirements of Type 4 construction.

2. Find a barn or mill built in the eighteenth or nineteenth century and sketch some typical connection details. How is the structure stabilized against wind loads?

3. Obtain a book on traditional Japanese construction from the library and compare Japanese timber joinery with eighteenth- or nineteenth-century American practice.

WOOD LIGHT FRAME CONSTRUCTION

A school of arts and crafts is housed in a cluster of small buildings of Wood Light Frame construction that cling to a dramatic mountainside site in New England. *(Architect: Edward Larrabee Barnes. Photograph by Joseph W. Molitor)*

Wood Light Frame construction is the most flexible of all building systems. There is scarcely a shape it cannot be used to construct, from a plain rectilinear box to cylindrical towers or to complex foldings of sloping roofs with dormers of every description. Over the century and a half since its invention, wood light framing has served to construct buildings in styles ranging from reinterpretations of nearly all the historical fashions to uncompromising expressions of every twentieth-century architectural philosophy. It has assimilated without difficulty during this same period a bewildering and unforeseen succession of technical improvements in building: central heating, gas lighting, electricity, thermal insulation, air conditioning, indoor plumbing, prefabricated components, and electronic communications. Light frame buildings are easily and swiftly constructed with a minimal investment in tools. Many observers of the building industry have criticized the sup-posed inefficiency of light frame construction, which is carried out largely by hand methods on the building site, yet it has successfully fought off competition from industrialized building systems of every sort, partly by incorporating their best features, to remain the least expensive form of durable construction. It is the common currency of residential and small commercial buildings in North America today.

Wood Light Frame construction also has its problems: If ignited, it burns rapidly; if exposed to dampness, it decays. It expands and contracts by significant amounts in response to changes in humidity, causing chronic difficulties with cracking plaster, sticking doors, and buckling floors. The framing is so unattractive to the eye that it is seldom left exposed in a building. But these problems can be controlled by clever design and careful workmanship, and there is no arguing with success: Frames made by the monotonous repetition of wooden joists, studs, and rafters are likely to remain the number one system of building in North America for a long time to come.

FIGURE 5.1
Carpenters apply plywood roof sheathing to a platform-framed apartment building. The ground floor is a concrete slab on grade. The edge of the wooden platform of the upper floor is clearly visible between the stud walls of the ground and upper floors. Most of the diagonal bracing is temporary, but permanent let-in diagonal braces occur between the two openings at the lower left, and immediately above in the rear building. The openings have been framed incorrectly, without supporting studs for the headers. (Courtesy of Southern Forest Products Association)

HISTORY

Wood Light Frame construction was the first uniquely American building system. It was invented in the 1830s in Chicago by George Washington Snow, an engineer by training and a lumber dealer and building contractor by vocation. Snow recognized that the closely spaced vertical members used to infill the walls of a heavy timber building frame were themselves sufficiently strong that the heavy posts of the frame could be eliminated. He also realized that boards and small framing members of wood had recently become inexpensive for the first time in history because of the advent of the water-powered sawmill; additionally, machine-made nails had also become remarkably cheap compared to the hand-forged nails that preceded them.

Snow's invention, labeled derisively by his critics the *balloon frame* because it seemed so thin and insubstantial, was a building system framed solely with small, closely spaced wooden members: *joists* for the floors, *studs* for the walls, *rafters* for the sloping roofs. Heavy posts and beams were completely eliminated, and with them, the difficult, expensive mortise-and-tenon joinery they required. There was no structural member in a frame that could not be handled easily by a single carpenter, and each of the hundreds of joints was made with lightning rapidity using two or three nails. The impact of this new building system was revolutionary: In 1865, G. E. Woodward could write that "A man and a boy can now attain the same results, with ease, that twenty men could on an old-fashioned frame...the Balloon Frame can be put up for forty per cent less money than the mortise and tenon frame."

The balloon frame (Figure 5.2) used full-length studs running two stories from foundation to roof. In time, these were recognized as being too long to erect efficiently, and the hollow spaces between studs acted as multiple chimneys in a fire, spreading the blaze rapidly to the upper floors, unless closed

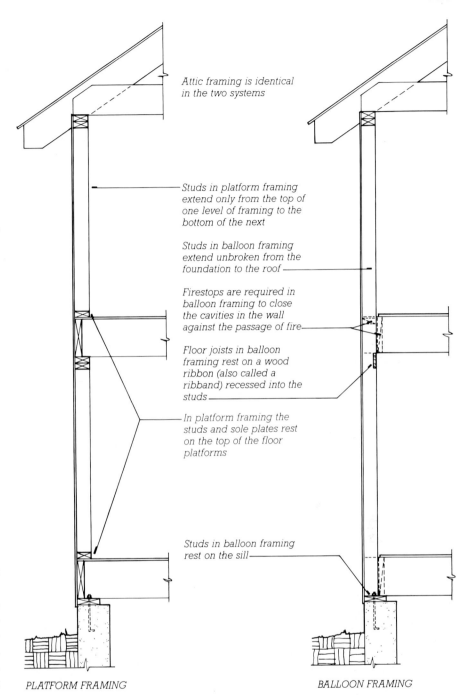

Attic framing is identical in the two systems

Studs in platform framing extend only from the top of one level of framing to the bottom of the next

Studs in balloon framing extend unbroken from the foundation to the roof

Firestops are required in balloon framing to close the cavities in the wall against the passage of fire

Floor joists in balloon framing rest on a wood ribbon (also called a ribband) recessed into the studs

In platform framing the studs and sole plates rest on the top of the floor platforms

Studs in balloon framing rest on the sill

PLATFORM FRAMING

BALLOON FRAMING

FIGURE 5.2

Comparative framing details for platform framing (left) and balloon framing (right). Platform framing is easier to erect but settles considerably as the wood dries and shrinks. If nominal 12-inch (300-mm) joists are used to frame the floors in these examples, the total amount of loadbearing cross-grain wood between the foundation and the attic joists is 33 inches (838 mm) for the platform frame, and only 4½ inches (114 mm) for the balloon frame. Interior partitions in a balloon frame building are essentially platform framed, however, which can result in tilting of floors.

off with wood or brick *firestops* at each floor line. Several modified versions of the balloon frame were subsequently developed in an attempt to overcome these difficulties, and the most recent of these, the *platform frame*, is now the universal standard.

THE PLATFORM FRAME

While complex in its details, the platform frame is very simple in concept. A floor platform is built. Loadbearing walls are erected upon it. A second floor platform is built upon these walls, and a second set of walls upon this platform. The attic and roof are then built upon the second set of walls. There are, of course, many variations: A concrete slab is often substituted for the ground floor platform; a building may be one or three stories tall instead of two; and several types of roofs are frequently built that do not incorporate attics. But the essentials remain: A floor platform is completed at each level, and the walls bear upon the platform rather than directly upon the walls below.

The advantages of the platform frame over the balloon frame are that it uses short, easily handled lengths of lumber for the wall framing; its hollow spaces are automatically firestopped at each floor; and its platforms are convenient working surfaces for the carpenters who build the frame. The major disadvantage of the platform frame is that each platform constitutes a thick layer of wood whose grain runs horizontally. This leads inevitably to a relatively large amount of vertical shrinkage in the frame as the excess moisture dries from the wood, which can lead to distress in the exterior and interior finish surfaces, especially around multistory spaces such as stairwells and tall rooms.

A platform frame is made entirely of nominal 2-inch members, actually $1\frac{1}{2}$ inches (38 mm) in thickness. These are ordered and delivered cut to the nearest 2-foot (600-mm) length and then are measured and sawn to exact length on the building site. All connections are made with nails, using either face nailing, end nailing, or toe nailing (Figures 3.35, 5.20) as required by the characteristics of each joint. Nails are driven either by hammer or by pneumatic nailing machines. In either case, the connection is quickly made because the nails penetrate the wood without the need for drilling holes or otherwise preparing the joint.

Each plane of structure in a platform frame is made by aligning a number of pieces of framing lumber parallel to one another at specified intervals, nailing these to crosspieces that *head off* the framing at either end to maintain its spacing and flatness, then covering the plane of framing with *sheathing* of boards or panels that join and stabilize the pieces into a single structural unit, ready for the application of finish materials inside and out. In a floor structure, the parallel pieces are the floor joists, and the crosspieces at the ends of the joists are the *headers* or *band joists*. The sheathing on a floor is known as the *subfloor*. In a wall structure, the parallel pieces are the studs, the crosspiece at the bottom of the wall is the *sole plate*, and the crosspiece at the top (which is doubled for strength if the wall bears a load from above) is called the *top plate*. In a sloping roof, the rafters are headed off by the top plates at the lower edge of the roof, and by the *ridge board* at the peak.

Openings are required in all these planes of structure: for windows and doors in the walls, for stairs and chimneys in the floors, and for chimneys and dormers in the roofs. In each case these are made by heading off the opening: Openings in floors are framed with *headers and trimmers* (Figure 5.14). In walls, *sills* head off the bottoms of openings, with strong *window headers* and *door headers* across the tops (Figure 5.29). Headers and trimmers must be doubled or otherwise strengthened to support the higher loads placed on them by the presence of the opening.

Sheathing is the component of platform framing that enables the concept to work. The end nails that connect the plates to the studs have little holding power against uplift of the roof by wind, but the sheathing connects the frame into a single, strong unit from foundation to roof. The rectilinear geometry of the parallel framing members has no useful resistance to *wracking* by lateral forces such as wind, but rigid sheathing panels or diagonal sheathing boards brace the building effectively against these forces. Sheathing also furnishes a surface to which shingles, boards, and flooring are nailed for finish surfaces. In buildings without sheathing, or with sheathing materials that are too weak to tie and brace the frame, diagonal bracing must be used for lateral stability.

FOUNDATIONS FOR LIGHT FRAME STRUCTURES

Foundations for light framing, originally of stone or brick, are now made of sitecast concrete, concrete block masonry, or preservative-treated wood (Figures 5.3 through 5.13). Concrete and masonry foundations are highly conductive of heat and must usually be insulated to meet code requirements concerning energy conservation (Figures 5.6, 5.7, 5.12, 5.13). A wood foundation is easily insulated in the same manner as the frame of the house it supports. Further advantages claimed for wood foundations are that they can be constructed in any weather by the same crew of carpenters that will frame the building, and that they allow for easy installation of electrical wiring, plumbing, and interior finish materials in the basement. A basement of any material needs to be carefully dampproofed and drained to avoid flooding with ground water and to prevent the buildup of water pressure that could cave in the walls (Figure 5.4).

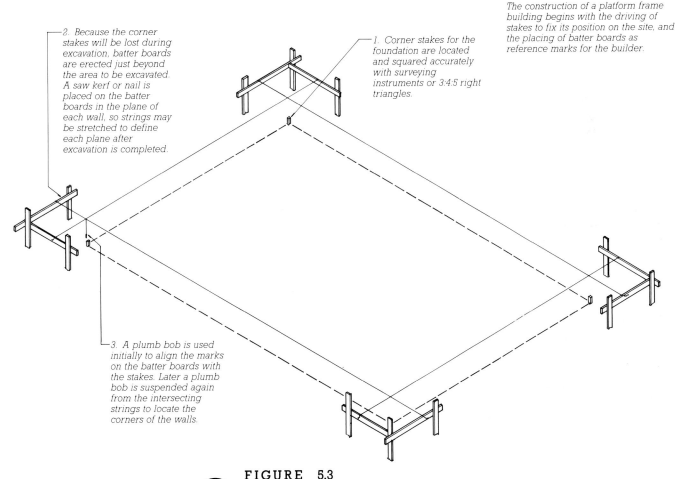

2. Because the corner stakes will be lost during excavation, batter boards are erected just beyond the area to be excavated. A saw kerf or nail is placed on the batter boards in the plane of each wall, so strings may be stretched to define each plane after excavation is completed.

1. Corner stakes for the foundation are located and squared accurately with surveying instruments or 3:4:5 right triangles.

The construction of a platform frame building begins with the driving of stakes to fix its position on the site, and the placing of batter boards as reference marks for the builder.

3. A plumb bob is used initially to align the marks on the batter boards with the stakes. Later a plumb bob is suspended again from the intersecting strings to locate the corners of the walls.

FIGURE 5.3

① *Step One in the construction of a simple platform-frame building: establishing the position, shape, and size of the building on the site. This drawing begins a series of isometric drawings that will follow the erection of this building step by step throughout the course of this chapter.*

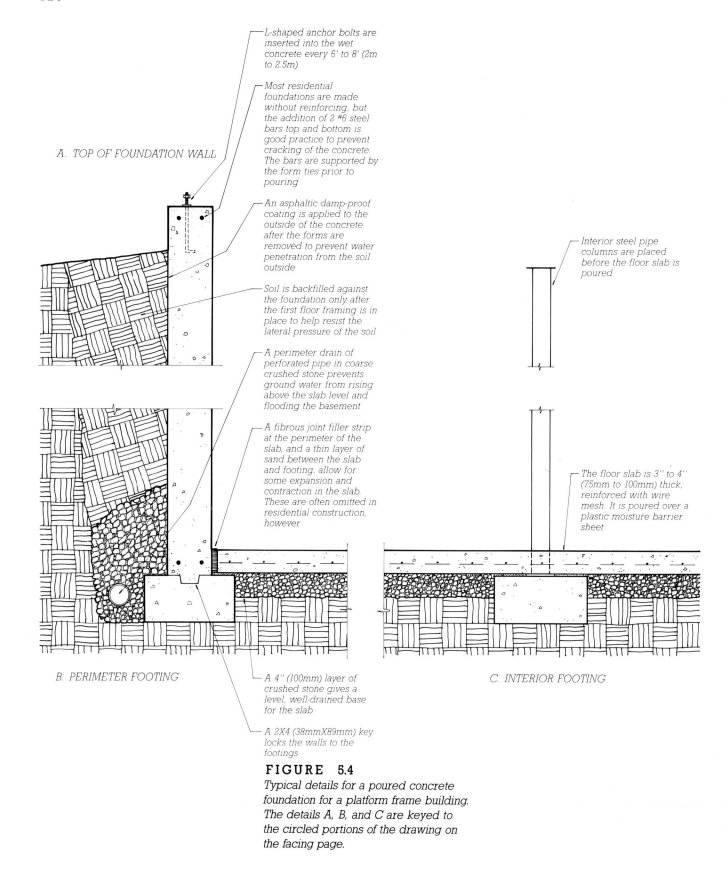

A. TOP OF FOUNDATION WALL

— L-shaped anchor bolts are inserted into the wet concrete every 6' to 8' (2m to 2.5m)

— Most residential foundations are made without reinforcing, but the addition of 2 #6 steel bars top and bottom is good practice to prevent cracking of the concrete. The bars are supported by the form ties prior to pouring

— An asphaltic damp-proof coating is applied to the outside of the concrete after the forms are removed to prevent water penetration from the soil outside

— Soil is backfilled against the foundation only after the first floor framing is in place to help resist the lateral pressure of the soil

— A perimeter drain of perforated pipe in coarse crushed stone prevents ground water from rising above the slab level and flooding the basement

— A fibrous joint filler strip at the perimeter of the slab, and a thin layer of sand between the slab and footing, allow for some expansion and contraction in the slab. These are often omitted in residential construction, however

— A 4" (100mm) layer of crushed stone gives a level, well-drained base for the slab

— A 2X4 (38mmX89mm) key locks the walls to the footings

— Interior steel pipe columns are placed before the floor slab is poured

— The floor slab is 3" to 4" (75mm to 100mm) thick, reinforced with wire mesh. It is poured over a plastic moisture barrier sheet

B. PERIMETER FOOTING

C. INTERIOR FOOTING

FIGURE 5.4
Typical details for a poured concrete foundation for a platform frame building. The details A, B, and C are keyed to the circled portions of the drawing on the facing page.

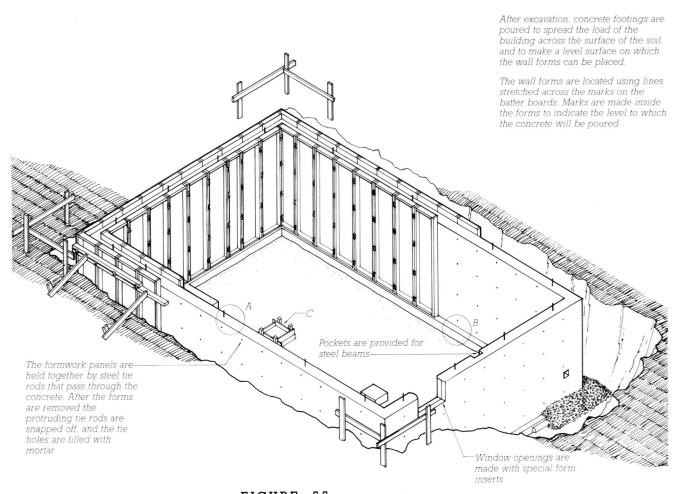

After excavation, concrete footings are poured to spread the load of the building across the surface of the soil, and to make a level surface on which the wall forms can be placed.

The wall forms are located using lines stretched across the marks on the batter boards. Marks are made inside the forms to indicate the level to which the concrete will be poured.

A

C

B

Pockets are provided for steel beams

The formwork panels are held together by steel tie rods that pass through the concrete. After the forms are removed the protruding tie rods are snapped off, and the tie holes are filled with mortar

Window openings are made with special form inserts

FIGURE 5.5

② *Step Two in the construction of a typical platform-frame building: excavation and foundations. The circles A, B, and C indicate portions of the foundation which are detailed in Figure 5.4 on the facing page.*

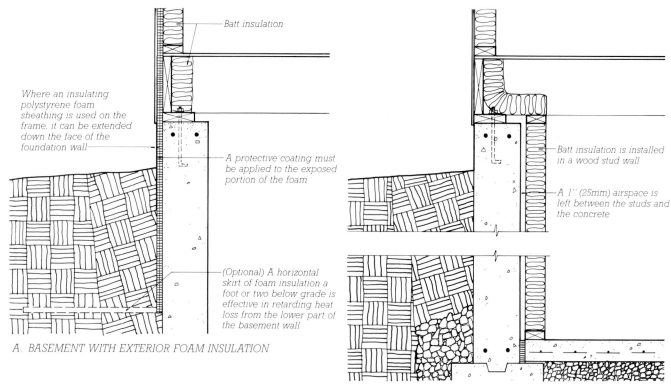

Where an insulating polystyrene foam sheathing is used on the frame, it can be extended down the face of the foundation wall

A protective coating must be applied to the exposed portion of the foam

(Optional) A horizontal skirt of foam insulation a foot or two below grade is effective in retarding heat loss from the lower part of the basement wall

A. BASEMENT WITH EXTERIOR FOAM INSULATION

Batt insulation

Batt insulation is installed in a wood stud wall

A 1" (25mm) airspace is left between the studs and the concrete

B. BASEMENT WITH INTERIOR BATT INSULATION

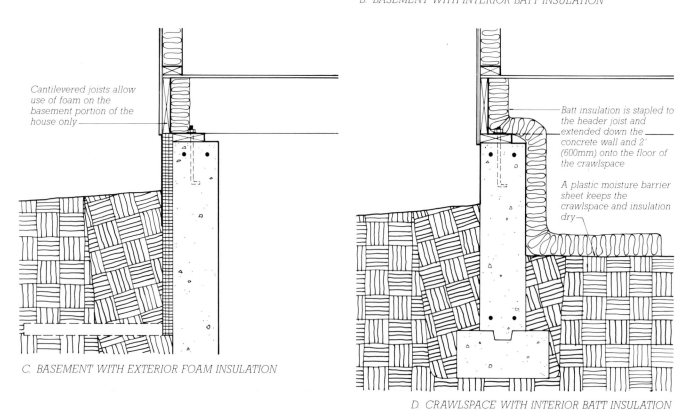

Cantilevered joists allow use of foam on the basement portion of the house only

C. BASEMENT WITH EXTERIOR FOAM INSULATION

Batt insulation is stapled to the header joist and extended down the concrete wall and 2' (600mm) onto the floor of the crawlspace

A plastic moisture barrier sheet keeps the crawlspace and insulation dry

D. CRAWLSPACE WITH INTERIOR BATT INSULATION

FIGURE 5.6

Most building codes require thermal insulation of the foundation. Shown here are three alternative ways of adding insulation to a poured concrete or concrete block basement, and one way of insulating a crawlspace foundation. The crawlspace may alternatively be insulated on the outside with panels of plastic foam. The interior batt insulation shown in Detail B is very commonly used but raises unanswered questions about how to avoid possible problems arising from moisture accumulating between the insulation and the wall.

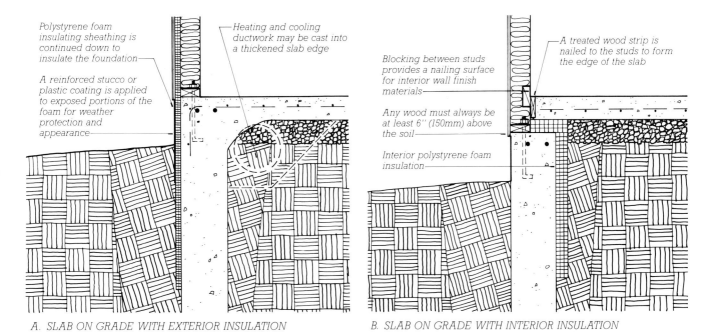

Polystyrene foam insulating sheathing is continued down to insulate the foundation

A reinforced stucco or plastic coating is applied to exposed portions of the foam for weather protection and appearance

Heating and cooling ductwork may be cast into a thickened slab edge

A. SLAB ON GRADE WITH EXTERIOR INSULATION

Blocking between studs provides a nailing surface for interior wall finish materials

Any wood must always be at least 6" (150mm) above the soil

Interior polystyrene foam insulation

A treated wood strip is nailed to the studs to form the edge of the slab

B. SLAB ON GRADE WITH INTERIOR INSULATION

2X6 (38mmX140mm) studs give a 3½" (89mm) bearing on the slab plus a 2" (51mm) projection to cover the edge of the foam insulation

2" (51mm) foam insulation with protective coating

C. SLAB ON GRADE WITH EXTERIOR INSULATION

The slab is thickened to at least 12" (300mm) under bearing partitions and posts

The width of the thickened slab is determined by the loadbearing capacity of the soil and the magnitude of the load

D. INTERIOR FOOTING FOR SLAB ON GRADE

FIGURE 5.7
Some typical concrete slab on grade details with thermal insulation.

FIGURE 5.8
Erecting formwork for a poured concrete foundation. The footing has already been poured and is visible at the lower left. (Photo by Joseph Iano)

FIGURE 5.9
Masons construct a foundation of concrete blocks. The first coat of parging, portland cement plaster used to help dampproof the foundation, has already been applied on the outside of the wall, and the drainage layer of crushed stone has been put in place. The projecting pilaster in the center of the wall will support a steel beam under the center of the main floor. After a second coat of parging, the outside of the foundation will be coated with an asphaltic dampproofing compound. (Courtesy of Portland Cement Association, Skokie, Illinois)

FIGURE 5.10
Erecting a preservative-treated wood foundation. One worker applies a bead of sealant to the edge of a panel of treated wood components, as another prepares to push the next panel into position. The entire foundation rests on a drainage layer of crushed stone. A flexible plastic perforated drainage pipe is partially visible at the lower right. A major advantage of a wood foundation is that it can be insulated in the same way as the superstructure of the building. (Courtesy of American Plywood Association)

FIGURE 5.11
The difference in color makes it clear where the treated wood foundation leaves off and the untreated superstructure begins. Heavy plastic moisture barrier sheets have been attached up to the grade line on portions of the foundation. (Courtesy of American Plywood Association)

FIGURE 5.12
External polystyrene foam insulation applied to the outside of a concrete block foundation. (Courtesy of Dow Chemical Company)

FIGURE 5.13
Following completion of the exterior siding, a worker staples a glass fiber reinforcing mesh to the foam insulation on the exposed portions of the basement wall. He will next trowel onto the mesh two thin coats of a cementitious, stuccolike material that will form a durable, attractive finish coating over the foam. (Courtesy of Dow Chemical Company)

BUILDING THE FRAME

Planning the Frame

While it is true that an experienced carpenter can frame a simple building from the most minimal of drawings, a platform frame of wood for a custom-designed building should be planned as carefully as a frame of steel or concrete for a larger building. The architect or engineer should determine an efficient layout and the appropriate sizes for joists and rafters, and communicate this information to the carpenters by means of *framing plans* (Figures 5.14, 5.46). For most purposes, the member sizes can be determined using standardized structural tables, an example of which is shown in Figure 5.19. Detailed section drawings, similar to those seen throughout this chapter, are also prepared for the major connections in the building. The *architectural floor plans* serve to indicate the locations and dimensions of all the walls, partitions, and openings, and the *elevations* are drawings that show side views of the building, with vertical framing dimensions indicated as required. For most buildings, *sections* are also drawn that cut completely through the building, showing the dimensional relationships of the various floor levels and roof planes, and the slopes of the roof surfaces.

FLOOR FRAMING PLAN

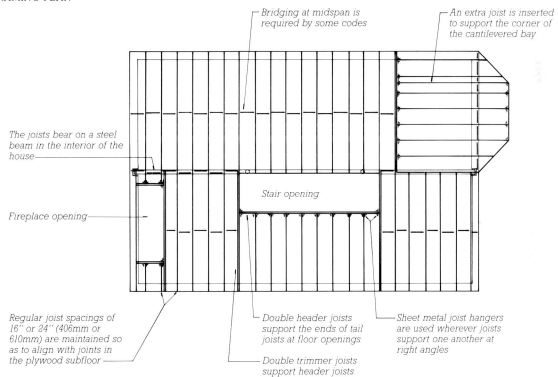

Bridging at midspan is required by some codes

An extra joist is inserted to support the corner of the cantilevered bay

The joists bear on a steel beam in the interior of the house

Stair opening

Fireplace opening

Regular joist spacings of 16" or 24" (406mm or 610mm) are maintained so as to align with joints in the plywood subfloor

Double header joists support the ends of tail joists at floor openings

Sheet metal joist hangers are used wherever joists support one another at right angles

Double trimmer joists support header joists

FIGURE 5.14
A framing plan for the ground floor platform of the building shown in Figure 5.16.

Erecting the Frame

The erection of a typical platform frame (referred to, inaccurately, as *rough carpentry* in architects' specifications) can best be understood by following the sequential isometric diagrams that begin with Figure 5.16. Notice the basic simplicity of the building process: A platform is built, walls are assembled horizontally on the platform and tilted up into place, and another platform or a roof is built on top of the walls. Most of the work is accomplished without the use of ladders or scaffolding, and

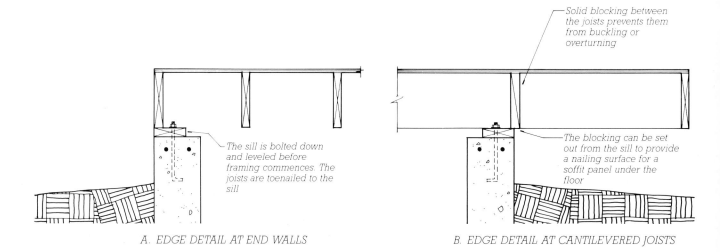

The sill is bolted down and leveled before framing commences. The joists are toenailed to the sill

A. EDGE DETAIL AT END WALLS

Solid blocking between the joists prevents them from buckling or overturning

The blocking can be set out from the sill to provide a nailing surface for a soffit panel under the floor

B. EDGE DETAIL AT CANTILEVERED JOISTS

Anchor bolts hold the frame to the foundation

A continuous sheet metal shield is required in areas with a high risk of termite infestation

C. EDGE DETAIL AT SIDE WALLS

Solid blocking prevents the joists from warping or overturning at the support, and transmits loads from the interior bearing partition above

A wood sill strip equalizes wood shrinkage with the perimeter of the house

D. DETAIL OF INTERIOR JOIST BEARING

FIGURE 5.15

Ground floor framing details, keyed to the lettered circles in Figure 5.16. A fibrous sill sealer material should be installed between the sill and the top of the foundation to reduce air leakage, but the sealer material is not shown on these diagrams. Using accepted architectural drafting conventions, continuous pieces of lumber are drawn with an X inside, and intermittent blocking with a single slash.

temporary bracing is needed only to support the walls until the next level of framing is installed and sheathed.

The details of a platform frame are not left to chance. While there are countless local and personal variations in framing details and techniques (not to mention terminology), the sizes, spacings, and connections of the members in a platform frame are closely regulated by building codes, down to the size and spacing of the studs (Figure 5.30), the size and number of nails for each connection (Figure 5.20), and the thickness of the sheathing panels.

When the foundation is complete, basement beams are placed, sills are bolted to the foundation, and the first floor joists and subfloor are installed.

Plywood sheets are considerably stiffer along their length than across their width, so they must be laid with their long dimension perpendicular to the joists. The end joints are staggered to avoid lines of weakness

Joist bridging is required by some building codes

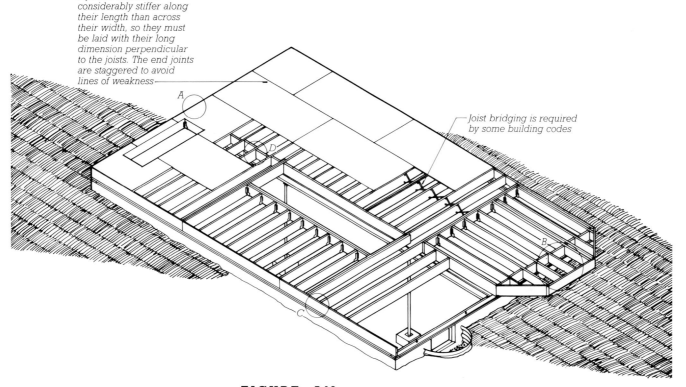

FIGURE 5.16

(3)

Step Three in erecting a typical platform-frame building: the ground floor platform. Compare this drawing with the framing plan in Figure 5.14. Notice that the direction of the joists must be changed to construct the cantilevered bay on the end of the building. A cantilevered bay on a long side of the building could be framed by merely extending the floor joists over the foundation.

A. STEEL BEAM UNDER JOISTS

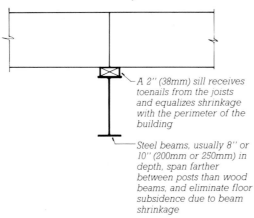

A 2'' (38mm) sill receives toenails from the joists and equalizes shrinkage with the perimeter of the building

Steel beams, usually 8'' or 10'' (200mm or 250mm) in depth, span farther between posts than wood beams, and eliminate floor subsidence due to beam shrinkage

B. STEEL BEAM RECESSED INTO JOISTS

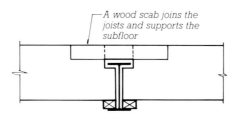

A wood scab joins the joists and supports the subfloor

This detail gives better headroom under the beam in a basement

C. WOOD BEAM UNDER JOISTS

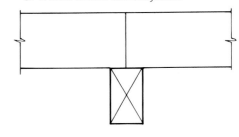

A wood beam in this position, though commonly used, causes noticeable subsidence of the floor as it shrinks

D. WOOD BEAM RECESSED INTO JOISTS

As an alternative to the scab shown in the detail to the left, the joist ends can be notched and lapped

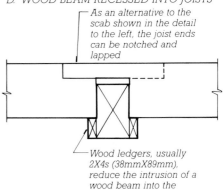

Wood ledgers, usually 2X4s (38mmX89mm), reduce the intrusion of a wood beam into the basement space, and reduce the effect of beam shrinkage

E. WOOD BEAM WITH JOIST HANGERS

A wood beam may be built up of several layers of joists nailed together

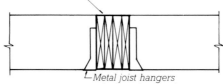

Metal joist hangers substitute for wood ledgers and allow the beam to be completely recessed in the basement ceiling. Subsidence is minimized by using joist hangers and a beam that is no deeper than the joists

FIGURE 5.17

Alternative ways of constructing an interior line of support for ground floor joists. Space is left over the tops of the beams in Details B and D to allow for drying shrinkage in the joists.

F. IN-LINE JOISTS In-line joists make a neatly-crafted frame that accepts plywood subflooring with a minimum of cutting. If a plywood joint occurs near the junction of joist ends, however, a metal strap is needed to tie the joists together

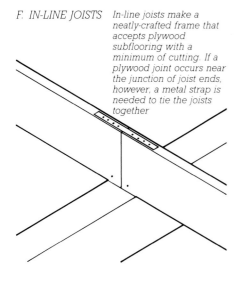

Species	Allowable Bending Stress		Modulus of Elasticity
	Floor Joists	Rafters	
Douglas fir/Larch	1450 psi (10.00 MN/m²)	1670 psi (11.51 MN/m²)	1,700,000 psi (11,721 MN/m²)
Eastern hemlock	1250 psi (8.62 MN/m²)	1440 psi (9.93 MN/m²)	1,100,000 psi (7,584 MN/m²)
Eastern spruce	950 psi (6.55 MN/m²)	1090 psi (7.52 MN/m²)	1,400,000 psi (9,653 MN/m²)
Hem-Fir	1150 psi (7.93 MN/m²)	1320 psi (9.10 MN/m²)	1,400,000 psi (9,653 MN/m²)
Southern pine	1400 psi (9.65 MN/m²)	1610 psi (11.10 MN/m²)	1,600,000 psi (11,032 MN/m²)
Spruce-Pine-Fir	1000 psi (6.89 MN/m²)	1150 psi (7.93 MN/m²)	1,300,000 psi (8,963 MN/m²)

FIGURE 5.18

Allowable bending stresses for some common species of framing lumber, to be used with the floor joist selection table in Figure 5.19. The allowable stresses for rafters are 15 percent higher than those for joists because of wood's greater ability to carry short-term loads such as snow and wind. These selected values, and the table in Figure 5.19, are taken from reference 4 in the list of references at the end of this chapter.

G. IN-LINE JOISTS

A wood scab can substitute for the metal strap

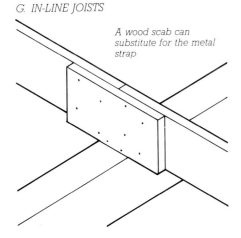

H. LAPPED JOISTS Lapped joists require no cutting to length, so are used more commonly than in-line joists, but they require more cutting of plywood panels for the subflooring

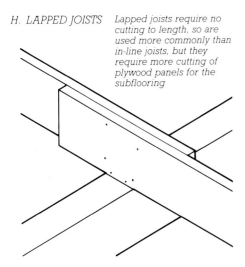

FLOOR JOISTS
40 Lbs. Per Sq. Ft. Live Load

DESIGN CRITERIA:
Deflection - For 40 lbs. per sq. ft. live load.
 Limited to span in inches divided by 360.
Strength - Live Load of 40 lbs. per sq. ft. plus
 dead load of 10 lbs. per sq. ft. determines the
 required fiber stress value.

Modulus of Elasticity, "E", in 1,000,000 psi

JOIST SIZE (IN)	SPACING (IN)	0.4	0.5	0.6	0.7	0.8	0.9	1.0	1.1	1.2	1.3	1.4	1.5	1.6	1.7	1.8	1.9	2.0	2.2	2.4
2x6	12.0	6-9 / 450	7-3 / 520	7-9 / 590	8-2 / 660	8-6 / 750	8-10 / 780	9-2 / 830	9-6 / 890	9-9 / 940	10-0 / 990	10-3 / 1040	10-6 / 1090	10-9 / 1140	10-11 / 1190	11-2 / 1230	11-4 / 1280	11-7 / 1320	11-11 / 1410	12-3 / 1490
	13.7	6-6 / 470	7-0 / 550	7-5 / 620	7-9 / 690	8-2 / 750	8-6 / 810	8-9 / 870	9-1 / 930	9-4 / 980	9-7 / 1040	9-10 / 1090	10-0 / 1140	10-3 / 1190	10-6 / 1240	10-8 / 1290	10-10 / 1340	11-1 / 1380	11-5 / 1470	11-9 / 1560
	16.0	6-2 / 500	6-7 / 580	7-0 / 650	7-5 / 720	7-9 / 790	8-0 / 860	8-4 / 920	8-7 / 980	8-10 / 1040	9-1 / 1090	9-4 / 1150	9-6 / 1200	9-9 / 1250	9-11 / 1310	10-2 / 1360	10-4 / 1410	10-6 / 1460	10-10 / 1550	11-2 / 1640
	19.2	5-9 / 530	6-3 / 610	6-7 / 690	7-0 / 770	7-3 / 840	7-7 / 910	7-10 / 970	8-1 / 1040	8-4 / 1100	8-7 / 1160	8-9 / 1220	9-0 / 1280	9-2 / 1330	9-4 / 1390	9-6 / 1440	9-8 / 1500	9-10 / 1550	10-2 / 1650	10-6 / 1750
	24.0	5-4 / 570	5-9 / 660	6-2 / 750	6-6 / 830	6-9 / 900	7-0 / 980	7-3 / 1050	7-6 / 1120	7-9 / 1190	7-11 / 1250	8-2 / 1310	8-4 / 1380	8-6 / 1440	8-8 / 1500	8-10 / 1550	9-0 / 1610	9-2 / 1670	9-6 / 1780	9-9 / 1880
	32.0					6-2 / 1010	6-5 / 1090	6-7 / 1150	6-10 / 1230	7-0 / 1300	7-3 / 1390	7-5 / 1450	7-7 / 1520	7-9 / 1590	7-11 / 1660	8-0 / 1690	8-2 / 1760	8-4 / 1840	8-7 / 1950	8-10 / 2060
2x8	12.0	8-11 / 450	9-7 / 520	10-2 / 590	10-9 / 660	11-3 / 720	11-8 / 780	12-1 / 830	12-6 / 890	12-10 / 940	13-2 / 990	13-6 / 1040	13-10 / 1090	14-2 / 1140	14-5 / 1190	14-8 / 1230	15-0 / 1280	15-3 / 1320	15-9 / 1410	16-2 / 1490
	13.7	8-6 / 470	9-2 / 550	9-9 / 620	10-3 / 690	10-9 / 750	11-2 / 810	11-7 / 870	11-11 / 930	12-3 / 980	12-7 / 1040	12-11 / 1090	13-3 / 1140	13-6 / 1190	13-10 / 1240	14-1 / 1290	14-4 / 1340	14-7 / 1380	15-0 / 1470	15-6 / 1560
	16.0	8-1 / 500	8-9 / 580	9-3 / 650	9-9 / 720	10-2 / 790	10-7 / 850	11-0 / 920	11-4 / 980	11-8 / 1040	12-0 / 1090	12-3 / 1150	12-7 / 1200	12-10 / 1250	13-1 / 1310	13-4 / 1360	13-7 / 1410	13-10 / 1460	14-3 / 1550	14-8 / 1640
	19.2	7-7 / 530	8-2 / 610	8-9 / 690	9-2 / 770	9-7 / 840	10-0 / 910	10-4 / 970	10-8 / 1040	11-0 / 1100	11-3 / 1160	11-7 / 1220	11-10 / 1280	12-1 / 1330	12-4 / 1390	12-7 / 1440	12-10 / 1500	13-0 / 1550	13-5 / 1650	13-10 / 1750
	24.0	7-1 / 570	7-7 / 660	8-1 / 750	8-6 / 830	8-11 / 900	9-3 / 980	9-7 / 1050	9-11 / 1120	10-2 / 1190	10-6 / 1250	10-9 / 1310	11-0 / 1380	11-3 / 1440	11-5 / 1500	11-8 / 1550	11-11 / 1610	12-1 / 1670	12-6 / 1780	12-10 / 1880
	32.0					8-1 / 990	8-5 / 1080	8-9 / 1170	9-0 / 1230	9-3 / 1300	9-6 / 1370	9-9 / 1450	10-0 / 1520	10-2 / 1570	10-5 / 1650	10-7 / 1700	10-10 / 1790	11-0 / 1840	11-4 / 1950	11-8 / 2070
2x10	12.0	11-4 / 450	12-3 / 520	13-0 / 590	13-8 / 660	14-4 / 720	14-11 / 780	15-5 / 830	15-11 / 890	16-5 / 940	16-10 / 990	17-3 / 1040	17-8 / 1090	18-0 / 1140	18-5 / 1190	18-9 / 1230	19-1 / 1280	19-5 / 1320	20-1 / 1410	20-8 / 1490
	13.7	10-10 / 470	11-8 / 550	12-5 / 620	13-1 / 690	13-8 / 750	14-3 / 810	14-9 / 870	15-3 / 930	15-8 / 980	16-1 / 1040	16-6 / 1090	16-11 / 1140	17-3 / 1190	17-7 / 1240	17-11 / 1290	18-3 / 1340	18-7 / 1380	19-2 / 1470	19-9 / 1560
	16.0	10-4 / 500	11-1 / 580	11-10 / 650	12-5 / 720	13-0 / 790	13-6 / 850	14-0 / 920	14-6 / 980	14-11 / 1040	15-3 / 1090	15-8 / 1150	16-0 / 1200	16-5 / 1250	16-9 / 1310	17-0 / 1360	17-4 / 1410	17-8 / 1460	18-3 / 1550	18-9 / 1640
	19.2	9-9 / 530	10-6 / 610	11-1 / 690	11-8 / 770	12-3 / 840	12-9 / 910	13-2 / 970	13-7 / 1040	14-0 / 1100	14-5 / 1160	14-9 / 1220	15-1 / 1280	15-5 / 1330	15-9 / 1390	16-0 / 1440	16-4 / 1500	16-7 / 1550	17-2 / 1650	17-8 / 1750
	24.0	9-0 / 570	9-9 / 660	10-4 / 750	10-10 / 830	11-4 / 900	11-10 / 980	12-3 / 1050	12-8 / 1120	13-0 / 1190	13-4 / 1250	13-8 / 1310	14-0 / 1380	14-4 / 1440	14-7 / 1500	14-11 / 1550	15-2 / 1610	15-5 / 1670	15-11 / 1780	16-5 / 1880
	32.0					10-4 / 1000	10-9 / 1080	11-1 / 1150	11-6 / 1240	11-10 / 1310	12-2 / 1380	12-5 / 1440	12-9 / 1520	13-0 / 1580	13-3 / 1640	13-6 / 1700	13-9 / 1770	14-0 / 1830	14-6 / 1970	14-11 / 2080
2x12	12.0	13-10 / 450	14-11 / 520	15-10 / 590	16-8 / 660	17-5 / 720	18-1 / 780	18-9 / 830	19-4 / 890	19-11 / 940	20-6 / 990	21-0 / 1040	21-6 / 1090	21-11 / 1140	22-5 / 1190	22-10 / 1230	23-3 / 1280	23-7 / 1320	24-5 / 1410	25-1 / 1490
	13.7	13-3 / 470	14-3 / 550	15-2 / 620	15-11 / 690	16-8 / 750	17-4 / 810	17-11 / 870	18-6 / 930	19-1 / 980	19-7 / 1040	20-1 / 1090	20-6 / 1140	21-0 / 1190	21-5 / 1240	21-10 / 1290	22-3 / 1340	22-7 / 1380	23-4 / 1470	24-0 / 1560
	16.0	12-7 / 500	13-6 / 580	14-4 / 650	15-2 / 720	15-10 / 790	16-5 / 860	17-0 / 920	17-7 / 980	18-1 / 1040	18-7 / 1090	19-1 / 1150	19-6 / 1200	19-11 / 1250	20-4 / 1310	20-9 / 1360	21-1 / 1410	21-6 / 1460	22-2 / 1550	22-10 / 1640
	19.2	11-10 / 530	12-9 / 610	13-6 / 690	14-3 / 770	14-11 / 840	15-6 / 910	16-0 / 970	16-7 / 1040	17-0 / 1100	17-6 / 1160	17-11 / 1220	18-4 / 1280	18-9 / 1330	19-2 / 1390	19-6 / 1440	19-10 / 1500	20-2 / 1550	20-10 / 1650	21-6 / 1750
	24.0	11-0 / 570	11-10 / 660	12-7 / 750	13-3 / 830	13-10 / 900	14-4 / 980	14-11 / 1050	15-4 / 1120	15-10 / 1190	16-3 / 1250	16-8 / 1310	17-0 / 1380	17-5 / 1440	17-9 / 1500	18-1 / 1550	18-5 / 1610	18-9 / 1670	19-4 / 1780	19-11 / 1880
	32.0					12-7 / 1000	13-1 / 1080	13-6 / 1150	13-11 / 1220	14-4 / 1300	14-9 / 1380	15-2 / 1450	15-6 / 1520	15-10 / 1580	16-2 / 1650	16-5 / 1700	16-9 / 1770	17-0 / 1830	17-7 / 1950	18-1 / 2070

Note: The required extreme fiber stress in bending, "F_b", in pounds per square inch is shown below each span.

FIGURE 5.19

A table for selecting floor joists for residential buildings. Most other building types have heavier floor loadings and must be designed with other tables. To use this table, start with allowable bending stress and modulus of elasticity values for the species and grade of lumber that will be used, as given in Figure 5.18. Read down the column that matches the modulus of elasticity value, looking at the upper figure in each square, until you find a number (expressed here in feet and inches) that equals or exceeds the required span. Then look to the left to see what size and spacing of joists are required. As an example, let us assume an apartment building with a required span of 13 feet 2 inches between supports, framed with Spruce-Pine-Fir, which has a modulus of elasticity of 1.3 million pounds per square inch (psi). Reading down the 1.3 column we find that 2 × 8 joists spaced 12 inches apart will apparently be satisfactory, but we must also check the lower figure in the

box, which is the required bending stress. In this case it is 990 psi, which is within the 1000 psi value found in Figure 5.18, so we could use 2 × 8 joists spaced 12 inches apart center to center. However, such a close spacing is usually not as economical as the more customary 16-inch spacing, so we will read further down the 1.3 column until we find a joist size that will be sufficiently stiff and strong at a 16-inch spacing. 2 × 10 joists can span 15 feet 3 inches on a 16-inch spacing, but require a bending stress of 1090 psi to do so, which is higher than the allowable stress (1000 psi) in the lumber we have selected. But we only need to span 13 feet 2 inches, so we can read to the left five squares to find by interpolation that for this span, the actual bending stress is only a little more than 800 psi, well within the 1000 psi allowable stress for our lumber. Thus we can adopt a design of 2 × 10 joists 16 inches on center. 2 × 10 joists at a spacing of 24 inches will not work because the actual

bending stress of 1250 psi would be higher than the allowable stress.

Similar tables are available in building codes and lumber literature for sizing ceiling joists and roof rafters. These are much more extensive because they must deal with high and low roof pitches, snow loads that vary according to the region in which the building is built, heavy and light roof coverings, and the presence or absence of a finish ceiling attached directly to the underside of the rafters. The reader is referred to reference 4 in the list of references at the end of this chapter, which includes both joist and rafter tables.

Joist and rafter tables available in the United States do not give metric equivalents, but the reader may make conversions using stress values from Figure 5.18, and conversion factors of 304.8 mm per foot, and 25.4 mm per inch. (Courtesy of National Forest Products Association)

Connection	Common Nail Size
Stud to sole plate	4–8d toe nails, or 2–16d end nails
Stud to top plate	2–16d end nails or toe nails
Double studs	10d face nails 12″ apart
Corner studs	16d face nails 24″ apart
Sole plate to joist or blocking	16d face nails 16″ apart
Double top plate	10d face nails 16″ apart
Lap joints in top plate	2–10d face nails
Rafter to top plate	3–8d toe nails
Rafter to ridge board	2–16d face nails or toe nails
Jack rafter to hip rafter	3–10d toe nails, or 2–16d face nails
Floor joists to sill or beam	3–8d toe nails
Ceiling joists to top plate	3–16d toe nails
Ceiling joist lap joint	3–10d face nails
Ceiling joist to rafter	3–10d face nails
Collar tie to rafter	3–10d face nails
Bridging to joists	2–8d face nails each end
Let-in diagonal bracing	2–8d face nails each stud or plate
Tail joists to headers	Use joist hangers
Headers to trimmers	Use joist hangers
Ledger to header	3–16d face nails per tail joist

FIGURE 5.20

Platform framing members are fastened according to this nailing schedule, which framing carpenters know by memory, and which is incorporated into most building codes.

Attaching the Frame to the Foundation

The *sill*, preferably of treated or naturally decay-resistant wood, is bolted to the foundation as a base for the wood framing. The top of the foundation is usually somewhat uneven, so the sill must be shimmed with wood shingle wedges to be able to transfer loads from the frame to the foundation. A compressible fibrous *sill sealer* should be inserted between the sill and the foundation to reduce air infiltration through the gap (Figures 5.21, 5.22). The normal foundation bolts are sufficient to hold most buildings on their foundations, but taller frames in areas subject to high winds or earthquakes may require more elaborate attachments (Figure 5.44).

Floor Framing and Bridging

Floor framing (Figure 5.16) is usually laid out in such a way that the joints of uncut subflooring panels will fall directly over joists; otherwise many panels will have to be cut, wasting both materials and time. Subflooring should be glued to the joists to prevent squeaking and increase floor stiffness (Figure 5.23). Plywood and OSB panels must be laid with the grain of their face plies perpendicular to the direction of the joists because these panels are considerably stiffer in this orientation. Panels should be spaced slightly at all their edges to prevent floor buckling during construction from the expansion of storm-wetted panels.

Bridging, which is crossbracing or solid blocking between joists at midspan, is a traditional feature of floor framing (Figure 5.24). Its function is to hold the joists straight and to help them share concentrated loads. Research has shown, however, that

FIGURE 5.21

Carpenters apply a treated wood sill to a concrete foundation. Glass fiber sill sealer has been placed on the top of the concrete wall, and the sill has been drilled to fit over the projecting anchor bolts. Before each section of sill is bolted tightly, it is leveled as necessary with wood shingle shims between the concrete and the wood. A length of completed sill is visible at the upper right. The basement windows were clamped into reusable steel form inserts and placed in the formwork before the concrete was poured, which causes the window to become integral with the basement wall. After the concrete was poured and the formwork was stripped, the steel inserts were removed, leaving a neatly formed concrete frame around each window. A pocket for a steel beam can be seen at the upper left. (Photo by the author)

FIGURE 5.22

Installing floor joists. Blocking will next be inserted between the joists over the two interior beams. (Photo by the author)

FIGURE 5.23
For stiffness and squeak resistance, subflooring should be glued to the joists. The adhesive is a thick mastic that is applied with a sealant gun. (Courtesy of American Plywood Association)

FIGURE 5.24
Bridging between joists, where required, may be solid blocks of joist lumber, wood crossbridging, or, as seen here, steel crossbridging. Steel crossbridging requires only one nail per piece, and no cutting, so is the fastest to install. (Courtesy of American Plywood Association)

bridging is of little use under most conditions, and many building codes no longer require it.

Where ends of joists are butted into supporting headers, as around stair openings and at changes of joist direction for projecting bays, end nails and toenails cannot carry the full weight of the joists, and sheet metal joist hangers must be used. Each provides a secure pocket for the end of the joist, and punched holes into which a number of special short nails are driven to make a safe connection.

Wall Framing, Sheathing, and Bracing

Wall framing, like floor framing, is laid out so a framing member occurs under each joint between sheathing panels. The head carpenter initiates wall fram-

FIGURE 5.25
Various types of manufactured joists are often used instead of dimension lumber, to achieve longer spans and lower installed costs. This I-beam joist is made entirely of plywood veneers and is manufactured in very long pieces. (Courtesy of Trus Joist Corporation)

FIGURE 5.26
Typical ground floor wall framing details, keyed by letter to Figure 5.27.

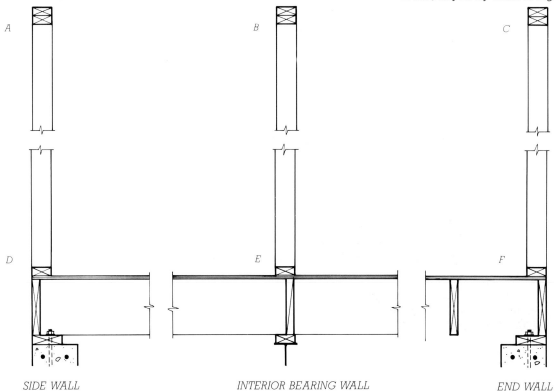

SIDE WALL

INTERIOR BEARING WALL

END WALL

The subfloor makes a convenient platform on which to assemble the first floor wall frames. The assembled frames are tilted up into place, nailed to the floor and to one another, and supported by temporary braces.

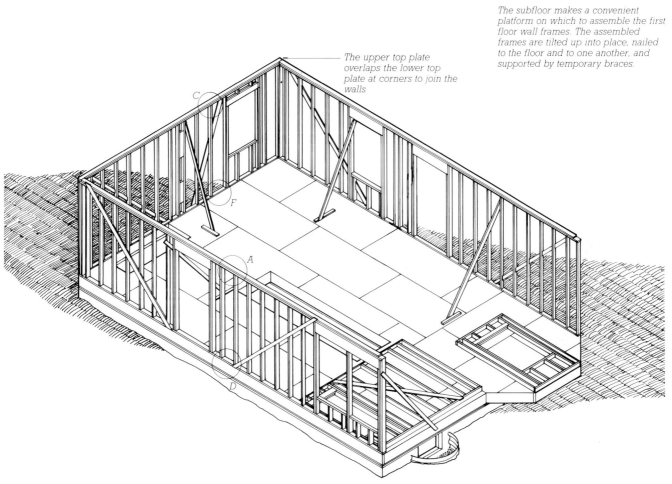

The upper top plate overlaps the lower top plate at corners to join the walls

FIGURE 5.27
Step Four in erecting a platform-frame building: The ground floor walls are framed.

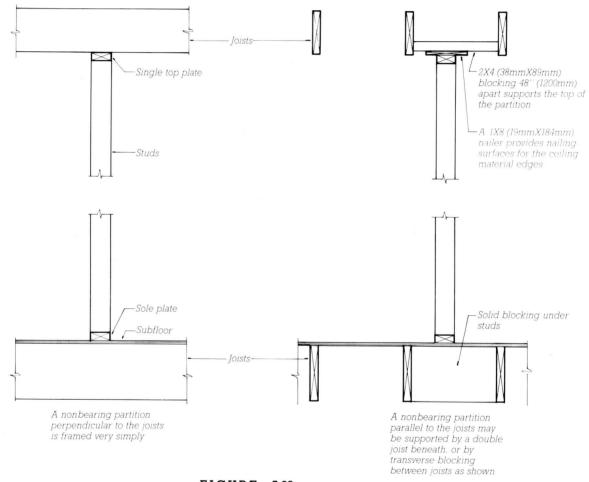

Joists

Single top plate

Studs

2X4 (38mmX89mm) blocking 48" (1200mm) apart supports the top of the partition

A 1X8 (19mmX184mm) nailer provides nailing surfaces for the ceiling material edges

Sole plate

Subfloor

Joists

Solid blocking under studs

A nonbearing partition perpendicular to the joists is framed very simply

A nonbearing partition parallel to the joists may be supported by a double joist beneath, or by transverse blocking between joists as shown

FIGURE 5.28
Framing details for nonloadbearing interior partitions.

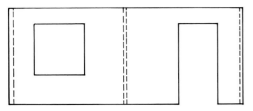

1. This is the layout of a typical exterior wall. It meets two other exterior walls at the corners, and a partition in the middle. It has two rough openings, one for a window and one for a door.

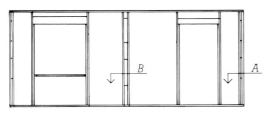

2. The framer begins by marking all the stud and opening locations on the sole plate and top plate. The "special" studs are cut and assembled first: two corner posts, a partition intersection, and full-length studs and supporting studs for the headers over the openings.

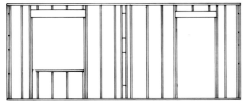

3. The wall is next filled with studs on a regular 16" (400mm) or 24" (600mm) spacing, to provide support for edges of sheathing panels.

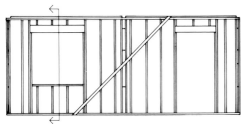

4. Diagonal bracing, usually 1X4 (19mmX89mm), is let into the face of the frame if the building will not have rigid sheathing. The second top plate may be added before the wall is tilted up, or after.

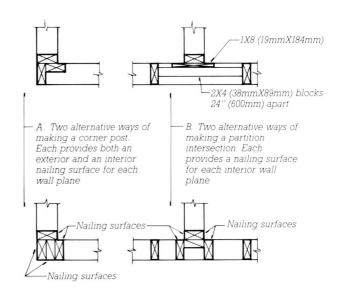

A. Two alternative ways of making a corner post. Each provides both an exterior and an interior nailing surface for each wall plane

B. Two alternative ways of making a partition intersection. Each provides a nailing surface for each interior wall plane

1X8 (19mmX184mm)

2X4 (38mmX89mm) blocks 24" (600mm) apart

Nailing surfaces

Nailing surfaces

Nailing surfaces

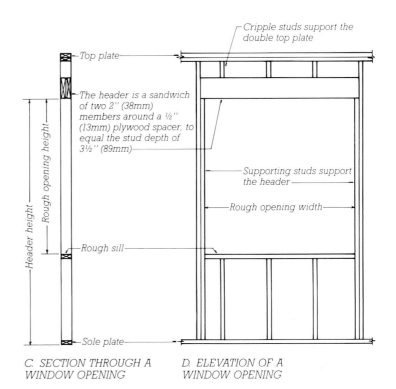

Top plate

The header is a sandwich of two 2" (38mm) members around a ½" (13mm) plywood spacer, to equal the stud depth of 3½" (89mm)

Rough sill

Sole plate

Header height

Rough opening height

Cripple studs support the double top plate

Supporting studs support the header

Rough opening width

C. SECTION THROUGH A WINDOW OPENING

D. ELEVATION OF A WINDOW OPENING

FIGURE 5.29
Procedure and details for wall framing.

Stud Size	Spacing	Maximum Stud Height	Maximum Stories Supported
2 × 4 (38 × 89 mm)	16" (406 mm)	14' (4.25 m)	One floor plus roof
	24" (610 mm)	14' (4.25 m)	Roof only
2 × 6 (38 × 140 mm)	16" (406 mm)	20' (6.10 m)	Two floors plus roof
	24" (610 mm)	20' (6.10 m)	One floor plus roof

FIGURE 5.30

Typical building code requirements for stud sizes and spacings. Notice that 2 × 4 (38 × 89 mm) studs are inadequate for walls that support two floors and a roof above, such as ground floor studs in a three-story building, or basement studs below a two-story building. Notice also that very tall studs may have to be increased in size.

Size of Wood Header	Supporting Roof Only	Supporting One Story Above	Supporting Two Stories Above	Not Supporting Floors or Roofs
2–2 × 4 (2–38 × 89 mm)	4' (1220 mm)	— —	— —	6' (1830 mm)
2–2 × 6 (2–38 × 140 mm)	4 to 6' (1220 to 1830 mm)	4' (1220 mm)	—	6 to 8' (1830 to 2440 mm)
2–2 × 8 (2–38 × 184 mm)	6 to 8' (1830 to 2440 mm)	4 to 6' (1220 to 1830 mm)	—	8 to 10' (2440 to 3050 mm)
2–2 × 10 (2–38 × 235 mm)	8 to 10' (2440 to 3050 mm)	6 to 8' (1830 to 2440 mm)	4 to 6' (1220 to 1830 mm)	10 to 12' (3050 to 3660 mm)
2–2 × 12 (2–38 × 286 mm)	10 to 12' (3050 to 3660 mm)	8 to 10' (2440 to 3050 mm)	6 to 8' (1830 to 2440 mm)	12 to 16' (3660 to 4880 mm)

FIGURE 5.31

Typical building code requirements for headers. These are rule-of-thumb values based on typical building designs and lumber of relatively low strength and stiffness; headers for unusual situations should be calculated using accepted structural engineering methods.

FIGURE 5.32

Laying out stud locations on the top and sole plates is the first step in constructing a wall frame. (Photo by the author)

FIGURE 5.33
Assembling studs to a plate, using a pneumatic nail gun. The triple studs are for a partition intersection. (Courtesy of Senco Products, Inc.)

FIGURE 5.34
Tilting an interior partition into position. The gap in the upper top plate will receive the projecting end of the upper top plate from another partition which intersects at this point. (Courtesy of American Plywood Association)

FIGURE 5.35

Fastening a wall to the floor platform. The studs have loosened slightly from the sole plate during the tilting operation and will have to be retightened by hammering downward *on the top plate. The blocks between studs are for a line of nailing to attach vertical wood siding. (Courtesy of Senco Products, Inc.)*

FIGURE 5.36

Ground floor wall framing is held up by temporary bracing until the upper floor framing is in place and wall bracing or sheathing is complete, after which the frame becomes completely self-bracing. The outer walls of this building are *framed with 2×6s (38 × 140 mm) to allow for a greater thickness of thermal insulation, while the interior partitions are made of 2×4s (38×89 mm). (Photo by Joseph Iano)*

ing by laying out the stud locations on the top plate and sole plate of each wall (Figure 5.32). Other carpenters follow behind to cut the studs and headers, and assemble the walls flat on the subfloor. As each wall frame is completed, the carpenters tilt it up and nail it into position, bracing it temporarily as needed (Figures 5.33 through 5.35).

Sheathing of rigid panels or diagonal boards acts as permanent bracing for the walls and is applied as soon as possible after the wall is framed. Some types of sheathing panels are intended only as insulation, however, and have no structural value. Where these are used, *let-in* diagonal bracing is inserted into each wall before it is erected (Figure 5.37).

Corners and partition intersections must furnish nailing surfaces for each plane of exterior and interior finish materials. This requires a minimum of three studs at each intersection, unless special metal clips are used to reduce the number to two (Figure 5.29).

Framing for Increased Thermal Insulation   The 2×4 (38×89 mm) has been the standard wall stud since light framing was invented. In recent years, however, pressures for heating fuel conservation have created a need to provide more thermal insulation than can be inserted in the cavities of a wall framed with 2×4s. Some designers have simply adopted the 2×6 (38×140 mm) as the standard stud, usually at a spacing of 24 inches (610 mm). Others have stayed with the 2×4 stud but have covered the wall either inside or out with insulating plastic foam sheathing, thus reaching an insulating value about the same as that of a 2×6 wall. Many use both 2×6 studs and insulating sheathing. Designers in some very cold climates are not satisfied even with these types of construction and use exterior walls framed in two separate layers, or with vertical truss studs made up of two ordinary studs joined at intervals by plywood plates. Some of these constructions are illustrated in Figures 7.10 and 7.11.

FIGURE 5.37
Applying a panel of insulating foam sheathing. Because this type of sheathing is incapable of bracing the frame, diagonal bracing is inserted at the corners of the building. T-shaped steel bracing, inserted into a saw cut and nailed at each stud, is used in this frame and is visible just to the right of the carpenter's leg. (Courtesy of The Celotex Corporation)

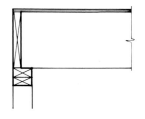

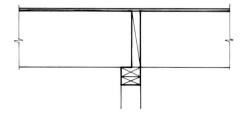

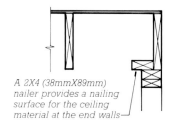

A 2X4 (38mmX89mm)
nailer provides a nailing
surface for the ceiling
material at the end walls—

A. SIDE WALL

B. INTERIOR BEARING WALL

C. END WALL

FIGURE 5.38

*Details of the second floor platform,
keyed to the letters on Figure 5.39. The
extra piece of lumber on top of the top
plate in Detail C, End Wall, is
continuous blocking whose function is to*
*provide a nailing surface for the edge of
the finish ceiling material, which is
usually either gypsum board or veneer
plaster base.*

*When the first floor walls are complete
and sheathed, much of the temporary
bracing can be removed. The second
floor platform is framed the same as
the first, with its joists resting on the
double top plates of the first floor
walls.*

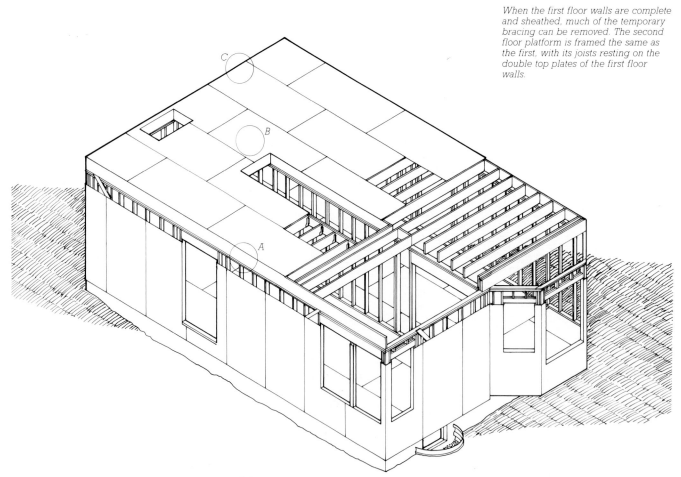

FIGURE 5.39

5

*Step Five in erecting a two-story
platform-frame building: Building the
upper floor platform.*

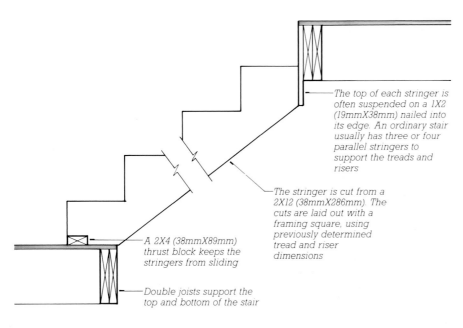

The top of each stringer is often suspended on a 1X2 (19mmX38mm) nailed into its edge. An ordinary stair usually has three or four parallel stringers to support the treads and risers

The stringer is cut from a 2X12 (38mmX286mm). The cuts are laid out with a framing square, using previously determined tread and riser dimensions

A 2X4 (38mmX89mm) thrust block keeps the stringers from sliding

Double joists support the top and bottom of the stair

A. STAIR FRAMING

FIGURE 5.40

Interior stairways are usually framed as soon as the upper floor platform is completed to give the carpenters easy up-and-down access during the remainder of the work. Temporary treads of joist scrap or plywood are fastened over the stringers. These will be replaced by finish treads after the wear and tear of construction are finished.

Wall framing procedures for the second floor are identical to those for the first floor.

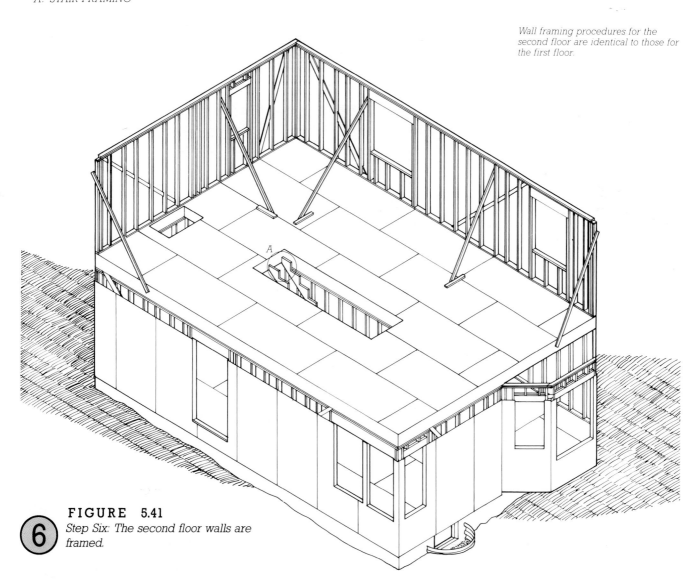

FIGURE 5.41

6 *Step Six: The second floor walls are framed.*

FIGURE 5.42
Nailing upper floor joists to the top plates. (Photo by the author)

FIGURE 5.43
Installing upper floor subflooring. The grade of the plywood panels used in this building is C-C Plugged, in which all surface voids are filled and the panel is lightly sanded to allow carpeting to be installed directly over the subfloor without additional underlayment. The long edges of the panels have interlocking tongue-and-groove joints to prevent excessive deflection of the edge of a panel under a heavy concentrated load such as a standing person or the leg of a piano. Notice the adhesive between the joists and the plywood. (Courtesy of American Plywood Association)

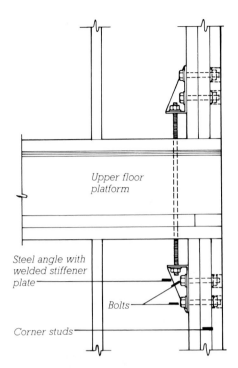

Upper floor platform

Steel angle with welded stiffener plate

Bolts

Corner studs

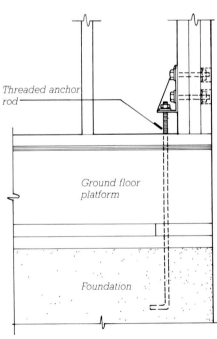

Threaded anchor rod

Ground floor platform

Foundation

FIGURE 5.44

Tall, narrow platform frame buildings in areas with high wind loads or earthquake risk must sometimes be reinforced against uplift at the corners. Hold-downs of the type shown here tie the entire height of the building securely to the foundation. To compensate for wood shrinkage, the nuts should be retightened after the first heating season, which can mean that access holes must be provided through the interior wall surfaces.

Roof Framing

The generic roof types for Wood Light Frame buildings are shown in Figure 5.45. These are often combined to make roofs suited to the covering of more complex plan shapes and building volumes.

For structural stability, gable and hip roofs must be securely tied by well-nailed ceiling joists to make what is in effect a series of triangular trusses. If

FIGURE   5.45
Basic roof shapes for Wood Light Frame buildings.

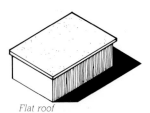

Flat roof

Flat and shed roofs exert no lateral thrust

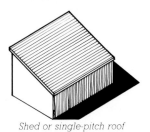

Shed or single-pitch roof

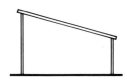

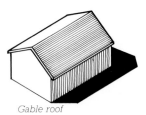

Gable roof

Ceiling joist   *Ridge beam*

Gable and hip rafters must be either tied with ceiling joists, or supported by a structural ridge beam

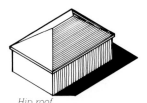

Hip roof

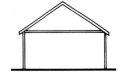

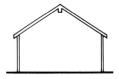

Gambrel roof

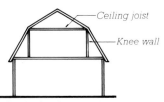

Ceiling joist

Knee wall

Gambrel and mansard roofs require both knee walls and ceiling joists for structural stability

Mansard roof

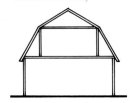

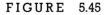

the designer wishes to eliminate the ceiling joists to expose the sloping underside of the roof as the finished ceiling surface, a beam or bearing wall must be inserted under the ridge unless a system of exposed ties is designed to replace the ceiling joists.

While the college-graduate architect or engineer would find it difficult to lay out the rafters for a pitched roof using trigonometry, the carpenter, without resorting to mathematics, has little problem making the layout if the *pitch* is specified as a ratio of *rise* to *run*. Rise is the vertical dimension and run is the horizontal. In the United States, pitch is usually given on the architect's drawings as inches of rise per foot (12 inches) of run. An old-time carpenter uses these two figures on the two sides of a framing square to lay out the rafter as shown in Figures 5.47 and 5.53. The actual length of the rafter is never figured, nor does it need

The balloon frame is closely connected with the level of industrialization which had been reached in America [in the early nineteenth century]. Its invention practically converted building in wood from a complicated craft, practiced by skilled labor, into an industry.... This simple and efficient construction is thoroughly adapted to the requirements of contemporary architects...elegance and lightness [are] innate qualities of the balloon-frame skeleton.

Sigfried Giedion, historian

to be, because all the measurements are made as horizontal and vertical distances with the aid of the square. Today, many carpenters prefer to do rafter layout with the aid of tables that give actual rafter lengths for various pitches and horizontal distances; these tables are stamped on the framing square itself or printed in pocket-size booklets.

Hips and valleys introduce another level of trigonometric complexity in rafter layout, but the experienced carpenter has little difficulty even here: Again, he or she can use published tables for hip and valley rafters, or do the layout the old-fashioned way, as illustrated in Figure 5.51.

The head carpenter lays out only one rafter of each type by these procedures. This then becomes the *pattern rafter* from which other carpenters trace and cut the remainder of the rafters (Figure 5.54).

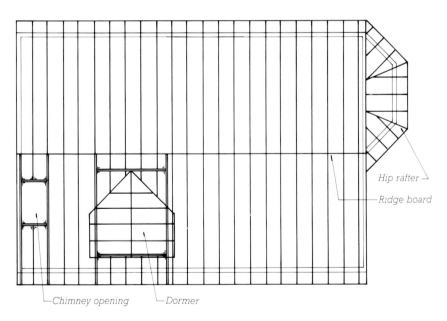

ROOF FRAMING PLAN

FIGURE 5.46

A roof framing plan for the building illustrated in Figure 5.48. The dormer and chimney openings are framed with doubled headers and trimmers. The dormer is then built as a separate structure nailed to the slope of the main roof.

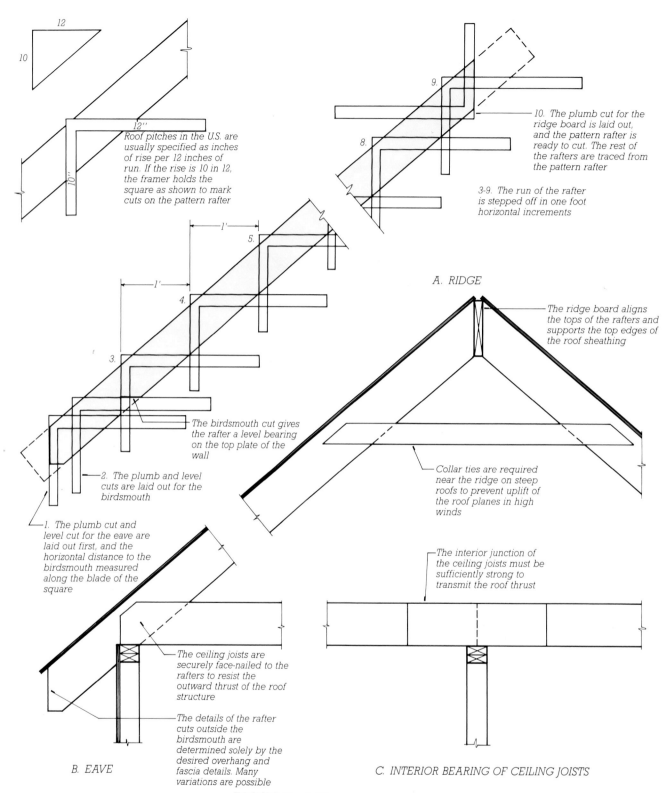

12

10

12"

10"

Roof pitches in the U.S. are usually specified as inches of rise per 12 inches of run. If the rise is 10 in 12, the framer holds the square as shown to mark cuts on the pattern rafter

9.

8.

10. The plumb cut for the ridge board is laid out, and the pattern rafter is ready to cut. The rest of the rafters are traced from the pattern rafter

3-9. The run of the rafter is stepped off in one foot horizontal increments

1'

5.

1'

4.

3.

A. RIDGE

The ridge board aligns the tops of the rafters and supports the top edges of the roof sheathing

The birdsmouth cut gives the rafter a level bearing on the top plate of the wall

Collar ties are required near the ridge on steep roofs to prevent uplift of the roof planes in high winds

2. The plumb and level cuts are laid out for the birdsmouth

The interior junction of the ceiling joists must be sufficiently strong to transmit the roof thrust

1. The plumb cut and level cut for the eave are laid out first, and the horizontal distance to the birdsmouth measured along the blade of the square

The ceiling joists are securely face-nailed to the rafters to resist the outward thrust of the roof structure

The details of the rafter cuts outside the birdsmouth are determined solely by the desired overhang and fascia details. Many variations are possible

B. EAVE

C. INTERIOR BEARING OF CEILING JOISTS

FIGURE 5.47

Roof framing: The lettered details are keyed to Figure 5.48. The remainder of the page shows how a framing square is used to lay out a pattern rafter, reading from the first step at the lower end of the rafter to the last step at the top.

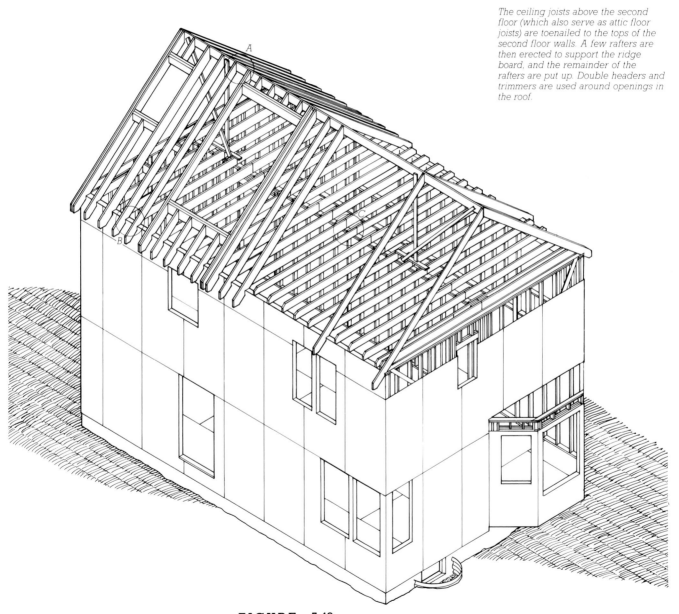

The ceiling joists above the second floor (which also serve as attic floor joists) are toenailed to the tops of the second floor walls. A few rafters are then erected to support the ridge board, and the remainder of the rafters are put up. Double headers and trimmers are used around openings in the roof.

FIGURE 5.48

7 *Step Seven: Framing the attic floor and roof.*

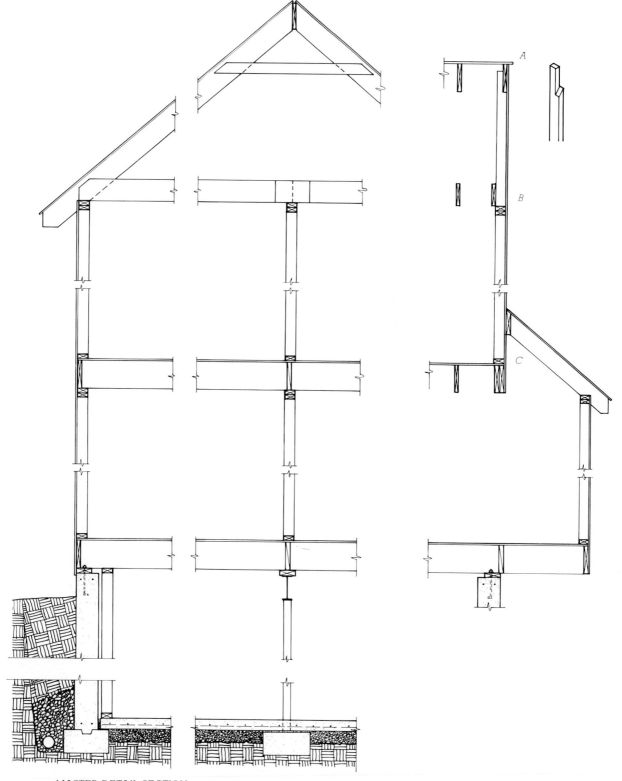

MASTER DETAIL SECTION

FIGURE 5.49

A summary of the major details for the structure shown in Figure 5.50, aligned in relationship to one another. The lettered details are keyed to Figure 5.50. The gable end studs are cut as shown in Detail A and face nailed to the rake rafter.

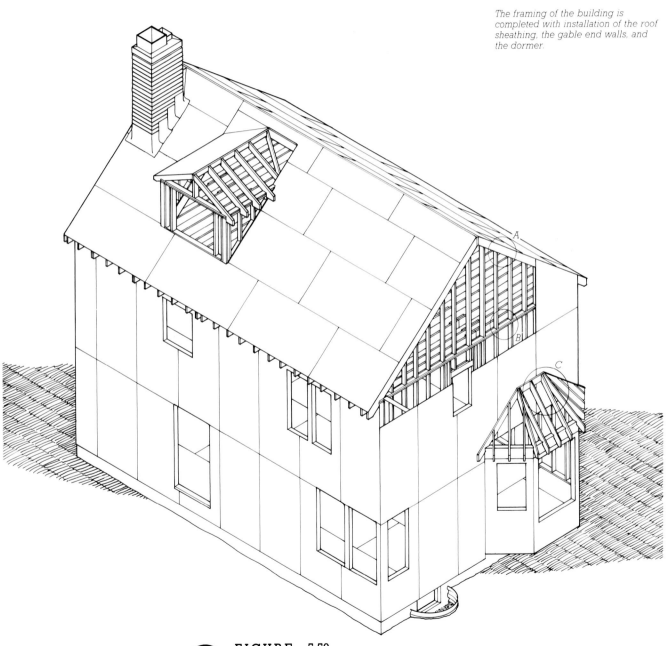

The framing of the building is completed with installation of the roof sheathing, the gable end walls, and the dormer.

FIGURE 5.50

Step Eight: The frame is completed.

8

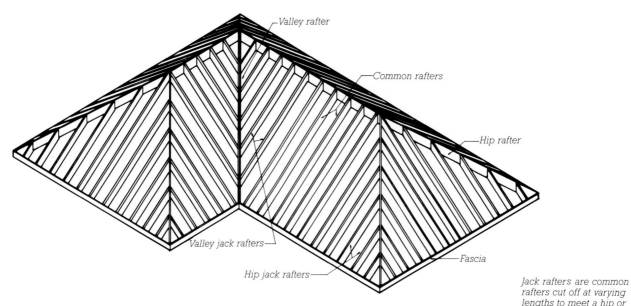

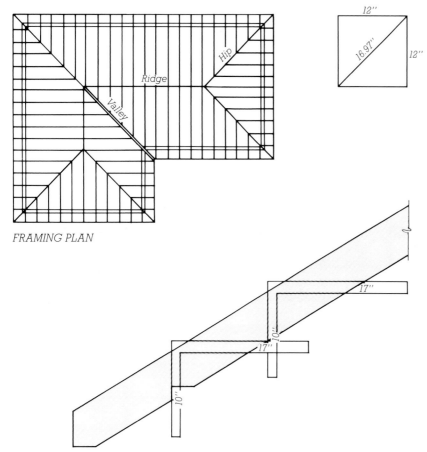

Jack rafters are common rafters cut off at varying lengths to meet a hip or valley rafter. The jacks meet the hip or valley at a compound angle that is easily laid out with a framing square

The diagonal of a 12" square is 16.97", or very nearly 17". In laying out a hip or valley rafter, the framer simply aligns the framing square to the rise per foot of a common rafter on the tongue, and 17" rather than 12" on the blade. The marking and stepping-off operations are otherwise identical to those for a common rafter

FRAMING PLAN

FIGURE 5.51
Framing for a hip roof. The difficult geometric problem of laying out the hip rafter is solved easily by using the framing square in the manner shown.

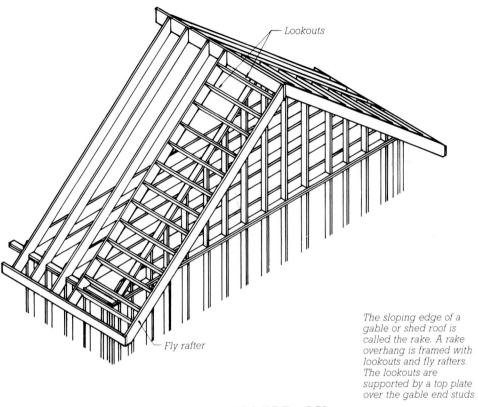

Lookouts

Fly rafter

The sloping edge of a gable or shed roof is called the rake. A rake overhang is framed with lookouts and fly rafters. The lookouts are supported by a top plate over the gable end studs

FIGURE 5.52
Framing for an overhanging rake.

FIGURE 5.53
A framing square being used to mark rafter cuts. The run of the roof, 12 inches, is aligned with the edge of the rafter on the blade of the square, and the rise, 7 inches in this case, is aligned on the tongue of the square. A pencil line along the tongue will be perfectly vertical when the rafter is installed in the roof, and one along the blade will be horizontal. True horizontal and vertical distances can be measured on the blade and tongue, respectively. (Photo by the author)

FIGURE 5.54
Tracing a pattern rafter to mark cuts for the rest of the rafters. The corner of the building behind the carpenters has let-in corner braces on both floors, and most of the rafters are already installed. (Courtesy of Southern Forest Products Association)

Prefabricated Framing Assemblies

Roof trusses and floor trusses find widespread use in platform frame buildings because of their speed of erection, economy of material usage, and long spans. Most are light enough to be lifted and installed by two carpenters (Figures 5.55, 5.56).

Manufactured wall framing panels have been adopted more slowly than roof and floor trusses, except with large builders who mass-market hundreds or thousands of houses per year. For the smaller builder, wall framing can be done on-site with the same amount of material as with panels and with little or no additional overall expenditure of labor, especially when the building requires walls of varying heights and shapes.

FIGURE 5.55
Roof framing with prefabricated trusses. Sheet metal clips and nails anchor the trusses to the top plate. (Courtesy of Gang-Nail Systems, Inc.)

FIGURE 5.56
Prefabricated trusses are trucked to the building site in bundles. Another set of trusses will roof the top story of this three-story apartment building. (Courtesy of Gang-Nail Systems, Inc.)

FIGURE 5.57
Applying plywood roof sheathing to a half-hipped roof. Blocking between rafters at the wall line has been drilled with large holes for attic ventilation. The line of horizontal blocking between studs is to support the edges of plywood siding panels applied in a horizontal orientation. (Courtesy of American Plywood Association)

FIGURE 5.58
Fastening roof sheathing with a pneumatic nail gun. (Courtesy of Senco Products, Inc.)

WOOD LIGHT FRAME CONSTRUCTION AND THE BUILDING CODES

It will be seen in the table in Figure 1.1 that the BOCA Basic/National Building Code allows buildings of every use group to be constructed with wood platform framing, which is classified as Type 5 construction, but there are severe restrictions on height and floor area. For some use groups, higher fire-resistance ratings are required for some components of the building. (These are usually achieved by using gypsum board or plaster finishes to add fire resistance to the components.) Additionally, zoning ordinances in most cities identify certain densely built areas in which wood platform frame construction is simply not permitted. And in some buildings, higher fire insurance premiums may discourage the owner from considering platform framing even if it is permitted by code.

Several examples serve to illustrate the limits within which wood platform framing can be used for different sorts of buildings. A nursing home would be classified as Use Group I-2, Institutional, incapacitated; Type 5B construction would not be permitted, and Type 5A could be used for a single-story building up to 7650 square feet (710 m²) in floor area. A meeting hall for a summer camp, Use Group A-3, can be only one story or 20 feet (6.1 m) tall, and 8925 square feet (830 m²) in area if protected (Type 5A) or 4200 square feet (390 m²) if unprotected (Type 5B). Retail stores (Use Group M) can be one or two stories tall with as much as 10,200 square feet (950 m²) of area per floor. Hotels, apartments, and single-family dwellings may be as tall as three stories, the maximum permitted for wood platform frame construction in any use group. These limits are broad enough that a surprisingly high percentage of all building projects can be built using this type of construction. And the economies of platform frame construction are such that most building owners will choose it over more fire-resistant types of construction if given the opportunity.

It would, however, be a mistake to assume from these facts that building codes are permissive when dealing with platform frame construction. On the contrary, wood platform framing is probably regulated more closely by codes than any other construction type, including detailed specifications for framing members and fasteners that have already been illustrated in this chapter, and detailed drawings that show exactly how certain parts of the structure must be assembled. Emergency exit requirements for residential structures generally include minimum size and maximum sill height dimensions for bedroom windows. An automatic fire alarm system including smoke detectors, heat detectors, or both is almost universally required in residential buildings to awaken the occupants and get them moving toward the exits before the building becomes fully involved in a fire. It seems likely, too, that automatic sprinkler systems of some type may soon become a requirement in many types of wood frame buildings, especially as simpler, cheaper systems are developed.

FIGURE 5.59
The W. G. Low house, built in Bristol, Rhode Island in 1887 to the design of architects McKim, Mead, and White, illustrates both the essential simplicity of wood light framing and the complexity of which it is capable. (Photo by Wayne Andrews)

FIGURES 5.60, 5.61
*Ordinary platform framing takes on
extraordinary qualities beneath the
translucent plastic roof of this Hawaiian
house designed by William Turnbull
Associates, Architects.* (Photos ©1982 by
David Franzen. Courtesy of the
architect)

FIGURE 5.62
*The Village Corner shopping center in
San Diego, California* (SPGA Planning &
Architecture, Architects) *makes skillful
use of the easy informality of wood
framing.* (Photo by Wes Thompson.
Courtesy of American Wood Council)

THE UNIQUENESS OF WOOD LIGHT FRAME CONSTRUCTION

Wood light framing in the form of platform framing is popular because it is an extremely flexible and economical way of constructing small buildings. Its flexibility stems from the ease with which carpenters with ordinary tools can create buildings of astonishing complexity in a variety of geometries. Its economy can be attributed in part to the relatively unprocessed nature of the materials from which it is made, and in part to mass-market competition among suppliers of components and materials and local competition among small builders.

Platform framing is the one truly complete and open system of construction we have. It incorporates structure, enclosure, thermal insulation, mechanical installations, and finishes into a single constructional concept. Thousands of products are made

FIGURE 5.63
In the late nineteenth century, the sticklike qualities of wood light framing often found expression in the exterior ornamentation of a house. (Photo by the author)

FIGURE 5.64
The Thorncrown Chapel in Arkansas, by Fay Jones and Associates, Architects, combines large areas of glass with special framing details of nominal 2" (38 mm) lumber to create a richly inspiring space. (Photo by Christopher Lark. Courtesy of American Wood Council)

to fit it: dozens of competing brands of windows and doors; a multitude of interior and exterior finish materials; scores of electrical, plumbing, and heating products. For better or worse, it can be dressed up to look like a building of wood or of masonry in any architectural style from any era of history. Architects have failed to exhaust its formal possibilities and engineers have failed to invent a new environmental control system it cannot assimilate.

At the other end of the scale, platform frame construction has led through sinister logic to the creation of countless horrors around the perimeters of our cities. To create the minimum-cost detached single-family house, one can begin by reducing the number of exterior wall corners to four because extra corners cost extra dollars for framing and finishing. Then the roof pitch can be made as shallow as asphalt shingles will permit, to allow roofers to work without scaffolding, and overhangs can be eliminated to save material and labor. Windows can be made as small as building codes will allow, because a square foot of window costs more than a square foot of wall. The thinnest, cheapest finish materials can be used inside and out, even if they will begin to look shabby after a short

period of time, and all ceilings can be made flat and as low as is legally permitted, to save materials and labor. The result is an uninteresting, squat, dark residence that weathers poorly, has little meaningful relation to its site, and contributes to a lifeless, uninspiring neighborhood. The same principles, with roughly the same result, can be applied to the construction of apartments and commercial buildings. In most cases things only get worse if one tries to dress up this minimal box building with inexpensive shutters, window boxes, and fake materials. The

single most devastating fault of Wood Light Frame construction is not its combustibility or its tendency to decay, but its ability to be reduced to the irreducibly minimal—scarcely human—building system. Yet one can look to examples of the Carpenter Gothic, Queen Anne, and Shingle Style buildings of the century past, or the Bay Region and Modern styles of our own time, to realize that wood light framing can be a system that gives to the designer the freedom to make a finely crafted building that nurtures life and elevates the spirit.

C.S.I./C.S.C. Masterformat Section Numbers for Wood Light Framing	
06100	**ROUGH CARPENTRY**
06105	Treated Wood Foundations
06110	Wood Framing Pre-assembled components
06115	Sheathing
06120	Structural Panels
06125	Wood Decking
06170	PREFABRICATED STRUCTURAL WOOD
06190	Wood Trusses
06195	Prefabricated Wood Beams and Joists

S E L E C T E D R E F E R E N C E S

1. Anderson, L. O. *Wood-Frame House Construction*.

A number of different publishers have issued versions of this straightforward, clearly illustrated volume, which was prepared by the U. S. Forest Products Laboratory and first published by the United States Government Printing Office as Agriculture Handbook No. 73.

2. Central Mortgage and Housing Corporation. *Canadian Wood-Frame House Construction*.

This book is based on Anderson (reference 1), with modifications to suit Canadian climates and Canadian building practices.

3. Dietz, Albert G. H. *Dwelling House Construction*, 4th Edition. Cambridge, Massachusetts, M.I.T. Press, 1974.

Dietz presents over 400 pages of lucid text and illustrations concerning light frame construction and is more detailed than either of the preceding texts.

4. National Forest Products Association. *Span Tables for Joists and Rafters*. Washington, D.C., 1977.

This is the standard reference for the structural design of platform-frame apartments and single-family residences. For use groups with heavier floor loadings, joist tables are available in *Wood Structural Design Data*, published by the same organization. (Address for ordering: 1619 Massachusetts Avenue N.W., Washington, D.C. 20036.)

K E Y   T E R M S
A N D   C O N C E P T S

Wood Light Frame construction	ridge board	bridging	gambrel
balloon frame	hip	stringer	mansard
platform frame	valley	let-in bracing	knee wall
firestop	trimmer	supporting stud	lookout
carpenter	sill	cripple stud	fly rafter
joist	wracking	pitch	eave
stud	framing plan	rise	plumb cut
rafter	elevations	run	level cut
header	sections	rafter	collar tie
band joist	rough carpentry	pattern rafter	dormer
sheathing	batter board	roof truss	fascia
subfloor	sill sealer	floor truss	framing square
sole plate	anchor bolt	prefabricated assemblies	hip rafter
top plate	blocking	shed	jack rafter
	joist hanger	gable	birdsmouth

R E V I E W   Q U E S T I O N S

1. Draw a series of very simple section drawings to illustrate the procedure for erecting a platform frame building, starting with the foundation and continuing with the ground floor, the ground floor walls, the second floor, the second floor walls, and the roof. Do not show details of connections but simply represent each plane of framing as a heavy line in your section drawing.

2. Draw from memory the standard detail sections for a two-story platform frame dwelling. **Hint:** The easiest way to draw a detail section is to draw the pieces in the order in which they are put in place during construction. If your simple drawings from Question 1 are correct, and if you follow this procedure, you will not find this question so difficult.

3. What are the differences between balloon framing and platform framing? What are the advantages and disadvantages of each? Why has platform framing become the method of choice?

4. Why is firestopping not usually required in platform framing?

5. Why is a steel beam preferred to a wood beam at the foundation level?

6. How is a platform frame building braced against wind and earthquake forces?

7. Light framing of wood is highly combustible. In what different ways does a typical building code take this fact into account?

E X E R C I S E S

1. Visit a building site where a wood platform frame is being constructed. Compare the details you see on the site with the ones shown in this chapter. Ask the carpenters why they do things the way they do. Where their details differ from the ones illustrated, make up your own mind about which is better, and why.

2. Develop floor framing and roof framing plans for a building you are designing. Figure the sizes of the floor joists using the table in Figure 5.19 if it is a residential building.

3. Make thumbnail sketches of twenty or more different ways of covering an L-shaped building with combinations of pitched roofs. Start with the simple ones (a single shed, two intersecting sheds, two intersecting gables) and work into the more elaborate ones. Note how the varying roof heights of some schemes could provide room for a partial second story or loft, or for high spaces with clerestory windows. How many ways do you think there are of covering an L-shaped building with pitched roofs? Look around you as you travel through areas with wood-frame buildings, especially older areas, and see how many ways designers and framers have roofed simple buildings in the past. Build up a collection of sketches of ingenious combinations of pitched roof forms.

4. Build a scale model of a platform frame from basswood or pine, reproducing accurately all its details, as a means of becoming thoroughly familiar with them. Better yet, build a small frame building for someone, perhaps a toolshed, playhouse, or kiosk.

EXTERIOR FINISHES FOR WOOD LIGHT FRAME CONSTRUCTION

Roofing
Finishing the Eaves and Rakes
Roof Shingling

Windows
Types of Windows   Materials for Windows
Installing Windows

Doors

Siding
Board Sidings
Plywood Sidings   Shingle Sidings
Metal and Plastic Sidings   Stucco
Masonry Veneer

Exterior Construction

Exterior Painting, Finish Grading, and Landscaping

Architects Hartman-Cox clad this small church with an economical but attractive siding of white-painted plywood and vertical wood battens. The roofs are of asphalt shingles and the windows of wood. The rectangular windows are double-hung. (Photo by Robert Lautman. Courtesy of American Wood Council)

As a platform frame nears completion, exterior finishing operations begin. First the roof edges are finished, then the roof is shingled, to offer as much weather protection as possible to subsequent operations. When the roof is tight to the weather, the windows, doors, and siding are applied, after which interior finishing work may take place completely sheltered from sun, rain, snow, and wind. The outside of the building is now complete for all practical purposes; exterior painting, finish grading, landscaping, and paving work may commence.

Roofing

Finishing the Eaves and Rakes

Before a roof can be shingled, the *eaves* (horizontal roof edges) and *rakes* (sloping roof edges) must be completed.

Several typical ways of doing this are shown in Figures 6.1 through 6.3. When designing roof edge details, the designer should keep several objectives in mind: The siding, which is not installed until after the eaves and rake, should be easy to join to them. The edges of the roof shingles should be supported in such a way that water flowing over them will drip free of the trim and siding below. The eaves must be ventilated to allow free circulation of air beneath the roof sheathing. And provision must be made to drain rainwater and snow melt from the roof without damaging the structure below.

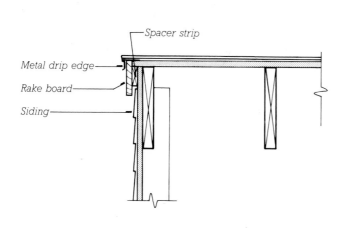
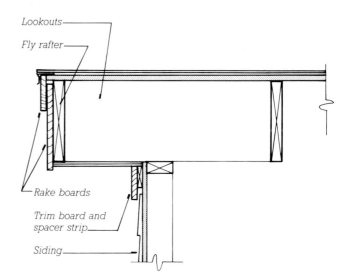

FIGURE 6.1

Two typical details for the rakes (sloping edges) of pitched roofs, the left with no overhang, and the right with an overhang supported on lookouts. A rake overhang is usually of little practical value because the amount of water that flows over a rake edge is inconsequential as compared to that which flows over an eave.

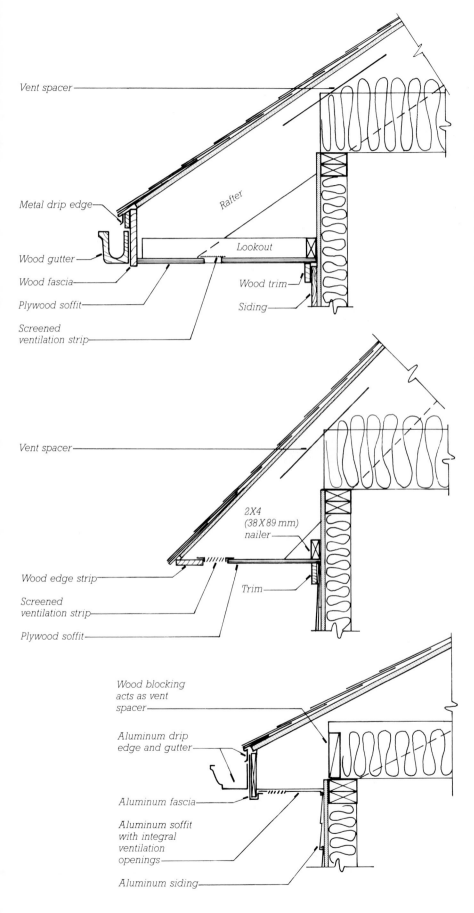

Vent spacer

Metal drip edge

Wood gutter

Wood fascia

Plywood soffit

Screened
ventilation strip

Rafter

Lookout

Wood trim

Siding

Vent spacer

2X4
(38 X 89 mm)
nailer

Trim

Wood edge strip

Screened
ventilation strip

Plywood soffit

Wood blocking
acts as vent
spacer

Aluminum drip
edge and gutter

Aluminum fascia

Aluminum soffit
with integral
ventilation
openings

Aluminum siding

FIGURE 6.2

Three ways, from among many, of finishing the eaves of a light frame building. The top detail has a wood fascia and a wood gutter. The gutter is spaced away from the fascia about $\frac{1}{4}$ inch (6 mm) on wood blocks, to help prevent decay. The width of the overhang can be varied by the designer, and a metal gutter could be substituted for the wood one. The sloping line at the edge of the ceiling insulation indicates a vent spacer as shown in Figure 6.7. The middle detail has no fascia or gutter; it works best for a steeply pitched roof with a sufficient overhang to avoid water damage to the walls below. The bottom detail is finished entirely in aluminum. It shows wood blocking as an alternative to vent spacers for maintaining free ventilation through the attic. Many designers entirely eliminate eave overhangs and gutters from their buildings for the sake of a "clean" appearance, but this is ill advised because it leads to staining, leaking, and premature deterioration of the windows, doors, and walls below.

FIGURE 6.3
A simple wood eave detail with an aluminum gutter and downspout. The wood bevel siding, corner boards, and eave are all painted white. The carpentry at the intersection of the eave and the rake shows some poorly fitted joints and badly finished ends, and the soffit is not ventilated at all, which may lead to the formation of ice dams on the roof. (Photo by the author)

Roof Drainage

Gutters and *downspouts* (downspouts are also called *leaders*) are often installed on a pitched roof to remove rainwater and snow melt without wetting the walls or causing splashing or erosion on the ground below. Gutters built into the roof, once popular with architects, are difficult to construct sufficiently well to prevent water from entering the building when the gutters become clogged with debris. They have been superseded almost completely by external gutters of wood, aluminum, or plastic (Figures 6.2, 6.3). These are sloped toward the points at which downspouts drain away the collected water. Spacings and flow capacities of downspouts are determined by formulas found in the building codes. At the bottom of each downspout, a means must be provided for getting the water away from the building. A simple precast concrete *splash block* can spread the water and direct it away from the foundation, or a system of underground piping can collect the water from all the downspouts and conduct it to a storm sewer, a *dry well* (an underground seepage pit filled with broken rock), or a drainage ditch.

Many codes allow rainwater gutters to be omitted if the overhang of the roof is sufficient, thus avoiding the problems of clogging and ice buildup commonly associated with gutters. Typical minimum overhangs for gutterless buildings are 1 foot (300 mm) for single-story buildings, and twice this for two-story buildings. To prevent soil erosion and mud spatter, the drip line below the eave must be protected with a bed of crushed stone.

Ice Dams and Roof Ventilation

Snow on a pitched roof may be melted by heat escaping through the roof from the heated space below. At the eave, however, the shingles and gutter, which are not directly above heated space, can be very cold, and the snow melt can freeze and begin to build up layers of ice. Soon an *ice dam* forms, and water accumulates above it, seeping between the shingles, through the roof sheathing, and into the building, causing damage to walls and ceilings (Figure 6.4). Ice dams can be largely elim-

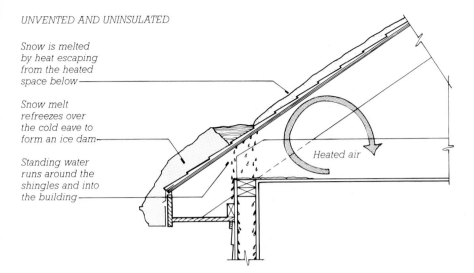

UNVENTED AND UNINSULATED

Snow is melted by heat escaping from the heated space below

Snow melt refreezes over the cold eave to form an ice dam

Standing water runs around the shingles and into the building

Heated air

FIGURE 6.4
Ice dams form because of inadequate insulation, combined with a lack of ventilation, as shown in the top diagram. The lower two diagrams show how insulation, soffit vents, and vent spacers are used to prevent snow from melting on the roof.

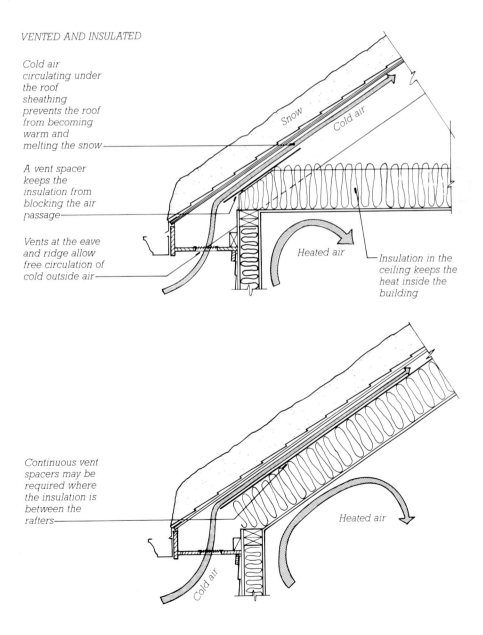

VENTED AND INSULATED

Cold air circulating under the roof sheathing prevents the roof from becoming warm and melting the snow

A vent spacer keeps the insulation from blocking the air passage

Vents at the eave and ridge allow free circulation of cold outside air

Snow

Cold air

Heated air

Insulation in the ceiling keeps the heat inside the building

Continuous vent spacers may be required where the insulation is between the rafters

Heated air

Cold air

FIGURE 6.5

A continuous soffit vent of perforated aluminum permits generous airflow to all the rafter spaces but keeps out insects. Both roof and walls of this building are covered with cedar shingles. (Photo by the author)

FIGURE 6.6

This building has both a louvered gable vent and a continuous ridge vent. The gable vent is of wood and has an insect screen on the inside. The aluminum ridge vent is designed to prevent snow or rain from entering even if blown by the wind. Vertical square-edged shiplap siding is used on this building. The roofing is a two-layered asphalt shingle designed to mimic the rough texture of a wood shake roof. (Photo by the author)

inated by keeping the entire roof cold enough so that snow will not melt. This is done by ventilating the roof continuously with cold outside air through vents at the eave and ridge. In buildings with an attic, the attic itself is ventilated and kept as cold as possible, with insulation in the ceilings below to retain the heat in the building. Where there is no attic, the spaces between the rafters should be ventilated by means of air passages between the insulation and the roof sheathing.

Soffit vents create the required ventilation openings at the eave; these usually take the form of a continuous slot covered either with insect screening or with a perforated aluminum strip made especially for the purpose (Figure 6.5). Ventilation openings at the ridge can be either *gable vents* just below the roof line in each of the end walls of the attic, or a *continuous ridge vent*, a screened cap that covers the ridge of the roof and draws air through gaps in the roof sheathing on either side of the ridge board (Figure 6.6). Building codes establish minimum area requirements for roof vents, based on the floor area of the building. Roof ventilation also keeps a structure cooler in summer by dissipating solar heat that has come through the shingles and roof sheathing.

Roof Shingling

Wood Light Frame buildings can be roofed with any of the systems described in Chapter 13. For reasons of economy and fire resistance, however, asphalt shingles are used in the majority of cases. These are applied either by the carpenters who build the frame, or by a roofing subcontractor. Figures 6.8 and 13.35 through 13.39 show how an asphalt shingle roof is applied. Wood shingle and shake roofs (Figures 13.32 through 13.34), clay and concrete tile roofs (Figure 13.42), and sheet metal roofs (Figures 13.43 through 13.48) are also fairly common. Flat roofing, either built-up or a single-ply membrane, is easily applied to a light frame building (Figures 13.4 through 13.31).

FIGURE 6.7

To maintain clear ventilation passages where thermal insulation materials come between the rafters, wood blocking may be inserted, or vent spacers of plastic foam or wood fiber may be installed as shown here. The positioning of the blocking or vent spacers is shown in Figures 6.2 and 6.4. Vent spacers are generally more economical than blocking and can be installed along the full slope of a roof to maintain free ventilation between the sheathing and the insulation where the interior ceiling finish material is applied directly to the bottom of the rafters. (Courtesy of Poly-Foam, Inc.)

FIGURE 6.8

Applying asphalt shingles to a roof. Shingles and other roof coverings are covered in detail in Chapter 13. (Photo by the author)

Windows

Windows were formerly made on the job site by highly skilled carpenters, but are now produced almost exclusively in factories. Some manufacturers make a range of standard sizes from which the designer can select, while others build windows to order. The rationale for factory production in either case is one of higher efficiency, lower cost, and most importantly, better quality. Windows need to be made to a very high standard of precision if they are to operate easily and maintain a high degree of weathertightness over a period of many years. In cold climates especially, a loosely fitted window with single glass and a frame that is highly conductive of heat will significantly increase heating fuel consumption for a building, cause noticeable discomfort to the people in the building, and create large quantities of condensate to stain and decay finish materials in and around the window.

Types of Windows

Figure 6.9 illustrates in diagrammatic form the window types used most commonly in Wood Light Frame buildings. *Fixed* windows are the least expensive and the least likely to leak air or water because they have no operable components. *Single-hung* and *double- hung* windows have one or two moving *sashes*, which are the frames in which the glass is mounted. The sashes slide up and down in tracks in the frame of the window. In older windows the sashes were held in position by cords and counterweights, but today's double-hung windows rely on a system of springs to counterbalance the weight of the sashes. A *sliding window* is essentially a single-hung window on its side, and shares with single-hung and double-hung windows the advantage that the sashes are always securely held in tracks in the frame. This allows the sashes to be more lightly built than those in *projected windows*,

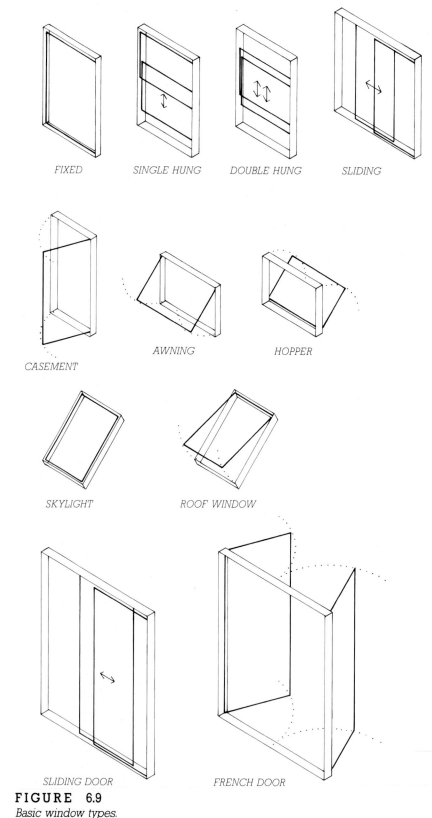

FIXED SINGLE HUNG DOUBLE HUNG SLIDING

CASEMENT AWNING HOPPER

SKYLIGHT ROOF WINDOW

SLIDING DOOR FRENCH DOOR

FIGURE 6.9
Basic window types.

which have sashes that pivot in their frames and must therefore have enough structural stiffness to resist wind loads. Windows with sliding sashes are also easier to make tight against the penetration of wind and water, but none can be opened to more than half its total area.

All projected windows can be opened to their full area. *Casement windows* are helpful in catching passing breezes and inducing ventilation through the building; they are generally narrow in width but can be joined to one another and to sashes of fixed glass to fill wider openings. *Awning windows* can be broad but are not usually very tall. Awning windows have the advantages of protecting an open window from water during a rainstorm and of lending themselves to a building-block approach to the design of window walls (Figure 6.15). *Hopper windows* are rare in residential construction but are more common in commercial work.

Windows in roofs are specially constructed and flashed for watertightness and may be either fixed (*skylights*) or openable (*roof windows*). Large glass doors may slide in tracks or swing open on hinges. The hinged *French door* opens fully and is a more generous type of door than the sliding door, but it cannot be used to regulate airflow through the room unless it is fitted with a plunger-type doorstop that can hold it open in any position.

Materials for Windows

Window Frames

The traditional frame material for windows, and in many ways still the best, is wood, though aluminum, steel, plastics, and combinations of these four materials have also come into widespread use. Wood is a fairly good thermal insulator and is easily worked into sash, but it shrinks and swells with changing moisture content and requires repainting every few years. Aluminum is inexpensive and easy to work as a sash material and requires

Everybody loves window seats, bay windows, and big windows with low sills and comfortable chairs drawn up to them...A room where you feel truly comfortable will always contain some kind of window place.

Christopher Alexander *et al*, in *A Pattern Language*, New York, Oxford University Press, 1977.

---

no periodic repainting. But unless *thermally broken* with plastic or synthetic rubber components to interrupt the flow of heat through the metal, aluminum conducts heat so well that condensation will form, and sometimes even frost, on interior window frame surfaces during cold winter weather. The chief advantage of steel is its strength—steel sash sections are very thin and unobtrusive compared to those of wood and aluminum. Unless permanently coated, steel windows need repainting at intervals. Plastic window frames, usually reinforced with a core of metal, were developed in Europe. At the time of this writing they do not account for a significant number of windows in the North American market. Much more common are frames that add a permanently finished plastic or aluminum outer skin to a wood structural core to combine the best features of wood and the second material.

Glazing and Screening

A number of glazing options for residential windows are shown in Figure 6.10. Single glazing is acceptable only in the mildest climates because of its low resistance to heat flow in winter. Storm windows or double glazing have become the minimum acceptable glazing under most building codes. The storm window has a slight thermal advantage over double glazing mounted

in the movable sash because it acts as a second seal against air leakage around the edges of the sash, which is typically a major source of heat loss. Removable glazing panels and storm windows must be removed periodically and cleaned, which is a nuisance that can be avoided by using sealed double or triple glazing. Sealed triple glazing is often heavy enough to cause problems with the operation of the sash but performs well from a thermal standpoint. Double glazing with an internal coating that is selectively reflective of longer-wave infrared radiation performs at least as well as triple glazing. It is the most satisfactory answer at this time to the problem of high heat loss through windows, though it still transmits heat five to ten times as rapidly as an equal area of well insulated wall.

Large windows that are near enough to the floor that they might be mistaken for open doorways must be glazed with breakage-resistant material under most building codes. Tempered glass is most often used for this purpose, but laminated glass and plastic glazing sheets are also permitted.

Insect screen arrangements vary from one type of window to another. Single-hung, double-hung, and sliding windows generally have removable or retractable screens on the outside. Projected windows must have interior screens. Sliding patio doors have an exterior sliding screen, while French doors require a pair of hinged screen doors on the exterior.

Installing Windows

Some catalog pages for windows are reproduced in Figures 6.11 through 6.16 to give an idea of the information available to the designer in window catalogs. Factory-made windows are extremely easy to install, often requiring only a few minutes per window. Figure 6.17 shows a typical procedure. Interior trimming of a window takes longer, as will be illustrated in the Chapter 7.

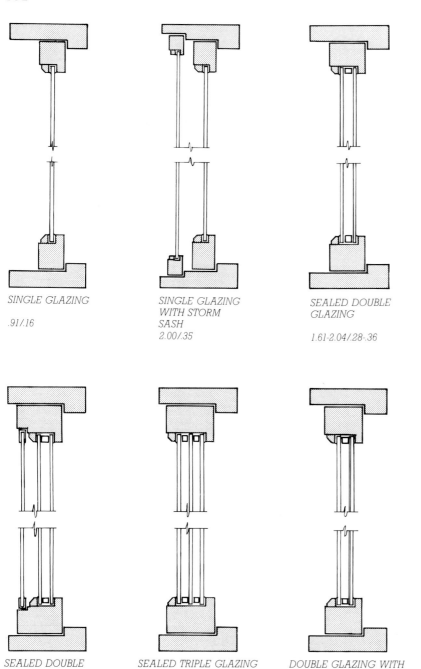

SINGLE GLAZING

.91/.16

SINGLE GLAZING
WITH STORM
SASH

2.00/.35

SEALED DOUBLE
GLAZING

1.61-2.04/.28-.36

SEALED DOUBLE
GLAZING WITH
REMOVABLE TRIPLE
GLAZING

2.56-3.23/.45-.57

SEALED TRIPLE GLAZING

2.56-3.23/.45-.57

DOUBLE GLAZING WITH
REFLECTIVE COATING

2.33-3.13/.41-.55

FIGURE 6.10

Glazing options for windows. A range of thermal resistance values (R) is given below each option to permit comparision. The first pair of numbers given in each case is in English units of hours-square feet-degrees Fahrenheit per Btu, and the second in SI units of square meters-degrees Kelvin per watt.

FIGURE 6.11

Details of a wood double-hung window as shown in a manufacturer's catalog. The left-hand detail is of a single-glazed window with a combination storm/ screen sash on the outside, mounted in a wood frame wall with wood siding. The right-hand detail shows double-glazed sash in a wood frame wall with brick veneer facing. Each of these top details is divided into three parts, which are, from top to bottom, a head detail showing how the top of the window meets the wall, a detail of the meeting rail where the upper and lower sashes come together, and a detail of the horizontal sill at the bottom of the window. The window unit is fundamentally the same in both instances, except for a difference in the size and shape of the wood stop that holds the glass in place. The bottom detail is a plan view of the jambs (vertical sides) of the two window installations and shows the extruded vinyl tracks on which the sashes slide. Not shown are the springs in the hollow spaces of the tracks, which counter-balance the weight of each sash for easier operation. (Courtesy of Marvin Windows, Warroad, Minnesota)

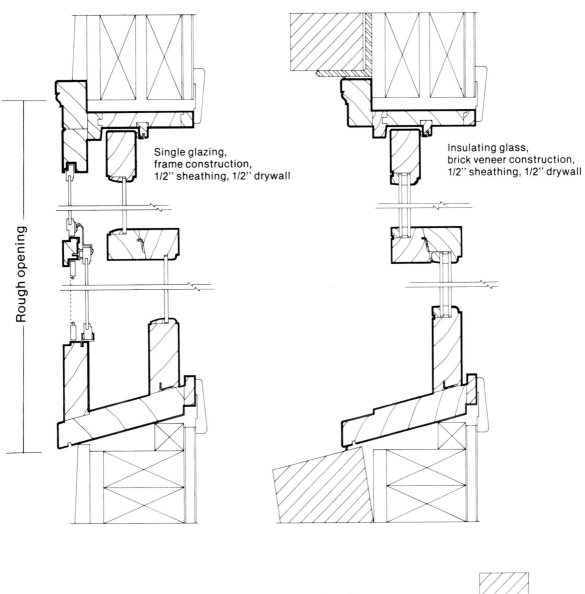

Single glazing,
frame construction,
1/2'' sheathing, 1/2'' drywall

Insulating glass,
brick veneer construction,
1/2'' sheathing, 1/2'' drywall

Rough opening

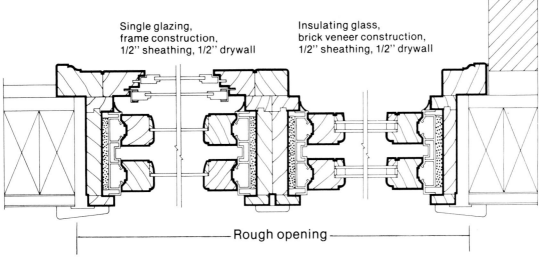

Single glazing,
frame construction,
1/2'' sheathing, 1/2'' drywall

Insulating glass,
brick veneer construction,
1/2'' sheathing, 1/2'' drywall

Rough opening

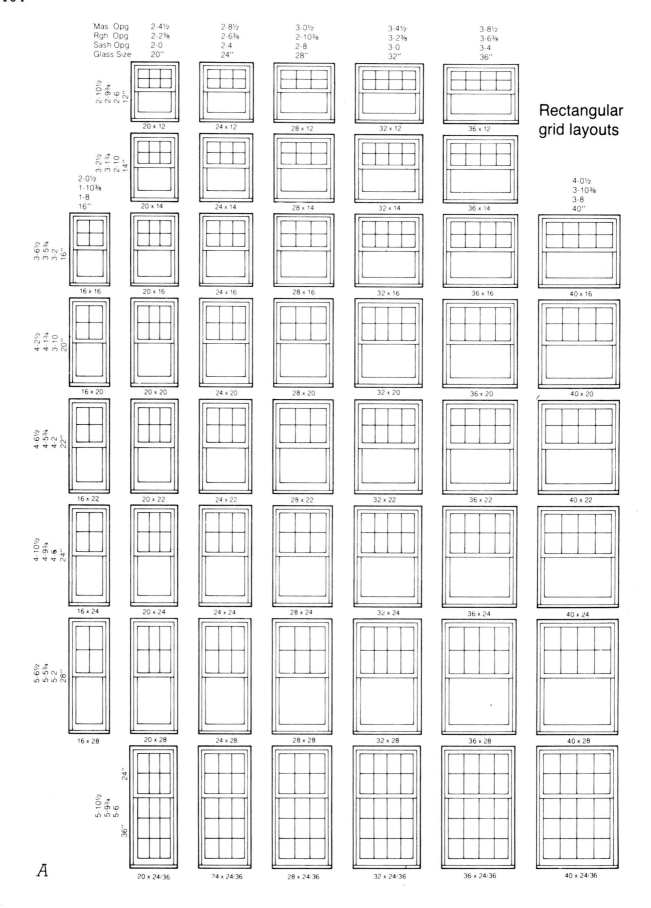

Mas. Opg
Rgh Opg
Sash Opg
Glass Size

Rectangular
grid layouts

A

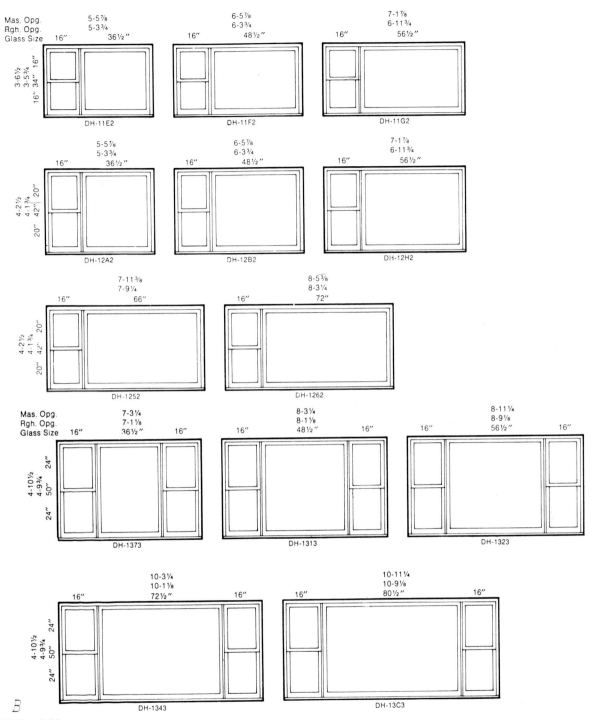

FIGURE 6.12

(a) *The details of each type of window in a manufacturer's catalog are typically accompanied by a chart such as this one, which shows the sizes of windows normally made and stocked. Four exact dimensions are given here for each height and width: the* masonry opening *that must be provided in a wall of brick, concrete block, or stone; the* rough opening *that must be provided in the framing of a wood stud wall; the size of the window unit itself, and the size of the glass. The windows may be ordered with or without the muntins (small wooden bars that divide the glass within a sash).* (b) *Combinations of double-hung windows and fixed windows are also available from stock. (Courtesy of Marvin Windows, Warroad, Minnesota)*

Frame construction
4 9/16 Jamb
½″ sheathing, ½″ drywall

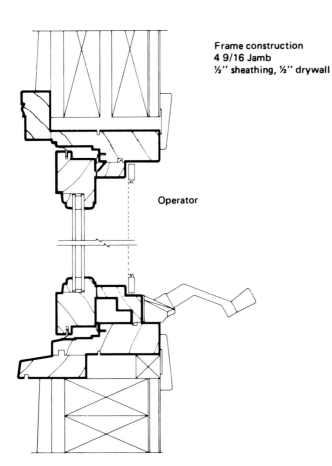

Operator

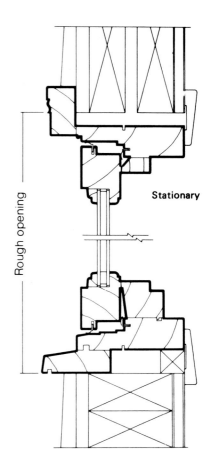

Stationary

Rough opening

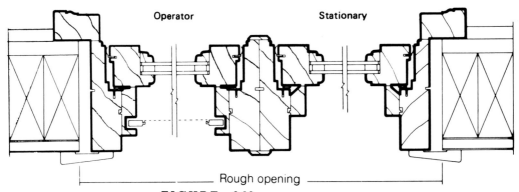

Operator   Stationary

Rough opening

FIGURE 6.13

*Details for wood casement windows,
also taken from a manufacturer's
catalog. The layout of these details
parallels that of Figure 6.11. (Courtesy of
Marvin Windows, Warroad, Minnesota)*

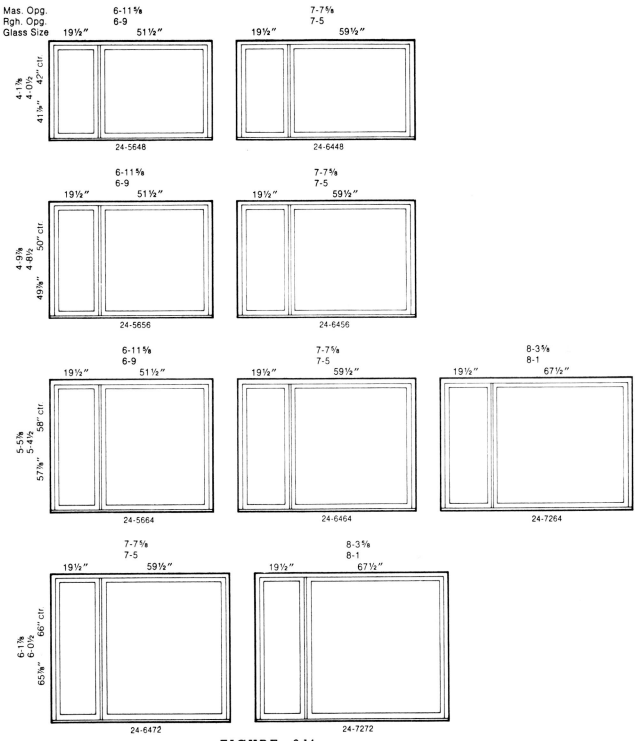

FIGURE 6.14

Casement windows may be combined in multiples to fill larger openings, or, as shown here, may be used with a flanking sash of fixed glass. (Courtesy of Marvin Windows, Warroad, Minnesota)

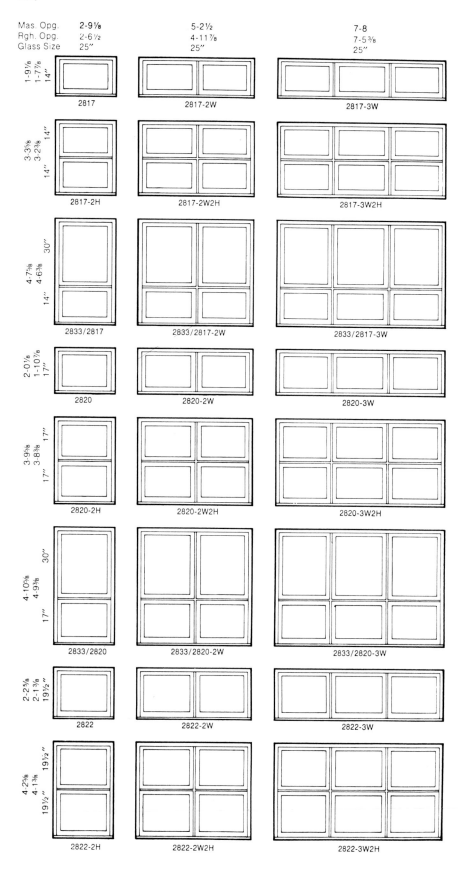

Mas. Opg.   2-9⅛   5-2½   7-8
Rgh. Opg.   2-6½   4-11⅞   7-5⅜
Glass Size   25″   25″   25″

FIGURE 6.15

Awning windows are easily combined with one another or with fixed glazing to form larger window units, up to the size of an entire wall. Details of these units are shown in Figure 6.16. (Courtesy of Marvin Windows, Warroad, Minnesota)

FIGURE 6.16

Details of wood awning windows. Each of the two vertical sections shows, from top to bottom, a head detail, a detail of a horizontal mullion (meeting of two window units), and a sill detail. The horizontal section at the bottom of the page details two jambs and a vertical mullion. (Courtesy of Marvin Windows, Warroad, Minnesota)

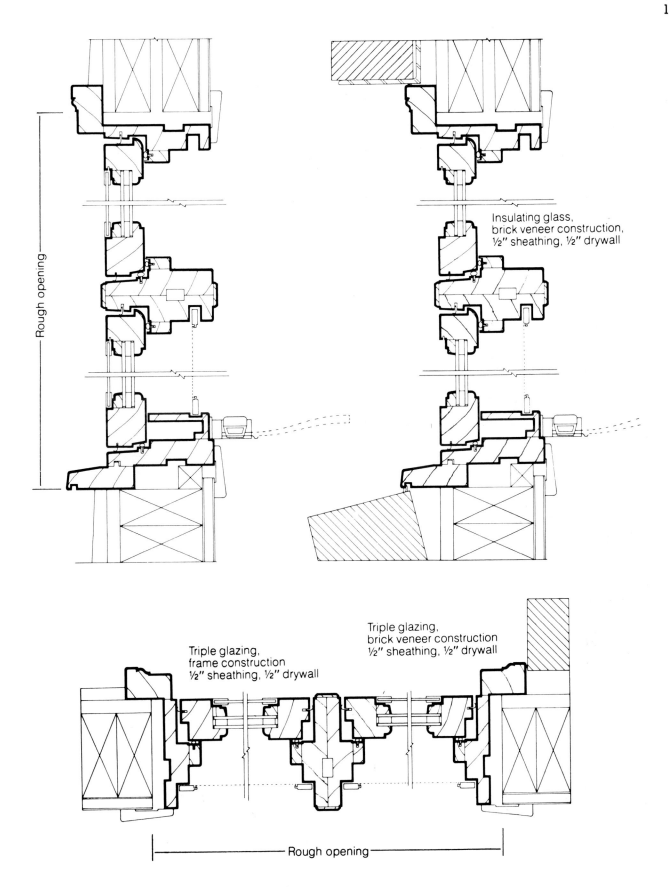

Rough opening

Insulating glass,
brick veneer construction,
½″ sheathing, ½″ drywall

Triple glazing,
frame construction
½″ sheathing, ½″ drywall

Triple glazing,
brick veneer construction
½″ sheathing, ½″ drywall

Rough opening

FIGURE 6.17

Installing a wood double-casement window unit in an existing building. (Installation in new construction is identical, except that the siding is installed after the window). (a) The carpenter checks the dimensions and squareness of the rough opening to be sure the window will fit. Small inaccuracies in the rough opening are usually of no consequence because of the large dimensional tolerance between the window unit and the frame. (b) The window unit, with its factory-applied exterior wood casing flange, is placed in the opening and centered. (c) A single finish nail is driven through a lower corner of the casing and into the wall framing. (d) The sill is checked with a spirit level to see if it is level. If it is not, it will be shimmed temporarily to make it level. (e) When the sill is level, a second nail is driven to maintain the window in a level position. (f) The level is used again to be sure that the jamb of the window unit is plumb. This is important because the unit may have been distorted during shipping or installation and will not operate properly unless it is brought back into square. (g) A nail halfway up the jamb holds the unit square in the opening. (h) Before inserting nails around the remainder of the unit, the jamb is checked again for verticality near the top. (i) Inside the building, wedges are used to support the sill snugly at several points. Most carpenters prefer to use pairs of wedges (usually pieces of wood shingle), laid over one another in opposing directions, to create a flat support at each point. (j) The jambs on both sides are shimmed at top, center, and bottom to keep the unit square. Again, wood shingle wedge pairs, which can easily be adjusted to fill any size gap, are preferable to the plywood blocks shown here. (k) Thermal insulation is stuffed lightly into the gaps between the window unit and the framing to reduce air leakage and heat loss. (l) Sealant is applied between the siding and the frame of the unit to reduce water leakage. If the head of the window is not well protected by a roof overhang, a Z-shaped aluminum flashing should be inserted behind the siding and over the window head to prevent water leakage. The installation is now complete except for the installation of interior casings, which will be shown in the next chapter. Installation procedures for other types of wood windows are similar. (Courtesy of Marvin Windows, Warroad, Minnesota)

W indows do more than let in light and air. The way they are placed in a wall affects our understanding of the whole house...

Charles Moore, Gerald Allen, and Donlyn Lyndon, in *The Place of Houses*, New York, Holt, Rinehart and Winston, 1974.

DOORS

Exterior doors must be well constructed and tightly weatherstripped if they are not to leak air and water. The problem can be compounded by the seasonal expansion and contraction of wood doors. Wood doors are discussed more fully in Chapter 7; properly installed and finished wood panel or solid-core doors are excellent for exterior use (Figures 6.18, 6.19). Wood doors exposed to direct sunlight should not be painted or stained in dark colors, which absorb so much solar heat that the wood in the door dries unevenly and warps.

Pressed sheet metal doors are becoming popular as an alternative to wood exterior doors. Their hollow cores are filled with insulating plastic foam.

FIGURE 6.18

A six-panel wood entrance door with flanking sidelights and a fanlight above. A number of elaborate traditional entrance designs such as this are available from stock for use in light frame buildings. (Courtesy of Morgan Products, Ltd., Oshkosh, Wisconsin)

FIGURE 6.19

Details of an exterior wood door installation. The door opens toward the inside of the building in this detail.

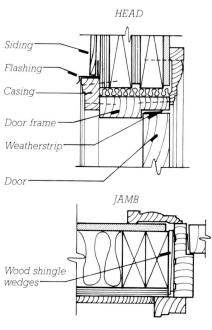

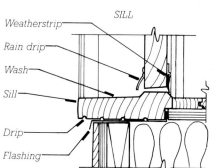

Metal doors do not suffer from moisture expansion and contraction and are often furnished *prehung*, meaning that they are already mounted on hinges in a frame, complete with weatherstripping, ready to install by merely nailing the frame in place. Wood doors can also be purchased prehung, although many are still hung and weatherstripped on the building site. The major disadvantage of metal exterior doors is that they do not have the satisfying appearance, feel, or sound of a wood door.

Glass in or near doors is *tempered* in order to strengthen it to meet build-ing code safety requirements. (Chapter 14 explains the tempering of glass.) Door manufacturers furnish tempered glass in doors as a matter of course, but the designer who specifies locally custom-fabricated doors may have to pay particular attention to see that this requirement is met.

SIDING

The exterior cladding material applied to the walls of a Wood Light Frame building is referred to as *siding*. Many different types of materials are used as siding: wood boards with various profiles, applied either horizontally or vertically; plywood; wood shingles; metal or plastic sidings; brick or stone; and stucco (Figures 6.20 through 6.39). Siding is applied over a paper or felt layer, which acts as an air barrier and backup waterproofing layer. This layer should allow water vapor to pass freely so it does not accumulate in the wall. The traditional material for this layer is asphalt-saturated felt, but perforated aluminum foils and airtight, vapor-permeable papers made of synthetic fibers are increasingly used.

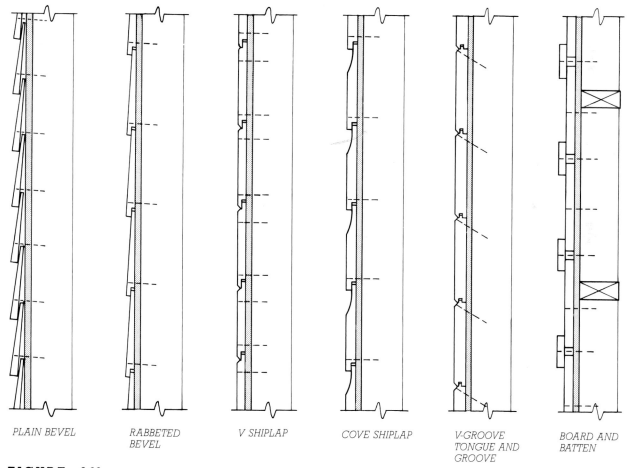

| PLAIN BEVEL | RABBETED BEVEL | V SHIPLAP | COVE SHIPLAP | V-GROOVE TONGUE AND GROOVE | BOARD AND BATTEN |

FIGURE   6.20

Six types of wood siding, from among many. The plain and rabbeted bevel sidings are designed to be applied in a horizontal orientation. Board and batten siding may only be applied vertically.

The other three sidings may be used either vertically or horizontally. The nailing pattern (shown with broken lines) for each type of siding is designed to allow for expansion and contraction of the boards. Nail penetration into the sheathing and framing should be a minimum of 1½ inches (38 mm) for a satisfactory attachment.

FIGURE   6.22
A carpenter applies V-groove tongue and groove redwood siding to an eave soffit, using a pneumatic nail gun. (Courtesy of Senco Products, Inc.)

FIGURE   6.21
Bevel siding on this late eighteenth century house is combined with wood quoins at the corners of the house, and elaborately molded casings and brackets. (Photo by the author)

FIGURE 6.23
A completed installation of tongue and groove siding, vertically applied in the foreground of the picture and diagonally in the area seen behind the chairs. The lighter colored streaks are sapwood in this mixed heartwood/sapwood grade of *redwood. The windows are framed in dark-colored aluminum. (Architect: Zinkhan/Tobey. Photo by Barbeau Engh. Courtesy of California Redwood Association)*

Board Sidings

Horizontally applied board sidings, made either of solid wood or of wood composition board, are usually nailed into the studs, so can be applied directly over insulating sheathing materials without requiring a *nail-base sheathing,* a material such as plywood or waferboard that accepts and holds nails. *Siding nails,* whose heads are intermediate in size between those of common nails and those of finish nails, are used to give the best compromise between holding power and appearance when attaching horizontal siding. Siding nails should be hot-dip galvanized, or made of aluminum or stainless steel, to prevent corrosion staining. Nailing is done in such a way that the individual pieces of siding may expand and contract freely without damage (Figure 6.20). Horizontal sidings are simply butted tightly to corner boards and window and door trim.

Vertically applied sidings are nailed at the top and bottom plates of the frame, and at one or more intermediate horizontal lines of wood blocking installed between the studs.

Heartwood redwood, cypress, and cedar sidings may be left unfinished if desired, to weather to various shades of gray. Other woods must be either stained or painted to prevent weathering and decay.

Plywood Sidings

Plywood siding materials (Figures 6.24 through 6.29) have become popular for their economy. The cost of the material per unit area of wall is usually somewhat less than for other siding materials, and labor costs tend to be relatively low because the large sheets of plywood are more quickly installed than equivalent areas of boards. In many cases the sheathing can be eliminated from the building if plywood is used for siding, leading to further cost savings. Most plywood sidings must be painted or stained, but some are made of decay-resistant heartwoods and can be left to weather without surface treatment. The most popular plywood sidings are grooved to imitate board sidings and conceal the vertical joints between sheets.

The largest problem in using plywood sidings for multistory buildings is how to detail the horizontal end joints between sheets. A Z-flashing of aluminum (Figure 6.25) is the usual solution, but like other solutions to the problem, it is clearly visible. The designer should include in the construction drawings a sheet layout for the plywood siding that organizes the horizontal joints in an acceptable manner. This will help avoid a random, unattractive pattern of joints on the face of the building.

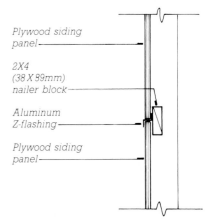

Plywood siding panel

2X4 (38 X 89mm) nailer block

Aluminum Z-flashing

Plywood siding panel

FIGURE   6.25
A detail of a simple Z-flashing, the device most commonly used to prevent water penetration at horizontal joints in plywood siding.

FIGURE   6.24
Grooved plywood siding is used vertically on this commercial building by Roger Scott Group, Architects. The horizontal metal flashings between sheets of plywood are purposely emphasized here with a special flashing detail that casts a dark shadow line.
(Courtesy of American Plywood Association)

FIGURE 6.26
Grooved plywood used in a horizontal orientation. The roof of this multifamily dwelling is made of cedar shakes. (Architect: Robert T. Steinberg. Photo by Karl Riek. Courtesy of American Plywood Association)

FIGURE 6.27
Townhouses of plywood siding and aluminum windows by architects Atkinson/Karius. The Z-flashings are clearly visible between the vertical sheets of plywood. (Photo by Karl Riek. Courtesy of American Plywood Association)

FIGURE 6.28
Dark-stained plywood and battens on a commercial building. (Architects: Hansen/Murakami/Eshima, Inc. Courtesy of American Plywood Association)

FIGURE 6.29
Smooth sheets of stained plywood are used here, beneath a pediment of horizontal siding and square-edged boards, for the formal front of a small museum. (Architect: Darrel Rippeteau. Photo by Harlan Hambright. Courtesy of American Wood Council)

Shingle Sidings

Wood shingles and shakes (Figures 6.30 through 6.35) require a nail-base sheathing material for installation. Either corrosion-resistant box nails or air-driven staples may be used for attachment. Most shingles are of cedar or redwood heartwood and need not be coated with paint or stain unless such a coating is desired for cosmetic reasons.

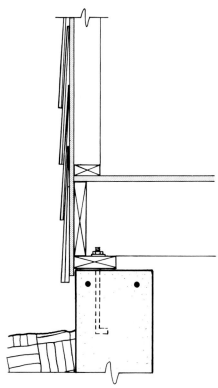

FIGURE 6.31

A detail of wood shingle siding at the sill of a wood platform frame building. The first course of shingles projects below the sheathing to form a drip, and is doubled so all the vertical joints between the shingles of the outside layer are backed up by the undercourse of shingles. Succeeding courses are single, but are laid so each course covers the open joints in the course below.

FIGURE 6.30

Applying wood shingles as siding, over asphalt-saturated felt building paper. (Courtesy of Red Cedar Shingle and Handsplit Shake Bureau)

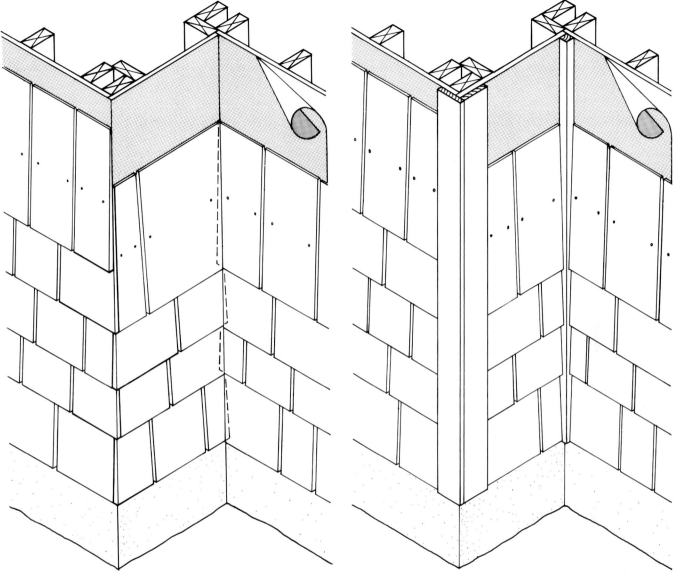

FIGURE 6.32
Wood shingles can be woven at the corners to avoid corner boards. Each corner shingle must be carefully trimmed to the proper line with a block plane, which is time consuming and relatively expensive, but the result is a more continuous, sculptural quality in the siding.

FIGURE 6.33
Corner boards save time when shingling walls, and become a strong visual feature of the building. Notice in this and the preceding diagram how the joints and nailheads in each course of shingles are covered by the course above.

FIGURE 6.34
Fancy cut wood shingles were often a featured aspect of shingle siding in the late nineteenth century. Notice the fish-scale shingles in the gable end, the serrated shingles at the lower edges of walls, and the sloping double shingle course along the rakes. (Photo by the author)

FIGURE 6.35
Fancy cut wood shingles stained in contrasting colors are used here on a contemporary restaurant. (Courtesy of Shakertown Corporation)

Metal and Plastic Sidings

Painted wood sidings are a constant maintenance problem, and are apt to decay unless they are carefully repainted every 3 to 6 years. Sidings formed of prefinished sheet aluminum or molded of vinyl plastic, usually designed to imitate wood sidings, will not decay and are generally guaranteed against needing repainting for long periods, typically 20 years (Figures 6.36, 6.37). Such sidings do, however, have their own problems, including the poor resistance of aluminum sidings to denting, and the tendency of plastic sidings to shatter on impact, especially in cold weather.

Stucco

Stucco, a portland cement plaster, is a strong, durable, economical, fire-resistant material for siding light frame buildings. Figure 6.38 shows the application of exterior stucco with a spray apparatus. Stucco is discussed in greater detail in Chapter 17.

FIGURE 6.37
Installing aluminum siding over insulating foam sheathing on an existing residence. Special aluminum pieces are provided for corner boards and window casings; each has a shallow edge channel to accept the cut ends of the siding. (Courtesy of Dow Chemical Company)

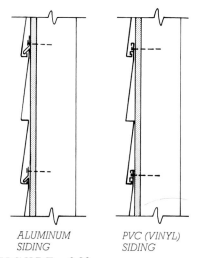

ALUMINUM SIDING   *PVC (VINYL) SIDING*

FIGURE 6.36
Aluminum and vinyl sidings are both intended to imitate horizontal lap siding in wood. Their chief advantage in either case is low maintenance. Aluminum siding must be grounded electrically to eliminate any hazard of accidental shock, and can cause interference with radio and television reception.

FIGURE 6.38
Applying exterior stucco over wire netting. The workers to the right hold a hose that sprays the stucco mixture onto the wall, while the man at the left levels the surface of the stucco with a straightedge. The small rectangular opening at the base of the wall is a crawlspace vent. (Courtesy of Keystone Steel and Wire Company)

Masonry Veneer

Light frame buildings can be faced with a single wythe of brick or stone in the manner shown in Figure 6.39. The corrugated metal ties support the masonry against falling away from the building, but allow for differential movement between the masonry and the frame. Masonry materials and detailing are covered in Chapter 8.

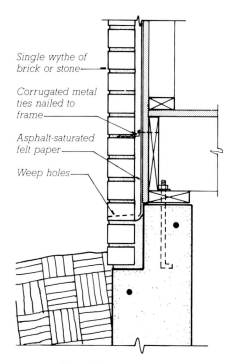

Single wythe of brick or stone

Corrugated metal ties nailed to frame

Asphalt-saturated felt paper

Weep holes

FIGURE 6.39

A detail of masonry veneer facing for a platform frame building. The weep holes drain any moisture that might collect in the cavity between the masonry and the sheathing.

FIGURE 6.40

Careful detailing is evident in every aspect of the exterior finishes of this commercial building. Notice especially the window casings, the purposeful use of both vertical boards and wood shingles for siding, and the neat junction between the sidewall shingles and the rake boards. (Architects: Woo and Williams. Photo by Richard Bonarrigo. Courtesy of the architects)

EXTERIOR CONSTRUCTION

Wood is often used outdoors for porches, decks, stairs, stoops, and retaining walls (Figure 6.41). Decay-resistant heartwoods and pressure-treated wood are suitable for these exposed uses. If nondurable woods are used, they will soon decay at the joints, where water is trapped and held by capillary action. Wood decking exposed to the weather should always be square-edged material with open, spaced joints rather than tongue-and-groove to allow for drainage of water through the deck and for expansion and contraction of the decking.

FIGURE 6.41
An exterior deck of redwood.
(Designer: John Matthias. Photo by Ernest Braun. Courtesy of California Redwood Association)

Exterior Painting, Finish Grading, and Landscaping

The final steps in finishing the exterior of a light frame building are painting or staining of exposed wood surfaces, finish grading of the ground around the building, installation of pavings for drives, walkways, and terraces, and seeding and planting of landscape materials. By the time these operations take place, interior finishing operations are well underway, having begun as soon as the roofing, sheathing, and windows and doors were in place.

C.S.I./C.S.C. Masterformat Section Numbers for Exterior Finishes for Light Frame Construction in Wood	
07300	SHINGLES AND ROOFING TILES
07400	PREFORMED ROOFING AND CLADDING/SIDING
07460	Cladding/Siding
08200	WOOD AND PLASTIC DOORS
08500	METAL WINDOWS
08510	Steel Windows
08520	Aluminum Windows
08560	Metal Storm Windows
08600	WOOD AND PLASTIC WINDOWS
08610	Wood Windows
08630	Plastic Windows

SELECTED REFERENCES

In addition to consulting the reference works previously listed in Chapter 5, the reader should acquire current catalogs from a number of manufacturers of wood and aluminum windows. Most lumber retailers also distribute unified catalogs of windows, doors, and millwork that can be an invaluable part of the designer's reference shelf.

KEY TERMS AND CONCEPTS

eave
rake
gutter
downspout or leader
splash block
dry well
ice dam
soffit vent
gable vent
continuous ridge vent
vent spacer
shingles
sash
fixed window
single-hung window
double-hung window
sliding window

projected window
casement window
awning window
hopper window
skylight
roof window
sliding door
French door
thermally broken
glazing
double glazing
triple glazing
masonry opening
rough opening
head
jamb
sill

mullion
muntin
storm window
insect screen
removable glazing panel
solid-core door
prehung door
nail-base sheathing
siding
siding nail
bevel siding
shiplap
tongue and groove
stucco
masonry veneer

REVIEW QUESTIONS

1. In what order are exterior finishing operations carried out on a platform-frame building, and why?

2. What are the relative advantages and disadvantages of wood windows and aluminum windows? How do these advantages and disadvantages change if the wood window is clad in plastic, and the aluminum window is thermally broken?

3. What are the basic types of windows, classified by mode of operation? What are some reasons that each type might be chosen by a designer?

4. Which types of siding require a nail-base sheathing? What are some nail-base sheathings?

5. What are the reasons for the relative economy of plywood sidings?

EXERCISES

1. For each of several completed wood frame buildings, make a list of the materials used for exterior finishes and sketch a set of details of the eaves, rakes, corners, and windows. Are there ways in which each could be improved?

2. Visit a lumber yard and look at all the alternative choices of sidings, windows, trim lumber, and roofing. Study one or more systems of gutters and downspouts. Look at eave vents, gable vents, and ridge vents. See what is available for exterior doors.

3. For a Wood Light Frame building of your design, list precisely the materials you would like to use for the exterior finishes. Sketch a set of typical details of how these finishes should be applied to achieve the appearance you desire, with special attention to the roof edge details.

## CHAPTER SEVEN

INTERIOR FINISHES FOR WOOD LIGHT FRAME CONSTRUCTION

———

———

In this Connecticut house architect Richard Meier produced uncommon results from the most common, least expensive interior wall finish material: gypsum board. *(Photo Ezra Stoller © 1971 ESTO)*

As the exterior roofing and siding of a platform frame building approach completion, the framing carpenters and roofers are joined by workers from a number of other building trades. Masons commence work on fireplaces and chimneys (Figures 7.1, 7.2). Plumbers begin *roughing in* their piping ("roughing in" refers to the process of installing the components of a system that will not be visible in the finished building): first the large *DWV (drain-waste-vent)* pipes, which drain by gravity and must therefore have right of way, then the small *supply* pipes, which bring hot and cold water to the fixtures, and the gas piping (Figures 7.3, 7.4). If the building is to have a *hydronic* (forced hot water) heating system, the plumbers put in the boiler and rough in the heating pipes and convectors at this time. If the building will have central warm air heating and/or air conditioning, sheet metal workers install the furnace and ductwork (Figures 7.5, 7.6). The electricians are the last of the mechanical and electrical trades to complete their roughing in because their wires are flexible and can generally be routed around the pipes and ducts without difficulty (Figure 7.7). When the plumbers, sheet metal workers, and electricians have completed their rough work, which consists of everything except the plumbing fixtures, electrical outlets, and air registers and grills, inspectors from the municipal building department check each of the systems for compliance with the plumbing, electrical, and sheet metal codes. Connections are then made to external sources of water, gas, electricity, and communications services, and to a means of sewage disposal.

Once these inspections have been passed, thermal insulation and a vapor retarder are added to the exterior ceilings and walls. At this point the general building inspector of the municipality is called in to examine the framing, exterior finish, and insulation of the building. Any deficiencies must be made good before the project is allowed to proceed further.

Now a new phase of construction begins, the interior finishing operations, during which the inside of the building undergoes a succession of radical transformations. The elaborate tangle of framing

members, ducts, pipes, wires, and insulation rapidly disappears behind the finish wall and ceiling materials. The interior *millwork*—doors, finish stairs, railings, cabinets, shelves, closet interiors, and door and window casings—is installed by a crew of *finish carpenters* who specialize in this exacting work. The finish flooring materials are installed as late in the process as possible to save them from damage by the passing armies of workers, and the finish carpenters follow behind the flooring installers to add the baseboards that cover the last of the rough edges in the construction.

Finally, the building hosts the painters who prime, paint, stain, varnish, and paper its interior surfaces. The plumbers, electricians, and sheet metal workers make brief return appearances on the heels of the painters to install the plumbing fixtures, the electrical receptacles, switches, and lighting fixtures, and the air grills and registers. At last, following a final round of inspections, and a last-minute round of repairs and corrections to remedy lingering defects, the building is ready for occupancy.

FIGURE **7.1** *(left)*
Insulated metal flue systems are often more economical than masonry chimneys for furnaces, boilers, water heaters, package fireplaces, and solid fuel stoves, and can be installed by carpenters. (Courtesy of Selkirk Metalbestos)

FIGURE **7.2**
A mason adds a section of clay flue liner to a chimney. The large flue is for a fireplace, and the three smaller flues are for a furnace and two woodburning stoves. (Photo by the author)

FIGURE **7.3**
At the basement ceiling, the plastic waste pipes are installed first to be sure that they are properly sloped to drain. The copper supply pipes for hot and cold water are installed next. The insulated duct seen at the lower right is for heating and air conditioning. (Photo by the author)

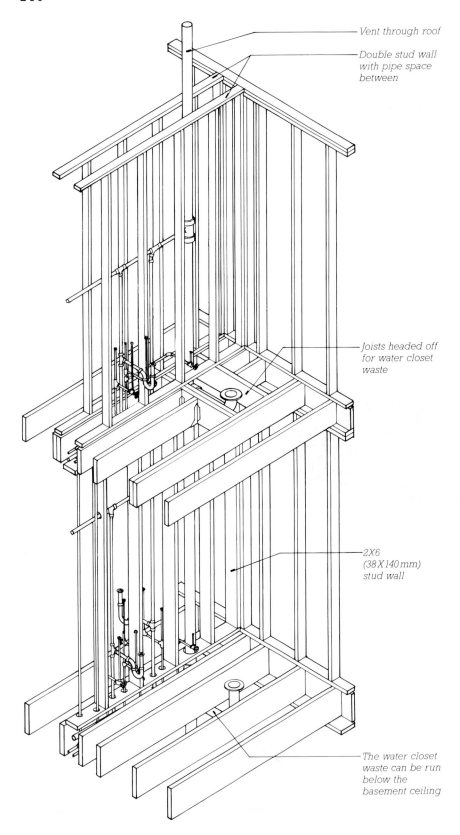

Vent through roof

Double stud wall with pipe space between

Joists headed off for water closet waste

2X6 (38 X 140 mm) stud wall

The water closet waste can be run below the basement ceiling

FIGURE 7.4

The plumber's work is easier and less expensive if the building is designed to accommodate the piping. The "stacked" arrangement shown here, in which a second-floor bathroom and a back-to-back kitchen and bath on the first floor share the same vertical runs of pipe, is economical and easy to rough in, as compared with plumbing that does not align vertically from one floor to the next. The double wall framing on the second floor allows plenty of space for the waste, vent, and supply pipes. The second floor joists are located to provide a slot through which the pipes can pass at the base of the double wall, and the joists beneath the water closet (toilet) are headed off to house its waste pipe. The first floor shows an alternative type of wall framing using a single layer of deeper studs, which must be drilled to permit the pipes to pass through.

FIGURE 7.5

The installation of this warm air furnace and air conditioning unit is almost complete. The metal pipe running diagonally to the left carries the exhaust gases from the oil burner to the masonry chimney. The ductwork is insulated to prevent moisture from condensing on it during the cooling season and to prevent excessive losses of energy from the ducts. (Photo by the author)

FIGURE 7.6
Ductwork and electrical wiring are installed conveniently through the openings in these floor trusses, making it easy to apply a finish ceiling if desired. (Courtesy of Trus Joist Corporation)

FIGURE 7.7
The electrician begins work by nailing the plastic boxes to the framing of the building in the locations shown on the electrical plan. Then holes are drilled through the framing, and the plastic-sheathed cable, which houses two insulated copper conductors and a bare ground wire, is pulled through the holes and into the boxes, where it is held by insulated staples driven into the wood. After the interior wall materials are in place, the electrician returns to connect outlets, lighting fixtures, and switches to the wires and the boxes and to affix cover plates to finish the installation. (Photo by the author)

Thermal insulation and vapor retarder

Thermal insulation helps keep a building cooler in summer and warmer in winter by retarding the passage of heat through the exterior surfaces of the building. It helps keep the occupants of the building more comfortable by raising the temperatures of the interior surfaces of the building and reducing convective drafts, and it reduces the energy consumption of the building for heating and cooling to a fraction of what it would be without insulation.

Thermal Insulating Materials

Figure 7.8 lists the most important types of thermal insulating materials for Wood Light Frame buildings, and gives some of their characteristics. Glass fiber batts are the most popular type of insulation for wall cavities in new construction. They are also widely used as attic and roof insulation. Some details of their installation are shown in Figure 7.9.

Increasing Levels of Thermal Insulation

Until a few years ago, the maximum resistance to heat flow (*R-value*) that could be achieved in a platform frame building was limited by the thickness of the studs and ceiling joists. With the rapid rise in energy costs in the mid-1970s, however, economic analyses indicated the strong desirability of insulating buildings to R-values well beyond those that could be reached by merely filling the available voids in the framing. At about the same time, building codes began to specify minimum insulating values for walls, ceilings, windows, and basement walls. These developments led to experimentation with thicker walls, insulating sheathing materials, and various schemes for insulating the concrete and masonry walls of basements. Figures 7.10 and 7.11 show some of the results of this experimentation. Insulating sheathing materials and 2×6 (38×140 mm) studs have become widely accepted in the building industry in North America, as have exterior and interior insulation for basements. Treated wood foundations (Figures 5.10, 5.11), which are easily insulated to high R-values, are gaining increasing acceptance. The other schemes illustrated here for adding extra insulation to platform frame buildings are not yet standard practice but are representative of current experimentation by designers and build-

Type	Material(s)	Method of Installation	R-value[a]	Combustibility	Advantages and Disadvantages
Batt or blanket	Glass wool; rock wool	The batt or blanket is installed between framing members and is held in place either by friction or by a facing stapled to the framing	3.3 23	The glass wool or rock wool is incombustible, but paper facings are combustible	Low in cost, fairly high R-value, easy to install
Loose fill	Glass wool; rock wool	The fill is blown onto attic floors, and into wall cavities through holes drilled in the siding	2.5–3.5 17–24	Incombustible	Good for retrofit insulation in older buildings. May settle somewhat in walls
Loose fibers with binder	Treated cellulose; glass wool	As the fill material is blown from a nozzle, a light spray of water activates a binder that adheres the insulation in place and prevents settlement	3.1–4.0 22–27	Incombustible	Low in cost, fairly high R-value
Foamed in place	Polyurethane and newer formulations	The foam is mixed from two components and sprayed or injected into place, where it adheres to the surrounding surfaces	5–7 27–49	Combustible in varying degrees, gives off toxic gases when burned	High R-value, high cost, good for structures that are hard to insulate by conventional means

(continued)

ing contractors in the colder regions of the United States and Canada. In the milder climates, the economic amounts of thermal insulation still fit easily in walls framed with ordinary 2 × 4s (38 × 89 mm).

Vapor Retarders

Vapor retarders (often called, less accurately, *vapor barriers*) have also received increased attention during this period of renewed interest in fuel economy. A vapor retarder is a membrane of metal foil, plastic, or treated paper placed on the warm side of thermal insulation to prevent water vapor from entering the insulation and condensing into liquid. The function of a vapor retarder is explained in detail on pages 517–519. Its role increases in importance as thermal insulation levels increase, and both designers and

contractors are becoming more concerned with installing high-quality vapor retarders.

Many batt insulation materials are furnished with a vapor retarder layer of treated paper or aluminum foil already attached. But most designers and builders in cold climates prefer to use unfaced batts and to apply a separate vapor retarder of polyethylene sheet, because a vapor retarder attached to batts has a seam at each stud that can leak significant quantities of air and vapor, while the separately applied sheet has few seams.

Air Infiltration and Ventilation

Much attention is given to reducing air infiltration between the indoors and outdoors in a residential building because this leakage often accounts for

the major portion of the fuel burned to heat the building in winter. This has led to higher standards of window and door construction, increased use of special vapor-permeable air barrier papers under the exterior siding, and greater care in sealing around electrical outlets and other penetrations of the exterior wall. This trend, coupled with the use of continuous vapor retarders inside the walls and ceilings, has sometimes resulted in houses and apartments with so little exchange of air with the outdoors that indoor moisture, odors, and chemical pollutants can build up to intolerable and often unhealthful levels. Opening a window to ventilate the dwelling in cold

FIGURE 7.8
Thermal insulation materials commonly used in light frame wood buildings. The R-values offer a direct means of comparing the relative effectiveness of the different types.

Type	Material(s)	Method of Installation	R-value[a]	Combustibility	Advantages and Disadvantages
Foamed in place	Urea formaldehyde	The foam is mixed from two components and sprayed or injected into place, where it adheres to the surrounding surfaces	3–5 *21–35*	Resistant to ignition, combustible	Low in cost, fairly high R-value. Gives off formaldehyde fumes if improperly formulated. Illegal in many areas
Rigid board	Polystyrene foam	The boards are applied over the wall framing members, either as sheathing on the exterior, or as a layer beneath the interior finish material	4–5 *27–35*	Combustible but self-extinguishing in most formulations	High R-value, can be used in contact with earth, moderate cost
Rigid board	Polyurethane foam	same	6.25 *43*	Combustible, gives off toxic gases when burned	Very high R-value, high cost
Rigid board	Newer formulations	same	4–7 *27–49*	Less combustible, less toxic combustion products	Very high R-value, high cost, safer against fire
Rigid board	Glass fiber	same	3.5 *24*	Incombustible	Moderate cost, vapor permeable
Rigid board	Cane fiber	same	2.5 *17*	Combustible	Low cost

[a]R-values are expressed first in units of hr-ft²-°F/Btu-in., then, in italics, in m-°K/W.

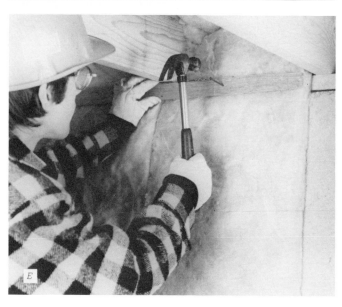

FIGURE 7.9

(a) *Installing a polyethylene vapor retarder over glass fiber batt insulation using a staple hammer, which drives a staple each time it strikes a solid surface. The batts are unfaced and stay in place between the studs by friction.* (b) *Stapling faced batts between roof trusses. Notice that the R value of the insulation is printed on the facing.* (c) *Placing unfaced batts between ceiling joists in an existing attic. Vent spacers should be used at the eaves (see Figure 6.7).* (d) *Insulating a floor over a*

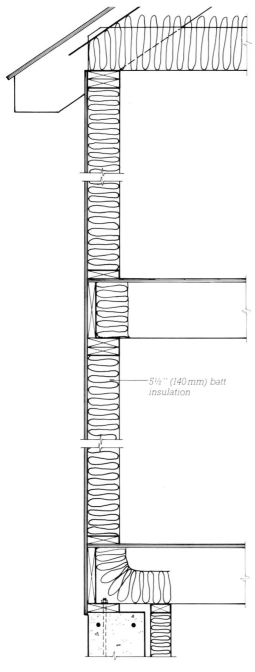

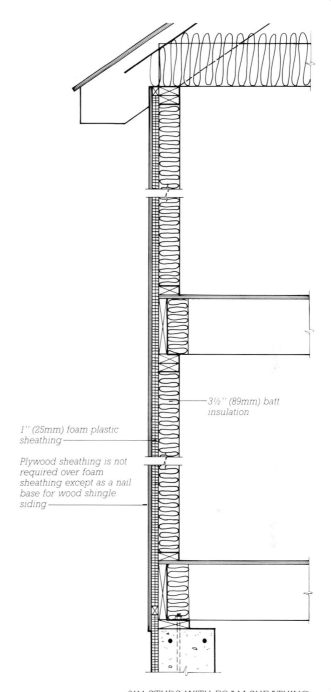

—5½″ (140mm) batt insulation

—3½″ (89mm) batt insulation

1″ (25mm) foam plastic sheathing——

Plywood sheathing is not required over foam sheathing except as a nail base for wood shingle siding——

2X6 STUDS

2X4 STUDS WITH FOAM SHEATHING

FIGURE 7.10

Insulation levels in walls of light frame buildings can be increased from the R12 (83 in SI units) of a normal stud wall to R19 (132) using either 2 × 6 framing and thicker batt insulation (left) or 2 × 4 framing with plastic foam sheathing in combination with batt insulation (right). The foam sheathing insulates the wood framing members as well as the cavities between, but can complicate the process of installing some types of siding.

crawlspace. Batts in this type of installation are usually retained in place by pieces of stiff wire cut slightly longer than the distance between joists and sprung into place at frequent intervals below the insulation. (e) Insulating crawlspace walls with batts of insulation suspended from the sill. The header space between the joist ends has already been insulated. (See also Figure 5.6.) (Photos courtesy of Owens-Corning Fiberglas Corporation)

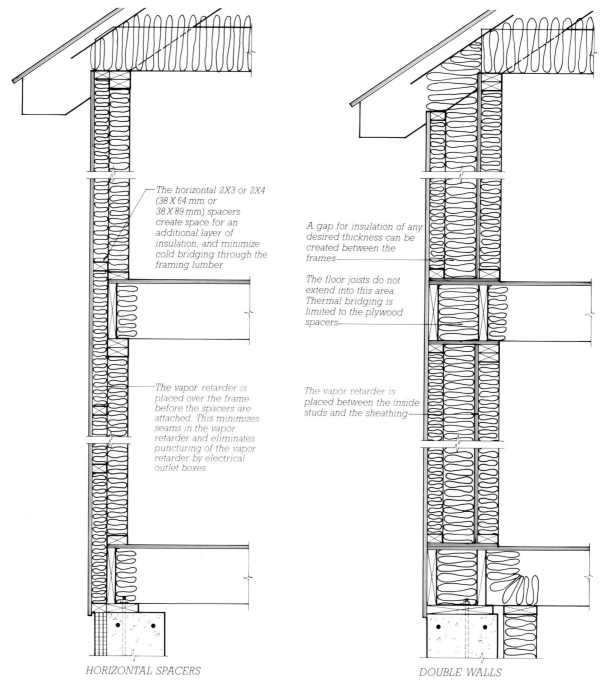

The horizontal 2X3 or 2X4 (38 X 64 mm or 38 X 89 mm) spacers create space for an additional layer of insulation, and minimize cold bridging through the framing lumber

The vapor retarder is placed over the frame before the spacers are attached. This minimizes seams in the vapor retarder and eliminates puncturing of the vapor retarder by electrical outlet boxes

A gap for insulation of any desired thickness can be created between the frames

The floor joists do not extend into this area. Thermal bridging is limited to the plywood spacers

The vapor retarder is placed between the inside studs and the sheathing

HORIZONTAL SPACERS

DOUBLE WALLS

FIGURE 7.11
With the two framing methods shown here, walls can be insulated to any desired level of thermal resistance.

weather, of course, wastes heating fuel, so many designers and builders install forced ventilation systems that employ *air-to-air heat exchangers*, devices that recover much of the heat from the air exhausted from the building and add it to the outside air being drawn in. Nonresidential buildings of Wood Light Frame construction generally require higher rates of ventilation than residential structures, and here, too, low-infiltration construction and air-to-air heat exchangers are of value in reducing heating and cooling costs.

WALL AND CEILING FINISH

Plaster-type finishes have always been the most popular for walls and ceilings in wood frame buildings. Their advantages include substantially lower installed costs than other types of finishes, adaptability to either painting or wallpapering, and, most importantly, a degree of fire resistance that offers considerable protection to the combustible frame. Three-coat *plaster* on wood strip *lath* was the prevalent wall and ceiling finish material until the Second World War, when *gypsum board* came into increasing use because of its lower cost, more rapid installation, and utilization of less skilled labor. More recently, *veneer plaster* systems have been developed that offer surfaces of a quality superior to that of gypsum board at comparable prices.

Plaster, veneer plaster, and gypsum board finishes are presented in detail in Chapter 17. Gypsum board remains the favored material for carpenters who do all the interior finish work in a building themselves because the skills and tools it requires are largely those of the carpenter. In geographic areas where there are plenty of skilled plasterers, veneer plaster captures a substantial share of the market. Almost everywhere in North America there are subcontractors who specialize in gypsum board installation and finishing and who are able to finish the interior surfaces of larger projects, such as apartments, retail stores, and rental office buildings, at highly competitive prices.

In most small buildings, all wall and ceiling surfaces are covered with plaster or gypsum board. Even wood paneling is best applied over a gypsum board backup layer for increased fire resistance. In buildings that require fire walls between dwelling units, or fire separation walls between different uses, a gypsum board wall of the required degree of fire resistance can be installed, eliminating the need to call in masons to put up a wall of brick or concrete masonry (Figure 17.34).

MILLWORK AND FINISH CARPENTRY

Millwork (so named because it originates in a planing and molding mill) includes all the wood interior finish components of a building. Millwork is produced from much higher quality wood than that used for framing: The softwoods used are those with fine, uniform grain structure and few defects. Flooring, stair treads, and millwork intended for transparent finish coatings such as varnish or shellac are customarily made of hardwoods and hardwood plywoods. Millwork is manufactured and delivered to the building site at a very low moisture content, typically in the range of 10 percent, so it is important to protect it from moisture and high humidity before and during installation to avoid swelling and distortions.

The humidity within the building is high at the conclusion of plastering or gypsum board work. The framing lumber and plywood, concrete work, and masonry mortar are still diffusing excess moisture into the interior air, and the wall finishing materials are giving off still more moisture. As much of this moisture as possible should be ventilated to the outdoors before *finish carpentry* (the installation of millwork) commences. Windows should be left open for a few days, and in cool or damp weather the building's heating system should be turned on to raise the interior air temperature and help drive off excess water.

Interior Doors

Figure 7.12 illustrates five doors that fall into three general categories: *Z-brace*, *panel*, and *flush*. Z-brace doors, built on-site, are used infrequently because they are subject to distortions and large amounts of moisture expansion and contraction in the broad surface of boards whose grain runs perpendicular to the width of the door. Panel doors were developed centuries ago to minimize dimensional changes and distortions caused by seasonal changes in the moisture content of the wood. They are widely available in ready-made form from millwork dealers. Flush doors may be either *solid core* or *hollow core*. Solid core doors consist of two veneered faces glued to a core of wood blocks or bonded wood chips (Figure 7.13). They are much heavier, stronger, and more resistant to the passage of sound than hollow core doors, and are also more expensive. In residential buildings their use is usually confined to entrance doors, but they are frequently installed throughout commercial and institutional buildings, where doors are subject to greater abuse. Hollow core doors have an interior grid of wood or paperboard spacers to which the veneers are bonded. Flush doors of either type are available in a variety of veneer species, the least expensive of which are intended to be painted.

For speed and economy of installation, most interior doors are furnished *prehung*, meaning that they have been hinged and fitted to frames at the mill. The carpenter on the site merely tilts the prehung door and frame unit up into the rough opening, plumbs

PROPORTIONING FIREPLACES

Ever since fireplaces were first developed in the Middle Ages, people have sought formulas for their construction that would ensure that the smoke from the fire would go up the chimney and the heat into the room, rather than the other way around, which is all too often the case. To this day, there is little scientific information on how fireplaces work and how best to design them, except for some measurements taken from fireplaces that seem to work reasonably well. These measurements have been arranged into tables of dimensions that enable designers to reproduce the critical features of these fireplaces as closely as possible.

Several general principles are clear: The chimney should be as tall as possible. The cross-sectional area of the flue should be about one-tenth the area of the front opening of the fireplace. A *damper* should be installed to close off the chimney when there is no fire burning, and to regulate the passage of air through the firebox when the fire *is* burning (Figure A). A *smoke shelf* above the damper reduces fireplace malfunctions caused by cold downdrafts in the chimney. Sloping sides and a sloping back in the firebox reduce smoking and to throw more heat into the room.

There are also a number of restrictions on fireplace and chimney construction in building codes. Typically these call for a 2-inch (51-mm) clearance between wood framing and the masonry of a chimney or fireplace, and clearances to combustible materials around the opening of the fireplace as shown on the accompanying diagram. Building codes usually specify also the minimum thicknesses of masonry around the firebox and flue, the minimum size of flue, and the minimum extension of the chimney above the roof. For an extended and knowledgeable discussion of fireplace design, see Albert G. H. Dietz, *Dwelling House Construction,* Cambridge, Massachusetts, M.I.T. Press, 1974, pages 147–163. For ready reference in proportioning a fireplace, use the values in the accompanying Figures B and C. In most cases the designer need not detail the internal construction of the fireplace beyond the information given in these dimensions, because masons are well versed in the intricacies of assembling a fireplace.

There are several alternatives to the conventional masonry fireplace. One is the steel fireplace liner that replaces the firebrick lining, damper, and smoke chamber, and which has internal passages to warm air drawn from the room and return it to the room. The liner is set in place on the underhearth, and built into a masonry facing and chimney by the mason. Another alternative is the "package" fireplace, a self-contained, fully insulated unit that needs no masonry whatsoever. It is set on the plywood subfloor and fitted with a chimney of prefabricated, insulated sections of metal pipe. It can be faced with any desired ceramic or masonry materials. A third alternative is a metal stove that burns wood, coal, or both. Stoves are available in hundreds of styles and sizes. Their principle functional advantage is that they provide more heat to the interior of the building per unit of fuel burned than a fireplace. A stove requires a hearth and a fire-protected wall that are rather large in extent—the designer should consult the local building code at an early stage of design to be sure the room is large enough to hold a stove of the desired dimensions.

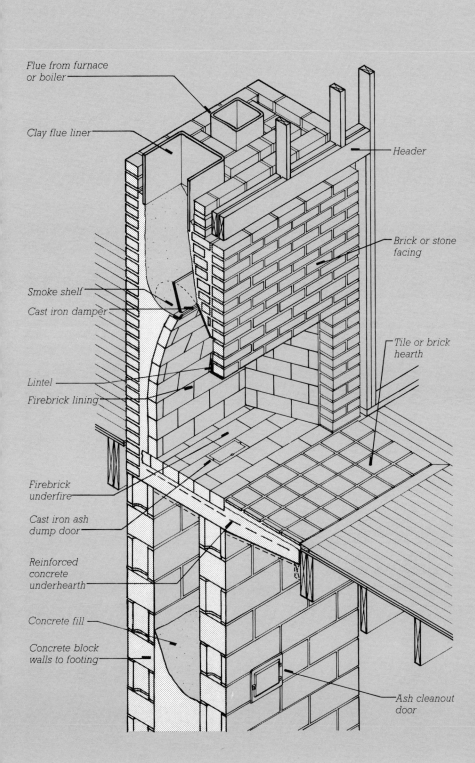

Flue from furnace or boiler

Clay flue liner

Header

Smoke shelf

Cast iron damper

Brick or stone facing

Tile or brick hearth

Lintel

Firebrick lining

Firebrick underfire

Cast iron ash dump door

Reinforced concrete underhearth

Concrete fill

Concrete block walls to footing

Ash cleanout door

FIGURE A

A cutaway view of a conventional masonry fireplace. Concrete block masonry is used wherever it will not show, to reduce labor and material costs. The damper and ash doors are prefabricated units of cast iron. The flue liners are made of fired clay and are highly resistant to heat. The flue from the furnace or boiler in the basement slopes as it passes the firebox until it adjoins the fireplace flue, to keep the chimney as small as possible.

(continued)

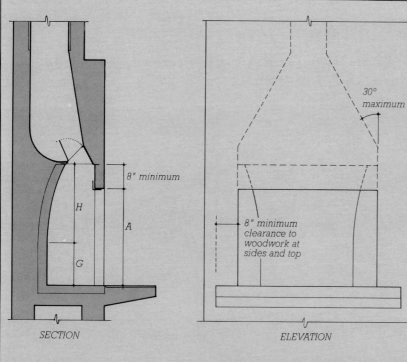

SECTION

ELEVATION

30° maximum

8" minimum

8" minimum clearance to woodwork at sides and top

H

A

G

FIGURE B

The critical dimensions of a conventional masonry fireplace, keyed to the table in Figure C. Dimension D, the depth of the hearth, is commonly required to be 16 inches (405 mm) for fireplaces with openings up to 6 square feet (0.56 m²) and 20 inches (510 mm) if the opening is larger. The side extension of the hearth, E, is usually fixed at 8 inches (203 mm) for smaller fireplace openings, and 12 inches (305 mm) for larger ones.

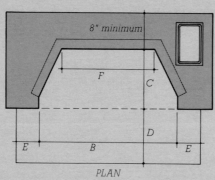

8" minimum

F

C

D

E   B   E

PLAN

FIGURE C

Recommended proportions for conventional masonry fireplaces, based largely on figures given in Ramsey/ Sleeper, Architectural Graphic Standards (7th Edition, Robert T. Packard, A.I.A., Editor), New York, John Wiley and Sons, 1981.

| Fireplace Opening | | | | Vertical Backwall Height (G) | Inclined Backwall Height (H) | Flue Lining | |
Height (A)	Width (B)	Depth (C)	Minimum Backwall Width (F)			Rectangular (outside dimensions)	Round (inside diameter)
24″ (610 mm)	28″ (710 mm)	16 to 18″ (405 to 455 mm)	14″ (355 mm)	14″ (355 mm)	16″ (405 mm)	8½ × 13″ (216 × 330 mm)	10″ (254 mm)
28 to 30″ (710 to 760 mm)	30″ (760 mm)	16 to 18″ (405 to 455 mm)	16″ (405 mm)	14″ (355 mm)	18″ (455 mm)	8½ × 13″ (216 × 330 mm)	10″ (254 mm)
28 to 30″ (710 to 760 mm)	36″ (915 mm)	16 to 18″ (405 to 455 mm)	22″ (560 mm)	14″ (355 mm)	18″ (455 mm)	8½ × 13″ (216 × 330 mm)	12″ (305 mm)
28 to 30″ (710 to 760 mm)	42″ (1065 mm)	16 to 18″ (405 to 455 mm)	28″ (710 mm)	14″ (355 mm)	18″ (455 mm)	13 × 13″ (330 × 330 mm)	12″ (305 mm)
32″ (815 mm)	48″ (1220 mm)	18 to 20″ (455 to 510 mm)	32″ (815 mm)	14″ (355 mm)	24″ (610 mm)	13 × 13″ (330 × 330 mm)	15″ (381 mm)

it carefully with a spirit level, shims it with pairs of wood shingle wedges between the finish and rough jambs, and nails it in place with finish nails through the jamb (Figure 7.14). *Casings* are then nailed around the frame on both sides of the partition to close the ragged gap between the door frame and the wall finish (Figure 7.15). To save the labor of casing doors, door units can also be purchased with *split jambs* that enable the door to be cased at the mill. At the time of installation, each door unit is separated into halves, and the halves are installed from opposite sides of the partition to telescope snugly together before being nailed in place (Figure 7.16).

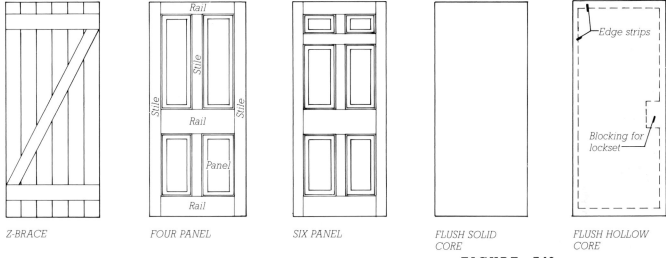

Z-BRACE FOUR PANEL SIX PANEL FLUSH SOLID CORE FLUSH HOLLOW CORE

FIGURE 7.12
Types of wood doors.

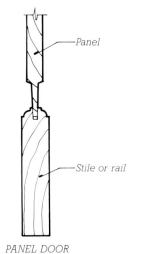

PANEL DOOR

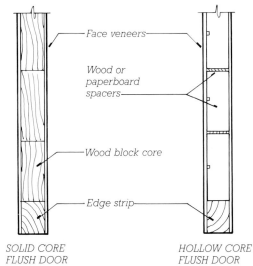

SOLID CORE FLUSH DOOR HOLLOW CORE FLUSH DOOR

FIGURE 7.13
Edge details of three types of wood doors. The panel is loosely fitted to the stiles and rails in a panel door to allow for moisture expansion of the wood. The spacers and edge strips in hollow core doors have ventilation holes to equalize air pressures inside and outside the door.

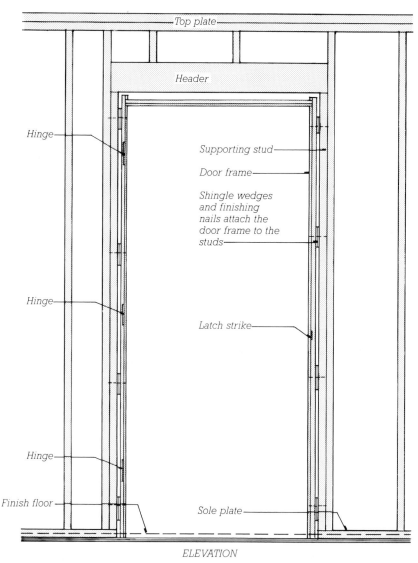

Top plate

Header

Hinge

Supporting stud

Door frame

Shingle wedges and finishing nails attach the door frame to the studs

Hinge

Latch strike

Hinge

Finish floor

Sole plate

ELEVATION

FIGURE 7.15

Casing a door frame. From top to bottom: The finish nails in the frame are set with a steel nail set. The top piece of casing, mitered to join the vertical casings, is ready to install, and glue is spread on the edge of the frame. The top casing is nailed into place. The nails are set below the surface of the wood, ready for filling. (Photos by Joseph Iano)

FIGURE 7.14

Installing a door frame in a rough opening. The shingle wedges at each nailing point are paired in opposing directions to create a flat, precisely adjustable shim to support the frame.

FIGURE 7.16

A split-jamb interior door arrives on the construction site prehung and precased. The halves of the frame are separated and installed from opposite sides of the partition.

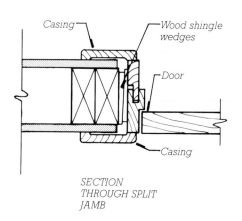

Casing

Wood shingle wedges

Door

Casing

SECTION THROUGH SPLIT JAMB

Window Casings and Baseboards

Windows are cased in much the same manner as doors (Figures 7.17, 7.18). After the finish flooring is in place, *baseboards* are installed at the junction of the floor and the wall to close the gap between the flooring and the

FIGURE 7.17

Casing a window, from left to right: Marking the length of a casing. Cutting the casing to length with a power miter saw. Nailing the casing. Coping the end of a molded edge of an apron with a coping saw, so the molding profile will terminate neatly at the end of the apron. Planing the edge of the apron, which *has been ripped (sawn lengthwise, parallel to the grain of the wood) from wider casing stock. Applying glue to the apron. Nailing the apron, which is wedged in position with a stick. The coped end of the apron, in place. The cased window, ready for filling, sanding, and painting. (Photos by Joseph Iano)*

FIGURE 7.18
Simple but carefully detailed and skillfully crafted window casings in a restaurant. (Architects: Woo and Williams. Photo by Richard Bonarrigo. Courtesy of the architects)

wall finish, and to protect the wall finish against damage by feet, furniture legs, and cleaning equipment (Figure 18.32).

The carpenter recesses the heads of finish nails below the surface of the wood using a *nail set*, a hardened steel punch. The first painting operation on millwork is to fill these nail holes with a paste filler and to sand the surface smooth after the filler has dried so the holes will be invisible in the painted woodwork. For transparent wood finishes, the filler must be selected carefully to match the color of the wood, or the nail holes may be left unfilled.

The nails in casings and baseboards must reach through the plaster or gypsum board to penetrate the framing members beneath in order to make a secure attachment. Eight- or ten-penny finish or casing nails are customarily used.

Cabinets

Cabinets for kitchens, bathrooms, and workrooms are nearly always fabricated in specialty cabinet shops and brought to the construction site fully finished. They are installed by shimming against wall and floor surfaces as necessary to make them level, and screwing through the backs of the cabinet units into the wall studs (Figures 7.19, 7.20). The tops are then attached with screws driven up from the cabinets beneath. Kitchen and bath countertops are cut out for built-in sinks and lavatories, which are subsequently installed by the plumber.

Finish Stairs

Finish stairs are either constructed in place (Figures 7.21 through 7.23) or shop built (Figure 7.24). The shop-built stairs tend to be more tightly constructed and to squeak less in use, but site-built stairs can be fitted more closely to the walls and are more adaptable to special situations and

FIGURE 7.19
Prepainted wood kitchen cabinets installed, but lacking shelves, drawers, doors, and countertops. (Photo by the author)

FIGURE 7.20
Finely wrought cabinets of redwood. (Architect: Alex Achimore. Photo by Jeff Weissman. Courtesy of California Redwood Association)

framing irregularities. Handrails for stairs are mounted on metal brackets that space them a convenient distance from the wall. These brackets are screwed through the wall finish material to the studs beneath.

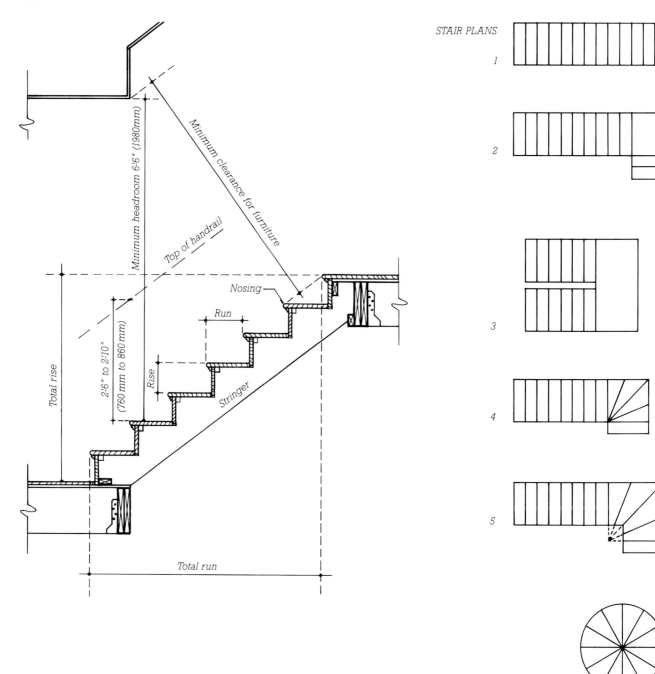

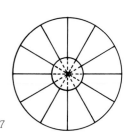

FIGURE 7.21

Left: *Stair terminology and clearances for wood frame residential construction.* Right: *Types of stairs.* **1.** *Straight run.* **2.** *L-shaped stair with landing.* **3.** *180 degree turn with landing.* **4.** *L-shaped stair with winders (triangular treads). Winders are helpful in compressing a stair into a much smaller space but are perilously steep where they converge, and their treads become much too shallow for comfort and safety. Many codes do not permit winders.* **5.** *L-shaped stair whose winders have an offset center. The offset center can* *increase the minimum tread dimension to within legal limits.* **6.** *A spiral stair (in reality a helix, not a spiral) consists entirely of winders and is generally illegal for any use but a secondary stair in a single-family residence.* **7.** *A spiral stair with an open center of sufficient diameter can have its treads dimensioned to legal standards.* (Adapted by permission from Albert G. H. Dietz, *Dwelling House Construction,* 4th Edition. Cambridge, Massachusetts, MIT Press, © 1974 Massachusetts Institute of Technology)

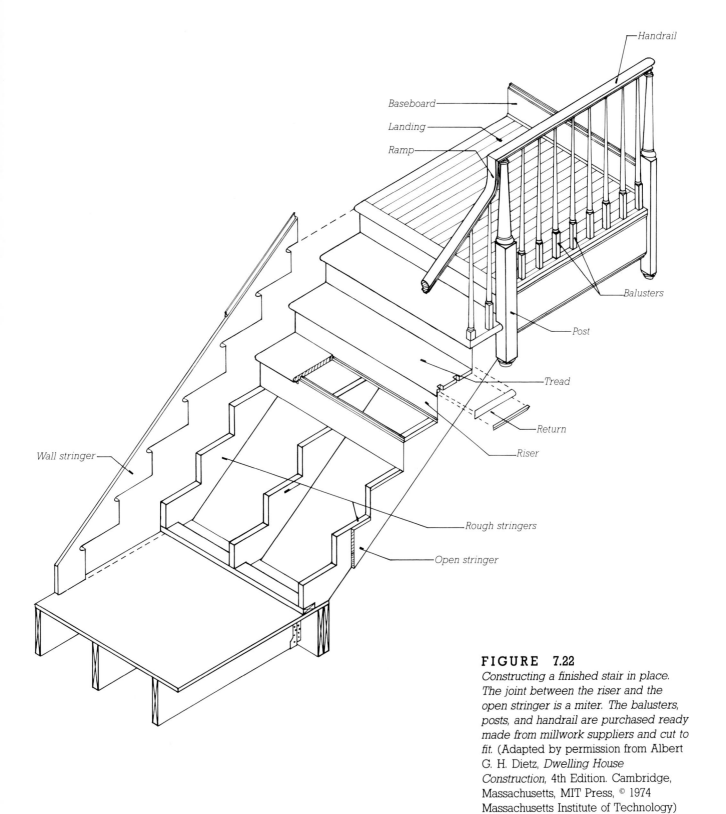

Handrail

Baseboard

Landing

Ramp

Balusters

Post

Tread

Return

Riser

Wall stringer

Rough stringers

Open stringer

FIGURE 7.22

Constructing a finished stair in place. The joint between the riser and the open stringer is a miter. The balusters, posts, and handrail are purchased ready made from millwork suppliers and cut to fit. (Adapted by permission from Albert G. H. Dietz, Dwelling House Construction, 4th Edition. Cambridge, Massachusetts, MIT Press, © 1974 Massachusetts Institute of Technology)

Three openings are required in stair-cases; the first is the door thro' which one goes up to the stair-case, which the less it is hid to them that enter into the house, so much the more it is to be commended. And it would please me much, if it was in a place, where before that one comes to it, the most beautiful part of the house was seen; because it makes the house (even tho' small) seem very large; but however, let it be manifest, and easily found. The second opening is the windows that are necessary to give light to the steps; they ought to be in the middle, and high, that the light may be spread equally every where alike. The third is the opening thro' which one enters into the floor above; this ought to lead us into ample, beautiful, and adorned places.

Andrea Palladio, 1508–1580

FIGURE 7.23
Millwork suppliers offer several different stair "packages" of coordinated parts in traditional designs. (Courtesy of Morgan Products Ltd.)

FIGURE 7.24
A shop-built stair. All the components are glued firmly together. (Adapted by permission from Albert G. H. Dietz, Dwelling House Construction, 4th Edition. Cambridge, Massachusetts, MIT Press, © 1974 Massachusetts Institute of Technology)

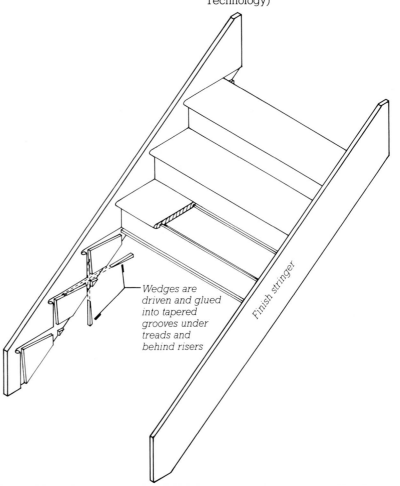

Wedges are driven and glued into tapered grooves under treads and behind risers

Finish stringer

FIGURE 7.25
Yohon Hadwiger, an itinerant carpenter, designed and built this spiral staircase of wood in 1877 for a chapel in New Mexico. It is held together with wooden pegs and has no central support. (Photo by Tony O'Brien. Courtesy of Fine Woodworking*)*

PROPORTIONING STAIRS

Building codes encourage the design of safe, comfortable stairs through a number of detailed requirements. The overall dimensional limitations for stairs as given in the BOCA Basic/ National Building Code are summarized in Figure A. Beyond these restrictions, the BOCA Code further specifies that the risers and treads shall be proportioned so that twice the riser dimension added to the tread dimension shall equal 24 to 25 inches (610 to 635 mm). This formula was derived in France two centuries ago from measurements of actual dimensions of comfortable stairs. Figure B gives an example of how it is used in designing a new stair, in this case for a nonresidential building. Because the code does not allow variations greater than $\frac{3}{16}$ inch (5 mm) between successive treads or risers, the floor-to-floor dimension should be divided equally into risers to an accuracy of 0.01 inch (0.5 mm) to avoid cumulative errors. The framing square used by carpenters in the United States to lay out stair stringers has a scale of hundredths of an inch, and riser dimensions should be given in these units rather than

fractions to achieve the necessary accuracy.

Most building codes do not allow a rise of more than 12 feet (3660 mm) between landings in a stair. Landings contribute to the safety of a stair by providing a moment's rest to the legs between flights of steps. (Architects also generally avoid designing flights of less than three risers because short flights, especially in public buildings, sometimes go unnoticed, leading to dangerous falls.) The width and depth of the landing must each be equal to the width of the stair. The width of a required exit stairway is calculated according to the number of occupants served by the stair, in accordance with formulas given in building codes.

Monumental outdoor stairs, such as those that lead to the entrances of public buildings, are designed with lower risers and deeper treads than indoor stairs. Many designers ease the 2R + T formula a bit for outdoor stairs, raising the sum to 26 or 27 inches (660 or 685 mm), but it is always best to make a full-scale mockup of a section of such a stair to be sure it is comfortable underfoot.

	Minimum Width	Maximum Riser Height	Minimum Tread Depth	Minimum Headroom
Residential stairs	36″	8¼″	9″	6′6″
	(915 mm)	(210 mm)	(230 mm)	(1980 mm)
Nonresidential stairs	44″	7½″	10″	6′8″
	(1120 mm)	(190 mm)	(255 mm)	(2030 mm)

FIGURE A

Dimensional limitations for stairs as established by the BOCA Basic/National Building Code.

Procedure	English	Metric
1. Measure the height (H) from finish floor to finish floor.	$H = 9'4\frac{3}{8}''$, or 112.375"	$H = 2854$ mm
2. Divide H by the approximate riser height desired, and round off to obtain a trial number of risers for the stair.	$\dfrac{H}{7''} = \dfrac{112.375''}{7''} = 16.05$ Try 16 risers	$\dfrac{H}{180 \text{ mm}} = \dfrac{2854 \text{ mm}}{180 \text{ mm}} = 15.86$ Try 16 risers
3. Divide H by the trial number of risers to obtain an exact trial riser height. Work to the nearest hundredth of an inch, or the nearest millimeter, to avoid any cumulative error that would result in one riser being substantially lower or higher than the rest. Check to make sure this trial riser height falls within the limits set by the building code.	$\dfrac{H}{16} = \dfrac{112.375''}{16} = 7.02''$ $R = 7.02''$ $7.02'' < 7.50''$ <u>OK</u>	$\dfrac{H}{16} = \dfrac{2854 \text{ mm}}{16} = 178$ mm $R = 178$ mm 178 mm < 190 mm <u>OK</u>
4. Substitute this trial riser height into the given formula and solve for the tread depth. The depth can be rounded down somewhat if desired, as long as $2R + T \geq 24$. Check the tread depth against the code minimum.	$2R + T = 25''$ $2(7.02) + T = 25''$ $T = 25'' - 14.04''$ $T = 10.96''$, say 10.9" $10.9'' > 10''$ <u>OK</u>	$2R + T = 635$ mm $2(178 \text{ mm}) + T = 635$ mm $T = 635 \text{ mm} - 356$ mm $T = 279$ mm, say 275 mm 275 mm > 255 mm <u>OK</u>
5. Summarize the results of these calculations. There are always one fewer treads than risers in a flight of stairs.	16 risers @ 7.02" 15 treads @ 10.9" Total run = (15) (10.9") = 163.5" = 13'7$\frac{1}{2}$"	16 risers @ 178 mm 15 treads @ 275 mm Total run = (15) (275 mm) = 4125 mm
6. If desired, a steeper or shallower stair can be tried as an alternative by subtracting or adding one riser and tread and recalculating riser and tread dimensions. Subtracting a riser and tread can reduce the overall run of the stair dramatically, which is helpful when designing a stair for a limited amount of space.	Try 15 risers: $R = \dfrac{112.375''}{15} = 7.49''$ $7.49'' < 7.50''$ <u>OK</u> $2(7.49) + T = 25''$ $T = 10.02''$, say 10" $10'' = 10''$ <u>OK</u> 15 risers @ 7.49" 14 treads @ 10" Total run = (14) (10") = 140" = 11'8" Saves almost 2' of run	Try 15 risers: $R = \dfrac{2854 \text{ mm}}{15} = 190$ mm = 190 mm <u>OK</u> $2(190) + T = 635$ mm $T = 255$ mm = 255 mm <u>OK</u> 15 risers @ 190 mm 14 treads @ 255 mm Total run = (14) (255 mm) = 3570 mm Saves 555 mm of run

FIGURE B
A procedure for proportioning stairs.

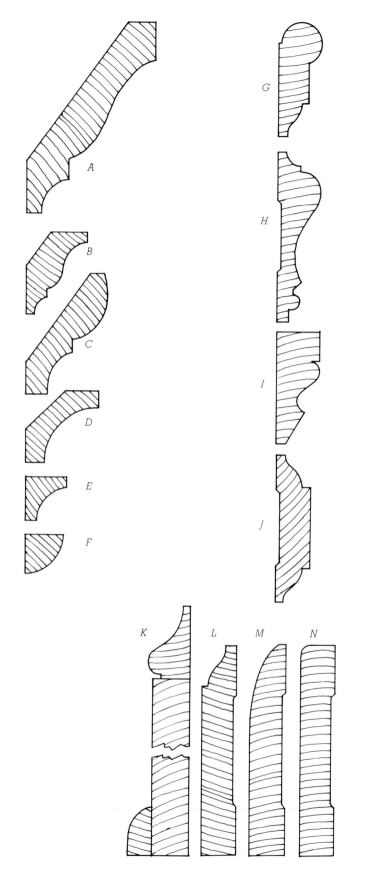

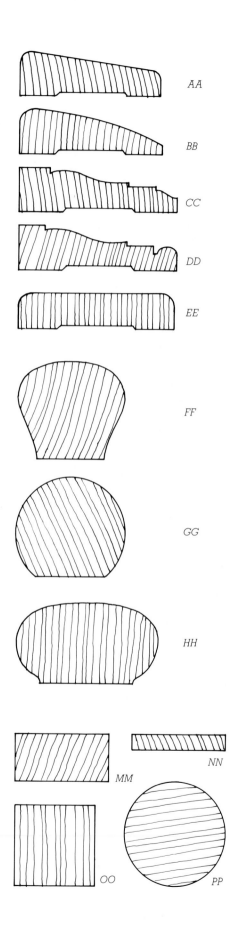

FIGURE 7.26

Some common molding patterns for wood interior trim. A and B are crowns, *C is a* bed, *D and E are* coves. *All are used to trim the junction of a ceiling and a wall. F is a* quarter-round, *for general purpose trimming of inside corners. Moldings G through J are used on walls—G is a* picture molding, *applied near the top of a wall so framed pictures can be hung from it at any point on special metal hooks that fit over the rounded portion of the molding. H is a* chair rail, *installed around dining rooms to protect the plaster from damage by the backs of chairs. I is a* panel molding *and J a* batten, *both used in traditional paneled wainscoting. Baseboards (K through N) include three single-piece designs, and one traditional design (K) using a separate* cap *molding and* shoe *in addition to a piece of S4S stock that is the baseboard itself (see also Figure 18.32).*

Designs AA through EE are standard casings for doors and windows. FF, GG, and HH are handrail stock. MM is representative of a number of sizes of S4S material available to the finish carpenter for miscellaneous uses. NN is lattice stock, *also used occasionally for flat trim. OO is* square stock, *used primarily for balusters. PP represents several available sizes of* round stock *for balusters, handrails, and closet poles. Wood moldings are furnished in either of two grades: N Grade, for transparent finishes, must be of a single piece. P Grade, for painting, may be finger jointed or edge glued from smaller pieces of wood. P Grade is less expensive because it can be made up of short sections of lower-grade lumber with the defects cut out. Once painted, it is indistinguishable from N Grade.*

Miscellaneous Finish Carpentry

Finish carpenters install dozens of miscellaneous items in the average building—closet shelves and poles, pantry shelving, bookshelves, wood paneling, chair rails, picture rails, ceiling moldings, mantelpieces, laundry chutes, door hardware, weatherstripping, doorstops, and bath accessories (towel bars, paper holders, and so on). Many of these items are available ready-made from millwork and hardware suppliers (Figures 7.26, 7.27), but others have to be crafted by the carpenter.

FIGURE 7.27

Fireplace mantels are available from millwork suppliers in a number of designs. Each is furnished largely assembled but detailed in such a way that it can easily be adjusted to fit any fireplace within a wide range of sizes. (Courtesy of Morgan Products Ltd.)

FIGURE 7.28
An existing room of a Victorian house before interior refinishing.

FIGURE 7.29
The same room finished in redwood paneling and casework. (Architect: Marshall Roath. Photos by Karl Riek. Courtesy of California Redwood Association)

FLOORING AND CERAMIC TILE WORK

Before finish flooring can be installed, the subfloor is cleaned of plaster droppings and swept thoroughly. Underlayment panels (in areas destined for resilient flooring materials and carpeting) are glued and nailed over the subfloor, their joints overlapped with those in the subfloor to eliminate weak spots. The thicknesses of the underlayment panels are chosen to make the finished floor surfaces as nearly equal in level as possible at junctions between different flooring materials.

Floor finishing operations require cleanliness and freedom from traffic, so other trades are banished from the area as the flooring materials are applied. Hardwood flooring is sanded level and smooth after installation, then vacuumed to remove the sanding dust. The finish coatings are applied in as dust-free an atmosphere as possible to avoid embedded specks. Resilient flooring installers vacuum the underlayment meticulously so particles of dust and dirt will not become trapped beneath the flooring and cause bumps in the finished surface. The finished floors are often covered with sheets of heavy paper to protect them during the final few days of construction activity. Carpet installation is less sensitive to dust, and the installed carpets are less prone to damage than hardwood and resilient floorings, but temporary coverings are applied as necessary to protect the carpet from paint spills and water stains.

The application of ceramic tile to a shower stall is illustrated in Figure 7.30. The installation of ceramic tile and finish flooring materials is covered in more detail in Chapters 17 and 18.

FINISHING TOUCHES

When flooring and painting are finished, the plumbers install and activate the lavatories, water closets, tubs,

FIGURE 7.30

Setting sheets of ceramic tile in a shower stall. The base coat of portland cement plaster over metal lath has already been installed. Now the tilesetter applies a thin coat of mortar with a trowel and presses a sheet of tiles into it, taking care to align the tiles individually around the edges. A day or two later, after the mortar has hardened sufficiently, the joints will be grouted to complete the installation. (Photos by Joseph Iano)

FIGURE 7.31

Ceramic tile is used for the floor, countertops, and backsplash in this kitchen. The border was made by selectively substituting tiles of four different colors for the white tiles used for the field of the floor. (Designer: Kevin Cordes. Courtesy of American Olean Tile)

FIGURE 7.32
Varnished oak flooring, millwork, and casework. (Architects: Woo and Williams. Photo by Richard Bonarrigo. Courtesy of the architects)

sinks, and shower fixtures. Gas lines are connected to appliances and the main gas valve is opened. The electricians connect the wiring for the water heater and heating and air conditioning equipment, mount the lighting fixtures, and put metal or plastic cover plates on the switches and receptacles. The electrical circuits are energized and checked to be sure they work. The smoke alarms and heat alarms, required by most codes in residential structures, are also connected and tested by the electricians. The heating and air conditioning system is completed with the installation of air grills and registers, or with the mounting of metal convector covers, then turned on and tested. Last-minute problems are identified and corrected through cooperative effort by the contractors, the owner of the building, and the architect. The building inspector is called in for a final inspection and issuance of an occupancy permit. After a thorough cleaning, the building is ready for use.

C.S.I./C.S.C. Masterformat Section Numbers for Interior Finishes for Light Frame Wood Construction	
06200	**FINISH CARPENTRY**
06220	Millwork
	Cabinets
	Closet and Storage Shelving
	Molding and Trim
06240	Plastic Laminate
06260	Board Paneling
06265	Molded Architectural Ornamentation
07190	**VAPOR AND AIR RETARDERS**
07192	Vapor Retarders
07200	**INSULATION**
07210	Building Insulation
	Batt and Blanket Insulation
	Board Insulation
	Foamed-in-Place Insulation
	Loose Fill Insulation
	Sprayed Insulation
08200	**WOOD AND PLASTIC DOORS**
08210	Wood Doors
	Flush Wood Doors
	Panel Wood Doors

(See also the Masterformat section numbers in Chapter 17)

SELECTED REFERENCES

1. Albert G. H. Dietz. *Dwelling House Construction*, 4th Edition. Cambridge, Massachusetts, M.I.T. Press, 1974.

This classic text has extensive chapters with clear illustrations concerning chimneys and fireplaces, insulation, wallboard, lath and plaster, and interior finish carpentry.

KEY TERMS AND CONCEPTS

roughing in	polystyrene foam	butt	Z-brace
DWV pipes	polyurethane foam	baseboard	panel door
supply pipes	isocyanurate foam	nail set	flush door
gas pipes	urea formaldehyde foam	molding	solid core
hydronic heating system	air-to-air heat exchanger	baluster	hollow core
finish carpentry	flue	wall stringer	prehung door
R-value	hearth	rise	casing
thermal insulation	damper	run	baseboard
vapor retarder	smoke shelf	riser	split jamb
millwork	underfire	tread	panel
glass fiber batt	ash dump	nosing	stile
cellulose	miter	landing	rail
rock wool			

REVIEW QUESTIONS

1. List the sequence of operations required to complete the interior of a Wood Light Frame building and explain the order in which these operations occur.

2. What are some alternative ways of insulating the walls of a Wood Light Frame building to R-values beyond the range normally possible with ordinary 2 × 4 (38 × 89 mm) studs?

3. Why are plaster and gypsum board so popular as interior wall finishes in wood frame buildings? List as many reasons as you can.

4. What is the level of humidity in a building at the time the interior wall finishes are completed? Why? What should be done about this, and why?

5. Summarize the most important things to keep in mind when designing a stair.

EXERCISES

1. Design and detail a fireplace for a building you are designing, using the information provided on page 220 to work out the exact dimensions, and the information in Chapter 8 to help in detailing the masonry.

2. Design and detail a stairway for a building you are designing using the information provided on pages 230 and 231 to calculate the dimensions.

3. Visit a wood frame building you admire. Make a list of the interior finish materials and components used, including species of wood, where possible. How does each material and component contribute to the overall feeling of the building? How do they relate to one another?

4. Make measured drawings of millwork details in an older building you admire. Analyze each detail to discover its logic. What woods were used and how were they sawn? How were they finished?

MASONRY

*F*lemish bond brickwork and limestone lintels and sills are combined simply and directly in the exterior walls of this townhouse on Boston's Beacon Hill. *(Photo by the author)*

Masonry is the simplest of building techniques: The mason stacks pieces of material (bricks, stones, concrete blocks, called collectively *masonry units*) atop one another to make walls. But it is also the richest and most varied, with its endless selection of colors and textures. And because the pieces of which it is made are small, masonry can take any shape, from a planar wall to a sinuous surface that defies the distinction of roof from wall.

Masonry is the material of earth, taken from the earth and comfortably at home in foundations, pavings, and walls that grow directly from the earth. But with modern techniques of reinforcing, masonry can rise many stories from the earth, and in the form of arches and vaults, masonry can take wing and fly across space.

The most ancient of our building techniques, masonry remains labor intensive, requiring the patient skills of experienced and meticulous artisans to achieve a satisfactory result. But it has kept pace with the times and remains highly competitive technically and economically with other systems of structure and enclosure, the more so because one mason can produce in one operation a completely finished, insulated, loadbearing wall, ready for use.

Masonry is durable. The designer can select masonry materials that are almost unaffected by water, air, or fire, ones with brilliant colors that will not fade, ones that will stand up to heavy wear and abuse, and make from them a building that will last for generations.

Masonry is a material of the small entrepreneur. One can set out to build a building of brick with no more tools than a trowel, a shovel, a measuring rule, a level, a square of scrap plywood, and a piece of string. Yet many masons can work together, aided by mechanized handling of materials, to put up projects as large as the human mind can conceive.

HISTORY

Masonry began spontaneously in the creation of low walls from stones or pieces of caked mud from dried puddles. Mortar was originally the mud smeared into the rising wall to lend stability and weathertightness. Where stone lay readily at hand, it was preferred to bricks; where stone was unavailable, bricks were made from local clays and silts. Change came with the passing millenia: People learned to quarry, cut, and dress stone with increasing precision. Fires built against mud brick walls brought a knowledge of the advantages of burned brick, leading to the invention of the kiln. Masons learned the simple art of turning limestone into lime, and lime mortar began to replace mud.

By the fourth millennium B.C., the peoples of Mesopotamia were building palaces and temples of stone and sun-dried brick. In the third millenium, the Egyptians erected the first of their stone temples and pyramids. In the

FIGURE 8.1

The Parthenon, constructed of marble, has stood on the Acropolis in Athens for more than twenty-four centuries. (Photo by James Austin, Cambridge, England)

last centuries prior to the birth of Christ, the Greeks perfected their temples of limestone and marble (Figure 8.1), and control of the western world passed to the Romans, who made the first large-scale use of masonry arches and roof vaults in their basilicas, baths, palaces, and aqueducts. Medieval civilizations in both Europe and the Islamic world brought masonry vaulting to a very high plane of development. The Islamic craftsmen built magnificent palaces, markets, and mosques of brick and often faced them with brightly glazed clay tiles. The Europeans directed their efforts toward fortresses and cathedrals of stone, culminating in the pointed vaults and flying buttresses of the great Gothic churches (Figures 8.2, 8.3). In Central America, South America, and Asia, other civilizations were carrying on a simultaneous evolution of building techniques in cut stone.

During the Industrial Revolution in Europe and North America, machines were developed that quarried and worked stone, molded bricks, and sped the transportation of these heavy materials to the building site. Sophisticated mathematics were applied for the first time to analysis of the structure of masonry arches and to the art of stonecutting. Portland cement mortar came into widespread use, enabling the construction of masonry buildings of greater strength and durability.

In the late nineteenth century, masonry began to lose its primacy among the materials of construction. The very tall buildings of the central cities required frames of metal to replace the thick masonry bearing walls that had limited the heights to which one could build. Reinforced concrete, poured rapidly and economically into simple forms of wood, began to replace brick and stone masonry in foundations and walls. The heavy masonry vault was supplanted by lighter floor and roof structures of steel and concrete.

The nineteenth-century invention of the hollow concrete block helped to avert the extinction of masonry as a craft. The concrete block was much

FIGURE 8.2

Construction in ashlar limestone of the magnificent Gothic cathedral at Chartres, France was begun in 1194 A.D. and was not finished until several centuries later. Seen here are the flying buttresses that resist the lateral thrusts of the stone roof vaulting. (Photo by James Austin, Cambridge, England)

FIGURE 8.3
The Gothic cathedrals were roofed with lofty vaults of stone blocks. The ambulatory roof at Bourges (built 1195– 1275) evidences the skill of the medieval French masons in constructing vaulting to cover even a curving floor plan. (Photo by James Austin, Cambridge, England)

FIGURE 8.4
Despite the steady mechanization of construction operations in general, masonry construction in brick, concrete block, and stone is still entirely dependent on simple tools and the highly skilled hands that use them. (Courtesy of International Masonry Institute)

cheaper than cut stone and required much less labor to lay than brick. It could be combined with brick or stone facings to make lower-cost walls that were still satisfactory in appearance. The brick cavity wall, an early nineteenth-century British invention, also contributed to the survival of masonry, for it produced a warmer, more watertight wall that was later to adapt easily to the introduction of thermal insulation when appropriate insulating materials became available in the middle of the twentieth century.

Other twentieth-century contributions to masonry construction include the development of techniques for steel reinforcing in masonry, high-strength and high-adhesion mortars, masonry units (both bricks and concrete blocks) that are higher in structural strength, and masonry units of many types that reduce the amount of labor required for masonry construction.

If this book had been written as recently as a century ago, it would have had to devote little space to materials of construction other than masonry and wood. Because other materials of construction were so late in developing, most of the great works of architecture in the world, and many of the best-developed vernacular architectures, are built of masonry. We live amid a rich heritage of masonry buildings—there is scarcely a town in the world that is without a number of beautiful examples from which the serious student of masonry architecture can learn.

Mortar

Mortar is as much a part of masonry as the masonry units themselves. Mortar serves to cushion the masonry units, giving them full bearing against one another despite their surface irregularities. Mortar seals between the units to keep water and wind from penetrating; it adheres the units to one another to bond them into a monolithic structural unit; and, inevitably, it is important to the appearance of the finished masonry wall.

I remember the masons on my first house. I was not much older than the apprentice whom I found choking back tears of frustration with the clumsiness of his work and the rebukes of his boss.

Nearly thirty years later, we still collaborate on sometimes difficult masonry walls, fireplaces and paving patterns. I work with bricklayers...whose years of learning their craft paralleled my years of trying to understand my profession. We are friends and we talk about our work like pilgrims on a journey to the same destination.

Henry Klein, Architect

Mortar is made of portland cement, hydrated lime, an inert *aggregate* (sand), and water. The sand must be clean, and screened to eliminate particles that are too coarse or too fine. The portland cement is the bonding agent in the mortar, but a mortar made only with portland cement is "harsh" and does not flow well on the trowel or under the brick, so lime is added to impart workability. Lime is produced by burning limestone or seashells (calcium carbonate) in a kiln to drive off carbon dioxide and leave quicklime (calcium oxide). The quicklime is then *slaked* by allowing it to absorb as much water as it will hold, resulting in the formation of calcium hydroxide, called *slaked lime* or *hydrated lime*. The slaking process, which releases large quantities of heat, is usually carried out in the plant, where the hydrated lime is subsequently dried, ground, and bagged for shipment. Until fairly recently, mortar was made without portland cement, and the lime itself

was the bonding agent; it hardened by absorbing carbon dioxide from the air to become calcium carbonate, a very slow and uneven process.

Prepackaged *masonry cements* are also widely used for making mortar. Most are proprietary formulations that contain a wide variety of additives intended to contribute to the workability of the mortar. Two colors of masonry cement are commonly available: *light*, which cures to about the same light gray color as ordinary concrete blocks, and *dark*, which cures to a dark gray. Other colors are easily produced by the mason, either by adding pigments to the mortar at the time of mixing, or by purchasing dry mortar mix that has been custom colored at the plant. Mortar mix can be obtained in shades ranging from pure white to pure black, including all the colors of the spectrum. Because mortar makes up a considerable fraction of the exposed surface area of a brick wall, usually about 20 percent, the color of the mortar is extremely important in the appearance of a brick wall and is almost as important in the appearance of stone or concrete masonry walls. Small sample walls are often constructed before a major building goes under construction to view and compare different brick and mortar combinations and make a final selection.

Mortar composition is specified in ASTM C270. Five basic mortar types are defined, as summarized in Figures 8.5 and 8.6.

Portland cement mortar cures by hydration, not by drying. Mortar that has been mixed but not yet used can become too stiff for use either by drying out, or by commencing its hydration. If the mortar was mixed less than 90 minutes prior to its stiffening, it has merely dried and can safely be retempered with water to make it workable again. If the unused mortar is more than 2 hours old, it must be discarded because it has already begun to hydrate and cannot be worked without damage to its final strength.

To prevent premature drying of mortar, which would weaken it, masonry units that are highly absorptive

Mortar Type	Description	Construction Suitability	Average Compressive Strength at 28 days
M	High-strength mortar	Masonry subjected to high lateral or compressive loads or severe frost action; Masonry below grade	2500 psi (17.25 MN/m²)
S	Medium high-strength mortar	Masonry requiring high flexural bond strength, but subjected only to normal compressive loads	1800 psi (12.40 MN/m²)
N	Medium strength mortar	General use above grade	750 psi (5.17 MN/m²)
O	Medium low-strength mortar	Non-loadbearing interior walls and partitions	350 psi (2.40 MN/m²)
K	Low strength mortar	Interior non-loadbearing partitions, where permitted by building code	75 psi (0.52 MN/m²)

FIGURE 8.5
Mortar types as defined by ASTM C270.

of water should be dampened before laying. Less absorptive units can be laid dry.

Mortars with exceptionally high strengths and adhesive qualities can be created by adding epoxies or latexes to the mix. These high-cost formulations are useful in masonry constructions where tensile strength is important, especially in the prefabrication of brick panels to be lifted by crane and installed on the face of a concrete or structural steel building frame.

Brick masonry

Among the masonry materials, brick is special in two respects: fire resistance and size. As a product of fire, it is the most resistant to building fires of any masonry unit. Its size may account for much of the love that many people instinctively feel for brick: A brick is shaped and dimensioned to fit the human hand. Hand-sized bricks are less likely to crack during drying or firing than larger bricks, and they are easy for the mason to manipulate. This small unit size makes brickwork very flexible in adapting to small-scale geometries and patterns and gives a pleasing scale and texture to a brick wall or floor.

Bricks

Molding of Bricks

Bricks, because of their weight and bulk, are produced by a large number of relatively small and scattered fac-

tories from a variety of local clays and shales. The raw material is dug from pits, crushed, ground, and screened to reduce it to a fine consistency. It is then tempered with water to produce a plastic clay ready for forming into bricks.

There are three major methods used today for forming bricks: the soft mud process, the dry-press process, and the stiff mud process. The oldest is the *soft mud process*, in which a relatively moist clay (20 to 30 percent water) is pressed into simple rectangular molds, either by hand or with the aid of molding machines. To keep the sticky clay from adhering to the molds, the molds may be dipped in water immediately before being filled, producing bricks with a relatively smooth, dense surface that are known as *water-struck bricks*. If the wet mold is dusted with sand just

FIGURE 8.6
Formulas for mortar types M through K.

Mortar Type	Parts by Volume of Portland Cement	Parts by Volume of Masonry Cement	Parts by Volume of Hydrated Lime	Aggregate Measured in a Loose, Damp Condition
M	1	1 (Type II)	—	
	1	—	$\frac{1}{4}$	
S	$\frac{1}{2}$	1 (Type II)	—	
	1	—	over $\frac{1}{4}$ to $\frac{1}{2}$	Not less than $2\frac{1}{4}$ and not more than 3 times the sum of the volumes of the cements and lime used
N	—	1 (Type II)	—	
	1	—	over $\frac{1}{2}$ to $1\frac{1}{4}$	
O	—	1 (Type I or II)	—	
	1	—	over $1\frac{1}{4}$ to $1\frac{1}{2}$	
K	1	—	over $2\frac{1}{2}$ to 4	

before forming the brick, *sand-struck* or *sand-mold* bricks are produced, with a matte-textured surface.

The *dry-press* process is used for clays that shrink excessively during drying. Clay mixed with a minimum of water (up to 10 percent) is pressed into steel molds by a machine working at very high pressures.

The high-production *stiff mud* process is the one most widely used today. Clay containing 12 to 15 percent water is passed through a vacuum to remove any pockets of air, then extruded through a rectangular die (Figures 8.7, 8.8). As the clay leaves the die, textures or thin mixtures of colored clays can be applied to its surface as desired. The rectangular column of moist clay is pushed by the pressure of extrusion across a cutting table, where cutter wires automatically slice it into bricks.

After molding by any of these three processes, the bricks are dried for one to two days in a low-temperature dryer kiln. They are then ready for transformation into their final form by a process known as *firing* or *burning*.

Firing of Bricks

Before the advent of modern kilns, bricks were most often fired by stacking them in a loose array called a *clamp*, covering the clamp with earth or clay, building a wood fire under the clamp, and maintaining the fire for a period of several days. After cooling, the clamp would be disassembled and the bricks sorted according to the degree of burning each had experienced. Bricks adjacent to the fire were usually overburned and distorted, making them unattractive and therefore unsuitable for use in exposed brickwork. Bricks in a zone of the clamp near the fire would be fully burned but undistorted, suitable for exterior facing bricks with a high degree of resistance to weather. Bricks further from the fire would be softer and would be set aside for use as backup bricks, while some bricks from around the perimeter of the clamp would not be burned sufficiently for any purpose and would be discarded. In the days before mecha-

FIGURE 8.7
A column of clay emerges from the die in the stiff mud process of molding bricks. (Courtesy of Brick Institute of America)

FIGURE 8.8
Rotating groups of parallel wires cut the column of clay into individual bricks, ready for drying and firing. (Courtesy of Brick Institute of America)

nized transportation, bricks for a building were often produced from clay obtained from the building site and burned in clamps adjacent to the work.

Today, bricks are usually burned either in a *periodic kiln* or a *tunnel kiln*. The periodic kiln is a fixed structure of masonry that is loaded with bricks, fired, cooled, and unloaded. Bricks are passed continuously through a long tunnel kiln on special railroad cars to emerge at the far end fully burned. In either type of kiln, the first stages of burning are *water-smoking* and *dehydration*, which drive off the remaining water from the clay. The next stages are *oxidation* and *vitrification*, during which the temperature rises to approximately 2400 degrees Fahrenheit (1300° C) and the clay is transformed into a ceramic material. This may be followed by a stage called *flashing*, in which the fire is regulated to create a reducing atmosphere in the kiln and thereby create color variations in the bricks. Finally, the bricks are cooled under controlled conditions to achieve the desired color and avoid thermal cracking. The finished bricks are inspected, sorted, and stacked for shipment. The entire process of firing takes from 40 to 150 hours and is monitored continuously to maintain product quality. Considerable shrinkage takes place in the bricks during drying and firing; this must be taken into account when designing the molds for the brick. The higher the temperature, the greater the shrinkage and the darker the brick. Bricks are often used in a mixed range of colors, with the darker bricks inevitably being smaller than the lighter bricks. Even in bricks of uniform color, some size variation is to be expected, and bricks in general are subject to a certain amount of distortion from the firing process.

The color of a brick depends on the chemical composition of the clay or shale and the temperature and chemistry of the fire in the kiln. Higher temperatures, as noted in the previous paragraph, produce darker bricks. The iron that is prevalent in most clays turns red in an oxidizing fire, and purple in

Name	Thickness	Height	Length
Standard	$3\frac{3}{4}''$ (95 mm)	$2\frac{1}{4}''$ (57 mm)	$8''$ (203 mm)
Modular	$3\frac{5}{8}''$ (92 mm)	$2\frac{1}{4}''$ (57 mm)	$7\frac{5}{8}''$ (194 mm)
Three-Inch	$3''$ (76 mm)	$2\frac{5}{8}''$ (67 mm)	$9\frac{5}{8}''$ (244 mm)
Oversize	$3\frac{3}{4}''$ (95 mm)	$2\frac{3}{4}''$ (70 mm)	$8''$ (203 mm)
Roman	$3\frac{5}{8}''$ (92 mm)	$1\frac{5}{8}''$ (41 mm)	$11\frac{5}{8}''$ (295 mm)
Norman	$3\frac{5}{8}''$ (92 mm)	$2\frac{1}{4}''$ (57 mm)	$11\frac{5}{8}''$ (295 mm)
Six-Inch Jumbo	$5\frac{5}{8}''$ (143 mm)	$3\frac{5}{8}''$ (92 mm)	$11\frac{5}{8}''$ (295 mm)
Jumbo Utility	$3\frac{5}{8}''$ (92 mm)	$3\frac{5}{8}''$ (92 mm)	$11\frac{5}{8}''$ (295 mm)

FIGURE 8.9

Dimensions of some bricks commonly used in North America, based on a mortar joint thickness of ⅜ inch. This list is by no means complete, but does serve to give an idea of the diversity of sizes and shapes available, and the difficulty of generalizing about brick dimensions. Modular bricks are dimensioned so that three courses plus mortar joints add up to a vertical dimension of 8 inches (203 mm), and one brick length plus mortar joint is also 8 inches (203 mm).

FIGURE 8.10

Bricks are often custom molded to perform particular functions in a particular building. This water table course in an English Bond wall was molded to an ogee curve. Architects often design custom-molded bricks to make wall corners other than 90 degrees and to create special surface textures and ornamentation. (Photo by the author)

a reducing fire. Other chemical elements interact in a similar way to the kiln atmosphere to make still other colors. For bright colors, the faces of the bricks can be glazed like pottery, usually in an additional firing.

Brick Sizes

There is no truly standard brick. The nearest thing in the United States is the modular brick, dimensioned to construct walls in modules of 4 inches horizontally and 8 inches vertically, but the modular brick has not found ready acceptance in some parts of the country, and traditional sizes persist on a regional basis. In an effort to reduce bricklaying costs, many manufacturers produce various sizes and proportions of larger bricks that further confuse the picture. Figure 8.9 shows the brick sizes that are most common in the United States. In practice, the designer selecting brick for a building usually views actual samples before completing the drawings for the building, and dimensions the drawings in accordance with the size of the particular brick selected (Figure 8.18). For most bricks in the normal range of sizes, three courses of bricks plus the accompanying three mortar joints add up to a height of 8 inches (203 mm). Length dimensions must be calculated specifically for the brick selected and must include the thicknesses of the mortar joints.

Custom shapes and sizes of brick are often required for buildings with ornamentation or unusual geometries (Figure 8.10). These are readily produced by most brick manufacturers if sufficient lead time is given.

Choosing Bricks

We have already considered three important qualities that the designer must consider in choosing the bricks for a particular building: molding process, color, and size. Several other qualities are also important and are measured in accordance with standard ASTM testing procedures. ASTM C216 es-

tablishes three *grades* of brick based on resistance to weathering, and three *types* of facing bricks (bricks that will be exposed to view) based on the degree of uniformity in shape, dimension, texture, and color from one brick to the next (Figure 8.11). The use of the three grades is related to a map of weathering indices prepared from weather service data on winter rainfall and freezing cycles (Figure 8.12). Grade SW is recommended for use in contact

with the ground, in horizontal surfaces such as pavings, or in situations where the brickwork is likely to be saturated with water, in any of the three regions of the map. Grade MW may be used above ground in any of the regions, but SW will provide greater durability in the severe weather region. Grade NW is intended for use in sheltered or indoor locations.

The compressive strength of brickwork is of obvious importance in struc-

Grades for Building and Facing Bricks	
Grade SW	Severe weathering
Grade MW	Moderate weathering
Grade NW	Negligible weathering
Types of Facing Bricks	
Type FBX	High degree of mechanical perfection, narrow color range, minimum size variation per unit
Type FBS	Wide range of color and greater size variation per unit
Type FBA	Nonuniformity in size, color, and texture per unit

FIGURE 8.11
Grades and types of bricks as defined by ASTM C216.

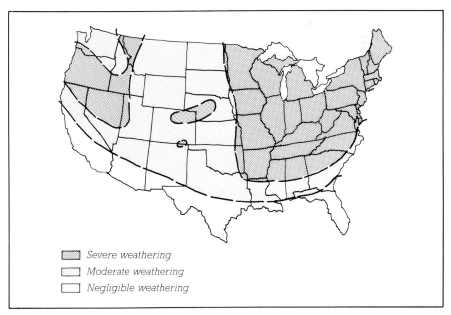

☐ Severe weathering
☐ Moderate weathering
☐ Negligible weathering

FIGURE 8.12
Weathering regions of the United States, as determined by winter rainfall and freezing cycles. Grade SW brick is *recommended for exterior use in the Severe Weathering region.* (Courtesy of Brick Institute of America)

tural walls and piers, and depends on the strengths of both the brick and the mortar. Typical allowable compressive stresses for unreinforced brick walls range from 75 to 400 pounds per square inch (0.52 to 2.76 MN/m^2).

For brickwork exposed to very high temperatures, such as the lining of a fireplace or a furnace, *firebricks* are used. These are made from special clays (*fireclays*) that produce bricks with refractory qualities. Firebricks are laid in very thin joints of fireclay mortar.

Laying Bricks

Figure 8.13 shows the basic vocabulary of bricklaying. Bricks are laid in the various positions for visual reasons, structural reasons, or both. The simplest brick wall is a single *wythe* of *stretchers*. For walls two or more wythes thick, *headers* are required to bond the wythes together into a structural unit. *Rowlock* courses are often used for caps on garden walls, and for sloping sills under windows. Architects frequently employ *soldier* courses for visual emphasis in such locations as window lintels or tops of walls.

The problem of bonding multiple wythes of brick has been solved in many ways in different regions of the world, often resulting in surface patterns that are particularly pleasing to the eye. Figures 8.14 and 8.15 show some *structural bonds* for brickwork, among which Common Bond, Flemish Bond, and English Bond are the most popular. On the exterior of buildings the *cavity wall*, with its single outside wythe, offers the designer little excuse to use anything but Running Bond. Inside a building, safely out of the weather, one may use solid brick walls in any desired bond. For fireplaces and other very small brick constructions, however, it is often difficult to create a long enough stretch of unbroken wall to justify the use of bonded brickwork.

The process of bricklaying is summarized in Figures 8.16 and 8.17. While conceptually simple, bricklaying requires both extreme care and

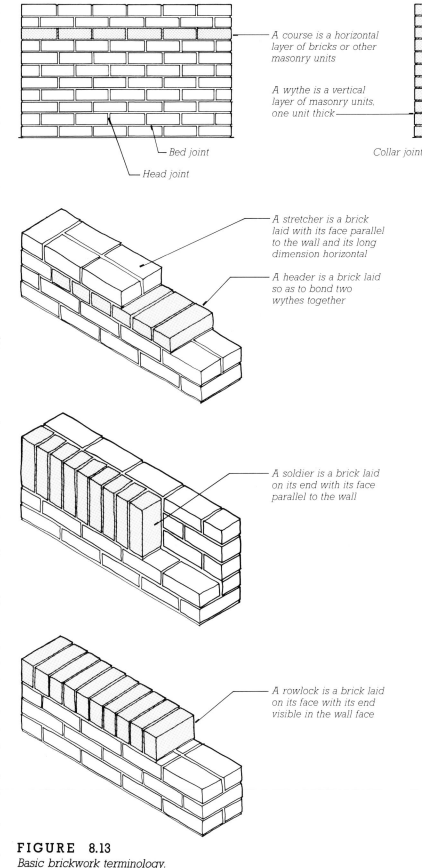

A course is a horizontal layer of bricks or other masonry units

A wythe is a vertical layer of masonry units, one unit thick

Bed joint

Head joint

Collar joint

A stretcher is a brick laid with its face parallel to the wall and its long dimension horizontal

A header is a brick laid so as to bond two wythes together

A soldier is a brick laid on its end with its face parallel to the wall

A rowlock is a brick laid on its face with its end visible in the wall face

FIGURE 8.13
Basic brickwork terminology.

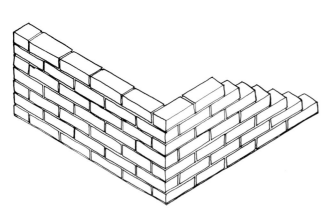

Running Bond consists entirely of stretchers

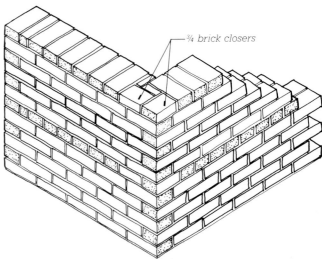

Common Bond (also known as American Bond) has a header course every sixth course. Notice how the head joints are aligned between the header and stretcher courses

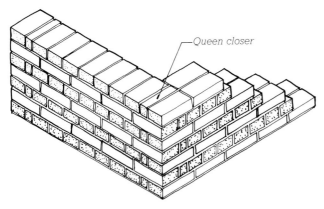

English Bond alternates courses of headers and stretchers

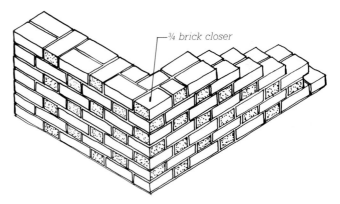

Flemish Bond alternates headers and stretchers in each course

FIGURE 8.14

Frequently used structural bonds for brick walls. Partial closer bricks are necessary at the corners to make the header courses come out even. The mason usually cuts the closers to length with a mason's hammer, but they are sometimes cut with a diamond saw.

FIGURE 8.15

Photographs of some brick bonds. In the left column, from top to bottom, are Running Bond, Common Bond, and English Garden Wall Bond with Flemish header courses. In the right column, English Bond, Flemish Bond, and Monk Bond, which is a Flemish Bond with two stretchers instead of one between headers. The Running Bond example shown here is from the late eighteenth century, with extremely thin joints, which require mortar made from very fine sand. Notice in the Common Bond wall (dating from the 1920s in this case)

that the header course began to fall out of alignment with the stretcher courses, so the mason inserted a partial stretcher to make up the difference; such small variations in workmanship account for some of the visual appeal of brick walls. Flemish header courses, such as those used in the English Garden Wall Bond at the lower left, are often used with bricks whose length, including mortar joint, is substantially more than twice their width; the Flemish header course avoids the thick joints between headers that would otherwise result. The English

Bond example at the upper right is from an early eighteenth-century New England church and uses rather small handmade bricks. The Flemish Bond example below it is modern, and is composed of modular sand mold bricks. The Monk Bond shown here has unusually thick bed joints, approximately ¾ inch (19 mm) high; these are difficult for the mason to lay unless the consistency of the mortar is very closely controlled. (Photos by the author)

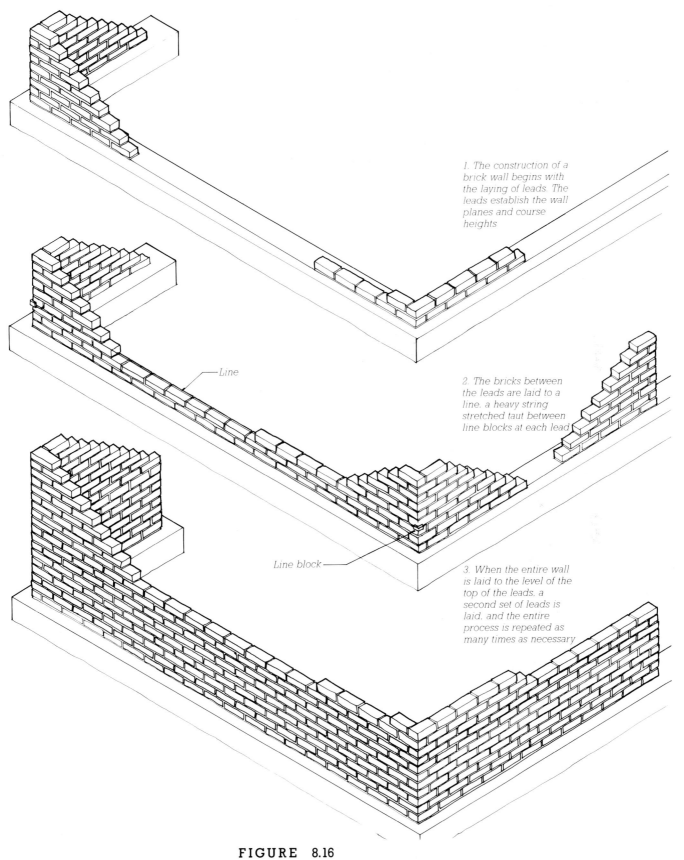

1. The construction of a brick wall begins with the laying of leads. The leads establish the wall planes and course heights

2. The bricks between the leads are laid to a line, a heavy string stretched taut between line blocks at each lead

3. When the entire wall is laid to the level of the top of the leads, a second set of leads is laid, and the entire process is repeated as many times as necessary

Line

Line block

FIGURE   8.16

*The procedure for building brick walls,
in this example of Running Bond.*

considerable experience to produce a satisfactory result, especially where a number of bricklayers working side by side must produce identical work on a major structure. Yet speed is essential to the economy of masonry construction. The work of a skilled mason is impressive both for its speed and for its quality. This level of expertise takes time and hard work to acquire, and the apprenticeship period for brickmasons is therefore both long and demanding.

The laying of leads (pronounced "leeds") is done with much laborious manipulation of a spirit level and a *story pole* to establish true course heights and flat, plumb surfaces. The laying of the infill bricks between the leads is much faster and easier because the mason needs only a trowel in one hand and a brick in the other to *lay to the line* and create a perfect wall. It follows that leads are expensive as compared to the wall surfaces between, so where economy is important the designer

should seek to minimize the number of corners in a brick structure.

Bricks may be cut as needed, either with sharp, well-directed blows of a mason's hammer, or for greater accuracy and more intricate shapes, with a power saw that utilizes a water-cooled diamond abrasive blade. Cutting of bricks slows the process of bricklaying considerably, however, and ordinary brick walls should be dimensioned to minimize cutting (Figure 8.18).

Mortar joints can vary in thickness

FIGURE 8.17

(a) *Laying a stretcher. Mortar for the bed joint has been laid with the trowel, and the end of the brick has been "buttered" with mortar for the head joint. The last three courses of the lead are visible to the right, with a line stretched across the face of the wall to guide the placement of the infill bricks. (b) The brick is pressed into place, compacting both the head joint and the bed joint simultaneously to create a dense, completely filled mortar joint. The stretched line serves as a guide for alignment of the upper front edge of the brick. The trowel is drawn across the face of the wall to remove the excess mortar that has squeezed out of the joints. The economy of brickwork is based on the rapid, skillful repetition of this simple process. Some masons do not butter the head joint but, instead, slide the brick across the bed joint to build up a ridge of mortar on the end of the brick. The wall being built here is of sand mold bricks over a backup of 4-inch (92-mm) concrete blocks. Every fourth course is bonded through the block wythe with Flemish headers to create English Garden Wall Bond.* (Photos by the author)

from about $\frac{1}{4}$ inch (6.5 mm) to more than $\frac{1}{2}$ inch (13 mm). Thin joints work only when the bricks are identical to one another within very small tolerances and the mortar is made with a very fine sand. Very thick joints require a stiff mortar that is difficult to work with. Mortar joints are usually standardized at $\frac{3}{8}$ inch (9.5 mm), which is easy for the mason and allows for considerable distortion and unevenness in the bricks.

The joints in brickwork are *tooled*

as the mortar begins to harden, to give a neat appearance, and to compact the mortar into a profile that meets the visual and weather-resistive requirements of the wall (Figures 8.19, 8.20). Outdoors, the *weathered joint, vee joint,* and *concave joint* shed water and resist freeze-thaw damage better than the others. The weathered joint casts a small shadow line under each brick that imparts more of a textured appearance to the wall than the concave joint. Indoors, a *raked* or *stripped joint*

can be used if desired to accentuate the pattern of bricks in the wall and deemphasize the mortar.

After joint tooling the face of the brick wall is swept with a soft brush to remove crumbs of mortar left by the tooling process. If the mason has worked cleanly, the wall is now finished, but most brick walls are later given a final cleaning by scrubbing with muriatic acid (HCl) and rinsing with water to remove mortar stains from the faces of the bricks.

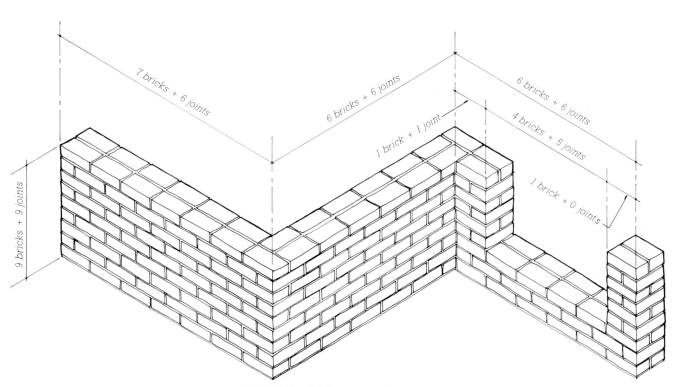

FIGURE 8.18
Dimensions for brick buildings are worked out in advance by the architect, based on the actual dimensions of the bricks and mortar joints to be used in the building. Bricks and mortar joints are carefully counted and converted to numerical dimensions for each portion of the wall.

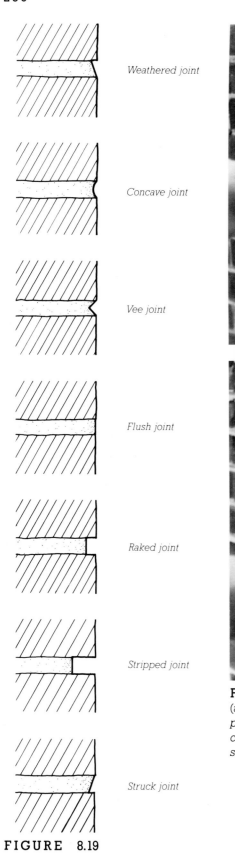

Weathered joint

Concave joint

Vee joint

Flush joint

Raked joint

Stripped joint

Struck joint

FIGURE   8.19
Joint tooling profiles for brickwork. The weathered joint, concave joint, and vee joint are suitable for outdoor use in severe climates.

FIGURE   8.20
(a)*Tooling horizontal joints to a concave profile. (b) Tooling vertical joints to a concave profile. The excess mortar squeezed out of the joints by the tooling* *process will be swept off with a brush, leaving a finished wall. (Courtesy of Brick Institute of America)*

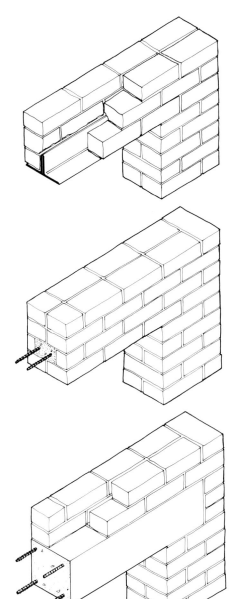

Spanning Openings in Brick Walls

Brick walls must be supported above openings for windows or doors. Lintels of reinforced concrete, reinforced brick, or steel angles (Figures 8.21, 8.22) are all equally satisfactory from a technical standpoint. The near-invisibility of the steel lintel is a source of delight to some designers but dissatisfies those who prefer that a building express visually its means of support. Wood is no longer used for lintels because of its tendency to shrink and allow the masonry above to settle and crack.

The *corbel* is an ancient structural device of limited spanning capability, but one that is often used for very small openings in brick walls, for beam brackets, and for ornament (Figures 8.23 through 8.25). If the successive projections are kept to a quarter of a brick length or less, a corbel is easy for a mason to lay.

The brick *arch* is a structural form so widely used and so powerful, both structurally and symbolically, that entire books could be devoted to it (Figure 8.26). Given a *centering* of wood or steel (Figure 8.27), a mason can lay a brick arch very rapidly, although the *spandrel*, with its numerous cut bricks, is much slower to make. In an arch of *gauged brick*, each brick is rubbed to the required wedge shape on an abrasive stone, which is laborious and ex-

FIGURE 8.21
Three types of lintels for spanning openings in brick walls. The double-angle steel lintel (top) requires some trimming of the first courses of brick but is scarcely visible in the finished wall. The reinforced brick lintel (center) works in the same manner as a reinforced concrete beam and gives no outward clues as to what supports the bricks over the opening. The precast reinforced concrete lintel (bottom) is clearly visible. For short spans, cut stone lintels without reinforcing can be used in the same manner as the concrete lintel.

FIGURE 8.22
Because of corbelling and arching action in the bricks, a lintel is considered to carry only the area of brickwork indicated by the shaded portion of this drawing. The broken line indicates a concealed steel angle lintel.

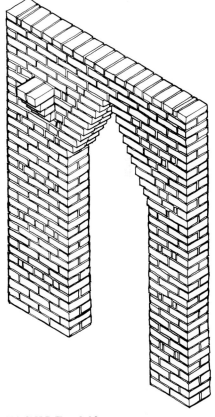

FIGURE 8.23
Corbelling has many uses in masonry construction. It is used in this example to span a door opening, and to create a bracket for support of a beam.

FIGURE 8.24
Corbelling creates a transition from the cylindrical tower to a hexagonal roof, and supports the interlocking arch ornamentation on the walls. Cut limestone is used for window sills, lintels, and arch intersections. The building is the Gothic cathedral in Albi, France. (Photo by the author)

FIGURE 8.25
All the skills of the nineteenth-century mason were called into play to create the corbels and arches of this brick cornice in Boston's Back Bay. (Photo by the author)

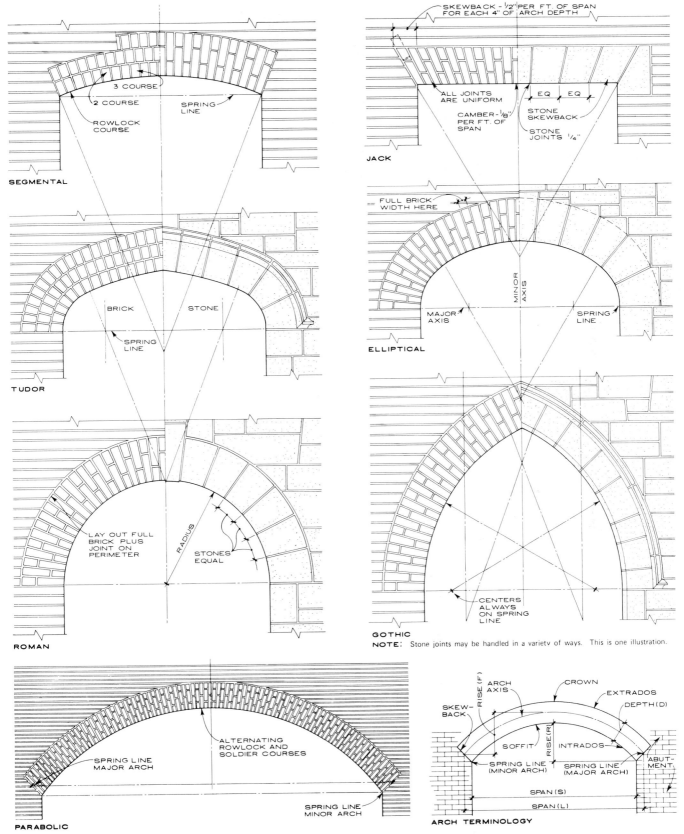

FIGURE 8.26

Arch forms and arch terminology in brick and cut stone. The spandrel is the area of wall that is bounded by the extrados of the arch. (Reprinted by permission of John Wiley and Sons, Inc., from Ramsey/Sleeper, Architectural Graphic Standards, 7th Edition, Robert T. Packard, A.I.A., Editor, copyright © 1981 by John Wiley and Sons, Inc.)

FIGURE 8.27

(a) *Two rough brick arches under
construction, each on its wooden
centering. (b) The brick locations were
marked on the centering in advance to
be sure that no partial bricks or unusual
mortar joint thicknesses will be required
to close the arch. (c) The brick arches
whose construction is illustrated in the
previous two photographs span a
barrel-vaulted fireplace room. The floor
is finished with quarry tiles. (Photos by
the author)*

pensive. The *rough arch*, which depends on wedge-shaped mortar joints for its curvature, is therefore much more usual in today's buildings (Figures 8.28, 8.29).

An arch translated along a line perpendicular to its plane produces a *barrel vault*. An arch rotated about its vertical centerline becomes a *dome*. From various intersections of these two basic roof shapes comes the infinite vocabulary of vaulted construction. Brick vaults and domes, if their lateral thrusts are sufficiently tied or buttressed, are strong, stable forms. In parts of the world where labor is inexpensive, they continue to be built on an everyday basis (Figure 8.30). In North America and most of Europe, where labor is more costly, they have been replaced almost entirely by less expensive, more compact structural elements such as beams and slabs of wood, steel, or concrete.

FIGURE 8.28
A rough brick triple rowlock arch spans a window opening. (Photo by the author)

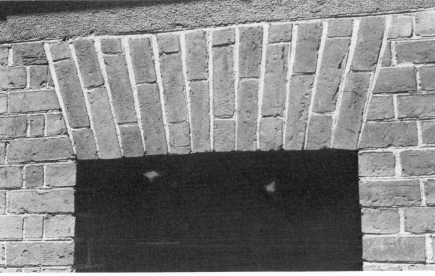

FIGURE 8.29
A jack arch (also called a flat arch) in a wall of Flemish Bond brickwork. (Photo by the author)

FIGURE 8.30

(a) *Masons in Mauritania, drawing on thousands of years of experience in masonry vaulting, build a dome for a patient room of a new hospital. The masonry is self-supporting throughout the process of construction; only a simple radius guide is used to maintain a constant diameter. The dome is double to insulate the room from the sun's heat. (b) The walls are buttressed with stack bond brick headers to resist the outward thrust of the domes.* (Courtesy of ADAUA, Geneva, Switzerland)

Reinforced Brick Masonry

Reinforced brick masonry (RBM) is analogous to reinforced concrete construction. The same type of deformed steel reinforcing bars used in concrete are placed in thickened collar joints to strengthen a brick wall or lintel. A reinforced brick wall (Figure 8.31) is created by constructing two wythes of brick the required distance apart, placing the reinforcing steel in the cavity, and filling the cavity with *grout*. Grout is a mixture of portland cement, aggregate, and enough water to cause the mixture to flow readily. ASTM C476 specifies the proportions and qualities of grout for use in filling masonry load-bearing walls.

There are two methods for grouting reinforced brick walls: low-lift and high-lift. In the *low-lift* method, the masonry is constructed to a height not greater than four feet (1200 mm) before grouting, taking care to keep the cavity free of mortar squeezeout and droppings, which might interfere with the placement of the reinforcing and grout. The vertical reinforcing bars are then inserted into the cavity and are left projecting at least thirty bar diameters above the top of the brickwork to transfer their loads to the steel in the next lift. The cavity is then filled with grout to within 1½ inches (38 mm) of the top, and the process is repeated for the next lift. In the *high-lift* method, the wall is grouted a story at a time. The cleanliness of the cavity is assured by temporarily omitting some of the bricks in the lowest course of masonry to create *cleanout* holes. As the bricklaying progresses the cavity is flushed periodically from above with water to drive debris down and out through the cleanouts. To resist the hydrostatic pressure of the wet grout, the wythes are held together by galvanized steel wire *ties* laid into the bed joints and across the cavity, usually at intervals of 24 inches (600 mm) horizontally and 16 inches (400 mm) vertically. After the cleanouts have been filled with bricks, and when the mortar has cured for at least three days, the reinforcing bars are placed and grout is pumped into the cavity from above in increments not more than 4 feet (1200 mm) high. Each increment is allowed to harden for an hour or so before the next increment is poured above it to minimize pressure on the brickwork.

The low-lift method is generally easier for small work where the grout is poured by hand. Where grout pumping equipment must be rented, the high-lift method is preferred because it minimizes rental costs.

While unreinforced brick walls are adequate for many structural purposes, RBM walls are much stronger against vertical loads, flexural loads from wind or earth pressure, and shear loads. With RBM it is possible to build bearing-wall buildings to heights formerly possible only with steel and concrete frames, and to do so with surprisingly thin walls (Figures 8.32, 8.33). RBM is also used for brick piers that are analogous to concrete columns, and, less commonly, for structural beams, slabs, and retaining walls.

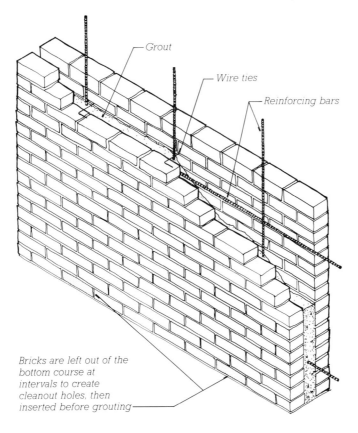

—Grout

—Wire ties

—Reinforcing bars

Bricks are left out of the bottom course at intervals to create cleanout holes, then inserted before grouting

FIGURE 8.31

A reinforced brick bearing wall is built by installing steel reinforcing bars in a thickened collar joint, then filling the joint with portland cement grout. The cleanout holes shown here are used in the high-lift method of grouting.

FIGURE 8.32
Twelve-inch (300-mm) reinforced brick walls bear the concrete floor and roof structures of a hotel. (Photo by the author)

FIGURE 8.33
The unreinforced brick walls of the sixteen-story Monadnock Building, built in Chicago in 1891, are 18 inches (460 mm) thick at the top, and 6 feet (1830 mm) thick at the base of the building. (Architects: Burnham and Root. Photo by William T. Barnum. Courtesy of Chicago Historical Society, IChi-18292)

FIGURE 8.34
Ornamental corbelled brickwork in an eighteenth-century New England chimney. Step flashings of lead sheet waterproof the junction between the chimney and the wood shingles of the roof. (Photo by the author)

FIGURE 8.35
In the gardens he designed at the University of Virginia, Thomas Jefferson used serpentine walls of brick that are only a single wythe thick. The shape of the wall makes it extremely stable against overturning despite its thinness. (Photo by Wayne Andrews)

FIGURE 8.36
Cylindrical bays of brick with stone lintels front these rowhouses. (Photo by the author)

FIGURE 8.37
Bricks were laid diagonally to create this mouse-tooth pattern. The window is spanned with a segmental arch of cut limestone. (Photo by the author)

FIGURE 8.38
Quoins *originated long ago as cut stone blocks used to form strong corners on walls of weak masonry materials such as mud bricks or round fieldstones. In more recent times, quoins (pronounced "coins") have been used largely for decorative purposes. At left, cut limestone quoins and a limestone water table dress up a Common Bond brick wall. The mortar joints between quoins are finished in a protruding beaded profile to emphasize the pattern of the stones. At right, brick quoins are used to make a graceful termination of a concrete block wall at a garage door opening. Notice that three brick courses match perfectly to one block course.* (Photos by the author)

FIGURE 8.39

Louis Sullivan's National Farmers' Bank in Owatonna, Minnesota, completed in 1908, rises from a red sandstone base. Enormous rowlock brick arches span the windows in the two street facades. Bands of glazed terra-cotta ornament in rich blues, greens, and browns outline the walls, and a flaring cornice of corbeled brick and terra-cotta caps the building. (Photo by Wayne Andrews)

FIGURE 8.40

Architect Edward Larrabee Barnes uses soldier arches of contrasting brick beneath a standing seam copper roof in this cathedral in Burlington, Vermont. (Photo by Nick Wheeler)

FIGURE 8.41
Frank Lloyd Wright used long, flat Roman bricks and cut limestone wall copings to emphasize the horizontality of the Robie House, built in 1906. (Photo by Mildred Mead. Courtesy of Chicago Historical Society, IChi-14191)

FIGURE 8.42
The Kline Science Center at Yale University rises on cylindrical piers of custom-made curved bricks. (Architects: Philip Johnson and Richard Foster. Photo courtesy of John Burgee Architects with Philip Johnson)

FIGURE **8.43**

Stone offers a wide range of expressive possibilities to the architect. The columns to the left are of polished granite, resting on a base of carved limestone. The limestone blocks have rough-pointed faces and tooth-axed edges. (Photo by the author)

STONE MASONRY

Types of Building Stone

Building stone is obtained by taking rock from the earth and reducing it to the required shapes and sizes for construction. Three types of rock are commonly quarried to produce building stone:

- *Igneous rock*, which is rock that was deposited in a molten state.
- *Sedimentary rock*, which is rock that was deposited by the action of water or wind.
- *Metamorphic rock*, which is either igneous or sedimentary rock that has been transformed by heat and pressure into a different type of rock.

Granite is the igneous rock most commonly quarried for construction in North America. It is a mosaic of mineral crystals, principally feldspar and quartz, and can be obtained in a range of colors that includes gray, black, pink, red, brown, buff, and green. Granite is nonporous, hard, strong, and durable, the most nearly permanent of building stones, suitable for use in contact with the ground or exposed to severe weathering. Its surface can be finished in any of a number of textures including a mirrorlike polish. In the United States it is quarried chiefly in the eastern mountains and the upper Midwest.

Limestone and *sandstone* are the principal sedimentary stones used in construction. Either may be found in a strongly stratified form, or in deposits that show little stratification (*freestone*). Neither will accept a high polish.

Limestone is quarried throughout North America, but the major quarries for large dimension stone are in Missouri and Indiana. Limestone may be composed either of calcium carbonate or a mixture of calcium and magnesium carbonates, originally furnished

His powers continually increased, and he invented ways of hauling the stones up to the very top, where the workmen were obliged to stay all day, once they were up there. Filippo had wineshops and eating places arranged in the cupola to save the long trip down at noon...He supervised the making of the bricks, lifting them out of the ovens with his own hands. He examined the stones for flaws and hastily cut model shapes with his pocketknife in a turnip or in wood to direct the men...

Giorgio Vasari (1511–1574), writing of Filippo Brunelleschi (1377–1446), the architect of the great masonry dome of the Cathedral of Santa Maria dei Fiori in Florence

in either case by the skeletons or shells of marine organisms. Its colors range from almost white through gray and buff to iron oxide red. It is a porous stone that contains considerable ground water (*quarry sap*) when quarried. While still saturated, most limestones are easy to work but are susceptible to frost damage. After seasoning in the air to evaporate the quarry sap, the stone becomes harder and is resistant to frost damage. *Travertine* is a richly patterned limestone, deposited by ancient springs, which is marblelike in its qualities.

Sandstone was formed from deposits of sand (silicon dioxide). It is quarried principally in New York, Ohio, and Pennsylvania. Two of its more familiar forms are *brownstone*, widely used in wall construction, and *bluestone*, a highly stratified stone espe-

cially suitable for pavings and wall copings.

Slate and *marble* are the major metamorphic stones utilized in construction. Slate was formed from clay, and is a dense, hard stone with closely spaced planes of cleavage, along which it is easily split into sheets, making it useful for paving stones, roof shingles, and thin wall facings. It is quarried in Vermont and Pennsylvania in black, gray, purple, blue, green, and red.

Marble is a recrystallized form of limestone. It is easily carved and polished and occurs in white, black, and nearly every color, often with beautiful patterns of veining. The marbles used in the United States come chiefly from Tennessee, Vermont, Georgia, Missouri, Italy, Greece, and Africa.

Quarrying and Milling of Stone

Some sedimentary rocks intended for rubble masonry can be broken out of the quarry strata with crowbars. Most building stone, however, must be cut from the quarry in large blocks. Figures 8.44 through 8.48 show some of the means by which this is done.

When preparing cut stone for a building, the stone producer works from the architect's drawings to make a set of shop drawings that show each individual stone and how it is to be dimensioned and shaped. After these drawings have been checked by the architect, they are used to guide the work of the mill in producing the stones. Rough blocks of stone are selected in the yard, brought into the mill, and sawed into slabs. The slabs are sawed, edged, planed flat, planed to a molding profile, turned on a lathe, or carved, as required for each piece, and given the desired surface finish. Holes for lewises and anchors are drilled as needed (Figures 8.49, 8.50). The finished pieces of stone are marked to correspond to their positions in the building as indicated on the shop drawings and shipped to the construction site.

FIGURE 8.44

A granite quarry is dominated by the guy derrick that hoists the blocks of stone to the surface. (Courtesy of Cold Spring Granite Company)

FIGURE 8.45

The wheels seen at the upper right corner of this granite quarry are part of a wire-sawing apparatus. (The large gantry saw at the lower right is a special-purpose machine for sawing channel-shaped chemical processing tanks directly from the quarry.) Drilling is also used to separate blocks of granite, as evidenced by the long parallel lines of drill holes on the blocks in the foreground. (Courtesy of Cold Spring Granite Company)

FIGURE 8.46
Separating a limestone quarry ledge into "cuts." In the foreground, a diamond core drill bores large-diameter holes downward into the rock. The holes will be joined to one another with a deep slot cut by the channeling machine, seen here on the level immediately above the core drill. Wire saw stanchions will then be inserted into each hole; one is visible just above the upper left corner of the channeling machine and appears to be an extension of the channeling machine. The wire saws use long loops of wire running between stanchions to slice down through the rock from the left to the right of the photograph from each hole, dividing the rock into long strips. An abrasive slurry of white silica sand and water is used as an abrasive under the fast-running wire. The slot between the holes carries the slurry out of the core holes when the wire saw is working. (Courtesy of the Indiana Limestone Institute)

FIGURE 8.47
Tipping out a key block of limestone, using a cable from a derrick. The wire-sawed face of a cut can be seen behind the block. (Courtesy of the Indiana Limestone Institute)

FIGURE 8.48
A key block is lifted out of the quarry by the derrick. A pneumatic drill cut holes in the block for the two steel "dogs," and the dogs are chained to the hook on the derrick line. When the cut has been completely cleared of key blocks, adjacent cuts will be simply tipped into the first cut, toward the camera, to free them from the ledge before lifting. (Courtesy of the Indiana Limestone Institute)

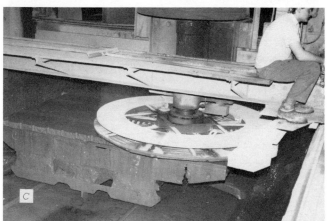

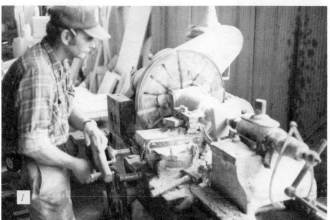

FIGURE 8.49

(a) *An overhead crane lifts a rough block of granite from the storage yard to transport it into the plant. The average block in this yard weighs 60 to 80 tons. (b) Inside the plant, a block of granite is sliced into slabs by a gang wire saw as the first step in processing it into building stone. Each wire is 1500 feet (450 m) long and moves at a speed of 60 miles per hour (95 km/hr) under a tension of 350 pounds (160 kg). An abrasive slurry of water and silicon carbide is applied continually to the* wire *to do the actual cutting. (c) A slab of granite is ground to produce a flat surface, prior to polishing operations. (d) If a textured* thermal finish *is desired, a propane-oxygen torch is passed across the slab under controlled conditions to cause small chips to explode off the surface. (e) A layout specialist, working from shop drawings for a specific building, marks a polished slab of granite for cutting. (f) The slab is cut into finished pieces with a large diamond saw that is capable of cutting 7* feet (2.1 m) per minute at a depth of 3 inches (76 mm). (g) Small pneumatic chisels with carbide-tipped bits are used for special details. (h) Hand polishers are used to finish edges that cannot be done by automatic machinery. (i) Cylindrical components are turned on a lathe. (j) A cylindrical column veneer is ground to true radius. (Photos a through h and j courtesy of Cold Spring Granite Company; photo i courtesy of the Indiana Limestone Institute)*

FIGURE 8.50

This 9-ton Corinthian column capital was carved from a single 30-ton block of Indiana limestone. Rough cutting took 400 hours, and carving another 500. Eight of these capitals were manufactured for a new portico on an existing church. (Architect: I. M. Pei & Partners. Photo courtesy of the Indiana Limestone Institute)

Stone Masonry

Stone is used in two fundamentally different ways in buildings: It may be laid in mortar, much like brick or concrete block, to make walls, arches, and vaults; or it may be mechanically attached in large sheets as a thin facing over the structural frame and walls of a building. This chapter deals only with stone masonry laid in mortar; thin facings of stone are covered in Chapter 15.

There are two simple distinctions that are useful in classifying patterns of stone masonry (Figures 8.51, 8.52):

- *Rubble* masonry, which is composed of unsquared pieces of stone, while *ashlar* is made up of squared pieces.
- *Coursed* stone masonry, which has continuous horizontal joint lines, while *uncoursed* or *random* does not.

Rubble can take many forms, from rounded river-washed stones to broken pieces from a quarry. It may be either coursed or uncoursed. Ashlar masonry may be coursed or uncoursed and may be in blocks the same size or several different sizes. Large squared blocks are known as *dimension stone*. The terms are obviously very general in their meaning, and the reader will find some variation in usage even among people experienced in the field of stone masonry.

Rubble stonework is laid very much like brickwork, except that the irregular shapes and sizes of the stones require the mason to select each stone carefully to fit the available space, and occasionally to trim a stone with a mason's hammer. Ashlar stonework, though similar in many ways to brickwork, has its unique problems. The stones are often too heavy to lift manually and must be lifted and lowered into place by a hoist. This requires a means of attaching the hoisting rope to the sides or top of the stone block so as not to interfere with the mortar joint, and several types of devices are commonly used for this purpose (Figure 8.53). Mortar joints in ashlar work are usually raked out after setting the stones, to avoid any uneven settling due to the more rapid drying and hardening of the mortar at the face of the wall. After the mortar in a wall has cured fully, the masons return to *point* the wall by filling the joints out to the face with mortar and tooling them to the desired profile. Stonework of either kind is laid with the *quarry bed* or grain of the stone running in the horizontal direction because stone is both stronger and more weather resistant in this orientation.

Some building stones, especially limestone and marble, deteriorate rapidly in the presence of acids. This restricts their outdoor use in regions whose air is heavily polluted and prevents their being cleaned with acid, as is often done with bricks. Exceptional care is taken during construction to keep stonework clean: Nonstaining mortars are used, high standards of workmanship are enforced, and the work is kept covered as much as possible. Flashings in stone masonry must be of plastic or nonstaining metal. Stonework may be cleaned only with mild soap, water, and a soft brush.

Random Rubble

Coursed Rubble

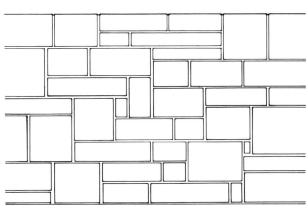

Random Ashlar

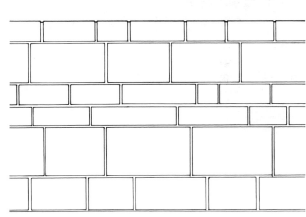

Coursed Ashlar

FIGURE 8.51
Rubble and ashlar stone masonry.

FIGURE 8.52
Uncoursed granite rubble masonry (above) *and uncoursed ashlar limestone* (below). (Photos by the author)

The rock ledges of a stone quarry are a story and a longing to me. There is a suggestion in the strata, and character in the formation. I like to sit and feel the stone as it is there. Often I have thought, were monumental great buildings ever given me to build I would go to the Grand Canyon of Arizona to ponder them.

Frank Lloyd Wright

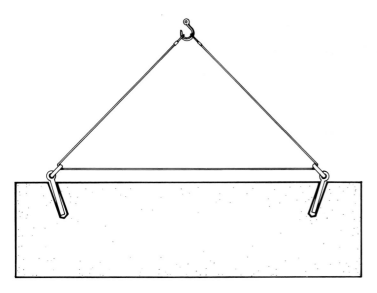

Lewis Pins

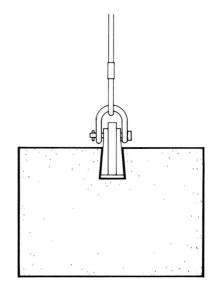

Box Lewis

FIGURE 8.53
Lewises, for lifting and placing blocks of building stone without interfering with the bed joint of mortar.

FIGURE 8.54
A method of attaching blocks of cut stone facing to a concrete block backup wall. For methods of attaching larger panels of stone to a building, see Figures 15.22, 15.24, and 15.25.

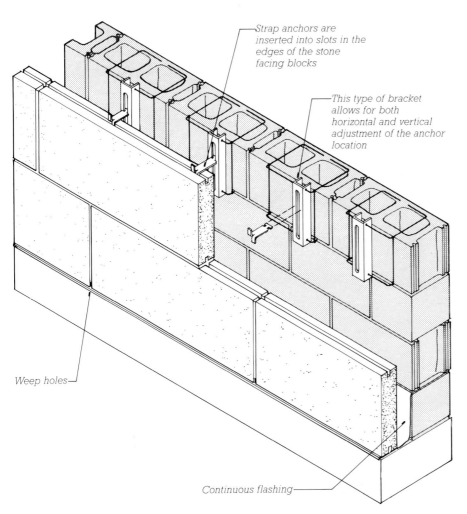

Strap anchors are inserted into slots in the edges of the stone facing blocks

This type of bracket allows for both horizontal and vertical adjustment of the anchor location

Weep holes

Continuous flashing

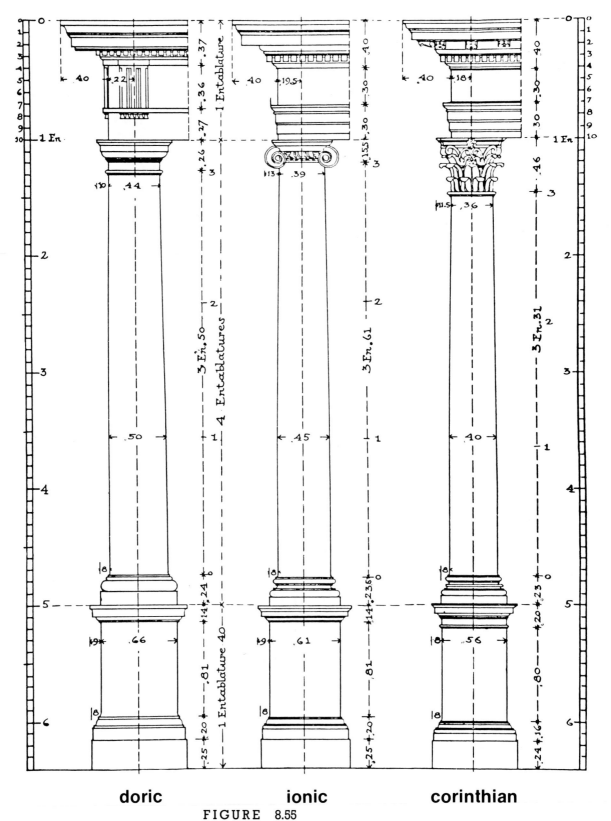

doric   **ionic**   **corinthian**

FIGURE   8.55

*Proportioning rules for the classical
orders of architecture.* (Courtesy of the
Indiana Limestone Institute)

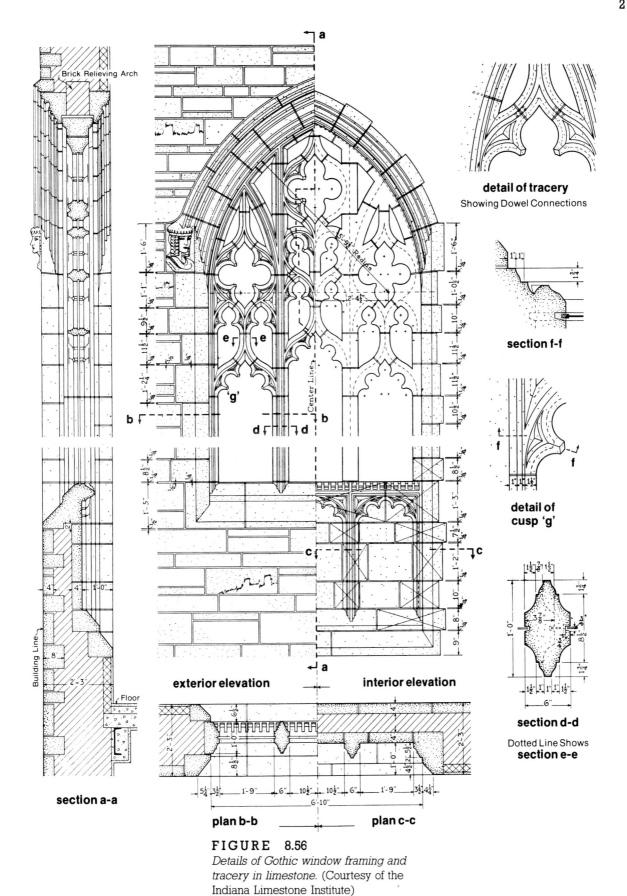

Brick Relieving Arch

detail of tracery
Showing Dowel Connections

section f-f

detail of
cusp 'g'

section d-d

Dotted Line Shows
section e-e

Building Line

Floor

section a-a

exterior elevation   interior elevation

plan b-b   plan c-c

FIGURE 8.56
*Details of Gothic window framing and
tracery in limestone.* (Courtesy of the
Indiana Limestone Institute)

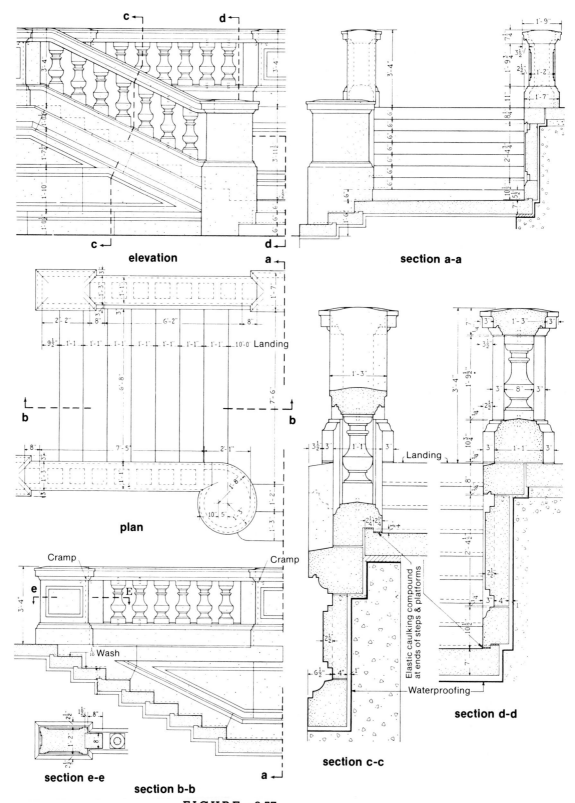

elevation

section a-a

plan

section c-c

section d-d

section e-e

section b-b

Cramp

Cramp

Wash

Landing

Landing

Elastic caulking compound at ends of steps & platforms

Waterproofing

FIGURE 8.57

*Details for a limestone stair and
balustrade in the classical manner. The
balusters are turned in a lathe.*
(Courtesy of the Indiana Limestone
Institute)

FIGURE 8.58
The Marshall Field Wholesale Store, built in 1885 in Chicago, rested on a two-story base of red granite. Its upper walls were built of red sandstone, and its interior was framed with heavy timber. (Architect: H. H. Richardson. Courtesy of Chicago Historical Society, IChi-01688)

FIGURE 8.59
Granite details from the rectory of H. H. Richardson's Trinity Church in Boston. (Photo by the author)

FIGURE 8.60

Cut stone detailing. Top row, reading from the upper left: *Random ashlar of broken face granite on H. H. Richardson's Trinity Church in Boston (1872–1877) still shows the drill holes of its quarrying. Dressed stonework in and around a window of the same church contrasts with the rough ashlar of the wall. Stonework on a nineteenth-century apartment building grows more and more refined as it distances itself from the ground and meets the brick walls* above. Bottom row, from the left: *The base of the Boston Public Library (1888–1895, McKim, Mead, and White, Architects) is constructed of pink granite, with strongly rusticated blocks between the windows. A chapel is simply detailed in limestone. A contemporary college library is clad in bands of limestone.* (Architects: Shepley, Bulfinch, Richardson and Abbott. All photos by the author)

FIGURE 8.61
Cut limestone loadbearing walls at the Wesleyan University Center for the Arts, designed by architects Kevin Roche, John Dinkeloo, and Associates. (Courtesy of Indiana Limestone Company)

FIGURE 8.62
The East Building of the National Gallery of Art in Washington, D. C. is constructed of pink marble. (Architects: I. M. Pei & Partners. Photo by Ezra Stoller © ESTO)

CONCRETE MASONRY

Concrete masonry units (CMUs) are manufactured in three basic forms: bricks, larger hollow units that will be referred to here as *concrete blocks*, and less commonly, larger solid units.

Manufacture of Concrete Masonry Units

Concrete masonry units are manufactured by vibrating a zero-slump concrete mixture into metal molds, then immediately turning out the wet blocks or bricks onto a rack so the mold can be reused, at the rate of a thousand or more units per hour. The racks of concrete masonry units are cured at an accelerated rate by subjecting them to steam, either at atmospheric pressure or, for faster curing, at higher pressure (Figure 8.63). After steam curing, the units are dried to a specified moisture content and bundled on wooden pallets for shipping to the construction site.

Concrete masonry units are made in a variety of sizes and shapes (Figures 8.64, 8.65) and in different densities of concrete, some of which use cinders, pumice, or expanded lightweight aggregates. Many colors and surface textures are available, and special shapes are relatively easy to produce if there is to be a sufficient number of units produced to amortize the expense of the mold. The major ASTM standards under which concrete masonry units are manufactured are C55 (concrete bricks), C90 (hollow loadbearing units), C129 (hollow nonloadbearing units), and C145 (solid loadbearing units). ASTM C90 establish two *Grades*, two *Types*, and three *Weights* of hollow loadbearing concrete masonry units (concrete blocks), as shown in Figure 8.66.

Hollow concrete block masonry is generally much more economical than that of brick or stone. The blocks themselves are cheaper on a volumetric basis and are made into a wall much more quickly because of their larger size (a single standard concrete block occupies the same volume as twelve modular bricks). Concrete blocks can be produced to any required degree

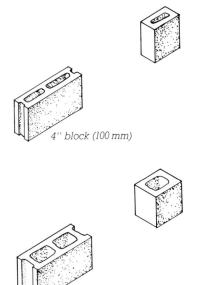

4" block (100 mm)

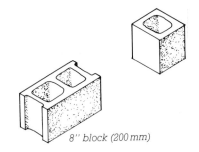

6" block (150 mm)

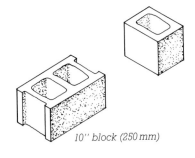

8" block (200 mm)

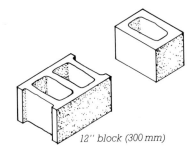

10" block (250 mm)

12" block (300 mm)

FIGURE   8.63
A forklift truck loads newly molded concrete blocks into an autoclave for steam curing. (Courtesy of Portland Cement Association, Skokie, Illinois)

FIGURE   8.64
American standard concrete blocks and half blocks. Each full block is nominally 8 inches (200 mm) high and 16 inches (400 mm) long.

of strength, and because their hollow cores allow for the easy insertion of reinforcing steel and grout, they are widely used in masonry bearing wall construction. Where surface qualities not available in block are desired on a masonry wall, concrete block is often used for the backup wythe behind a brick or stone facing. Block walls also accept plaster, stucco, or tile work directly.

Although innumerable special sizes, shapes, and patterns are available, American concrete blocks are standardized around an 8-inch (203.2-mm) cubic module. The most common block is nominally $8 \times 8 \times 16$ inches ($203.2 \times 203.2 \times 406.4$ mm). The actual size of the block is $7\frac{5}{8} \times 7\frac{5}{8} \times 15\frac{5}{8}$ inches ($193.7 \times 193.7 \times 396.9$ mm), which allows for a mortar joint $\frac{3}{8}$ inch (9.5 mm) thick. This size block is de-

signed to be laid conveniently with two hands (as compared to a brick, which is designed to be laid with one). Its double-cube proportions work well for running bond stretchers, for headers, and for corners. While concrete blocks can be cut with a diamond-bladed power saw (Figure 8.67), it is more economical and produces better results if the designer lays out buildings of block masonry in dimensional units

FIGURE 8.65

Other concrete masonry shapes. Concrete bricks are interchangeable with modular clay bricks. Header units accept the tails of a course of headers from a brick facing. The use of control joint units is illustrated in Figure 8.95. Bond beam units have space for horizontal reinforcing bars and grout and are used to tie a wall together horizontally, or for reinforced block lintels.

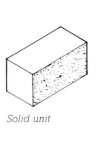
Solid unit

Cap or paving unit

Concrete brick

4" high unit (100 mm)

Capping unit

Header unit

Control joint unit

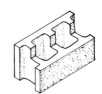
Channel bond beam
"LINTEL BLOCK"

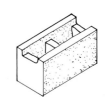
Low-web bond beam

Grades	
Grade N	General use above and below grade
Grade S	Above grade only, in exterior walls with weather-protective coatings, or in walls not exposed to weather
Types	
Type I	Moisture-controlled units, for use where drying shrinkage of units would cause cracking
Type II	Non-moisture-controlled units
Weights	
Normal weight	Made from concrete weighing more than 125 pcf (2000 kg/m³)
Medium weight	Made from concrete weighing between 105 and 125 pcf (1680–2000 kg/m³)
Lightweight	Made from concrete weighing 105 pcf or less (1680 kg/m³)

FIGURE 8.66

Grades, Types, and Weights of concrete masonry units, from ASTM C90.

that correspond to the module of the block (Figure 8.68). Nominal 4-inch, 6-inch, and 12-inch block thicknesses (101.6 mm, 152.4 mm, and 304.8 mm) are also common, as is a solid concrete brick that is exactly equal in size and proportion to a modular clay brick. A handy feature of the standard 8-inch block height is that it corresponds exactly to three courses of ordinary clay or concrete brickwork, making it easy to interweave blockwork and brickwork.

Laying Concrete Blocks

The accompanying photographic sequence (Figure 8.69) illustrates the technique of laying block walls. The mortar is identical to that used in brick walls, but in most walls only the face shells of the blocks are mortared, with the webs left unsupported.

Concrete masonry is often reinforced with steel to increase its load-bearing capacity and its resistance to cracking. Horizontal reinforcing is usually inserted in the form of grids of welded small-diameter steel rods that are laid into the bed joints of mortar at the desired vertical intervals (Figure 8.70). If stronger horizontal reinforcing is required, bond beam blocks (Figure 8.65) or special blocks with channeled webs (Figure 8.71) allow reinforcing bars to be placed in the horizontal direction. The bars are then embedded in grout before the next course is laid. The grout is contained

FIGURE 8.67
Concrete blocks and bricks can be cut as necessary on a water-cooled, diamond-bladed saw. For rougher sorts of cuts, a few skillful blows from the mason's hammer will suffice. (Courtesy of Portland Cement Association, Skokie, Illinois)

in the cores by a strip of metal mesh previously laid into the bed joint beneath to bridge across the core openings. Vertical block cores are easily reinforced by inserting bars and grouting, using either the low-lift or high-lift technique, as described earlier in this chapter. In most cases, only those cores that contain reinforcing bars are grouted, but sometimes all the vertical cores are filled, whether or not they contain bars (Figure 8.72).

Lintels for concrete block walls may be made of steel angles, combinations of rolled steel shapes, reinforced concrete, or bond beam blocks with grouted horizontal reinforcing (Figure 8.73).

In recent years, *surface bonding* of concrete block walls has found favor in certain low-rise building applications where the cost or availability of skilled labor is a problem. The blocks are laid without mortar, after which a thin layer of a special cementitious compound containing short fibers of alkali-resistant glass is plastered on each side of the wall. This surface bonding compound, when cured, joins the blocks securely to one another both in tension and in compression. It serves also as a surface finish whose appearance resembles stucco.

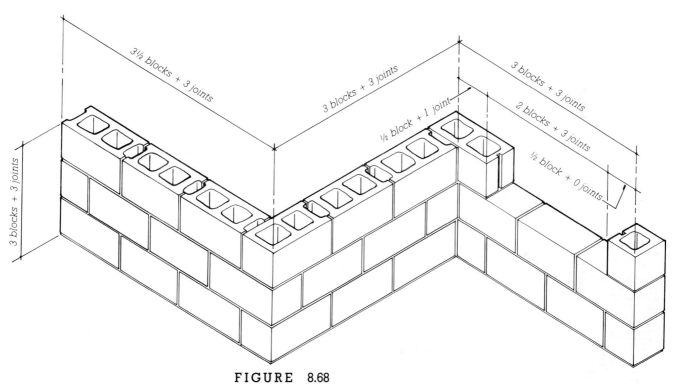

FIGURE 8.68
Concrete block buildings should be dimensioned to use uncut blocks except for special circumstances.

FIGURE 8.69

Laying a concrete block wall: (a) A bed of mortar is spread on the footing. (b) The first course of blocks for a lead is laid in the mortar. Mortar for the head joint is applied to the end of each block with the trowel before the block is laid. (c) The lead is built higher. Mortar is normally applied only to the face shells of the block and not to the webs. (d) As each new course is started on the lead, its height is meticulously checked with

either a folding rule or, as shown here, a story pole marked with the height of each course. (e,f) Each new course is also checked with a spirit level to be sure it is level and plumb. Time expended in making sure the leads are accurate is amply repaid in the accuracy of the wall and the speed with which blocks can be laid between the leads. (g) The joints of the lead are tooled. (h) A soft brush removes mortar

(continued)

crumbs after tooling. (i) *A mason's line is held taut between the leads on line blocks.* (j) *The courses of block between the leads are laid rapidly, and are aligned only with the line; no story pole or spirit level is necessary. The mason has laid bed joint mortar and buttered the head joints for a number of blocks.* (k) *Each course of infill blocks is completed with a closer, which must be inserted between blocks that have*

already been laid. *The head joints of the already-laid blocks are buttered.* (l) *Both ends of the closer block are also buttered with mortar, and the block is lowered carefully into place. Some touching up of the head joint mortar is often necessary.* (All photos courtesy of Portland Cement Association, Skokie, Illinois)

FIGURE 8.69 *(continued)*

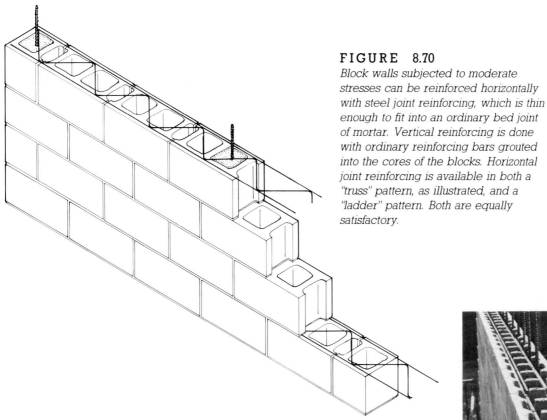

FIGURE 8.70
Block walls subjected to moderate stresses can be reinforced horizontally with steel joint reinforcing, which is thin enough to fit into an ordinary bed joint of mortar. Vertical reinforcing is done with ordinary reinforcing bars grouted into the cores of the blocks. Horizontal joint reinforcing is available in both a "truss" pattern, as illustrated, and a "ladder" pattern. Both are equally satisfactory.

FIGURE 8.71
In this proprietary system for building more heavily reinforced concrete block walls, the webs are grooved to allow the insertion of horizontal reinforcing bars into the wall. The cores of the blocks are then grouted to embed the bars. (Courtesy of G. R. Ivany and Associates, Inc.)

FIGURE 8.72
Grout is deposited in the cores of a reinforced block wall using a grout pump and hose. (Courtesy of Portland Cement Association, Skokie, Illinois)

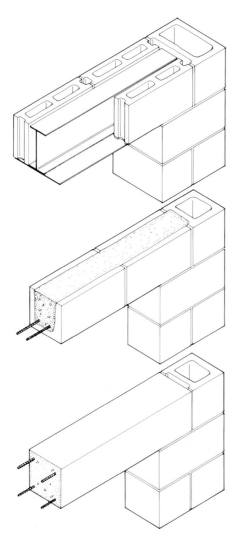

FIGURE 8.73

Lintels for openings in concrete block walls: At the top, a steel lintel made up of a wide-flange section welded to a plate. Steel angle lintels are also used. In the middle, a reinforced block lintel composed of bond beam units. At the bottom, a precast reinforced concrete lintel.

Decorative Concrete Masonry Units

Concrete masonry units are easily and economically manufactured in an unending variety of surface patterns, textures, and colors intended for exposed use in exterior and interior walls. A few such units are diagrammed in Figure 8.75, and some of the resulting surface textures are depicted in Figures 8.74, and 8.76 through 8.79. Mold costs for producing special units are low when spread across the number of units required for medium to large buildings, and many of the textured concrete masonry units that are now considered standard originated as special designs created by architects for particular buildings.

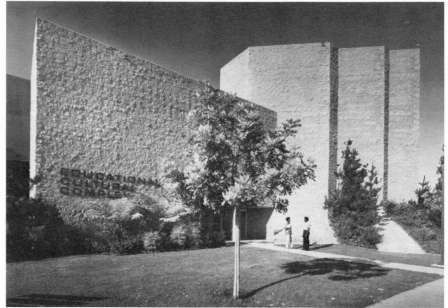

FIGURE 8.74

A facade of split concrete blocks. (Architects: Paderewski, Dean, Albrecht & Stevenson. Courtesy of National Concrete Masonry Association)

Scored-face unit

Ribbed-face unit

Ribbed-face unit

Fluted-face unit

Ribbed split-face unit

Angular-face unit

FIGURE 8.75
Some decorative concrete masonry units, representative of literally hundreds of designs currently in production. The scored-face unit, if the slot in the face is filled with mortar and tooled, produces a wall that looks as if it were made entirely of half blocks. The ribbed split-face unit is produced by casting "Siamese twin" blocks joined at the ribs, then shearing them apart.

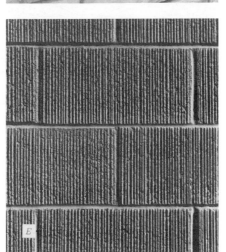

FIGURE 8.76
Some walls of decorative concrete blocks: (a) Split block. (b) Slump block, which is allowed to sag slightly after molding and before curing. (c) Split blocks of varying sizes laid in a random ashlar pattern. (d) Ribbed split-face blocks. (e) Striated blocks. (Courtesy of National Concrete Masonry Association)

FIGURE 8.77
Split blocks are used indoors in this high-school auditorium. (Architects: Marcel Breuer and Associates. Courtesy of National Concrete Masonry Association)

FIGURE 8.78
Ribbed split-face blocks. (Architects: Collins and Kronstadt. Courtesy of National Concrete Masonry Association)

FIGURE 8.79
High rise apartments constructed with reinforced bearing walls of fluted-face units. Specially cast blocks were used to produce the curved balcony fronts. (Architect: Paul Rudolph. Courtesy of National Concrete Masonry Association)

OTHER TYPES OF MASONRY UNITS

Bricks, stones, and concrete blocks are the most commonly used types of masonry units. In the past, hollow tiles of cast gypsum or fired clay were often used for partition construction (Figure 17.37), but both have been supplanted in the United States by concrete blocks. *Structural glazed tiles* of clay remain in use, especially for partitions where their durable, easily cleaned surfaces are advantageous, as in public corridors, institutional kitchens, locker and shower rooms, and industrial plants (Figure 17.38). *Glass blocks*, available in many textures and in clear, heat-absorbing, and reflective glass, are enjoying a second era of popularity after nearly disappearing from the market (Figures 8.80 and 8.81). *Structural terra cotta*, glazed or unglazed molded decorative units of fired clay, was widely used until a few decades ago and is often seen on the facades of late nineteenth-century masonry buildings in the United States (Figure 8.39). It is a material worthy of revival but is currently manufactured only by one or two companies.

MASONRY BEARING WALL CONSTRUCTION

Walls of brick, stone, concrete block, or combinations of masonry units can be used to carry roof and floor structures of wood, steel, concrete, or masonry vaulting. Because these masonry *bearing walls* also serve as exterior walls and interior partitions, they are often a very economical system of construction as compared to systems that carry their structural loads on columns of wood, steel, or concrete.

Wall Types

A masonry loadbearing wall can be classified in three different ways, ac-

FIGURE 8.80
Glass blocks used to enclose a conference room in a corporate office. (Courtesy of Pittsburgh Corning Corporation)

FIGURE 8.81
Exterior use of glass blocks in the wall of a stairwell. (Courtesy of Pittsburgh Corning Corporation)

cording to how it is constructed: It may be *reinforced* or *unreinforced;* it may be constructed entirely of one type of masonry unit, or it may be a *composite wall,* made of two or more types of units; and it may be a solid masonry wall, or a *cavity wall.*

Reinforced Masonry Walls

Loadbearing walls may be built with or without reinforcing. Unreinforced masonry walls cannot carry such high stresses as reinforced walls and may be unsuitable for use in regions with high seismic risk, but such walls have been used in this country to support buildings as tall as sixteen stories (Figure 8.33). In a multistory building of masonry loadbearing wall construction, the top floor walls carry only the load of the roof. The walls on the floor below carry the loads of the roof, the top floor, and the top floor walls. Each succeeding story, counting from top to bottom, carries a greater load than the one above, and an unreinforced masonry bearing wall must therefore grow progressively thicker. With reinforcing, this thickening can be reduced or eliminated entirely (Figure 8.32), with substantial savings in labor and materials, and much taller buildings may be constructed. Methods of reinforcing brick and concrete masonry walls have been described earlier in this chapter.

Composite Masonry Walls

For greater economy, masonry walls are often constructed with an outer wythe of stone or face brick and a backup wythe of hollow concrete masonry. The two are bonded together either by steel horizontal joint reinforcing, or by headers from the outer wythe that penetrate the backup wythe. Headers may penetrate completely through the backup (Figure 8.17) or may interlock with courses of header blocks (Figure 8.65).

Composite walls seldom have problems, but the designer should be sure that the differences in the thermal or moisture expansion characteristics of the two materials are not sufficient to cause warping or cracking of the wall. In composite loadbearing walls, the different strengths and elasticities of the two materials must be taken into account in locating the centroid of the wall and calculating the deformations it will experience under load.

Cavity Walls

A wall that is totally contained within a building is not exposed to the weather and is almost always constructed as a solid masonry wall, but masonry walls on the exterior of a building must resist water penetration and heat transfer and, for these reasons, are usually built with internal cavities. A masonry *cavity wall* consists of an inner, structural wythe and an outer wythe of masonry facing separated by a continuous airspace that is spanned only by corrosion-resistant metal ties that hold the wythes together (Figures 8.82 *b, c, d, e, f;* Figures 8.85, 8.88, 8.89).

Every masonry wall is porous to some degree, and especially as the wall ages and its mortar joints start to deteriorate, some water will find its way through the masonry if the wall is wetted for a sustained period. A cavity wall prevents the water from reaching the interior of the building by interposing the cavity between the outside and inside wythes of the wall. When penetrating water reaches the cavity, it has no place to go but down. When it reaches the bottom of the cavity, it is caught by a thin, impervious membrane of metal or plastic called a *flashing* and channeled through *weep holes* to the exterior of the building. To maintain the watertightness of a cavity wall, it is essential that no accidental bridges be created by which water can cross the cavity to the inner wythe. The metal ties should be fashioned in such a way that drops of water cannot cling to them and work their way across, and the cavity must be kept clear of mortar droppings during the laying of the masonry. Strips of wood suspended on wires in the cavity and pulled up periodically during the construction process catch falling mortar and debris and remove them from the cavity.

Detailing Masonry Walls
Flashings and Weep Holes

The purpose of flashings in masonry construction is to exclude moisture or to conduct any moisture that has penetrated the masonry back to the outside of the wall. There are two types of flashings: *External flashings* are used to prevent the penetration of water at intersections of the wall with roofs or decks; *internal flashings* catch water that has penetrated the masonry and redirect it to the outside of the wall. Internal flashings are used wherever water might penetrate or accumulate, in such locations as below masonry copings and sills, above lintels and shelf angles, at intermediate floors supported by the masonry wall, and at the base of the wall. Flashings are installed by the mason as the wall is constructed (Figures 8.83, 8.84).

An internal flashing is always accompanied by a line of *weep holes* to drain moisture to the outdoors as it accumulates. Weep holes are created with short pieces of oiled rope laid in the mortar joints and later pulled out (Figure 8.83), by plastic or metal tubes laid in the mortar, or by simply leaving head joints unmortared at intervals in the courses where weep holes are required (Figure 8.94). Weep holes are usually spaced 18 to 24 inches apart (450 to 600 mm) in the horizontal direction.

Flashings are made of copper, bituminous membranes, flexible plastic sheets, or combinations of these materials. Aluminum flashings react chemically with mortar and are therefore unsuitable for masonry work.

Thermal Insulation of Masonry Walls

A solid masonry wall is a good conductor of heat, which is another way

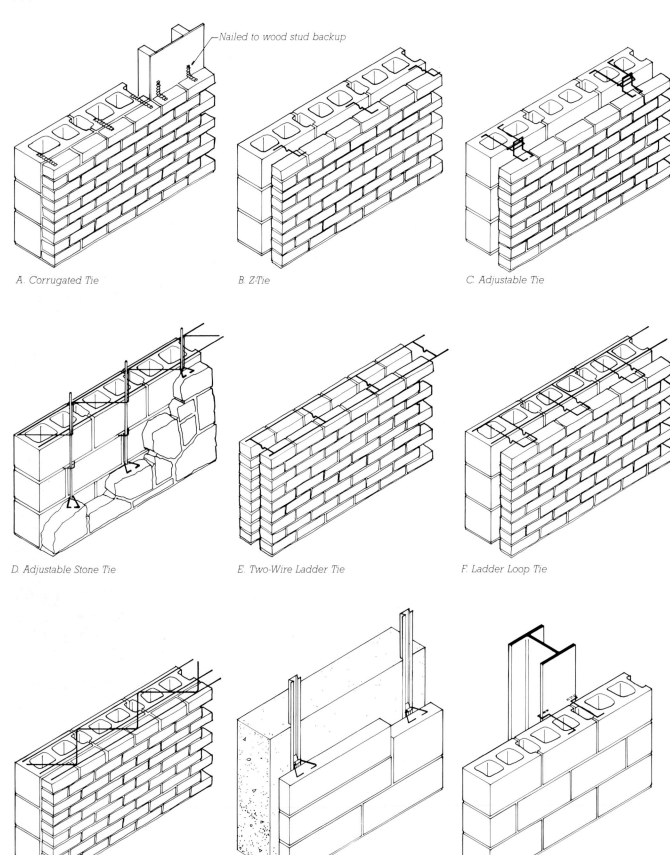

—Nailed to wood stud backup

A. Corrugated Tie

B. Z-Tie

C. Adjustable Tie

D. Adjustable Stone Tie

E. Two-Wire Ladder Tie

F. Ladder Loop Tie

G. Three-Wire Truss Tie

H. Dovetail Anchors for Concrete Backup

I. Steel Column Anchor

FIGURE 8.82
Masonry ties. These are only a few examples from among dozens of types available from a number of manufacturers to meet every conceivable need to tie wythes of masonry to one another or to supporting structures of wood, concrete, or steel. The ties shown in B, E, and F have drips to prevent drops of water from running across the tie to the inner wythe.

FIGURE 8.83
Laying a rowlock brick window sill. The flashing has already been installed over the face bricks and backup blocks, the bed joint of mortar has been applied, and the first rowlock bricks have been laid at the two corners of the sill. A mason's line has been stretched across the corner bricks as a guide for laying the remainder of the sill. The mason's hand holds a brick buttered with mortar, ready to be laid. The piece of cord in the bed joint will be pulled out as the mortar hardens to form a weep hole. (Courtesy of Brick Institute of America)

FIGURE 8.84
A steel angle lintel in place across a window head, with a plastic membrane flashing ready for laying of the first course of bricks across the lintel. (Courtesy of Brick Institute of America)

of saying that it is a poor insulator. In many hot, dry climates the capacity of an uninsulated masonry wall to store heat works effectively to keep the inside of the building cool during the hot day and warm during the cold night, but in climates with sustained cold or hot seasons, measures must be taken to improve the insulating qualities of masonry walls. The introduction of a cavity into a wall improves its thermal properties considerably, but not to a level fully sufficient for cold climates.

There are three general ways of insulating masonry walls: on the outside face, within the wall, and on the inside face. Insulation on the outside face is a relatively recent development. It is usually accomplished by adhering panels of plastic foam insulation to the outside surface of the masonry, and covering the foam with a thin, continuous layer of a polymeric stucco reinforced with glass fiber mesh. The masonry is completely concealed, and can be of inexpensive materials and workmanship. Exterior insulation is frequently used for insulating existing masonry buildings in cases where the exterior appearance of the masonry does not need to be retained. An advantage of exterior insulation is that the masonry is protected from temperature extremes, and can function effectively to stabilize the interior temperature of the building. A disadvantage is that the thin stucco coatings are usually not very resistant to denting or penetration damage, although repair is easy. Figure 8.90 details a building insulated in this manner, and photographs of the installation of exterior insulation are shown in Figure 15.46.

Insulation within the wall can take several forms. If the cavity in a wall is made sufficiently thick, the masons can insert slabs of plastic foam insulation against the inside wythe of masonry as the wall is built (Figure 8.85). The hollow cores of a concrete block wall can be filled with loose granular insulation (Figure 8.86) or with special molded-to-fit liners of foam plastic. Insulating the cores of concrete blocks does not retard the passage of heat through the webs of the blocks, however, and works

FIGURE 8.85

An insulated cavity wall. From left to right it is composed of a wythe of face brick, an airspace, two layers of polystyrene foam insulation, and an interior backup wythe of large bricks. Wire joint reinforcing ties the two wythes together. (Courtesy of Brick Institute of America)

FIGURE 8.87

A worker installs foam plastic insulation on the interior of a concrete masonry wall, using a patented system of steel furring strips in which only isolated clips, clearly visible in this photograph, contact the masonry wall, to minimize thermal conduction through the metal.

The furring strips serve as a base to which interior finish panels such as gypsum board can be attached. Other furring systems for masonry walls are shown in Figures 17.5 through 17.7. (Courtesy of W. R. Grace & Co.)

FIGURE 8.86

Insulating the cores in a concrete block wall with a dry fill insulation. The insulation is incombustible, inorganic, nonsettling, and treated to repel water that might be present in the block core from condensation or leakage. (Courtesy of W. R. Grace & Co.)

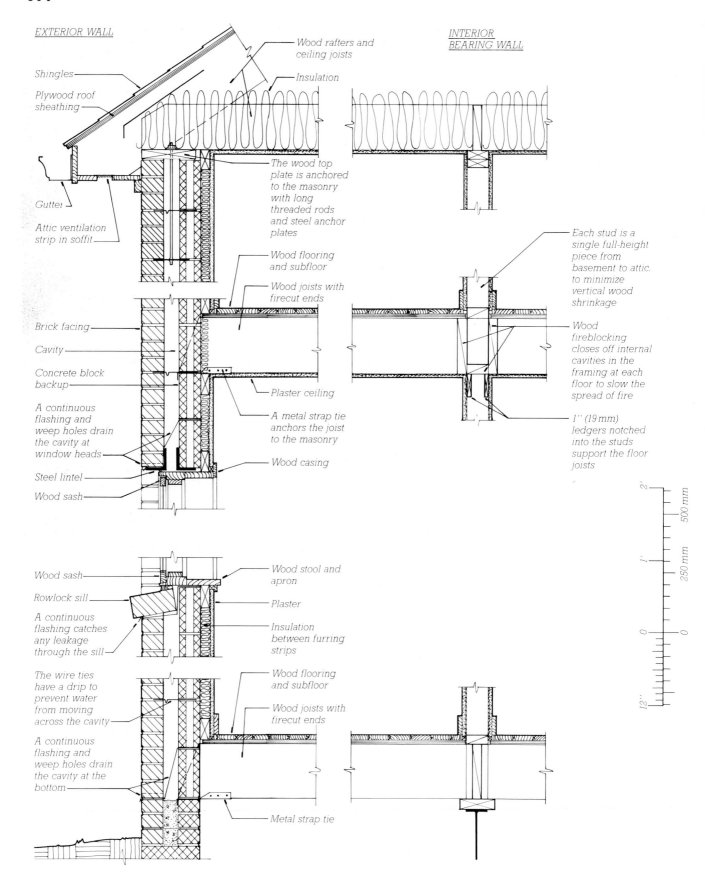

EXTERIOR WALL

Shingles

Plywood roof sheathing

Gutter

Attic ventilation strip in soffit

Wood rafters and ceiling joists

Insulation

The wood top plate is anchored to the masonry with long threaded rods and steel anchor plates

Wood flooring and subfloor

Wood joists with firecut ends

Brick facing

Cavity

Concrete block backup

A continuous flashing and weep holes drain the cavity at window heads

Steel lintel

Wood sash

Plaster ceiling

A metal strap tie anchors the joist to the masonry

Wood casing

INTERIOR BEARING WALL

Each stud is a single full-height piece from basement to attic, to minimize vertical wood shrinkage

Wood fireblocking closes off internal cavities in the framing at each floor to slow the spread of fire

1" (19 mm) ledgers notched into the studs support the floor joists

Wood sash

Rowlock sill

A continuous flashing catches any leakage through the sill

The wire ties have a drip to prevent water from moving across the cavity

A continuous flashing and weep holes drain the cavity at the bottom

Wood stool and apron

Plaster

Insulation between furring strips

Wood flooring and subfloor

Wood joists with firecut ends

Metal strap tie

2'

500 mm

1'

250 mm

0

0

12"

best when coupled with an unbroken layer of insulation in the cavity or on one face of the wall.

The inside surface of a masonry wall can easily be insulated by adhering slabs of plastic foam to the wall and applying plaster directly to the foam. More usual is the practice of attaching wood or metal *furring strips* to the inside of the wall with masonry nails or power-driven metal fasteners (Figures 8.87, 17.5 through 17.7). Furring strips are analogous to wood or metal studs in light frame construction and are insulated and finished in much the same manner. They can be of any desired depth to house the necessary thickness of fibrous or foam insulation. Furring strips can also solve another chronic problem of masonry construction by creating a space in which electrical wiring and plumbing can easily be concealed.

Spanning Systems for Masonry Bearing Wall Construction

"Ordinary" Joisted Construction

So-called *"Ordinary"* construction, in which the floors and roof are framed with wood joists and rafters, is the fabric of which American center cities were largely built in the nineteenth century. It still finds use today in a small percentage of new buildings and is listed as Type 3A and Type 3B construction in the building code table in Figure 1.1. Ordinary construction is essentially balloon framing (Figure 5.2) in which the outer walls of wood are replaced with masonry bearing walls. Balloon framing of the interior load-bearing partitions is preferable to platform framing because it minimizes the sloping of floors that might be caused by wood shrinkage along the interior lines of support. Figure 8.88 shows the essential features of Ordinary construction. Notice two very important details: the firecut ends of the joists, and the metal anchors used to tie the

wood framing and the masonry wall together. As might be expected from the combustibility of its interior construction, the size and extent of a building that can be built of Ordinary construction is severely restricted by code (Figure 1.1).

"Mill" Construction

"Mill" construction, listed as Type 4 in the code table, is similar in concept to Ordinary construction, but uses heavy timbers rather than light joists, rafters, and studs. Because heavy timbers are slower to catch fire and burn than nominal 2-inch (38-mm) framing members, Mill construction receives more favorable treatment under the code than Ordinary construction. Mill construction is discussed in detail in Chapter 4 and is illustrated in Figures 4.5 through 4.13.

Steel and Concrete Decks With Masonry Bearing Walls

Spanning systems of structural steel, sitecast concrete, and precast concrete are frequently used in combination with masonry bearing walls. Figures 8.89 and 8.90 show representative details of some of these combinations. Depending on the degree of fire resistance of the spanning elements, these constructions may be classified under Type 1 or Type 2 in the table in Figure 1.1.

SOME SPECIAL PROBLEMS OF MASONRY CONSTRUCTION

Expansion and Contraction

Masonry walls expand and contract slightly in response to changes in both temperature and moisture content. Thermal movement is relatively easy to quantify (Figure 8.91). Moisture movement is more difficult: Clay masonry units tend to absorb water and

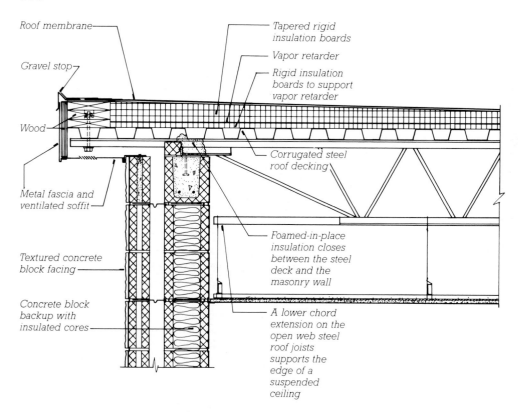

Roof membrane

Gravel stop

Wood

Metal fascia and
ventilated soffit

Textured concrete
block facing

Concrete block
backup with
insulated cores

Tapered rigid
insulation boards

Vapor retarder

Rigid insulation
boards to support
vapor retarder

Corrugated steel
roof decking

Foamed-in-place
insulation closes
between the steel
deck and the
masonry wall

A lower chord
extension on the
open web steel
roof joists
supports the
edge of a
suspended
ceiling

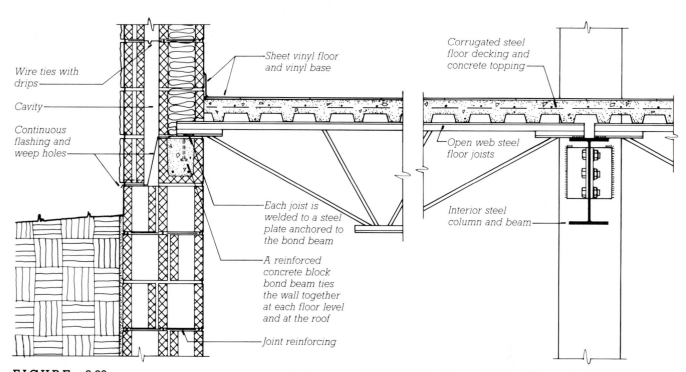

Wire ties with
drips

Cavity

Continuous
flashing and
weep holes

Sheet vinyl floor
and vinyl base

Each joist is
welded to a steel
plate anchored to
the bond beam

A reinforced
concrete block
bond beam ties
the wall together
at each floor level
and at the roof

Joint reinforcing

Corrugated steel
floor decking and
concrete topping

Open web steel
floor joists

Interior steel
column and beam

FIGURE 8.89

*An example of concrete block exterior
bearing walls with roof and floor of steel
joists and corrugated steel decking. In
areas with very severe winters,
insulation in the wall cavity or on the
inside surface of the wall would offer*
*better cold-weather performance than
insulation within the cores as shown
here. In areas of high seismic risk,
vertical reinforcing would be included
in the cores of the concrete block
backup. With vertical reinforcing, and*
*suitable fire protection for the steel floor
and roof structures, this type of
construction could be built many stories
high.*

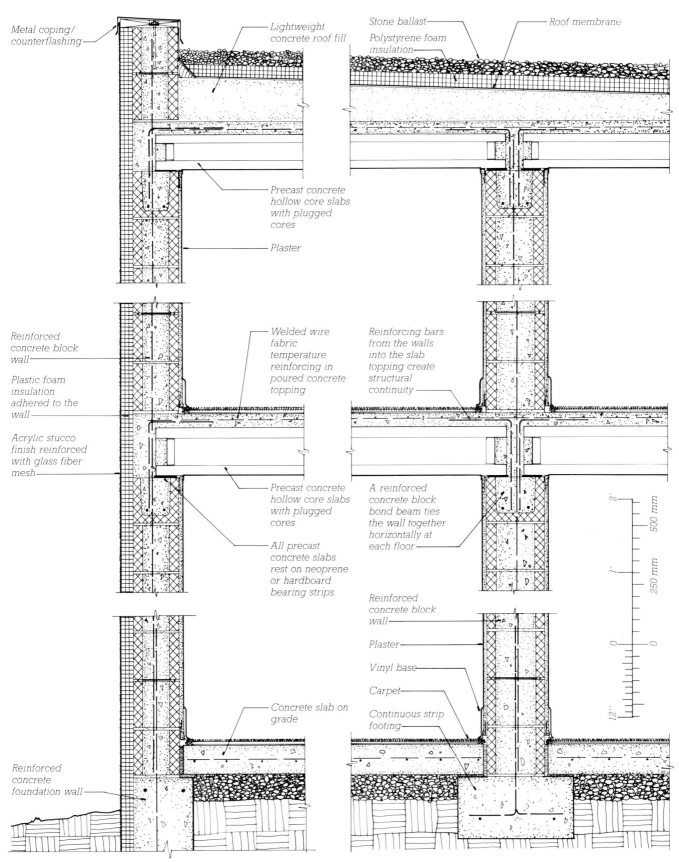

Metal coping/counterflashing

Lightweight concrete roof fill

Stone ballast

Roof membrane

Polystyrene foam insulation

Precast concrete hollow core slabs with plugged cores

Plaster

Reinforced concrete block wall

Plastic foam insulation adhered to the wall

Acrylic stucco finish reinforced with glass fiber mesh

Welded wire fabric temperature reinforcing in poured concrete topping

Reinforcing bars from the walls into the slab topping create structural continuity

Precast concrete hollow core slabs with plugged cores

All precast concrete slabs rest on neoprene or hardboard bearing strips

A reinforced concrete block bond beam ties the wall together horizontally at each floor

Reinforced concrete block wall

Plaster

Vinyl base

Carpet

Continuous strip footing

Concrete slab on grade

Reinforced concrete foundation wall

2'

500 mm

1'

250 mm

0

0

12"

FIGURE 8.90

An example of concrete block exterior and interior bearing walls with precast concrete hollow-core slabs spanning the roof and floors. This system can be used for multistory buildings. Full wall reinforcing is shown, with thermal insulation applied to the outside of the building beneath a thin layer of acrylic stucco. For further details of this exterior finishing system, see Chapter 15, especially Figure 15.46.

expand under moist conditions. Concrete masonry units usually shrink somewhat as they give off excess water following manufacture, which is the reason Type I units (moisture controlled) should be used where shrinkage must be minimized. Some types of building stone also shrink after quarrying.

To accommodate these movements, as well as foundation settlement and structural deflections, long walls and large buildings need to be broken at intervals with *expansion joints* (Figure 8.92). Expansion joints are also critical in masonry facings applied over multistory structural frames of steel or concrete, to prevent fracture of the masonry when the frame deflects under load, as discussed in Chapter 15.

In addition to expansion joints, it is wise to provide *control joints* in locations where masonry is likely to crack because of stress concentrations due to configurational changes in a wall (Figures 8.93, 8.94). Unless very large amounts of movement are anticipated, expansion and control joints can be constructed using details of the type shown in Figure 8.95. If the wall is reinforced, the horizontal reinforcing must be interrupted at expansion and control joints.

	in/in/°F	mm/mm/°C
Clay or shale brick masonry	0.0000036	0.0000065
Normal weight concrete masonry	0.0000052	0.0000094
Lightweight concrete masonry	0.0000043	0.0000077
Granite	0.0000047	0.0000085
Limestone	0.0000044	0.0000079
Marble	0.0000073	0.0000131
Normal weight concrete	0.0000055	0.0000099
Structural steel	0.0000065	0.0000117

FIGURE 8.91
Coefficients of thermal expansion for some masonry materials.

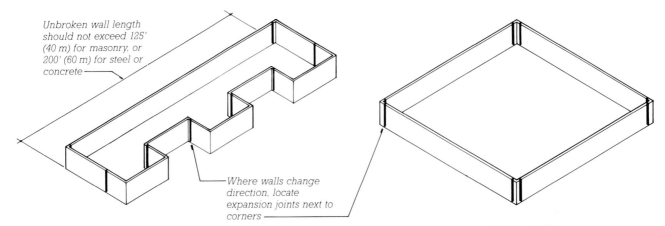

FIGURE 8.92
Placing expansion joints in masonry walls.

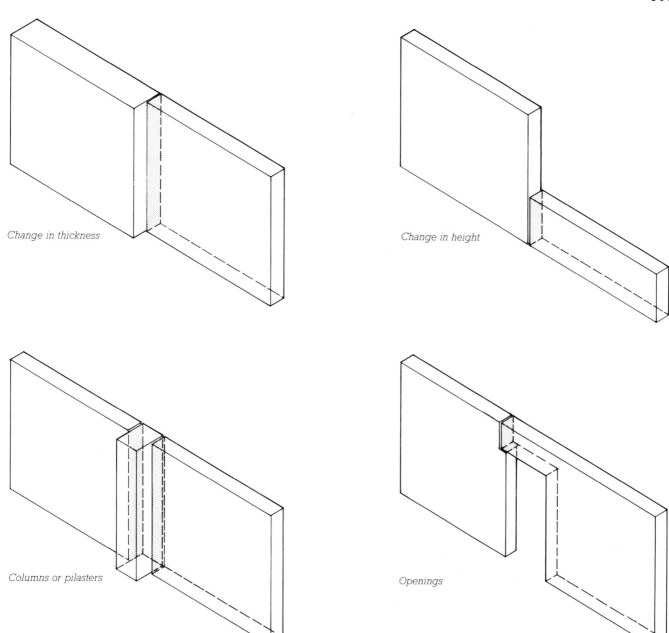

Change in thickness

Change in height

Columns or pilasters

Openings

FIGURE 8.93

*Control joints in masonry walls should
be located at discontinuities in the wall,
where cracks tend to form.*

FIGURE 8.94
*A window head and control joint. The
soldier course is supported on a steel
angle lintel. Two weep holes are visible
at the bottom of the soldier course.*
(Photo by the author)

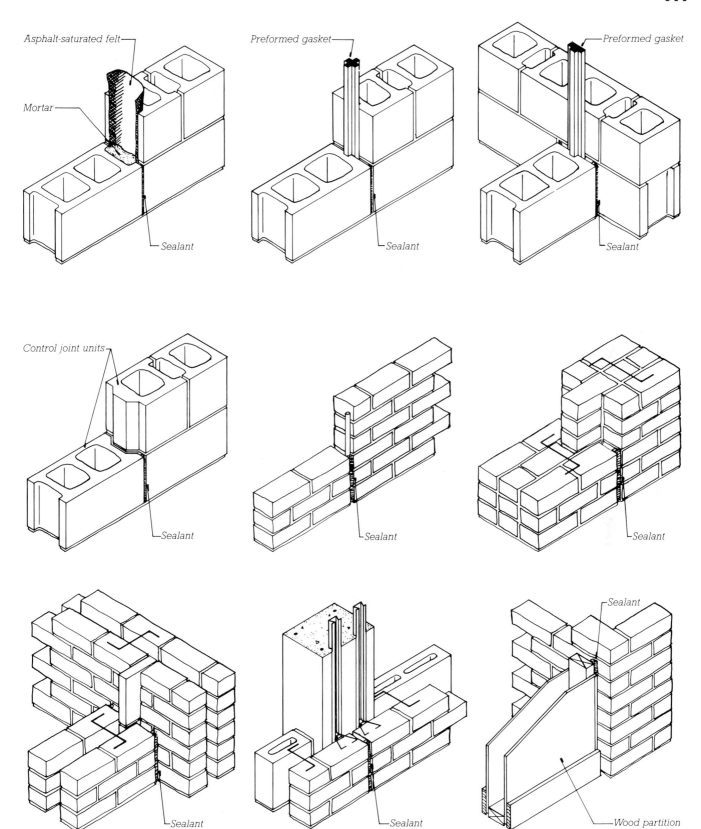

FIGURE 8.95

Some ways of making expansion and control joints in masonry. Detailing of sealant joints is covered in Chapter 15.

Efflorescence

Efflorescence is a fluffy powder, usually white, that sometimes appears on the surface of a wall of brick, stone, or concrete masonry (Figure 8.96). It consists of one or more water-soluble salts that were present either in the masonry units or in the mortar. These were brought to the surface and deposited there by water that had first seeped into the masonry, then migrated to the surface and evaporated. Efflorescence can usually be avoided by choosing masonry units that have been shown by laboratory testing not to contain water-soluble salts, and by using clean ingredients in the mortar. Most forms of efflorescence that form soon after the completion of construction are easily removed with water and a brush, and although efflorescence is likely to reappear, it will diminish and finally disappear with time as the salt is gradually leached out of the wall. Efflorescence that forms for the first time after a period of years is an indication that water is entering the wall, and can only be controlled by investigating and correcting the source of leakage.

Mortar Joint Deterioration

Mortar joints are the weakest link in most masonry walls. Water running down a wall tends to accumulate in the joints, where cycles of freezing and thawing weather can gradually *spall* (split off flakes of) the mortar in an accelerating process of destruction that eventually creates water leaks and loosens the masonry units. To forestall this process as long as possible, a suitably weather-resistant mortar formulation must be used, and joints must be well filled and tightly compacted with a concave, vee, or weathered tooling at the time the masonry is laid. Even with all these precautions, a masonry wall in a severe climate will show some joint deterioration after a few decades of weathering, and may require *tuck pointing*, a process of raking and cutting out the defective mortar and replacing it with fresh.

Moistureproofing of Masonry

Masonry materials, including mortar, are porous, and under some circumstances can transmit water from the outside of the wall to the inside. To prevent water from entering a building through a masonry wall, the designer should begin by specifying appropriate types of masonry units, mortar, and joint tooling. Above ground, masonry walls should be protected against excessive wetting of the exterior surface of the wall insofar as practical, through proper roof drainage and overhangs. Cavity wall construction should be utilized rather than solid wall construction. Beyond these measures, consideration can also be given to coating the wall with a sealer or paint. It is important that any exterior coating be highly permeable to water vapor to avoid blistering and rupture of the coating from outward vapor migration (see pages 517–519). *Silicone-based clear masonry sealers* are highly effective in most situations because they do not fill the pores of the wall, are highly permeable to water vapor, and do not alter the wall's appearance. They are sprayed onto the masonry and work by destroying the adhesive force between water and the microscopic interior surfaces of the masonry pores. Primer-sealers and paints based on portland cement fill the pores of the wall and render it more waterproof, without obstructing the outward passage of water vapor. Most other masonry paints are based on latex formulations and are also permeable to water vapor.

Below grade, masonry should first be *parged* (plastered on the outside) with two coats of Type M mortar to a total thickness of $\frac{1}{2}$ inch (13 mm). The parging serves to seal the cracks and pores in the masonry. After the parging has cured and dried, it can be coated with a bituminous dampproofing compound. If a truly watertight wall is re-

FIGURE 8.96
Efflorescence on a wall of Flemish Bond brickwork. (Photo by the author)

quired below grade, it can be covered with a built-up or elastomeric sheet membrane, or a layer of bentonite clay, as discussed in Chapter 2.

Cold Weather Construction

Mortar cannot be allowed to freeze before it has cured, or its strength and watertightness may be seriously damaged. In severe climates, special precautions are necessary if masonry work is carried through the winter months. These include such measures as protecting the masonry units and sand against water and freezing temperatures prior to use, warming the mixing water (and sometimes the sand as well) to produce a mortar mix at an optimum temperature for workability and curing, using a Type III (high early strength) cement to accelerate the curing of the mortar, and mixing the mortar in smaller quantities in order that it not cool excessively before it is used. The masons' workstations should be protected from wind with temporary enclosures and heated if temperatures inside the enclosures cannot be kept above freezing. The finished masonry must be protected against freezing for at least 2 to 3 days after it is laid, and tops of walls should be kept protected from rain and snow. Chemical accelerators and so-called "antifreeze" admixtures are, in general, harmful to mortar and reinforcing steel, and their use should be discouraged.

MASONRY AND THE BUILDING CODES

The utilization of masonry bearing walls in Type 1, Type 2, Type 3 (Ordinary) and Type 4 (Mill, or Heavy Timber) construction has been discussed in earlier sections of this chapter. Masonry walls are frequently constructed for interior partitions, fire separation walls, and fire walls in buildings of all construction types. Figure 8.97 gives rule-of-thumb values for fireresistance

Wall Type	Fireresistance (hours)	STC
4″ (100 mm) brick	1	45
8″ (200 mm) brick	4	52
10″, 12″ (250 mm, 300 mm) brick	4	59
4″ concrete block	$\frac{1}{2}$ to 1	44[a]
8″ concrete block	1 to 2	55[a]

[a]painted or plastered on both sides

FIGURE 8.97
Rule-of-thumb Fireresistance and Sound Transmission Class for some masonry partitions.

	Ultimate Compressive Strength	Density
Bricks	2000–14,000 psi (14–96 MN/m²)	100–140 lbs/ft³ (1600–2240 kg/m³)
Concrete masonry units	1500–6000 psi (10–41 MN/m²)	75–135 lbs/ft³ (1200–2160 kg/m³)
Limestone	2600–21,000 psi (18–147 MN/m²)	130–170 lbs/ft³ (2080–2720 kg/m³)
Sandstone	4000–28,000 psi (28–195 MN/m²)	140–165 lbs/ft³ (2240–2640 kg/m³)
Marble	9000–18,000 psi (62–123 MN/m²)	165–170 lbs/ft³ (2640–2720 kg/m³)
Granite	15,600–30,800 psi (108–212 MN/m²)	165–170 lbs/ft³ (2640–2720 kg/m³)

FIGURE 8.98
Ranges of strength and density for some masonry materials, to allow comparison of the properties of various bricks, blocks, and building stones. In practice, the allowable compressive stresses used in structural calculations for masonry are much lower than the values given, to take into account the strength of the mortar and a substantial factor of safety.

Material	Allowable Tensile Strength	Allowable Compressive Strength	Density	Modulus of Elasticity
Wood (average)	700 psi (4.83 MN/m²)	1,100 psi (7.58 MN/m²)	30 pcf (480 kg/m³)	1,200,000 psi (8,275 MN/m²)
Brick masonry (average)	0	250 psi (1.72 MN/m²)	120 pcf (1,920 kg/m³)	1,200,000 psi (8,275 MN/m²)
Steel (ASTM A36)	22,000 psi (151.69 MN/m²)	22,000 psi (151.69 MN/m²)	490 pcf (7,850 kg/m³)	29,000,000 psi (200,000 MN/m²)
Concrete (average)	0	1,350 psi (9.31 MN/m²)	145 pcf (2,320 kg/m³)	3,150,000 psi (21,720 MN/m²)

FIGURE 8.99

Comparative properties of four common structural materials.

FIGURE 8.100

Some masonry paving patterns in brick and granite. All six of these examples are laid without mortar in a bed of sand.

Outdoor pavings of masonry are also laid in mortar over reinforced concrete slabs. (Photos by the author)

ratings of some common types of masonry walls and partitions, along with Sound Transmission Class ratings to allow comparison of the acoustical isolation capabilities of these walls with those of the partition systems shown in Chapter 17.

THE UNIQUENESS OF MASONRY

Masonry is often chosen as a material of construction for its association in people's minds with qualities of permanence and solidity, and with buildings and architectural styles of the past. It is often chosen for its unique colors, textures, and patterns; for its fire resistance and its compliance with building code requirements; and for reasons of economy. Although masonry is labor intensive, it creates a high-performance structure and enclosure in a single operation by a single trade, bypassing the difficulties frequently encountered in managing the numerous trades and subcontractors needed to erect a comparable building of other materials.

Masonry, like Wood Light Frame construction, is a construction process carried out with small, relatively inexpensive tools and machines on the construction site. Unlike steel and concrete construction, it does not require (except in the case of ashlar stonework) a large and expensively equipped shop operation to process the major materials prior to erection. It shares with sitecast concrete construction the long construction schedule that requires special precautions and can encounter delays during periods of very hot, very cold, or very wet weather. But generally it does not require an extensive period of preparation and fabrication in advance of the beginning of construction because it uses standardized units and materials that are put into final form as they are placed in the building.

From the beginning of human civilization, masonry has been the me-

FIGURE 8.101
A detail of the porch of H. H. Richardson's First Baptist Church, Boston, built in 1871. (Photo by the author)

dium from which we have created our most carefully crafted and highly prized buildings. It has given us the massiveness of the pyramids, the inspirational elegance of the Parthenon, the light-filled loftiness of the great European cathedrals, as well as the reassuring coziness of the fireplace, the brick cottage, and the walled garden. Masonry can express our highest aspirations and our deepest yearnings for a rootedness in the earth. It reflects both the tiny scale of the human hand, and the boundless power of that hand to create.

C.S.I./C.S.C. Masterformat Section Numbers for Masonry	
04100	MORTAR
04150	MASONRY ACCESSORIES
	Anchors and Tie Systems
	Control Joints
	Joint Reinforcement
04200	UNIT MASONRY
04210	Brick Masonry
04220	Concrete Unit Masonry
04230	Reinforced Unit Masonry
04245	Structural Clay Facing Tile
04270	Glass Unit Masonry
04400	STONE
04410	Rough Stone
04420	Cut Stone
04450	Stone Veneer
04455	Marble
04460	Limestone
04465	Granite
04470	Sandstone
04475	Slate
04500	MASONRY RESTORATION AND CLEANING
04550	REFRACTORIES
04555	Flue Liners
04565	Firebrick

SELECTED REFERENCES

1. National Concrete Masonry Association. *Architectural and Engineering Concrete Masonry Details for Building Construction*. McLean, Virginia, 1976.

A treasury of typical concrete masonry details is contained in this 112-page book. (Address for ordering: P. O. Box 781, 2302 Horse Pen Road, Herndon, VA 22071.)

2. Brick Institute of America. *BIA Technical Notes on Brick Construction*. McLean, Virginia, various dates.

This is a large ring binder that contains the current set of *Technical Notes* with the latest information on every conceivable topic concerning bricks and brick masonry. (Address for ordering: 11490 Commerce Park Drive, Reston, VA 22091.)

3. Indiana Limestone Institute of America, Inc. *Indiana Limestone Institute of America, Inc., Handbook*. Bedford, Indiana, 1982–83.

Updated frequently, this large booklet tells of the history and provenance of Indiana limestone, recommended standards and details for its use, and architectural case histories. (Address for ordering: Suite 400, Stone City Bank Building, Bedford, IN 47421.)

4. Italian Institute for Foreign Trade. *Italian Marble Technical Guide*. Rome, 1982.

The first volume of this boxed set is an extended, heavily illustrated treatment of the quarrying and milling of Italian marble, with details of its installation in buildings. The second is a sumptuous full-color catalog of all the available colors and patterns of Italian marble. (Address for ordering: Italian Institute for Foreign Trade, Ministry of Foreign Trade, Rome, Italy.)

5. Brick Institute of America. *Principles of Clay Masonry Construction*, 2nd Edition. McLean, Virginia, 1973.

The seventy pages of this booklet present a complete curriculum in clay masonry construction for the student of building construction. (Address for ordering: See reference 2.)

6. Randall, Frank A., and William C. Panarese. *Concrete Masonry Handbook for Architects, Engineers, and Builders*, 4th Edition. Skokie, Illinois, Portland Cement Association, 1976.

This is a clearly written, beautifully illustrated guide to every aspect of concrete masonry. (Address for ordering: 5420 Old Orchard Road, Skokie, IL 60076.)

K E Y   T E R M S
A N D   C O N C E P T S

masonry unit
brick
concrete masonry unit
  (CMU)
mortar
mason
trowel
lime
quicklime
slaked or hydrated lime
masonry cements, light
  and dark
soft mud process
dry-press process
stiff mud process
water-struck
sand-struck, sand-mold
firing, burning
face bricks
header
stretcher

rowlock
soldier
wythe
course
bed joint
head joint
collar joint
bond
Common Bond
Flemish Bond
English Bond
cavity wall
leads
weathered joint
vee joint
concave joint
raked or stripped joint
lintel
corbel
arch
centering

spandrel
gauged brick
barrel vault
dome
reinforced brick masonry
  (RBM)
grout
low-lift method
high-lift method
igneous rock
sedimentary rock
metamorphic rock
granite
limestone
travertine
sandstone
slate
marble
lewis
rubble

ashlar
dimension stone
coursed
random
point
quarry bed
concrete block
surface bonding
structural glazed tiles
terra cotta
bearing wall
flashing
weep hole
tie
furring strip
Ordinary construction
firecut
Mill construction
efflorescence
tuck pointing

R E V I E W   Q U E S T I O N S

1. How many syllables are in the word "masonry?" (**Hint:** There cannot be more syllables in a word than there are vowels. Many people, even masons and building professionals, mispronounce this word.)

2. What are the most common types of masonry units?

3. List the functions of mortar.

4. How is the process of laying a brick different from that of laying a concrete block?

5. Why are mortar joints tooled?

6. What is the function of a structural brick bond such as Common or Flemish Bond?

7. Describe how a cavity wall works, and sketch its major constructional features.

8. In what ways is the laying of stone masonry different from the laying of brick?

9. Where should flashings be installed in a masonry wall?

10. Where should weep holes be provided? Describe the function of a weep hole and indicate several ways in which it may be constructed.

11. What are the differences between Ordinary construction and Mill construction? What features of each are related to fire resistance?

E X E R C I S E S

1. What are the allowable height and floor area for a restaurant in a building of Mill construction (Type 4)? How do these figures change if unprotected Ordinary construction (Type 3B) is used instead? What if unprotected steel joists are substituted for the wood joists? Or precast concrete plank floors with a two hour fire rating?

2. What is the exact length of a wall of standard $8 \times 8 \times 16$ concrete blocks that is 17 blocks long?

3. What is the exact length of an opening that is $4\frac{1}{2}$ blocks long in a concrete block wall?

4. What is the exact height of a brick wall that is 44 courses high, when three courses of brick plus their three mortar joints are 8 inches (203.2 mm) high?

5. Design a masonry gateway for one of the entrances to a college campus with which you are familiar. Choose whatever type of masonry you feel is appropriate, and make the fullest use you can of the decorative and structural potentials of the material.

STEEL FRAME CONSTRUCTION

Ironworkers place open-web steel joists on a frame of steel wide-flange sections as a crane lowers bundles of joists from above. (Photo by Balthazar Korab. Courtesy of Vulcraft Division of Nucor Corporation)

Steel, strong and stiff, is a material of slender towers and soaring spans. Precise and predictable, light in proportion to its strength, it is also well suited to rapid construction, highly repetitive building frames, and architectural details that satisfy the eye with a clean, mechanical elegance. Among the metals it is uniquely plentiful and inexpensive. If its weaknesses—a tendency to corrode in certain environments and a loss of strength during severe building fires—are held in check by intelligent construction measures, it offers the designer possibilities that exist in no other material.

FIGURE 9.1
*Landscape architect Joseph Paxton
designed the Crystal Palace, an
exposition hall of cast iron and glass,
which was built in London in 1851.*
(Bettmann Archive)

FIGURE 9.2
*Allied Bank Plaza, designed by
Architects Skidmore, Owings, and
Merrill.* (Courtesy of American Institute
of Steel Construction)

HISTORY

Prior to the beginning of the nineteenth century, metals had little structural role in buildings except in connecting devices. The Greeks and Romans used hidden cramps of bronze to join blocks of stone, and architects of the Renaissance countered the thrust of masonry vaults with wrought iron chains and rods. The first all-metal structure, a cast iron bridge, was built in the late eighteenth century in England and still carries traffic across the Severn River more than two centuries after its construction. Cast iron and wrought iron were used increasingly for framing industrial buildings in Europe and North America in the first half of the nineteenth century, but their usefulness was limited by the unpredictable brittleness of cast iron and the relatively high cost of wrought iron.

Until this time, steel was a rare and expensive material, produced only in small batches for such applications as weapons and cutlery. Plentiful, inexpensive steel first became available in the 1850s with the introduction of the Bessemer process, in which air was blown into a vessel of molten iron to burn out the impurities. By this means a large batch of iron could be made into steel in about twenty minutes, and the resulting steel was vastly superior in structural properties to the cast iron. Another economical steel-making process, the open hearth method, was developed in Europe in 1868 and was soon adopted in America. By 1889, when the Eiffel Tower was built of

FIGURE 9.3
Engineer Gustav Eiffel's magnificent tower of wrought iron was constructed in Paris from 1887 to 1889. (Photo by James Austin, Cambridge, England)

FIGURE 9.4
The first true steel frame skyscraper was the Home Insurance Company Building, designed by William LeBaron Jenney and built in Chicago in 1883. The steel framing was fireproofed with masonry, and the exterior masonry facings were supported on the steel frame. (Photo by Wm. T. Barnum. Courtesy of Chicago Historical Society ICHi-18293)

wrought iron (Figure 9.3), several steel frame skyscrapers had already been erected in the United States (Figure 9.4). A new material of construction had been born.

THE MATERIAL STEEL

Steel

Steel is any of a range of alloys of iron and carbon that contain less than 2 percent carbon. Ordinary structural steel, called *mild steel*, contains less than three-tenths of 1 percent carbon, plus traces of detrimental impurities such as phosphorus, sulfur, oxygen, and nitrogen, and of beneficial elements such as manganese and silicon. Ordinary *cast iron*, by contrast, contains 3 percent to 4 percent carbon and considerable quantities of impurities. Carbon content is a crucial determinant of the properties of a ferrous metal: Too much carbon makes a hard but brittle metal, while too little produces a soft, weak material. Thus mild steel is iron whose properties have been optimized for structural purposes by controlling the amount of carbon and other elements in the metal.

Iron is produced in a blast furnace charged with alternating layers of iron ore (oxides of iron), coke (coal whose volatile constituents have been distilled out, leaving only carbon), and crushed limestone (Figure 9.6). The coke is burned by large quantities of air forced into the bottom of the furnace to produce carbon monoxide,

FIGURE 9.5
Molten iron is poured into a crucible to begin its conversion to steel in the basic oxygen process. (Courtesy of United States Steel Corporation)

The gap between stone and steel-and-glass was as great as that in the evolutionary order between the crustaceans and the vertebrates.

Lewis Mumford, *The Brown Decades* (New York: Dover Publications, Inc., 1955), pages 130–131.

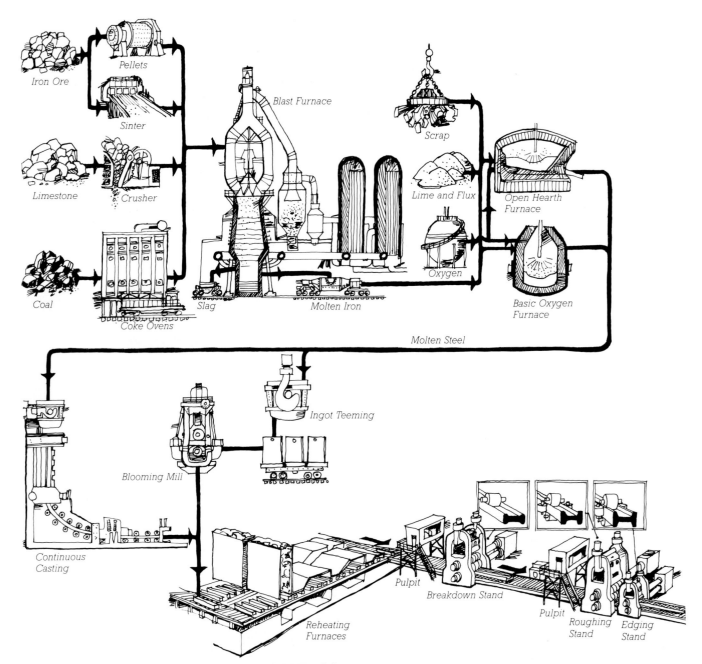

Iron Ore

Pellets

Sinter

Limestone

Crusher

Coal

Coke Ovens

Slag

Blast Furnace

Molten Iron

Scrap

Lime and Flux

Oxygen

Open Hearth Furnace

Basic Oxygen Furnace

Molten Steel

Continuous Casting

Blooming Mill

Ingot Teeming

Reheating Furnaces

Pulpit

Breakdown Stand

Pulpit

Roughing Stand

Edging Stand

FIGURE 9.6

The steel-making process, from iron ore to structural shapes. Notice particularly the steps in the evolution of a wide-flange shape as it progresses through the various stands in the rolling mill. (Adapted from "Steelmaking Flowlines," by permission of the American Iron and Steel Institute)

which reacts with the ore to reduce it to elemental iron. The limestone forms a slag with various impurities, but large amounts of carbon and other elements are inevitably incorporated into the iron. The molten iron is drawn off at the bottom of the furnace and held in a liquid state for processing into steel. The manufacture of a ton of iron requires about $1\frac{3}{4}$ tons of iron ore, $\frac{3}{4}$ ton of coke, $\frac{1}{4}$ ton of limestone, and 4 tons of air.

Today most structural steel is produced by the *basic oxygen process*, in which a water-cooled lance is lowered into a container of molten iron and recycled steel scrap until it is just above the surface of the metal. A stream of pure oxygen at very high pressure is blown from the hollow lance into the metal to burn off the excess carbon and impurities. A flux of lime and fluorspar reacts with other impurities, particularly phosphorus, to form a slag that is discarded. New metallic elements may be added to the container at the end of the process to adjust the composition of the steel as desired: Manganese gives resistance to abrasion and impact, molybdenum gives strength, vanadium imparts strength and toughness, nickel and chromium give toughness and stiffness. The entire process takes place with the aid of careful sampling and laboratory analysis techniques to assure the finished quality of the steel and takes less than an hour from start to finish.

Molten steel from the furnace is cast into molds to produce *ingots*, which are then maintained at an elevated temperature in a heated "soaking" pit preparatory to rolling, or it is cast continuously into slabs or blooms, ready for the structural mill.

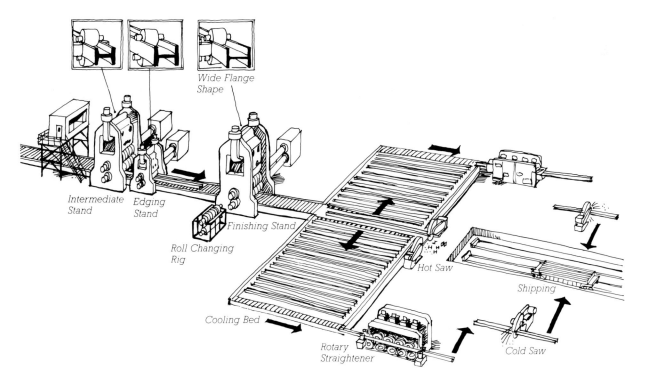

Wide Flange Shape

Intermediate Stand *Edging Stand*

Roll Changing Rig

Finishing Stand

Cooling Bed

Rotary Straightener

Hot Saw

Cold Saw

Shipping

FIGURE 9.7
A glowing steel ingot is worked by the rolls of a blooming mill. Next it will pass to the stands that roll it into standard structural shapes. (Courtesy of Bethlehem Steel Corporation)

Production of Structural Shapes

The rolling process for ingots commences in the *blooming mill*, where the hot ingot is reduced in size by squeezing it between steel rollers, producing a rectangular prism of steel of smaller cross section known as a *bloom* (Figures 9.6, 9.7). The bloom is then taken to the *structural mill*, still at a temperature of about 2200 degrees Fahrenheit (1200° C). Here it passes through a succession of rollers

FIGURE 9.8
A hot saw cuts pieces of wide-flange stock from a continuous length that has just emerged from the finishing stand in the background. Workers in the booth control the process. (Courtesy of United States Steel Corporation)

that press the metal into progressively more refined approximations of the desired shape and size, until it exits from the last set of rollers ready for cooling. A typical wide flange shape makes forty passes through sets of rollers before it reaches its final size and shape, gaining in speed each time its cross-sectional area is reduced, until it is traveling at about 15 miles per hour (25 km/h) through the last roller stand. After cooling, a straightening machine takes out such minor crookedness as may have been rolled into each piece, and a saw cuts the lengths of steel into shorter pieces (Figure 9.8). Each piece is then labeled with its shape designation and the number of the batch of steel from which it was rolled; later, when the piece is shipped to a fabricator, it will be accompanied by a certificate that gives the chemical analysis of that particular batch, as evidence that the steel meets standard structural specifications.

The roller spacings in the structural mill are adjustable; by varying the spacings between rollers a number of different shapes of the same nominal dimensions can be produced (Figure 9.9). This provides the architect and structural engineer with a finely graduated selection of shapes from which to select each structural member in a building and assures that little steel will be wasted through the specification of shapes that are larger than required.

Wide-flange shapes are used for most beams and columns, superseding the older *American Standard* (I-beam) shapes. American Standard shapes are

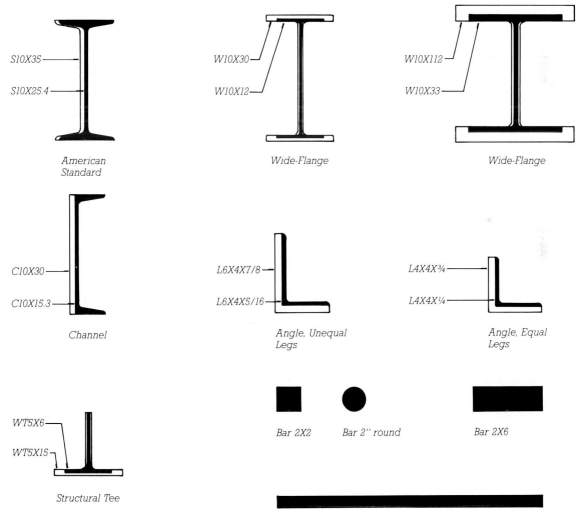

S10X35
S10X25.4
American Standard

W10X30
W10X12
Wide-Flange

W10X112
W10X33
Wide-Flange

C10X30
C10X15.3
Channel

L6X4X7/8
L6X4X5/16
Angle, Unequal Legs

L4X4X¾
L4X4X¼
Angle, Equal Legs

WT5X6
WT5X15
Structural Tee

Bar 2X2 *Bar 2" round* *Bar 2X6*

Plate 24X½

FIGURE 9.9
Examples of the standard shapes of structural steel, showing how different weights of the same section are produced by varying the spacing of the rollers in the structural mill.

less efficient structurally than wide-flanges because the roller arrangement that produces them is incapable of increasing the amount of steel in the flanges without also adding steel to the web, where it does little to increase the load-carrying capacity of the member. Wide-flanges are available in a vast range of sizes and weights. The smallest available depth in the United States is a nominal 4 inches (100 mm), and the largest is 36 inches (900 mm). Weights per foot range from 12 to 730 pounds (18 to 1080 kg/m), the latter for a nominal 14-inch (360-mm) shape with flanges nearly 5 inches (130 mm) thick. Wide-flanges come in two basic proportions: tall and narrow for beams,

and squarish for columns and foundation piles. The accepted nomenclature for wide-flange shapes begins with a letter W, followed by the nominal depth of the shape, a multiplication sign, and the weight of the shape in pounds per foot. Thus, a W12 × 26 is a wide-flange shape nominally 12 inches (305 mm) deep weighing 26 pounds per foot of length (38.5 kg/m). More information about this shape is contained in a table of dimensions and properties (Figure 9.11): Its actual depth is 12.22 inches (310.4 mm) and its flange is 6.49 inches (164.9 mm) wide. These proportions indicate that the shape is mainly intended for use as a beam or girder. Reading across

the table square by square, the designer can learn everything there is to know about this section, from its thicknesses and the radii of its fillets to various quantities that are useful in computing its structural behavior under load. At the upper end of the portion of the table dealing with 12-inch (305-mm) wide-flanges we find shapes weighing up to 336 pounds per foot (501 kg/m), with actual depths of almost 17 inches (432 mm). These heavier shapes have flanges nearly as wide as the shapes are deep, suggesting that they are intended for use as columns.

Steel *angles* (Figure 9.12) are extremely versatile. They can be used for very short beams supporting small

Shape	Sample Designation	Explanation	Range of Available Sizes
Wide-flange	W21 × 83	W denotes a wide-flange shape. 21 is the nominal depth in inches, and 83 is the weight per foot of length in pounds	Nominal depths from 4 to 18″ to 2″ increments, and from 18 to 36″ in 3″ increments
American Standard	S18 × 70	S denotes American Standard. 18 is the nominal depth in inches, 70 is the weight per foot of length in pounds	Nominal depths of 3″, 4″, 5″, 6″, 7″, 8″, 10″, 12″, 15″, 18″, 20″, and 24″
Angle	L4 × 3 × $\frac{3}{8}$	L denotes an angle. The first two numbers are the nominal depths of the two legs, and the last is the thickness of the legs	Leg depths of 2″, 2$\frac{1}{2}$″, 3″, 3$\frac{1}{2}$″, 4″, 5″, 6″, 7″, 8″, and 9″. Leg thicknesses from $\frac{1}{8}$″ to 1$\frac{1}{8}$″
Channel	C9 × 13.4	C denotes a channel. 9 is the nominal depth in inches, 13.4 is the weight per foot of length in pounds	Nominal depths of 3″, 4″, 5″, 6″, 7″, 8″, 9″, 10″, 12″, and 15″
Structural tee	WT13.5 × 47	This is a tee made by splitting a W27 × 94. It is 13.5 inches deep and weighs 47 pounds per foot of length. Tees split from American Standard shapes are designated ST rather than WT	Nominal depths of 2 to 9″ in 1″ increments, and 10$\frac{1}{2}$ to 18″ in 1$\frac{1}{2}$″ increments

FIGURE 9.10
Commonly used steel shapes.

FIGURE 9.11

A portion of the table of dimensions and properties of wide-flange shapes from the Manual of Steel Construction *of the American Institute of Steel Construction, reproduced by permission of the AISC. One inch equals 25.4 mm.*

W SHAPES — Properties

Nominal Wt. per Ft. (Lb.)	$b_f/2t_f$	F_y' (Ksi)	d/t_w	F_y'' (Ksi)	r_T (In.)	d/A_f	X-X I (In.⁴)	X-X S (In.³)	X-X r (In.)	Y-Y I (In.⁴)	Y-Y S (In.³)	Y-Y r (In.)	Torsional const. J (In.⁴)	Z_x (In.³)	Z_y (In.³)
336	2.3	—	9.5	—	3.71	0.43	4060	483	6.41	1190	177	3.47	243	603	274
305	2.4	—	10.0	—	3.67	0.46	3550	435	6.29	1050	159	3.42	185	537	244
279	2.7	—	10.4	—	3.64	0.49	3110	393	6.16	937	143	3.38	143	481	220
252	2.9	—	11.0	—	3.59	0.53	2720	353	6.06	828	127	3.34	108	428	196
230	3.1	—	11.7	—	3.56	0.56	2420	321	5.97	742	115	3.31	83.8	386	177
210	3.4	—	12.5	—	3.53	0.61	2140	292	5.89	664	104	3.28	64.7	348	159
190	3.7	—	13.6	—	3.50	0.65	1890	263	5.82	589	93.0	3.25	48.8	311	143
170	4.0	—	14.6	—	3.47	0.72	1650	235	5.74	517	82.3	3.22	35.6	275	126
152	4.5	—	15.8	—	3.44	0.79	1430	209	5.66	454	72.8	3.19	25.8	243	111
136	5.0	—	17.0	—	3.41	0.87	1240	186	5.58	398	64.2	3.16	18.5	214	98.0
120	5.6	—	18.5	—	3.38	0.96	1070	163	5.51	345	56.0	3.13	12.9	186	85.4
106	6.2	—	21.1	—	3.36	1.07	933	145	5.47	301	49.3	3.11	9.13	164	75.1
96	6.8	—	23.1	—	3.34	1.16	833	131	5.44	270	44.4	3.09	6.86	147	67.5
87	7.5	—	24.3	—	3.32	1.28	740	118	5.38	241	39.7	3.07	5.10	132	60.4
79	8.2	62.6	26.3	—	3.31	1.39	662	107	5.34	216	35.8	3.05	3.84	119	54.3
72	9.0	52.3	28.5	—	3.29	1.52	597	97.4	5.31	195	32.4	3.04	2.93	108	49.2
65	9.9	43.0	31.1	—	3.28	1.67	533	87.9	5.28	174	29.1	3.02	2.18	96.8	44.1
58	7.8	55.9	33.9	57.6	2.72	1.90	475	78.0	5.28	107	21.4	2.51	2.10	86.4	32.5
53	8.7	—	35.0	54.1	2.71	2.10	425	70.6	5.23	95.8	19.2	2.48	1.58	77.9	29.1
50	6.3	—	32.9	60.9	2.17	2.36	394	64.7	5.18	56.3	13.9	1.96	1.78	72.4	21.4
45	7.0	—	36.0	51.0	2.15	2.61	350	58.1	5.15	50.0	12.4	1.94	1.31	64.7	19.0
40	7.8	—	40.5	40.3	2.14	2.90	310	51.9	5.13	44.1	11.0	1.93	0.95	57.5	16.8
35	6.3	—	41.7	38.0	1.74	3.66	285	45.6	5.25	24.5	7.47	1.54	0.74	51.2	11.5
30	7.4	—	47.5	29.3	1.73	4.30	238	38.6	5.21	20.3	6.24	1.52	0.46	43.1	9.56
26	8.5	57.9	53.1	23.4	1.72	4.95	204	33.4	5.17	17.3	5.34	1.51	0.30	37.2	8.17
22	4.7	—	47.3	29.5	1.02	7.19	156	25.4	4.91	4.66	2.31	0.847	0.29	29.3	3.66
19	5.7	—	51.7	24.7	1.00	8.67	130	21.3	4.82	3.76	1.88	0.822	0.18	24.7	2.98
16	7.5	—	54.5	22.2	0.96	11.3	103	17.1	4.67	2.82	1.41	0.773	0.10	20.1	2.26
14	8.8	54.3	59.6	18.6	0.95	13.3	88.6	14.9	4.62	2.36	1.19	0.753	0.07	17.4	1.90

W SHAPES — Dimensions

Designation	Area A (In.²)	Depth d (In.)	Depth d (In.)	Web t_w (In.)	Web t_w (In.)	$t_w/2$ (In.)	Flange Width b_f (In.)	Flange Width b_f (In.)	Flange Thick. t_f (In.)	Flange Thick. t_f (In.)	T (In.)	k (In.)	k_1 (In.)
W 12×336	98.8	16.82	16 7/8	1.775	1 3/4	7/8	13.385	13 3/8	2.955	2 15/16	9 1/2	3 11/16	1 1/2
×305	89.6	16.32	16 3/8	1.625	1 5/8	13/16	13.235	13 1/4	2.705	2 11/16	9 1/2	3 7/16	1 7/16
×279	81.9	15.85	15 7/8	1.530	1 1/2	3/4	13.140	13 1/8	2.470	2 1/2	9 1/2	3 3/16	1 3/8
×252	74.1	15.41	15 3/8	1.395	1 3/8	11/16	13.005	13	2.250	2 1/4	9 1/2	2 15/16	1 5/16
×230	67.7	15.05	15	1.285	1 5/16	11/16	12.895	12 7/8	2.070	2 1/16	9 1/2	2 3/4	1 1/4
×210	61.8	14.71	14 3/4	1.180	1 3/16	5/8	12.790	12 3/4	1.900	1 7/8	9 1/2	2 5/8	1 1/4
×190	55.8	14.38	14 3/8	1.060	1 1/16	9/16	12.670	12 5/8	1.735	1 3/4	9 1/2	2 7/16	1 3/16
×170	50.0	14.03	14	0.960	15/16	1/2	12.570	12 5/8	1.560	1 9/16	9 1/2	2 1/4	1 1/8
×152	44.7	13.71	13 3/4	0.870	7/8	7/16	12.480	12 1/2	1.400	1 3/8	9 1/2	2 1/8	1 1/16
×136	39.9	13.41	13 3/8	0.790	13/16	3/8	12.400	12 3/8	1.250	1 1/4	9 1/2	1 15/16	1
×120	35.3	13.12	13 1/8	0.710	11/16	3/8	12.320	12 3/8	1.105	1 1/8	9 1/2	1 13/16	1
×106	31.2	12.89	12 7/8	0.610	5/8	5/16	12.220	12 1/4	0.990	1	9 1/2	1 11/16	15/16
×96	28.2	12.71	12 3/4	0.550	9/16	5/16	12.160	12 1/8	0.900	7/8	9 1/2	1 5/8	7/8
×87	25.6	12.53	12 1/2	0.515	1/2	1/4	12.125	12 1/8	0.810	13/16	9 1/2	1 1/2	7/8
×79	23.2	12.38	12 3/8	0.470	1/2	1/4	12.080	12 1/8	0.735	3/4	9 1/2	1 7/16	7/8
×72	21.1	12.25	12 1/4	0.430	7/16	1/4	12.040	12	0.670	11/16	9 1/2	1 3/8	7/8
×65	19.1	12.12	12 1/8	0.390	3/8	3/16	12.000	12	0.605	5/8	9 1/2	1 5/16	13/16
W 12×58	17.0	12.19	12 1/4	0.360	3/8	3/16	10.010	10	0.640	5/8	9 1/2	1 3/8	13/16
×53	15.6	12.06	12	0.345	3/8	3/16	9.995	10	0.575	9/16	9 1/2	1 1/4	13/16
W 12×50	14.7	12.19	12 1/4	0.370	3/8	3/16	8.080	8 1/8	0.640	5/8	9 1/2	1 3/8	13/16
×45	13.2	12.06	12	0.335	5/16	3/16	8.045	8	0.575	9/16	9 1/2	1 1/4	13/16
×40	11.8	11.94	12	0.295	5/16	3/16	8.005	8	0.515	1/2	9 1/2	1 1/4	3/4
W 12×35	10.3	12.50	12 1/2	0.300	5/16	3/16	6.560	6 1/2	0.520	1/2	10 1/2	1	9/16
×30	8.79	12.34	12 3/8	0.260	1/4	1/8	6.520	6 1/2	0.440	7/16	10 1/2	15/16	1/2
×26	7.65	12.22	12 1/4	0.230	1/4	1/8	6.490	6 1/2	0.380	3/8	10 1/2	7/8	1/2
W 12×22	6.48	12.31	12 1/4	0.260	1/4	1/8	4.030	4	0.425	7/16	10 1/2	7/8	1/2
×19	5.57	12.16	12 1/8	0.235	1/4	1/8	4.005	4	0.350	3/8	10 1/2	13/16	1/2
×16	4.71	11.99	12	0.220	1/4	1/8	3.990	4	0.265	1/4	10 1/2	3/4	1/2
×14	4.16	11.91	11 7/8	0.200	3/16	1/8	3.970	4	0.225	1/4	10 1/2	11/16	1/2

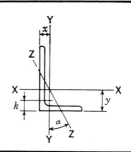

ANGLES
Equal legs and unequal legs
Properties for designing

Size and Thickness	k	Weight per Foot	Area	AXIS X-X				AXIS Y-Y				AXIS Z-Z	
				I	S	r	y	I	S	r	x	r	Tan α
In.	In.	Lb.	In.²	In.⁴	In.³	In.	In.	In.⁴	In.³	In.	In.	In.	α
L 4 x3 x 5/8	1 1/16	13.6	3.98	6.03	2.30	1.23	1.37	2.87	1.35	0.849	0.871	0.637	0.534
1/2	15/16	11.1	3.25	5.05	1.89	1.25	1.33	2.42	1.12	0.864	0.827	0.639	0.543
7/16	7/8	9.8	2.87	4.52	1.68	1.25	1.30	2.18	0.992	0.871	0.804	0.641	0.547
3/8	13/16	8.5	2.48	3.96	1.46	1.26	1.28	1.92	0.866	0.879	0.782	0.644	0.551
5/16	3/4	7.2	2.09	3.38	1.23	1.27	1.26	1.65	0.734	0.887	0.759	0.647	0.554
1/4	11/16	5.8	1.69	2.77	1.00	1.28	1.24	1.36	0.599	0.896	0.736	0.651	0.558
L 3 1/2 x3 1/2 x 1/2	7/8	11.1	3.25	3.64	1.49	1.06	1.06	3.64	1.49	1.06	1.06	0.683	1.000
7/16	13/16	9.8	2.87	3.26	1.32	1.07	1.04	3.26	1.32	1.07	1.04	0.684	1.000
3/8	3/4	8.5	2.48	2.87	1.15	1.07	1.01	2.87	1.15	1.07	1.01	0.687	1.000
5/16	11/16	7.2	2.09	2.45	0.976	1.08	0.990	2.45	0.976	1.08	0.990	0.690	1.000
1/4	5/8	5.8	1.69	2.01	0.794	1.09	0.968	2.01	0.794	1.09	0.968	0.694	1.000
L 3 1/2 x3 x 1/2	15/16	10.2	3.00	3.45	1.45	1.07	1.13	2.33	1.10	0.881	0.875	0.621	0.714
7/16	7/8	9.1	2.65	3.10	1.29	1.08	1.10	2.09	0.975	0.889	0.853	0.622	0.718
3/8	13/16	7.9	2.30	2.72	1.13	1.09	1.08	1.85	0.851	0.897	0.830	0.625	0.721
5/16	3/4	6.6	1.93	2.33	0.954	1.10	1.06	1.58	0.722	0.905	0.808	0.627	0.724
1/4	11/16	5.4	1.56	1.91	0.776	1.11	1.04	1.30	0.589	0.914	0.785	0.631	0.727
L 3 1/2 x2 1/2 x 1/2	15/16	9.4	2.75	3.24	1.41	1.09	1.20	1.36	0.760	0.704	0.705	0.534	0.486
7/16	7/8	8.3	2.43	2.91	1.26	1.09	1.18	1.23	0.677	0.711	0.682	0.535	0.491
3/8	13/16	7.2	2.11	2.56	1.09	1.10	1.16	1.09	0.592	0.719	0.660	0.537	0.496
5/16	3/4	6.1	1.78	2.19	0.927	1.11	1.14	0.939	0.504	0.727	0.637	0.540	0.501
1/4	11/16	4.9	1.44	1.80	0.755	1.12	1.11	0.777	0.412	0.735	0.614	0.544	0.506
L 3 x3 x 1/2	13/16	9.4	2.75	2.22	1.07	0.898	0.932	2.22	1.07	0.898	0.932	0.584	1.000
7/16	3/4	8.3	2.43	1.99	0.954	0.905	0.910	1.99	0.954	0.905	0.910	0.585	1.000
3/8	11/16	7.2	2.11	1.76	0.833	0.913	0.888	1.76	0.833	0.913	0.888	0.587	1.000
5/16	5/8	6.1	1.78	1.51	0.707	0.922	0.865	1.51	0.707	0.922	0.865	0.589	1.000
1/4	9/16	4.9	1.44	1.24	0.577	0.930	0.842	1.24	0.577	0.930	0.842	0.592	1.000
3/16	1/2	3.71	1.09	0.962	0.441	0.939	0.820	0.962	0.441	0.939	0.820	0.596	1.000

Angles in shaded rows may not be readily available. Availability is subject to rolling accumulation and geographical location, and should be checked with material suppliers.

FIGURE 9.12

A portion of the table of dimensions and properties of angle shapes from the Manual of Steel Construction of the American Institute of Steel Construction, reproduced by permission of the AISC. One inch equals 25.4 mm.

loads and are frequently found playing this role as lintels spanning door and window openings in masonry construction. In steel frame buildings, they are most often seen cut into short pieces and used to connect wide-flange shapes. They also find use as diagonal braces in steel frames, and as members of steel trusses, where they are doubled back-to-back to connect conveniently to flat *gusset plates* at the joints of the truss (Figure 9.75). *Channel* sections are also used as truss members and bracing, as well as for short beams and lintels. *Tees, plates,* and *bars,* all have their various roles in a steel frame building, as shown in the diagrams that accompany this text.

Steel Alloys

Mild structural steel, known by its ASTM designation of A36, is by far the predominant type used in steel building frames, but other steels are also widely employed (Figure 9.13). With small additions of other elements to the molten metal, high-strength, low-alloy steels are produced. The increased tensile strength of these steels makes them economical for use in tension members, or in columns whose cross-sectional areas are restricted by architectural considerations. But because the elastic modulus of these steels is not increased by the added alloys, their use for a beam is justified only if the deflection of the beam is not the controlling factor in its design. Certain of the low-alloy steels have another useful property besides higher strength: When exposed to the atmosphere, a coating of tenacious oxide forms to protect them from further corrosion. These *weathering steels,* joined with bolts and welding electrodes of similar alloys, can be left exposed to the weather without painting. This offers significant savings in maintenance costs, and the deep, warm hue of the oxide coating can be attractive on the face of a building. In other locations where corrosion is expected to be a problem, steel structural members are often coated with zinc, a self-protecting metal, in a process known as *galvanizing*.

Open-Web Steel Joists

Among the many structural steel products fabricated from hot- and cold-rolled shapes, the most common is the *open-web steel joist*, a mass-produced truss used in closely spaced arrays to support floor and roof decks (Figures 9.14, 9.15). Standard depths range from 8 to 72 inches (200 to 1800 mm), and spans up to 144 feet (44 m) are possible. Most buildings utilize joists that are less than 2 feet (600 mm) deep to achieve spans of up to 40 feet (13 m). The spacings between joists commonly range from 2 to 10 feet (610 to 3050 mm), depending upon the magnitude of the applied loads and the spanning capability of the decking. *Joist girders*, which are prefabricated steel trusses designed for use as primary framing members, can be used to replace wide-flange beams and girders in situations such as industrial roof structures where the greater depth of the truss is not objectionable. Open web joists and joist girders are invariably made of high-strength steel.

Alloy	Yield Strength	Allowable Stress in Bending	Modulus of Elasticity
ASTM A36 Carbon steel	36,000 psi (248.22 MN/m²)	22,000 psi (151.69 MN/m²)	29,000,000 psi (200,000 MN/m²)
ASTM A242 High strength low alloy; corrosion resisting	42,000–50,000 psi in 3 grades (289.59–344.75 MN/m²)	25,200–30,000 psi (173.75–206.85 MN/m²)	29,000,000 psi (200,000 MN/m²)
ASTM A441 High strength low alloy; structural manganese-vanadium	40,000–50,000 psi in 3 grades (275.80–344.75 MN/m²)	24,000–30,000 psi (165.48–206.85 MN/m²)	29,000,000 psi (200,000 MN/m²)
ASTM A572 High strength low alloy; columbium-vanadium steels of structural quality	42,000–65,000 psi in 4 grades (289.59–448.18 MN/m²)	25,200–39,000 psi (173.75–268.91 MN/m²)	29,000,000 psi (200,000 MN/m²)
ASTM A588 High strength low alloy; corrosion resisting	42,000–50,000 psi in 3 grades (289.59–344.75 MN/m²)	25,200–30,000 psi (173.75–206.85 MN/m²)	29,000,000 psi (200,000 MN/m²)

FIGURE 9.13
Properties of some steels used in construction.

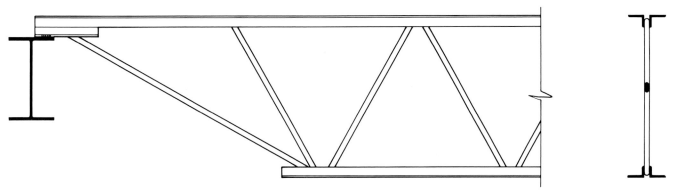

FIGURE 9.14
An elevation (left) of the end of an open-web steel joist, and (right) a section through it showing the double- *angle top and bottom chords with the steel rod diagonal chords welded between.*

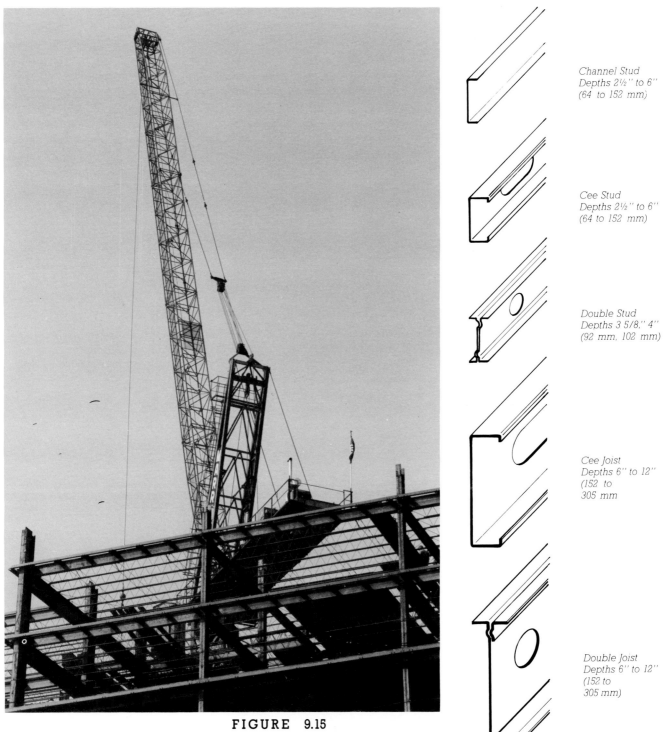

FIGURE 9.15
Open-web steel joist floor framing, with a climbing crane above. (Photo by the author)

*Channel Stud
Depths 2½'' to 6''
(64 to 152 mm)*

*Cee Stud
Depths 2½'' to 6''
(64 to 152 mm)*

*Double Stud
Depths 3 5/8,'' 4''
(92 mm, 102 mm)*

*Cee Joist
Depths 6'' to 12''
(152 to
305 mm*

*Double Joist
Depths 6'' to 12''
(152 to
305 mm)*

FIGURE 9.16
Cold-formed steel light framing members.

Cold-Worked Steel

Steel can be formed in a cold state as well as in a hot state, by rolling or bending. Steel sheet is bent into C-shaped and Z-shaped sections, and bent and welded into I-shaped sections, to make short-span framing members that are frequently used in partitions and exterior walls of larger buildings, and in floor structures of smaller buildings (Figure 9.16). Steel sheet stock is also rolled into corrugated configurations utilized as floor and roof decking in steel framed structures (Figure 9.56). Heavier sheet stock is cold-formed or hot-formed into square, rectangular, and round cross sections, which are then welded along the longitudinal seam to form *structural tubing*. Structural tubing is often used for columns and for members of welded steel trusses and space frames. Tube shapes are especially useful for members subjected to torsional stresses.

Steel that is cold-rolled gains considerably in strength, through a realignment of its crystalline structure. The normal range of wide-flange shapes is too large to be cold-rolled with today's machinery, but cold rolling is used to produce steel rods and steel components for open web joists, where the higher strength can be utilized to good advantage. Steel can be cold-drawn through dies to produce the very high-strength wires used in wire ropes, bridge cables, and concrete prestressing strands.

Joining Steel Members

Rivets

Steel shapes can be joined into a building frame with any of three fastening techniques—rivets, bolts, or welds—or by combinations of these. A *rivet* is a fastener consisting of a cylindrical body and a formed head, which is brought to a white heat, inserted through holes in the members to be joined, and hot-worked with a pneumatic hammer to produce a second head opposite the first (Figure 9.17). As the rivet cools, it

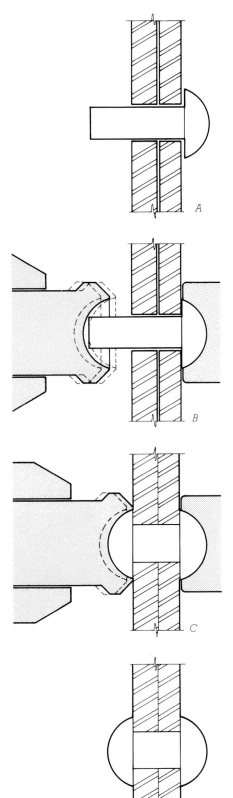

FIGURE 9.17
A. *A hot steel rivet is inserted in holes drilled in the two members to be joined.* **B, C.** *Its head is then held with a hand hammer with a cup-shaped depression, while a pneumatic hammer drives a rivet set repeatedly against the body of the rivet to form the second head.* **D.** *The rivet shrinks as it cools, drawing the members tightly together.*

shrinks, clamping the joined pieces together and forming a tight joint. Riveting was for many decades the predominant fastening technique in steel frame buildings, but it has been almost entirely replaced in recent times by the less labor-intensive techniques of bolting and welding.

Bolts

The bolts commonly used in steel frame construction fall into two general categories: *carbon steel bolts* (ASTM A307), and *high-strength bolts* (ASTM A325 and A490). Carbon steel bolts (also called *unfinished* or *common* bolts) are similar to the ordinary machine bolts that can be purchased in hardware stores. Their "in-place" cost is less than that of high-strength bolts, so they are used in many structural joints where their lower strength is sufficient to carry the necessary loads; they act primarily in shear. High-strength bolts are heat treated during manufacture to develop the necessary strength. They derive their connecting ability either from their shear resistance, which is much higher than that of carbon steel bolts, or from being tightened to the point that the members they join are kept from slipping by the friction between them, producing what is known as a *friction* or *slip-critical* connection.

High-strength bolts are inserted into holes slightly larger than the shank diameter of the bolt. Washers may or may not be used; a washer is required in certain cases to spread the load of the bolt over a larger area of the shapes being joined. Sometimes a washer is inserted under the head or nut, whichever is turned to tighten the bolt, to prevent *galling* of the softer parent

FIGURE 9.18
An ironworker tightens high-strength bolts with a pneumatic impact wrench. (Courtesy of Bethlehem Steel Corporation)

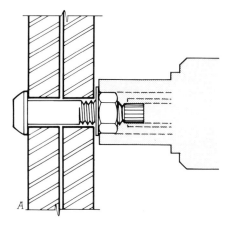

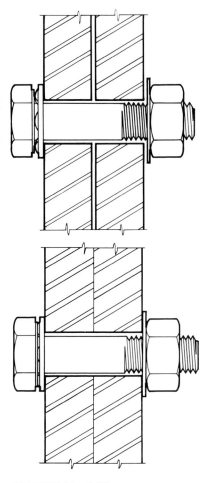

FIGURE 9.19

At top, *an untightened high-strength bolt with a load indicator washer under the head. At* bottom, *the bolt and washer after tightening.*

metal. The bolt is usually tightened using a pneumatic or electric *impact wrench* (Figure 9.18). If the bolt is to act only in shear, the amount of tension in the bolt is not critical; but if it is to connect by friction, it must be tightened reliably to at least 70 percent of its ultimate tensile strength. The extreme tension in the bolt produces the high clamping force necessary to allow the surfaces of the two members to transfer the load between them entirely by friction.

A major problem in high-strength bolting of friction-type connections is how to verify that the necessary tension has been achieved in all the bolts in a connection. This can be accomplished in any of several ways. In the *turn-of-nut method,* each bolt is tightened snug, then turned a specified additional fraction of a turn. Depending on bolt length, bolt alloy, and other factors, the additional tightening required will range from one-third of a turn to a full turn. In another method, *load indicator washers* are placed under the head or nut of the bolt. As the bolt is tightened, the protrusions on the washer are progressively flattened in proportion to the tension in the bolt (Figure 9.19). Inspection for proper bolt tension then becomes a simple matter of inserting a feeler gauge to determine whether the protrusions have flattened sufficiently to indicate the required tension. Yet another method employs *tension control bolts* with either hex or button heads and protruding, splined ends that extend beyond the threaded portion of the body of the bolt (Figure 9.20). The nut is tightened by a special power-driven wrench that grips both the nut and the splined end simultaneously, turning the one against the other (Figure 9.21). The splined end is formed in such a way that when the required torque has been reached, the end twists off. Installation of this type of bolt is accomplished by a single worker, while conventional high-strength bolts require a second person with a wrench to prevent the other end of the bolt assembly from turning.

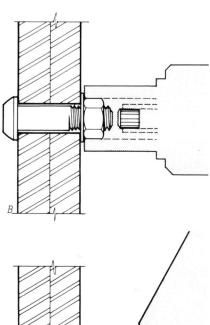

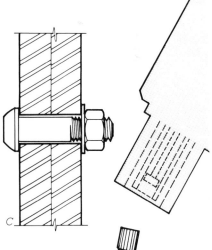

FIGURE 9.20

Tightening a tension-control bolt. **A.** *The wrench holds both the nut and the splined body of the bolt, and turns them against one another to tighten the bolt.* **B.** *When the required torque is achieved, the splined end twists off in the wrench.* **C.** *A plunger inside the wrench discharges the splined end into a container.*

FIGURE 9.21
The compact design of the electric wrench for tightening tension-control bolts makes it easy to reach bolts in tight situations. (Courtesy of Ingersoll-Rand Corporation)

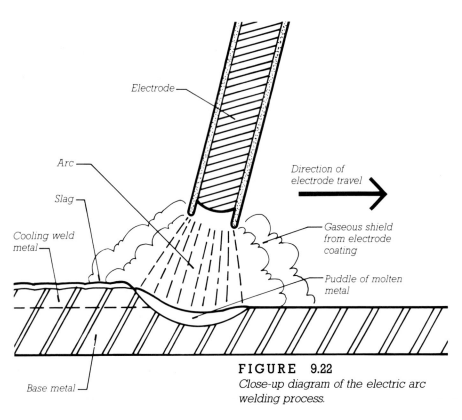

FIGURE 9.22
Close-up diagram of the electric arc welding process.

Welding

Welding offers a unique and valuable capability to the structural designer: It can join the members of a steel frame as if they were a monolithic whole. Welded connections, properly designed and executed, are stronger than the members they join in resisting both shear and moment forces. While it is possible to achieve this same performance with high-strength bolted connections, such connections are often cumbersome compared to equivalent welded joints. Bolting, on the other hand, has its own advantages: It is quick and easy for field connections that need only resist shearing forces and can be accomplished under conditions of adverse weather or difficult physical access that would make welding impossible. Sometimes welding and bolting are combined in the same connections to take advantage of the unique qualities of each: Welding may be used in the fabricator's shop for its inherent economies, and in the field for its structural continuity, while bolting is often employed in the simpler field connections and to hold connections in alignment for welding. The choice between bolting and welding is often dictated by the designer, but such a choice may also be influenced by considerations such as the fabricator's and erector's equipment and expertise, availability of electric power, climate, and location.

Electric *arc welding* is conceptually simple. An electrical potential is established between the steel pieces to be joined and a metal *electrode* held either by a machine (as in some shop welding processes) or a person. When the electrode is held close to the steel members, a continuous electric arc is established that generates sufficient heat to melt both a localized area of the steel members and the tip of the electrode (Figure 9.22). The molten steel from the electrode merges with that of the members to form a single puddle. The electrode is drawn slowly along the seam being welded, leaving behind a continuous bead of metal that

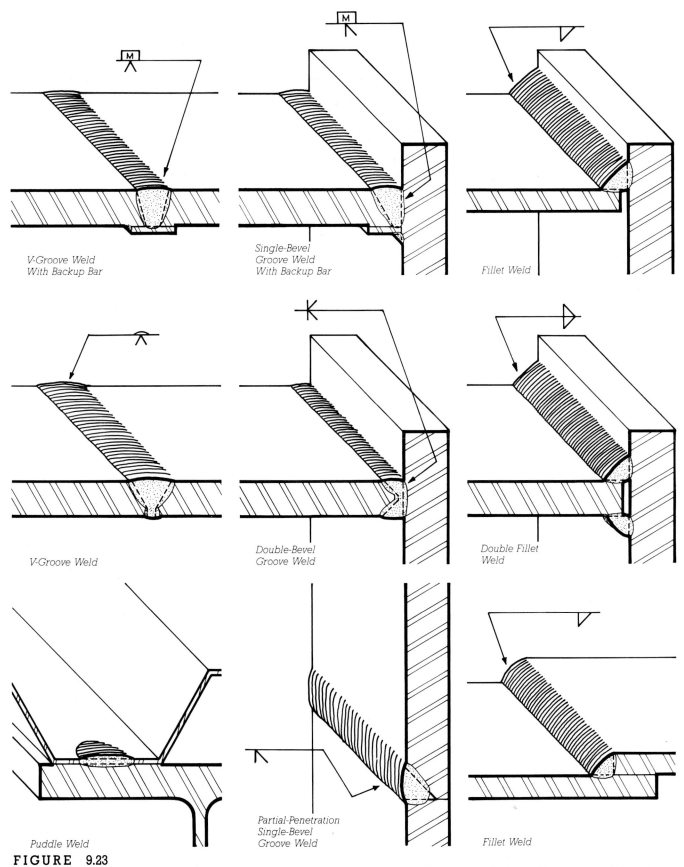

V-Groove Weld
With Backup Bar

Single-Bevel
Groove Weld
With Backup Bar

Fillet Weld

V-Groove Weld

Double-Bevel
Groove Weld

Double Fillet
Weld

Puddle Weld

Partial-Penetration
Single-Bevel
Groove Weld

Fillet Weld

FIGURE 9.23

Typical welds used in steel frame construction. Fillet welds are more economical to make because they require no advance preparation of the joint, but the full-penetration groove welds are stronger. The standard weld symbols used here are explained in Figure 9.24.

THE ARROW

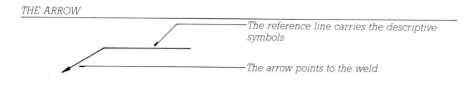

The reference line carries the descriptive symbols

The arrow points to the weld

THE BASIC SYMBOLS

The basic weld symbol is located on either side of the reference line as follows:

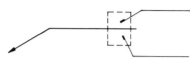

Symbols on the top of the reference line refer to welds on the side of the joint opposite the arrow

Symbols on the bottom of the reference line refer to welds on the same side of the joint as the arrow

The basic symbols are

BACK	FILLET	PLUG or SLOT	GROOVE or BUTT						
			SQUARE	V	BEVEL	U	J	FLARE V	FLARE BEVEL
⌒	◺	▭	‖	∨	∨	⋃	⊳	⋎	⋁

SUPPLEMENTARY SYMBOLS

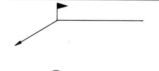

FIELD WELD
This weld must be done in the field during erection. Other welds are done earlier in the fabricator's shop

WELD ALL AROUND
This indicates that the weld should be carried fully around the perimeter of the joining pieces

BACKUP BAR
As indicated in this example, a backup bar to support the first pass of the weld must be placed on the side of the joint opposite the arrow

SPACER
Small metal spacers are used to maintain a gap between the pieces to be joined, prior to welding

A sharp bend near the end of the arrow indicates that the arrowhead is pointing toward the grooved side of a bevel or J-grooved joint

FIGURE   9.24

Standard weld symbols, as used on steel connection details.

cools and solidifies to form a strong connection between the members. For small members, a single pass of the electrode may suffice to make the connection. For larger members, a number of passes are made in order to build up a weld of the required depth.

In practice, welding is a complex science. The metallurgy of both the structural steel and the welding electrodes must be carefully coordinated. Voltage, amperage, and polarity of the electric current are selected to achieve the right heat and penetration for the weld. Air must be kept away from the electric arc to prevent rapid oxidation of the liquid steel; this is accomplished in simple welding processes by a thick coating on the electrode that melts to create a liquid and gaseous shield around the arc, or by a core of vaporizing flux in a tubular steel electrode. It may also be done by means of a continuous flow of inert gas around the arc, or with a dry flux that is heaped over the end of the electrode as it moves across the work.

The required thickness and length of each weld are calculated by the designer to match them to the forces to be transmitted between the members. For deep welds, the edges of the members are beveled to permit access of the electrode to the full thickness of the piece (Figure 9.23). Small strips of steel called *backup bars* are welded beneath the connection prior to beginning the actual weld, to prevent the molten metal from flowing out the bottom of the groove. In some cases, *run-off bars* are required at the ends of a groove weld to facilitate the formation of a full thickness of weld metal at the edges of the member (Figure 9.43).

Workers doing structural welding should be methodically trained and periodically tested to ensure that they have the required level of skill and knowledge. When an important weld is completed, it should be inspected to see that it is of the required size and quality; often this involves sophisticated magnetic particle, dye penetrant, ultrasonic, or radiographic testing procedures that search for hidden voids and flaws within each weld.

DETAILS OF STEEL FRAMING

Typical Connections

Most steel frame connections use angles, plates, or tees as transitional elements between the members being connected. A simple bolted beam-to-column-flange connection requires two angles and a number of bolts (Figures 9.25, 9.26). The angles are cut to length and the holes are made in all the components prior to assembly. The angles are usually bolted to the web of the beam in the fabricator's shop, and the bolts through the flange of the column are added after the beam is in place. This type of connection, which joins only the web of the beam and not the flanges, is known as a *framed connection*. It is capable of transmitting all the vertical forces (*shear*) from a beam to a column. Because it does not connect the beam flanges to the column, it is of no value in transmitting bending moment from one to the other. To produce a moment-transmitting connection, it is necessary to connect the flanges strongly across the joint, either with heavy angles, plates, and numerous bolts top and bottom, or by means of full-penetration welds across the beam flanges (Figures 9.27, 9.28). If the column flanges are insufficiently strong to accept the bending moment transmitted from the beam, *stiffener plates* must be installed inside the column flanges.

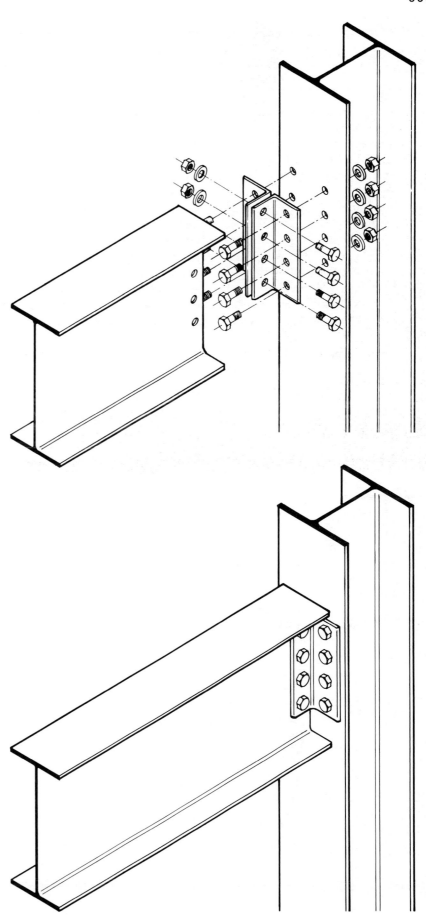

FIGURE 9.25

Exploded and assembled views of a framed, bolted beam-to-column-flange connection. The size of the angles and the number and size of the bolts are determined by the magnitude of the load the connection must transmit from the beam to the column.

Shear Connections and Moment Connections

In order to understand the respective roles of shear connections and moment connections in a building frame, it is necessary to understand the means by which buildings may be made stable against the lateral forces of wind and earthquake. Three basic mechanisms are commonly used: *diagonal bracing*, *shear panels*, and *moment connections* (Figure 9.29). Diagonal bracing works by creating stable triangular configurations within the unstable rectilinear geometry of a steel building frame. The connections within a diagonally braced frame need not

transmit moments; they can behave like pins or hinges, which is another way of saying that they can be *shear connections* such as the one in Figure 9.25. Shear panels, which may be made of steel or concrete, act much like braced rectangles within the building frame, and also do not require moment connections. Moment connections are capable of stabilizing a frame against lateral forces without the use of either diagonal bracing or shear panels. A large number of the connections in a frame stabilized in this manner must be moment connections, but many may be shear connections.

To illustrate, consider two common methods of stabilizing the frame of a tall building (Figure 9.30). One is to

provide a stable core area in the center of the building. The core, the area that contains the elevators, stairs, mechanical chases, and washrooms, is structured as a rigid tower, using diagonal braces, shear panels, or moment connections. The floors of the structure act as horizontal shear panels that make the outer bays of the building rigid by connecting them to the rigid core. The outer bays may then be structured with shear connections only, although a nominal number of moment connections may be included to provide additional rigidity or restrict *drift*, which is a lateral displacement of the frame caused by wind or earthquake loads.

A second method of achieving stability is to make the perimeter of the

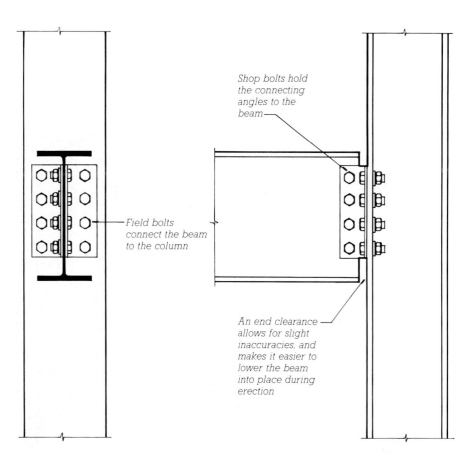

Shop bolts hold the connecting angles to the beam

Field bolts connect the beam to the column

An end clearance allows for slight inaccuracies, and makes it easier to lower the beam into place during erection

FIGURE 9.26
Two elevation views of the framed, bolted beam-to-column-flange connection shown in Figure 9.25. This is a shear connection only (AISC Type 2), because the flanges of the beam are not rigidly connected to the column.

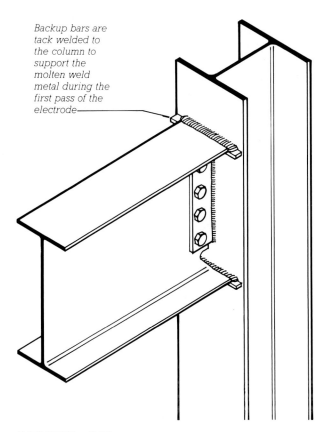

Backup bars are tack welded to the column to support the molten weld metal during the first pass of the electrode—

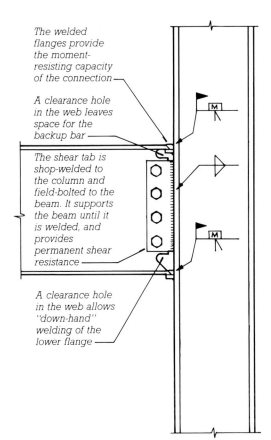

The welded flanges provide the moment-resisting capacity of the connection—

A clearance hole in the web leaves space for the backup bar—

The shear tab is shop-welded to the column and field-bolted to the beam. It supports the beam until it is welded, and provides permanent shear resistance—

A clearance hole in the web allows ''down-hand'' welding of the lower flange—

FIGURE 9.27

A welded moment connection (AISC Type 1) for joining a beam to a column. The groove welds develop the full strength of the flanges of the beam, allowing the connection to transmit moments between the beam and the column. If the column flanges are not stiff enough to accept the moments from the beam, stiffener plates similar to those in Figure 9.35 are welded between the column flanges in the plane of each of the beam flanges. The weld symbols are those explained in Figure 9.24. Note that some are field welds, and some are shop welds.

FIGURE 9.28

A welder works on the upper flange of a welded moment connection similar to the one shown in Figure 9.27. The lower flange weld is already complete. The welder stood on the small wood platform, called a float, while welding the lower flange. These exceptionally large beams are too large to have been rolled and had to be shop welded from steel plates. (Courtesy of United States Steel Corporation)

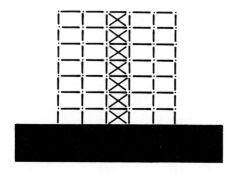

Diagonal Bracing

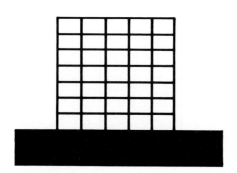

*Moment
Connections*

Shear Panels

FIGURE 9.29
*Elevation views of the basic
mechanisms for imparting lateral
stability to a frame. The dots at
connection points indicate that they are
shear connections only, and solid
intersections indicate moment
connections.*

building rigid, again using diagonal braces, shear panels, or moment connections. When this is done, the entire interior of the structure can be assembled with shear connections, provided there is enough continuity in the floor system to attain *diaphragm action*, which is the rigidity possessed by a thin plate of material such as a welded steel deck with a concrete topping. Moment connections are much more costly to make than shear connections, so the object in designing a scheme for lateral stabilization of a building is to use shear connections wherever possible, and moment connections only where necessary.

Shear connections, then, are sufficient for most purposes in buildings braced by diagonal bracing or shear panels and for many of the joints in buildings braced by moment connections. Moment connections may be used as the means for imparting lateral stability to a building instead of diagonal bracing or shear panels. Moment connections are also utilized to connect cantilevered beams to columns (and sometimes to beams or girders). The American Institute of Steel Construction (AISC) defines three types of Steel Frame construction, classified according to the manner in which they achieve stability against lateral forces. *AISC Type 1*, Rigid Frame construction, assumes that beam-to-column connections are sufficiently rigid that the geometric angles between members will remain virtually unchanged under loading. *AISC Type 2* construction, Simple Frame construction, assumes shear connections only and requires diagonal bracing or shear panels for lateral stability. *AISC Type 3* construction is defined as semi-rigid, in which the connections are not as rigid as those required for AISC Type 1 construction, but possess a dependable and predictable moment-resisting capacity that can be used to stabilize the building.

A series of simple, fully bolted shear (AISC Type 2) connections are shown as a beginning basis for understanding steel connection details (Figures 9.25,

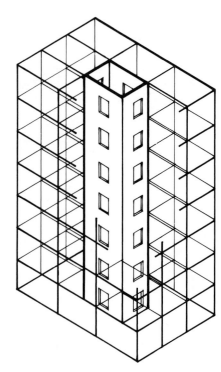

Rigid Core

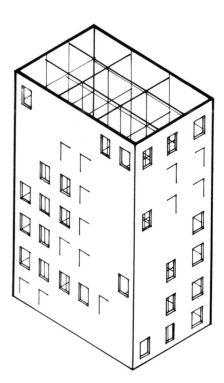

Rigid Perimeter
FIGURE 9.30
Rigid core versus rigid perimeter.

9.26, 9.31, 9.33, and 9.34). These are followed by a series of welded moment-resisting (AISC Type 1 and AISC Type 3) connections (Figures 9.27, 9.35, 9.36, 9.38). Welding is also widely used for making shear connections, an example of which is shown in Figure 9.37. In practice, there are a number of different ways of making any of these connections, using various kinds of connecting elements and different combinations of bolting and welding. The object is economy. The text and drawings in the accompanying sidebar illustrate some of the ways in which steel connections can be made more economical while still performing their designated function in the structure.

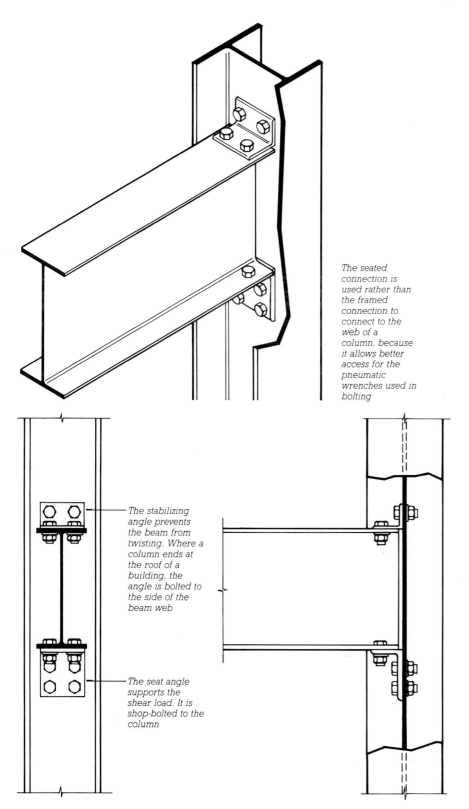

The seated connection is used rather than the framed connection to connect to the web of a column, because it allows better access for the pneumatic wrenches used in bolting

The stabilizing angle prevents the beam from twisting. Where a column ends at the roof of a building, the angle is bolted to the side of the beam web

The seat angle supports the shear load. It is shop-bolted to the column

FIGURE 9.31

A seated beam-to-column-web connection. Although the beam flanges are connected to the column, this is an AISC Type 2 (shear) connection, not a moment connection, because the two bolts are incapable of developing the full strength of the beam flanges.

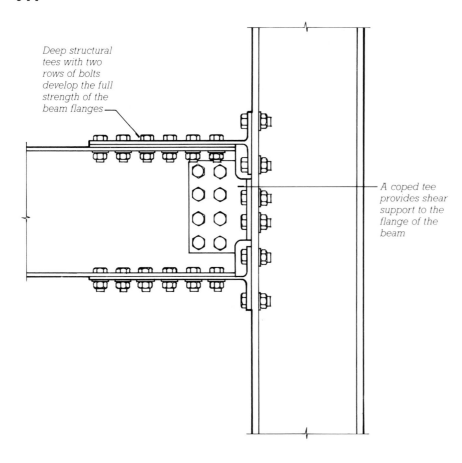

Deep structural tees with two rows of bolts develop the full strength of the beam flanges

A coped tee provides shear support to the flange of the beam

FIGURE 9.32
This detail of an all-bolted AISC Type 1 (moment) connection illustrates the difficulty of developing the full strength of the beam flanges with bolts. This type of connection is so troublesome and expensive to make that it is very seldom used.

FIGURE 9.33
A coped beam-girder connection (AISC Type 2). A girder is a beam that supports other beams, as shown on the framing plan in Figure 9.39.

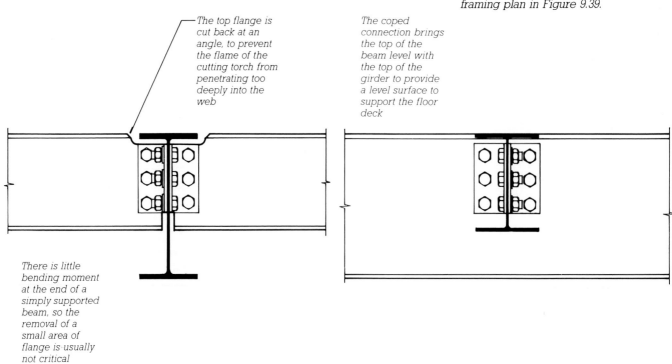

The top flange is cut back at an angle, to prevent the flame of the cutting torch from penetrating too deeply into the web

The coped connection brings the top of the beam level with the top of the girder to provide a level surface to support the floor deck

There is little bending moment at the end of a simply supported beam, so the removal of a small area of flange is usually not critical

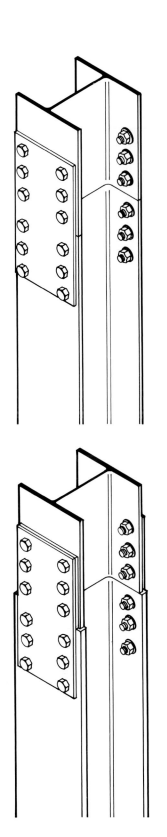

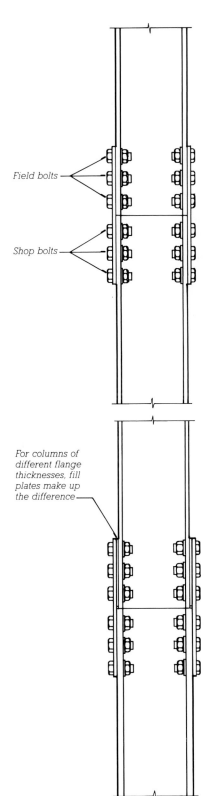

Field bolts

Shop bolts

For columns of
different flange
thicknesses, fill
plates make up
the difference

FIGURE 9.34
*Bolted column-column connections.
Column sizes diminish as the building
rises, requiring frequent use of the
lower detail.*

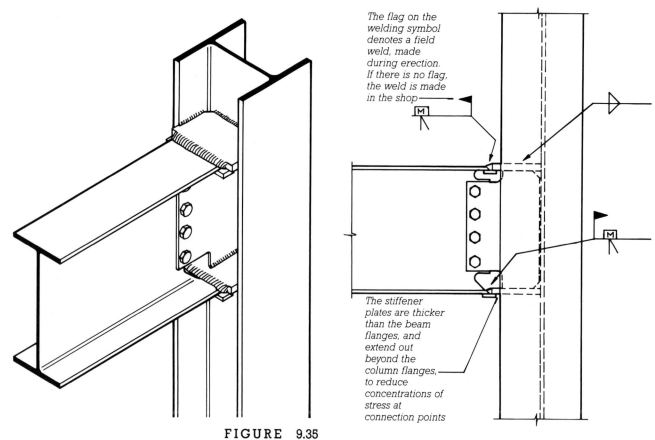

The flag on the welding symbol denotes a field weld, made during erection. If there is no flag, the weld is made in the shop

The stiffener plates are thicker than the beam flanges, and extend out beyond the column flanges, to reduce concentrations of stress at connection points

FIGURE 9.35
A welded beam-to-column-web connection (AISC Type 1).

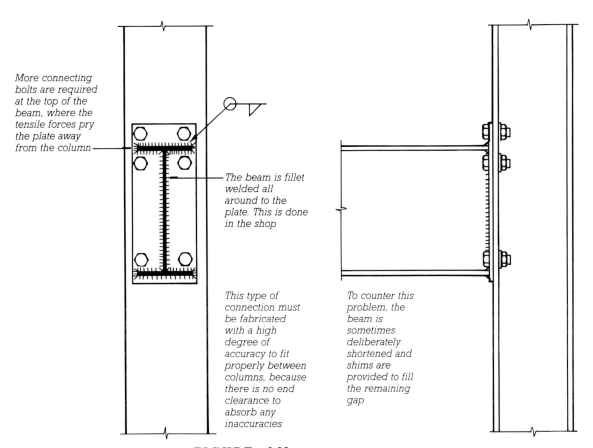

More connecting bolts are required at the top of the beam, where the tensile forces pry the plate away from the column

The beam is fillet welded all around to the plate. This is done in the shop

This type of connection must be fabricated with a high degree of accuracy to fit properly between columns, because there is no end clearance to absorb any inaccuracies

To counter this problem, the beam is sometimes deliberately shortened and shims are provided to fill the remaining gap

FIGURE 9.36
A welded/bolted end plate beam-column connection. The connection shown is AISC Type 3 (semi-rigid); with more bolts an AISC Type 1 end plate connection is possible, although not common.

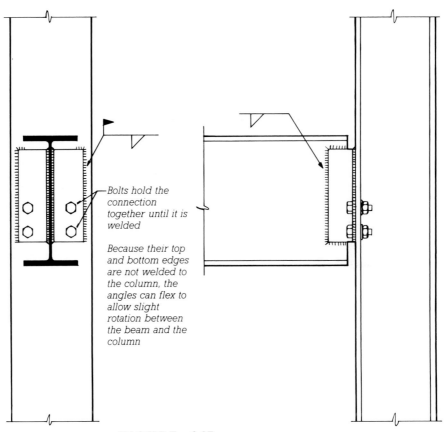

Bolts hold the connection together until it is welded

Because their top and bottom edges are not welded to the column, the angles can flex to allow slight rotation between the beam and the column

FIGURE 9.37

AISC Type 2 (shear) connections are also made by welding. The beam flanges are not connected to the column, and the angles are welded in such a way that they can flex to allow the beam to rotate away from the column.

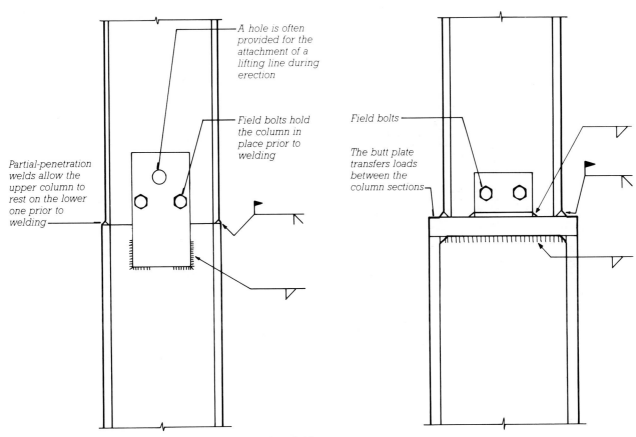

A hole is often provided for the attachment of a lifting line during erection

Field bolts hold the column in place prior to welding

Partial-penetration welds allow the upper column to rest on the lower one prior to welding

Field bolts

The butt plate transfers loads between the column sections

FIGURE 9.38
Welded column-column connections. The butt plate connection is used when a column changes from one nominal size of wide-flange to another.

DESIGNING STEEL CONNECTIONS

by

D A V I D T. R I C K E R, P.E.

Chief Engineer,
The Berlin Steel Construction Company,
Berlin, Connecticut

Steel connections account for a very substantial portion of the design time and construction cost of a steel building frame. The factors that bear on the efficient design of a connection are many; they can be summarized as follows:

1. The connection must perform with complete structural reliability. It must carry safely the loads indicated by the structural engineer, and it must offer the required degree of rigidity or flexibility.

2. The connection must be aesthetically acceptable to the engineer, the architect, and the owner of the building, particularly if it will be exposed in the finished structure.

3. The connection must keep within its allotted space in the building. It must avoid conflict with other steel members coming into the same joint, and it must keep out of the way of piping, ductwork, wiring, and finish surfaces.

4. The connection should use the most economical, readily available material, to minimize cost and avoid delays. ASTM A36 steel usually satisfies this requirement.

5. The connection should utilize the fabricator's equipment and manpower to best advantage. Some fabricators are set up to do punching and bolting economically; others are better equipped for welding; some can do both equally well.

6. The connection should require the minimum number of shop operations, and the minimum of materials handling in the shop. Each shipping piece should be designed to require only one method of connecting, either bolting or welding, but not both, so as to avoid having to move the piece about the shop to several different machines and work areas.

7. The connection should have no superfluous parts, such as stiffener plates that are not structurally required, or extra connecting plates and angles.

8. The connection should utilize fasteners to best advantage. Bolts in shear carry a greater load than bolts in friction; bolts in double shear carry more than bolts in single shear; bolting and welding can often be used in combination to bring out the best characteristics of each.

9. Fillet welds, which require no edge preparation, should be used instead of groove welds wherever possible.

10. The connection should be designed with normal mill, fabrication, and erection tolerances in mind. Adjustments must be provided for, but unnecessary precision should be avoided.

11. The connection should be designed to permit rapid, simple, safe erection.

12. The connection should be designed to require a feasible minimum of inspection. Bolts in shear, for example, can be inspected visually for tightness, while welds, and bolts in friction-type connections, may require more elaborate inspection procedures.

13. In general, simpler is better, and usually less expensive.

Figures A through E illustrate the application of some of these principles.

FRAMED CONNECTION—SHOP WELDED TO COLUMN

ADVANTAGES

1. No holes in column flange.
2. Very safe, simple, fast to erect.
3. Bay length can be accurately controlled.
4. Ample erection clearance.
5. Does not interfere with connections to other sides of column.
6. Bolts are in double shear, which doubles their capacity.
7. Beam requires only routine web punching.

DISADVANTAGES

1. Requires coping the bottom flange so beam can be inserted from above.
2. Burrs must be removed from beam web so it will slide easily between the angles.

THIS CONNECTION IS COMMONLY KNOWN AS A "KNIFE" CONNECTION

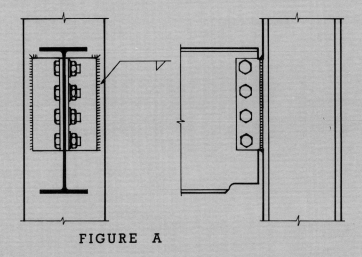

FIGURE   A

FRAMED CONNECTION—SHOP WELDED TO BEAM

ADVANTAGES

1. Bottom flange is not coped.
2. Useful for a wide range of beam sizes and loads.

DISADVANTAGES

1. Requires holes in column flanges, which might not otherwise require punching.
2. Bolts through flange may foul beams connecting to web of column.
3. Requires twice as many field bolts as a "knife" connection.
4. Has no tolerance for variation in column spacing or fabrication tolerances, other than expensive shimming.

FRAMED BEAM CONNECTION TO COLUMN WEB

ADVANTAGES

1. Useful for a large range of beam sizes and loads.
2. Beam flanges do not require punching, unless for erection seat.
3. Doesn't require a subsequently installed piece such as the stabilizing angle clip for a seated connection.

DISADVANTAGES

1. Difficult to use with small columns.
2. May have to interact with a connection on the far side of the web, as shown.
3. Beam may be difficult to erect if it must be tilted into place.
4. Bolts may be hard to install if there are also bolts in the column flanges.
5. An erection seat is usually required (as shown) for temporary support of one beam, if two beams share the same bolts.

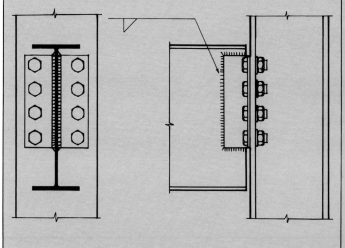

FIGURE B

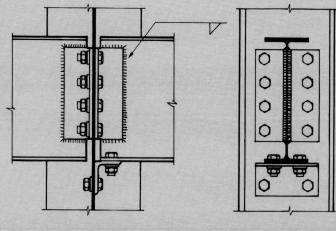

FIGURE C

SHOP WELDED SEATED BEAM CONNECTION

ADVANTAGES

1. Column shaft does not require punching.
2. Beam requires only routine flange punching.
3. Ample erection clearance.
4. Erection is fast, safe, and simple.
5. Connection parts are clean and require only routine shop operations.
6. Requires only 2 field bolts.
7. Useful for a wide range of beam and column sizes.
8. Does not affect connections on other 3 sides of column.
9. Column spacing is easy to maintain.
10. Simple to detail.

DISADVANTAGES

1. Delivers load to column with a slight eccentricity.
2. The stabilizer clip angle must be installed after the plumbing up of the structure.

SINGLE ANGLE FRAMED BEAM-GIRDER CONNECTION

ADVANTAGES

1. Ample erection clearance.
2. Erection is fast, simple, and safe.
3. Simple to detail.
4. The connection angle may also be shop welded to the girder.
5. The bay spacing is easily maintained. Short slotted holes may be used in the angle leg against the beam to facilitate this.
6. The beam requires only routine web punching.
7. A second connection angle may be added to increase the load capacity.

DISADVANTAGES

1. The load is delivered to the girder with a slight eccentricity.
2. Should not be used if the beam is subject to severe torsion at the connection.
3. Bolts may have to be staggered to assure driving clearances.

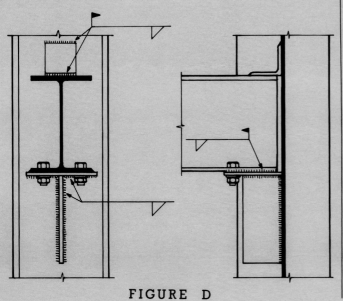

FIGURE D

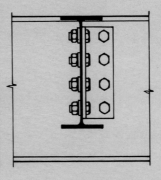
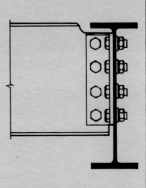

FIGURE E

THE CONSTRUCTION PROCESS

A steel building frame begins as a rough sketch on the drafting board of an architect or engineer. As the building design process progresses, the sketch evolves through many stages of drawings and calculations to become a finished set of structural drawings. These show accurate column locations and the shapes and sizes of all the members of the frame, but they do not give the exact length to which each member must be cut to mate with the members it joins, and they do not give details of the more routine connections of the frame. These are left to be worked out by a subsequent recipient of the drawings, the *fabricator*.

The Fabricator

The fabricator's job is to deliver steel components to the construction site ready to be assembled without further processing. This work begins with the preparation in the fabricator's shop of detailed drawings that show exactly how each piece will be made, and what its precise dimensions will be. Connections are designed to transmit the loads indicated by the engineer's drawings. The fabricator is free, within the limits of accepted engineering practice, to design the connections to be made as economically as possible, using various combinations of welding and bolting that best suit available equipment and expertise. Drawings are also prepared by the fabricator to show the general contractor exactly where and how to install foundation anchor bolts to connect to the columns of the building, and to guide the erector in assembling the steel frame on the building site. When completed, the fabricator's *shop drawings* are submitted to the engineer and the architect for review and approval to be sure that they conform exactly to the intentions of the design team. Meanwhile, the fabricator places an order with a producer

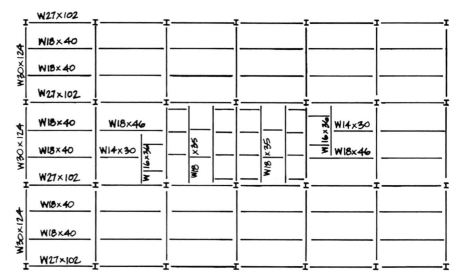

FIGURE 9.39

A typical steel framing plan for a multistory building, showing beam size designations. Notice how this frame requires beam-to-column-flange connections where the W30 girders meet the columns, beam-to-column-web connections where the W27 beams meet the columns, and coped beam-girder connections where the W18 beams meet the W30 girders. The small squares in the middle of the plan are openings for elevators, stairways, and mechanical shafts.

FIGURE 9.40

An automatic punching machine, for punching bolt holes in duplicate plates to match those in a pattern plate on the table to the right. (Courtesy of W. A. Whitney Corporation)

of steel for the stock from which the structural steel members will be fabricated. (The major beams, girders, and columns are usually ordered cut to exact length by the mill.) When the approved shop drawings, with corrections and comments, are returned by the design team, revisions are made as necessary, and full-size templates of cardboard or wood are prepared as required to assist the shop workers in laying out the various connections on the actual pieces of steel.

Plates, angles, and tees for connections are brought into the shop and cut to size and shape with oxyacetylene cutting torches, power shears, and saws. With the aid of the templates, bolt hole locations are marked. If the plates and angles are not unusually thick, the holes are made rapidly and economically with a punching machine, an automatic version of which is pictured in Figure 9.40. In very thick stock, or in pieces that will not fit conveniently into the punching machines, holes are drilled rather than punched.

Pieces of steel stock for the beams, girders, and columns are brought into the fabricator's shop with an overhead traveling crane or conveyor system. Each is stenciled or painted with a code that tells which building it is intended for and exactly where it will go in the building. With the aid of the shop drawings, each piece is measured and marked for its exact length and for the locations of all holes, stiffeners, connectors, and other details. Cutting to length, for those members not already cut to length at the mill, is done with a power saw or an oxyacetylene cutting torch. The ends of column sections that must bear fully on base plates or on one another are then squared and made perfectly flat by sawing, milling, or facing. In cases where the columns will be welded to one another, and for beams and girders that are to be welded, the ends of the flanges are beveled as necessary. Beam flanges are coped as required. Bolt holes are punched or drilled (Figure 9.41).

Where called for, beams and girders are *cambered* (curved slightly in a up-

Т here are 175,000 ironworkers in this country . . . and apart from our silhouettes ant-size atop a new bridge or skyscraper, we are pretty much invisible.

Mike Cherry, *On High Steel: The Education of an Ironworker* (New York: Quadrangle/ The New York Times Book Co., 1974), page xiii.

ward direction) so they will deflect into a straight line under load. Cambering is done by heating local areas of one flange of the member with a large oxyacetylene torch. As each area is heated to a cherry-red color, the metal softens, expands, and deforms to make a slight bulge in the width and thickness of the flange because the surrounding steel, which is cool, prevents the heated flange from lengthening. As the heated flange cools, the metal contracts, pulling the member into a slight bend at that point. By repeating this process at several points along the beam, a camber of the desired shape and magnitude is produced.

As a last step in fabricating beams, girders, and columns, stiffener plates are arc welded to each piece as required, and connecting plates, angles, and tees are welded or bolted at the appropriate locations (Figure 9.42). As much connecting as possible is done in the shop, where tools are handy and

FIGURE 9.41
Punching bolt holes in a wide-flange beam. (Courtesy of W. A. Whitney Corporation)

FIGURE 9.42
Welders attach connector plates to an exceptionally heavy column section in a fabricator's shop. The twin channels bolted to the end of the column will be used to attach a lifting line for erection, after which they will be removed and reused. (Courtesy of United States Steel Corporation)

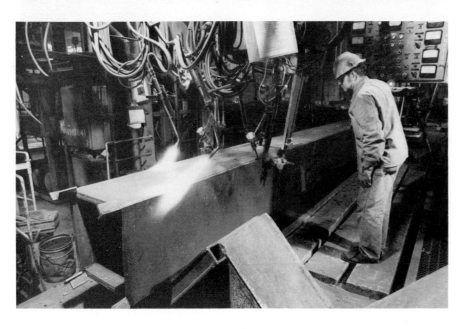

FIGURE 9.43
Machine-welding plates together to form a box column. The torches to the left preheat the metal to help avoid thermal distortions in the column. Mounds of powdered flux around the electrodes in the center indicate that this is the submerged arc process of welding. The small steel plates tacked onto the corners of the column at the extreme left are runoff tabs, which are used to allow the welding machine to go past the end of the column to make a complete weld. These will be cut off as soon as welding is complete. (Courtesy of United States Steel Corporation)

access is easy. This saves time and money during erection, when tools and working conditions are less optimal and total costs per manhour are higher.

Plate girders, built-up columns, trusses, and other large components are assembled in the shop in as large units as can practically be transported to the construction site, whether by truck, railway, or barge (Figure 9.43). Intricate assemblies such as large trusses are usually preassembled in their entirety in the shop, to be sure they will go together smoothly in the field, then broken down again into transportable components.

As the members are completed, each is straightened, cleaned and painted as necessary, and inspected for quality and for conformance to the job specifications and shop drawings. The members are then taken from the shop to the fabricator's yard by crane, conveyor, trolley, or forklift, where they are organized in stacks according to the order in which they will be needed on the building site.

The Erector

Where the fabricator's job ends, the *erector's* begins. Some companies both fabricate and erect, but in many cases the two operations are done by separate companies. The erector is responsible for assembling the steel components furnished by the fabricator on the building site. The erector's workers, by tradition, are called *ironworkers*.

Erecting the First Tier

Erection of a multistory steel building frame starts with assembly of the first two-story tier of framing. Lifting of the steel components is begun with a truck-mounted or crawler-mounted crane. In accordance with the erection drawings prepared by the fabricator, the columns for the first *tier*, usually in sections two stories high, are picked up from organized piles on the site and

If nobody plumbed-up, all the tall buildings in our cities would lean crazily into each other, their elevators would scrape and bang against the shaft walls, and the glaziers would have to redress all the windowglass into parallelograms....What leaned an inch west on the thirty-second floor is sucked back east on the thirty-fourth, and a column that refused to quit leaning south on forty-six can generally be brought over on forty-eight, and by the time the whole job is up the top is directly over the bottom.

Mike Cherry, *On High Steel: The Education of an Ironworker* (New York: Quadrangle/The New York Times Book Co., 1974), pages 110–111.

lowered carefully over the anchor bolts and onto the foundations, where the ironworkers bolt them down.

Foundation details for steel columns vary (Figure 9.44). Steel baseplates, which distribute the concentrated loads of the steel columns across a larger area of the concrete foundation, are shop welded to all but the largest of columns. The foundations and anchor bolts have been put in place previously by the general contractor, following the plan prepared by the fabricator. The contractor may, if requested, provide thin steel *leveling plates* set perfectly level and to the proper height on a bed of *grout* atop each concrete foundation. The baseplate of the column rests upon the leveling plate and is held down with the protruding anchor bolts. Alterna-

tively, especially for larger baseplates with four anchor bolts, the leveling plate is omitted. The column is supported at the proper elevation on stacks of steel shims inserted between the baseplate and the foundation, or on leveling nuts placed beneath the baseplate on the anchor bolts. After the first tier of framing is plumbed up as described below, the baseplates are grouted and the anchor bolts tightened. For very large, heavy columns, baseplates are shipped independently of the columns. Each is leveled in place with shims, wedges, or shop-attached leveling screws, then grouted prior to column placement.

After the first tier of columns is in place, the beams and girders for the first two stories are bolted in place (Figures 9.46, 9.47). The two-story tier of framing is then *plumbed up* (straightened and squared) using diagonal cables and turnbuckles, checking the alignment with plumb bobs and transits. When the tier is plumb, connections are tightened, baseplates are grouted if necessary, welds are made, and permanent diagonal braces, if called for, are rigidly attached.

At this level a temporary working surface of 2- or 3-inch (50- or 75-mm) wood planks or corrugated steel decking may be laid over the steel framing. Similar platforms will be placed every second story as the frame rises, unless safety nets are used instead, or the permanent floor decking is installed as erection progresses. The platforms protect the ironworkers from falling long distances down through the frame, and they protect workers on lower levels of the building from falling objects. They also furnish a convenient working surface for tools, materials, and derricks. Column splices are made at waist level above this platform, both as a matter of convenience and as a way of avoiding conflict between the column splices and the beam-to-column connections. Columns are generally fabricated in two-story lengths, a transportable size that also corresponds to the two-story spacing of the plank surfaces.

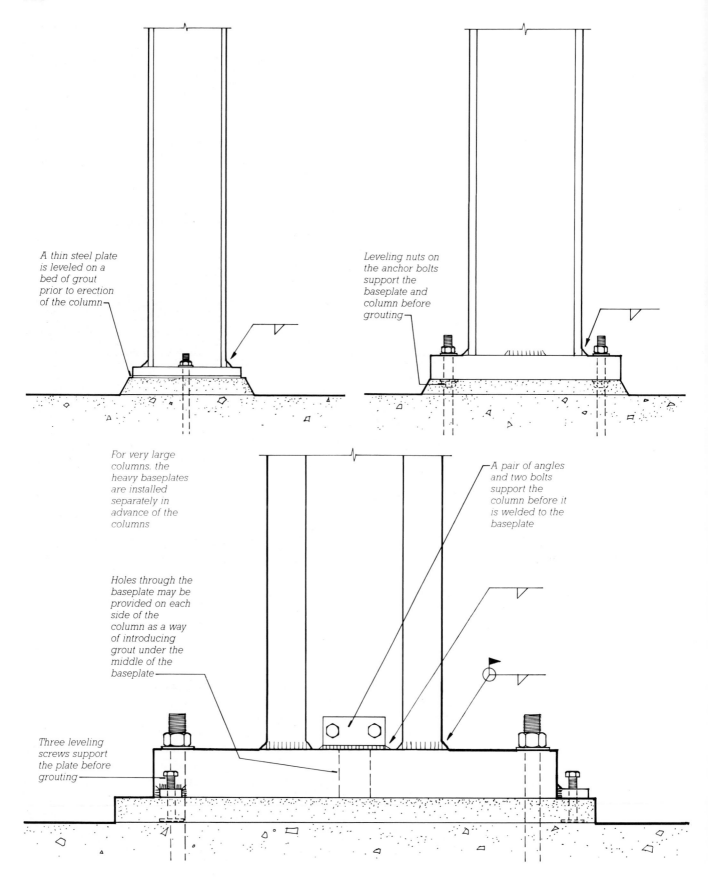

A thin steel plate is leveled on a bed of grout prior to erection of the column

Leveling nuts on the anchor bolts support the baseplate and column before grouting

For very large columns, the heavy baseplates are installed separately in advance of the columns

A pair of angles and two bolts support the column before it is welded to the baseplate

Holes through the baseplate may be provided on each side of the column as a way of introducing grout under the middle of the baseplate

Three leveling screws support the plate before grouting

FIGURE 9.44
*Three typical column base details.
(Upper left:) A small column with
welded baseplate set on a steel leveling
plate. (Upper right:) A larger column
with welded baseplate set on leveling
nuts. (Below:) A heavy column field-
welded to a loose baseplate that has
been previously leveled and grouted.*

FIGURE 9.45
*Ironworkers guide the placement of a
heavy column fabricated from two
rolled wide-flange sections and two
thick steel plates. It will be bolted to its
baseplate through the holes in the small
plate welded between the flanges on
either side.* (Courtesy of Bethlehem
Steel Corporation)

FIGURE 9.46
*Placing the steel for the first tier. The
two ironworkers near the top of the
picture are placing a girder.* (Photo by
Architectural Camera. Courtesy of
American Institute of Steel Construction)

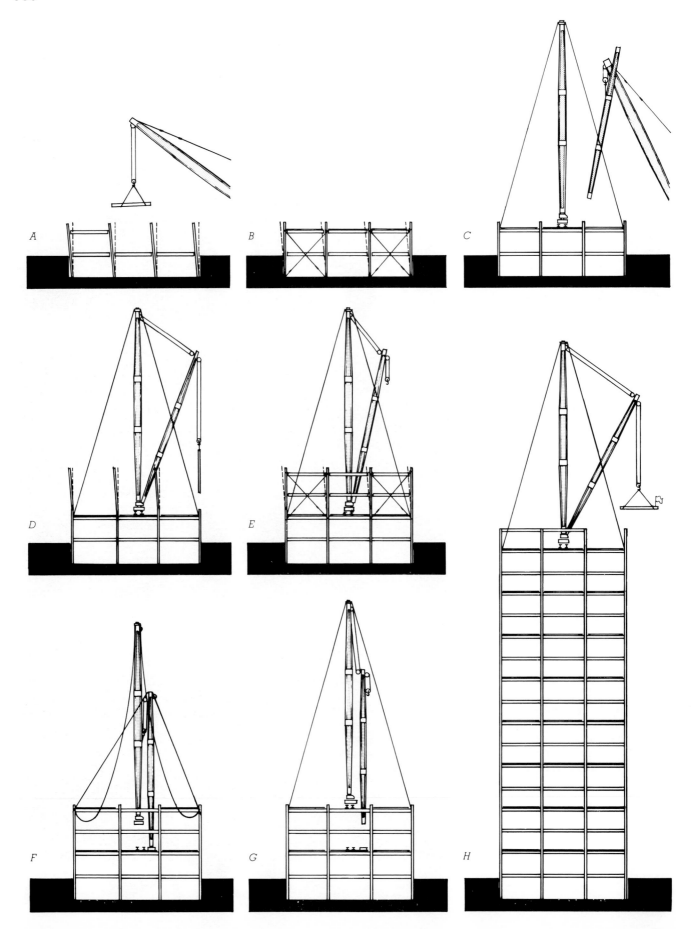

Erecting the Upper Tiers

Erection of the second tier proceeds much like that of the first. Two-story column sections are hoisted into position and connected by splice plates to the first tier of columns. The beams and columns for the two floors are set, the tier is plumbed and tightened up, and another layer of planks, decking, or safety netting is installed.

If the building is not too tall, the ground-based crane will do the lifting for the entire building. For a taller building, the ground-based crane does the work until it gets to the maximum height to which it can lift a climbing crane or derrick. A building-mounted crane or derrick takes over from there,

lifting itself tier by tier as the frame rises.

The *guy derrick* has been the workhorse of steel erection for many decades, but new and more efficient types of climbing cranes have replaced it in all but the tallest of frames. The guy derrick is supported by the building frame and hoists itself upward a tier at a time in a process called "jumping the derrick" (Figure 9.47). After the building is topped out, the derrick is disassembled and lowered to the ground with smaller hoisting equipment, which in turn can be taken down the construction elevator. The *climbing crane*, such as the one in Figure 9.50, works somewhat differently, building itself an independent tower

FIGURE 9.47

The steps in erecting a tall steel building frame, using a guy derrick. **A.** *The first tier is erected with a ground-based crane.* **B.** *The first tier is plumbed up, and a temporary plank floor is placed.* **C.** *The ground-based crane erects the guy derrick. This is usually done at the highest level the ground-based crane can reach but is shown here at the second floor to keep the drawing to a reasonable size.* **D, E.** *An upper tier of framing is erected and plumbed.* **F.** *The boom of the guy derrick is disconnected from the mast, mounted on a temporary block, and braced with guy cables. The boom then lifts the derrick mast to the next tier.* **G.** *The mast is secured in its new position and lifts the boom up to be reconnected to the mast. This jumping of the derrick is repeated every two stories.* **H.** *The frame is topped out, after which the derrick is disassembled and lowered to the ground in pieces by a small hoist. The small hoist, in turn, is disassembled and taken down the elevator.*

FIGURE 9.48

Rather than install temporary plank floors, the erector of this building has chosen to lay the corrugated steel decking as soon as each floor has been framed. (Photo by Architectural Camera. Courtesy of American Institute of Steel Construction)

as the building rises, usually within an elevator shaft or a vertical space temporarily left open in the frame.

As each piece of steel is lowered toward its final position in the frame, it is guided by an ironworker holding a rope called a *tagline*, the other end of which is attached to the piece. Other ironworkers in the raising gang guide the piece by hand as soon as they can reach it, until its bolt holes align with those in the mating pieces (Figures 9.51, 9.52). Sometimes crowbars or hammers must be used to pry, wedge, or drive components until they fit properly, and bolt holes may on occasion have to be reamed larger to admit bolts through slightly misaligned pieces. When an approximate alignment is achieved, tapered steel *drift pins* from the ironworker's tool belt are shoved into enough bolt holes to hold the pieces together until a few bolts can be inserted. A gang of bolters

FIGURE 9.49
Three guy derricks raise the steel for this forty-one-story office building in New York City. Two temporary outside elevators transport workers and small materials up and down the building. (Courtesy of Bethlehem Steel Corporation)

follows behind the raising gang, filling the remaining holes with bolts from leather carrying baskets, tightening them first with hand wrenches and then with impact wrenches. Field-welded connections are initially held in alignment with bolts, then welded when the frame is plumb.

The placing of the last beam at the top of the building is carried out with a degree of ceremony appropriate to the magnitude of the building. At the very least a small evergreen tree, a national flag, or both are attached to the beam before it is lifted (Figure 9.53). For major buildings, assorted digni-

taries are likely to be invited to a building-site *topping-out* party that includes music and refreshments. After the party, work goes on as usual, for although the frame is complete, the building is not. Cladding and finishing operations will continue for many months.

FIGURE 9.50
A climbing crane being used for steel erection. The tower of the crane grows as the building grows around it. (Courtesy of United States Steel Corporation)

FIGURE 9.51
Connecting a beam to a column.
(Courtesy of Bethlehem Steel
Corporation)

FIGURE 9.52
Ironworkers attach a girder to a box column. Each carries two wrench/drift pin combination tools in a holster on his belt, inserting the tapered drift pins into each connection to hold it until a few bolts can be added. Bundles of corrugated steel decking are ready to be opened and distributed over the beams to make a floor deck. (Courtesy of Bethlehem Steel Corporation)

FIGURE 9.53
Topping out: The last beam in a steel frame is special. (Courtesy of United States Steel Corporation)

Floor and Roof Decking

If plank decks are used during erection of the frame, they must be replaced with permanent floor and roof decks of incombustible materials. In early Steel Frame buildings, shallow arches of brick or tile were often built between the beams, tied with steel tension rods, and filled over with concrete to produce level surfaces (Figure 9.54). These were both heavier, necessitating larger framing members to carry their weight, and more consumptive of labor, than the metal deck systems commonly used today.

Metal Decking   *Metal decking* at its simplest is a sheet of steel that has been corrugated to increase its stiffness. The spanning capability of the deck is determined mainly by the thickness of the sheet from which it is made and the depth of the corrugations. Single corrugated sheets are commonly used for roof decking, where concentrated loads are not expected to be great and deflection criteria are not as stringent as in floors (Figures 9.55, 9.56). They are also used as permanent formwork for concrete floor decks, with a wire-fabric-reinforced concrete slab supported by the steel decking until the slab can support itself and its live loads. *Cellular decking* is manufactured by welding together two sheets, one corrugated and one flat, and can be made sufficiently stiff to support normal floor loads without structural assistance from the concrete fill that is poured over it to produce a level floor. Cellular decking offers the important side benefit of providing spaces for running electrical and communications wiring. Metal decking is usually puddle-welded to the joists, beams, and girders at intervals by melting through the decking to the supporting members below with a welding electrode. If the deck is required to act as a diaphragm, the longitudinal edges of the decking panels must be connected to one another at frequent intervals with screws or welds.

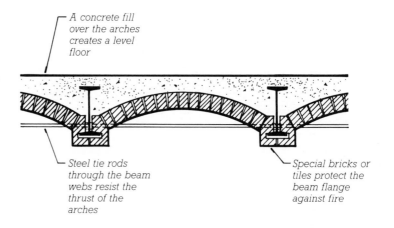

FIGURE   9.54
Tile or brick arch flooring will be found in many older steel frame buildings.

A concrete fill over the arches creates a level floor

Steel tie rods through the beam webs resist the thrust of the arches

Special bricks or tiles protect the beam flange against fire

FIGURE   9.55
Workers install corrugated steel form decking over a floor structure of open-web steel joists supported by joist girders. (Photo by Balthazar Korab. Courtesy of Vulcraft Division of Nucor Corporation)

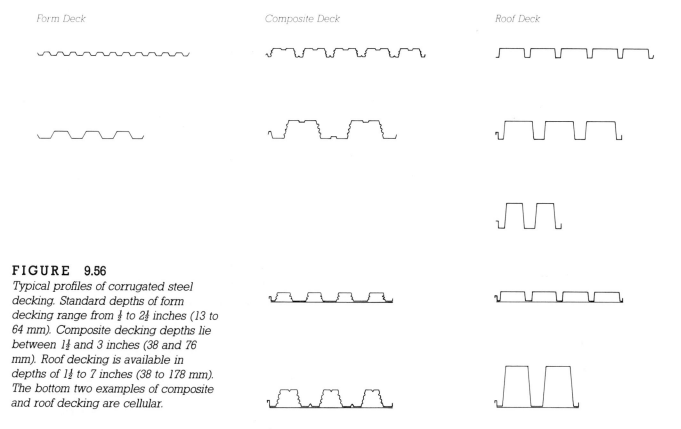

Form Deck

Composite Deck

Roof Deck

FIGURE 9.56
Typical profiles of corrugated steel decking. Standard depths of form decking range from $\frac{1}{2}$ to $2\frac{1}{2}$ inches (13 to 64 mm). Composite decking depths lie between $1\frac{1}{2}$ and 3 inches (38 and 76 mm). Roof decking is available in depths of $1\frac{1}{2}$ to 7 inches (38 to 178 mm). The bottom two examples of composite and roof decking are cellular.

Composite Construction *Composite metal decking* (Figure 9.57) is designed to work together with its concrete fill to make a stiff, lightweight, economical deck. The decking itself serves as tensile reinforcing for the concrete, which is bonded to it by special rib patterns or by small steel rods or wire fabric welded to the tops of the corrugations.

Composite construction is often carried a step beyond the decking to include the beams and girders of the floor.

Before the concrete is poured over the metal deck, *shear studs* are welded through the decking to the top of each beam and girder, using a special electric welding gun (Figures 9.58, 9.59). It would be more economical to attach the shear studs in the shop rather than in the field, but the danger of tripping up ironworkers during the erection process delays their installation until the steel decking is in place. The purpose of the studs is to create a strong shear connection between the con-

crete slab and the steel beam. A strip of the slab can then be assumed to act together with the top flange of the steel shape to resist compressive forces. The result of composite design is a steel member whose loadbearing capacity has been greatly enhanced at relatively low cost by taking advantage of the unused strength of the concrete topping that must be present in the construction anyway. The payoff is a stiffer, lighter, less expensive building.

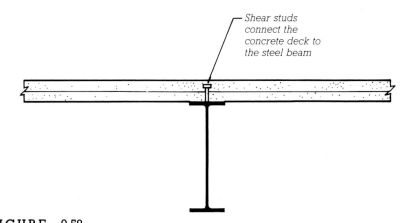

Shear studs connect the concrete deck to the steel beam

FIGURE 9.58
Composite beam construction.

FIGURE 9.59
Pouring a concrete fill on a steel roof deck, using a concrete pump to deliver the concrete from the street below to the point of the pour. Shear studs are plainly visible over the lines of the beams below. The welded wire fabric reinforces the concrete against cracking. (Courtesy of Schwing America, Inc.)

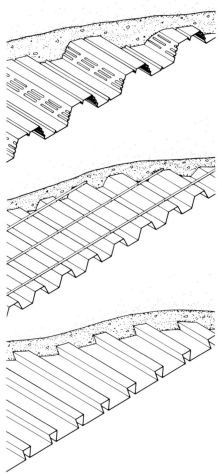

FIGURE 9.57
Composite decking acts as steel reinforcing for the concrete topping installed over it. The top example bonds to the concrete with deformed ribs, and the middle example with welded steel rods. The bottom type makes an attractive ceiling texture if left exposed and furnishes dovetail channels for the insertion of special fastening devices to hang ductwork, piping, conduits, and machinery from the ceiling.

Concrete Decks Concrete floor and roof slabs are often used in steel building frames without metal decking. Concrete may be poured in place over removable plywood forms, or it may be erected as precast planks lifted into place much like the metallic elements of the building (Figure 9.60). Precast decks are relatively light in weight and are quick to erect, even under weather conditions that would preclude the pouring of concrete, but they usually require the addition of a thin, poured-in-place concrete topping to produce a smooth floor.

Roof Decking For roofs of low steel-framed buildings, many different types of decks are available. Corrugated metal may be used, with or without a concrete fill; many types of rigid insulation boards are capable of spanning the corrugations to make a flat surface for the roof membrane. Some corrugated decks are finished with a weather-resistant coating that allows them to serve as the water-resistive surface of the roof. A number of different kinds of insulating deck boards are produced of such fibers as wood and glass bonded with various substances. Many of the insulating boards are designed as permanent formwork for reinforced, poured slabs of gypsum or lightweight concrete; in this application, they are usually supported on steel *subpurlins* (Figures 9.61, 9.62). Heavy timber decking, or even wood joists and plywood, are also used over steel framing

FIGURE 9.60
A tower crane installs precast concrete hollow-core planks for floor decks in an apartment building. Precast concrete is also used for exterior cladding of the building. The steel framing is a design known as the staggered truss system, *in which story-height steel trusses at alternate levels of the building support the floors. The trusses are later enclosed with interior partitions.*
(Courtesy of Blakeslee Prestress, Inc.)

FIGURE   9.61
*Welding truss-tee subpurlins to open-
web steel joists for a roof deck.
(Courtesy Keystone Steel and Wire)*

FIGURE   9.62
*Installing insulating formboard over
truss-tee subpurlins. A light wire
reinforcing mesh and poured gypsum
fill will be installed over the formboards.
The trussed top edges of the subpurlins
will be embedded in the gypsum slab
to form a composite deck. (Courtesy
Keystone Steel and Wire)*

in situations where building codes
permit combustible materials.

Corrugated steel sheets are also used
for siding of industrial buildings, where
they are supported on *girts*, which are
horizontal zees or channels that span
between the outside columns of the
building.

FIREPROOFING OF STEEL FRAMING

Building fires are not hot enough to
melt steel, but many are able to weaken
it sufficiently to cause structural failure (Figures 9.63, 9.64). For this reason, building codes generally limit the
use of exposed steel framing to buildings of one to three stories, where escape in case of fire is rapid and where
collapse of the building is unlikely to
endanger people or other buildings.
For taller buildings, it is necessary to
protect the steel frame from heat for
a length of time sufficient that the
building can be fully evacuated and
the fire extinguished.

Fireproofing (fire protection might
be a more accurate term) of steel framing was originally done by encasing
steel beams and columns in brick masonry or poured concrete (Figures 9.65,
9.66). These heavy encasements were
effective, absorbing heat into their great
mass and dissipating some of it through
dehydration of the mortar and concrete, but their weight added considerably to the load that the steel frame
had to bear and, therefore, to the
weight and cost of the frame. The search
for lighter-weight fireproofing led first
to thin enclosures of metal lath and
plaster around the steel members
(Figure 9.67). These derive their effectiveness from the large amounts of
heat needed to dehydrate the water of
crystallization from the gypsum plaster. Plasters using lightweight aggregates such as vermiculite instead of sand
have come into use to further reduce
the weight and to add thermal insulating properties to the plaster.

Today's designers can also choose

from a group of fireproofing techniques that are lighter still. Plaster fireproofing has largely been replaced by beam and column enclosures made of boards or slabs of gypsum or other fire-resistive materials (Figures 9.68 through 9.70). These are fastened mechanically around the steel shapes, and in the case of the gypsum board fireproofing, they can also serve as the finished surface on the interior of the building. Where the fireproofing material need not serve as a finished surface, spray-on materials have become the most prevalent type. These generally consist of a fiber and a binder, or of a cementitious mixture, and are sprayed over the steel to the required thickness (Figure 9.71). These products are available in weights of about 12 to 40 pounds per cubic foot (190 to 640 kg/m³). The lighter materials are, as a group, fragile and unattractive, and must be covered with finish materials. The denser materials are generally more durable and attractive. All these materials act primarily by insulating the steel from high temperatures for long periods of time. They are generally the least expensive form of fireproofing.

The newest generation of fireproofing techniques for steel offers new possibilities to the designer. *Intumescent mastics and paints* allow steel structural elements to remain exposed in situations of low to moderate fire risk; these coatings expand when exposed to fire to form a stable char that insulates the steel from the heat of the fire for varying lengths of time, depending on the thickness of the coat-

FIGURE 9.63
An exposed steel structure following a prolonged fire in the highly combustible contents of a warehouse. (Courtesy of National Fire Protection Association)

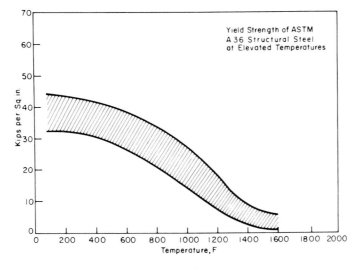

FIGURE 9.64
The relationship between temperature and strength in structural steel. (Courtesy American Iron and Steel Institute)

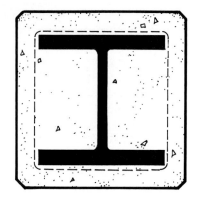

A

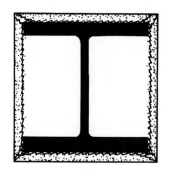

B

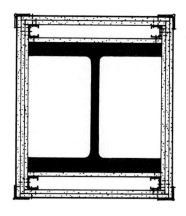

C

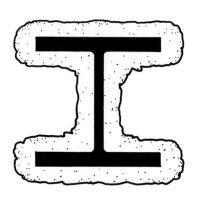

D

E

F

FIGURE 9.65

Some methods for fireproofing steel columns: **A.** *Encasement in reinforced concrete.* **B.** *Enclosure in metal lath and plaster.* **C.** *Enclosure in multiple layers of gypsum board.* **D.** *Spray-on fireproofing.* **E.** *Loose insulating fill inside a sheet metal enclosure.* **F.** *Water-filled box column made of a wide-flange shape with added steel plates.*

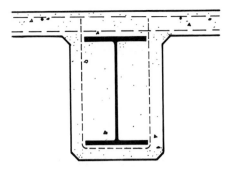

A

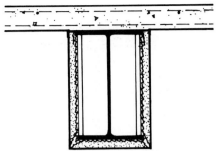

B

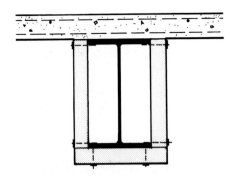

C

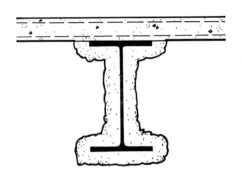

D

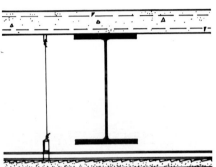

E

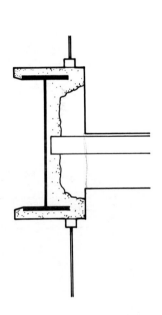

F

FIGURE 9.66

Some methods for fireproofing steel beams and girders: **A.** *Encasement in reinforced concrete.* **B.** *Enclosure in metal lath and plaster.* **C.** *Rigid slab fireproofing.* **D.** *Spray-on fireproofing.* **E.** *Suspended plaster ceiling.* **F.** *Flame-shielded exterior spandrel girder with spray-on fireproofing inside.*

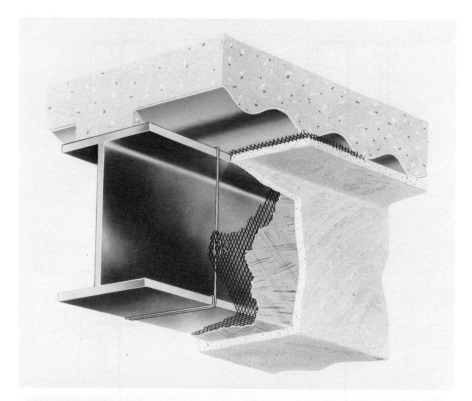

FIGURE 9.67
Lath-and-plaster fireproofing around a steel beam. (Courtesy of United States Gypsum Company)

FIGURE 9.68
Gypsum board fireproofing around a steel column. The gypsum board layers are screwed to the four cold-formed steel C-channels at the corners of the column, and finished with steel corner bead and drywall compound on the corners. (Courtesy of United States Gypsum Company)

FIGURE 9.69
Attaching slab fireproofing made of mineral fiber to a steel column, using welded attachments. (Courtesy of United States Gypsum Company)

FIGURE 9.70
Slab fireproofing on a steel beam and deck. (Courtesy of United States Gypsum Company)

FIGURE 9.71
Applying spray-on fireproofing to a steel beam, using a gage to measure the depth. (Courtesy of W. R. Grace Company)

ing. Most intumescent coatings can be obtained in an assortment of colors and can also serve as a base coat under ordinary paints if another color is desired. Another rather specialized technique, applicable only to steel box or tube columns exposed on the exterior of buildings, is to fill the columns with water and antifreeze. Heat applied to a region of a column by a fire is dissipated throughout the column by convection in the liquid filling. New mathematical techniques have been developed for calculating temperatures that will be reached by steel members in various situations during a fire; these allow the designer to experiment with a variety of ways of protecting the members, including metal flame shields that allow the compo-

nent to be left exposed on the exterior of a building (Figure 9.66*f*).

LONGER SPANS IN STEEL

Beams of standard wide-flange shapes are suitable for the range of structural spans normally encountered in offices, schools, hospitals, apartments, hotels, retail stores, warehouses, and other buildings in which columns may be brought to earth at intervals without obstructing the activities that take place within. For many other types of buildings—athletic buildings, certain types of industrial buildings, aircraft hangars, auditoriums, theaters, religious

buildings, transportation terminals—longer spans are required than can be accomplished with wide-flange beams. A rich assortment of longer-span structural devices is available in steel for these uses.

Improved Beams

One general class of longer-span devices might be called improved beams. The *castellated beam* (Figure 9.72) is produced by flame-cutting the web of a wide-flange section along a zigzag path, then reassembling the beam by welding it point-to-point, thus increasing its depth without increasing its weight. This greatly increases the spanning potential of the beam, pro-

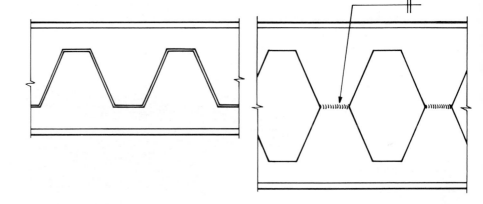

FIGURE 9.72
Manufacture of a castellated beam.

FIGURE 9.73
Erecting a welded steel plate girder. Notice how the girder is custom made with cutouts for the passage of pipes and ductwork. The section being erected is 115 feet (35 m) long, 13 feet (4 m) deep, and weighs 192,000 pounds (87,000 kg). (Courtesy of Bethlehem Steel Corporation)

vided the superimposed loads are not exceptionally heavy. For longspan beams tailored to any loading condition, *plate girders* are custom designed and fabricated. Steel plates and angles are assembled by bolting or welding in such a way as to put the steel exactly where it is needed: The flanges often become thicker in the middle of the span where bending forces are higher, more web stiffeners are provided near the ends where shear stresses are high, and so on. Almost any depth can be manufactured as needed, and very long spans are possible, even under heavy loads (Figure 9.73). Often these members are tapered, having greater depth where the bending moment is largest. *Rigid frames* are easily produced by welding

together steel wide-flange sections or plate girders. They may be set up in a row to roof a rectangular space (Figure 9.74), or rotated about a point to cover a circular area. Castellated beams, plate girders, and rigid frames share the characteristic that they must be braced laterally by purlins, decking, or diagonal bracing to prevent them from buckling.

Trusses

Steel *trusses* (Figures 9.75 through 9.78) are generally deeper and lighter than improved beams and can span correspondingly longer distances. They can be designed to carry light or heavy loads. Earlier in this chapter, one class

of steel trusses, open web joists and joist girders, were presented. These are light members for light loadings but are capable of fairly long spans, and they are usually less expensive than custom-made trusses. Roof trusses for light loadings are most often made of double-angle *chords* with gusset plate connectors and may be either welded or bolted. Trusses for heavier loadings, such as the transfer trusses that are used in some building frames to transmit column loads from floors above across a wide meeting room or lobby in a building, can be made of wide-flange or tubular shapes.

A steel *space truss* (more popularly called a *space frame*) is a truss made three-dimensional (Figures 9.79 and 9.80). It carries its load by bending

FIGURE 9.74
A farm building spanned with steel rigid frames. (Courtesy of Varco-Pruden Building Systems)

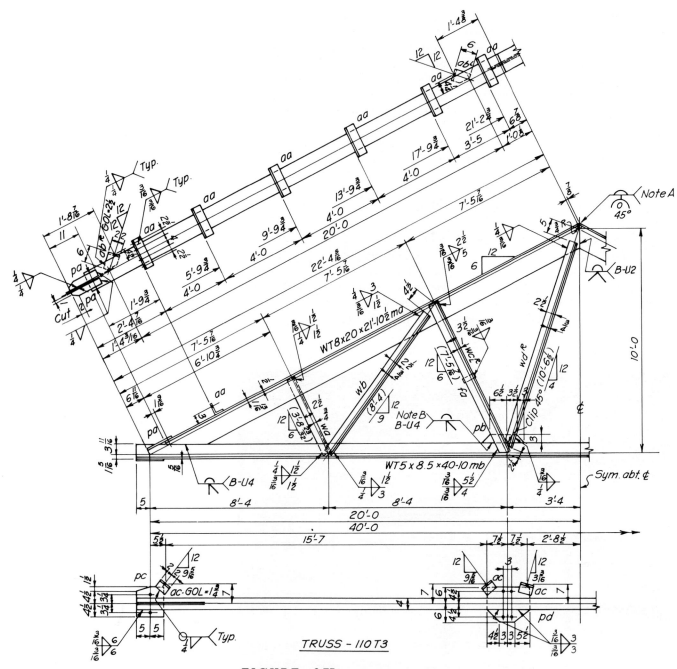

FIGURE 9.75

A fabricator's shop drawing of a welded steel roof truss made of tees and double-angle chords. (From Detailing for Steel Construction, *Chicago, AISC, 1983. Reproduced by permission of the American Institute of Steel Construction)*

FIGURE 9.76
*Bolted steel roof trusses over a
shopping mall support steel purlins
which carry the corrugated steel roof
deck. Automatic sprinklers provide fire
protection and permit a larger building
under the building code.* (Courtesy of
American Institute of Steel Construction)

FIGURE 9.77
Ironworkers seat the end of a heavy roof truss made of wide-flange sections. (Courtesy of American Institute of Steel Construction)

FIGURE 9.78
Tubular steel trusses support the roof of a convention center. (Courtesy of American Institute of Steel Construction)

along both its axes, much like a two-way concrete slab (Chapter 11). It must therefore be supported by columns that are approximately equally spaced in both directions.

Arches

Steel *arches*, produced by bending standard wide-flange shapes or by joining plates and angles, can be made into cylindrical roof vaults or circular domes of considerable span (Figure 9.81). For greater spans still, the arches may be built of trusswork. Lateral thrusts are produced at the base of an

FIGURE 9.79

Assembling a space truss. (Courtesy of Unistrut Space-Frame Systems, GTE Products Corporation)

FIGURE 9.80

A space truss roof over a ferry terminal. (Architects: Braccia/DeBrer/Heglund. Structural engineer: Kaiser Engineers. Photo by Barbeau Engh. Courtesy of American Institute of Steel Construction)

The spider web is a good inspiration for steel construction.

Frank Lloyd Wright, "In the Cause of Architecture: The Logic of the Plan," *Architectural Record,* January 1928.

FIGURES 9.82, 9.83
The Olympic roof in Munich, Germany, made of steel cables and transparent acrylic plastic panels. For scale, notice the worker seen through the roof at the upper left of Figure 9.83. (Architects: Frei Otto, Ewald Bubner, and Benisch and Partner. Photo courtesy of Institute for Lightweight Structures, Stuttgart)

arch and must be resisted by the foundations or some other part of the building structure.

Tensile Structures

High-tensile-strength wires of cold drawn steel, made into cables, are the material for a fascinating variety of tent-like roofs that can span almost unlimited distances (Figures 9.82, 9.83). With *anticlastic* (saddle-shaped) curvature, cable *stays*, or other ways of restraining the cable net, hanging roofs can be made rigid against wind uplift and flutter.

Industrialized Systems

Steel adapts well to industrialized systems of construction. The two most successful prefabrication systems in the United States are probably the mobile home and the "package" industrial building. The mobile home, built largely of wood, is made possible by a rigid undercarriage welded together from steel shapes. The package building is most commonly based on a structure of welded steel rigid frames supporting an enclosure of corrugated metal sheets. The mobile home is founded on steel because of steel's matchless stiffness and strength. The package building depends on steel for these qualities, for the repeatable precision with which components can be produced, and for the ease with which the relatively light steel components can be transported and assembled. It is but a short step from the usual process of steel fabrication and erection to the serial production of repetitive building components.

STEEL AND THE BUILDING CODES

Steel Frame construction appears in the typical building code tables in Figures 1.1 and 1.2 as five different construction types—1A, 1B, 2A, 2B, and 2C—the exact classification depending on the degree of fireproofing treatment applied to the various members of the frame. With a high degree of fireproofing, especially on members supporting more than one floor, unlimited building heights and areas are permitted for most types of occupancies. With no fireproofing whatsoever of steel members, building heights and areas are severely restricted, but many building types can easily meet these restrictions. Another section of the typical building code allows the height and area of a building to be extended significantly if an automatic fire suppression system (sprinklers) is installed throughout. Except for high hazard occupancies, the allowable area of a sprinklered building under this code may be increased by 200 percent for one- and two-story buildings, or by 100 percent for taller buildings.

THE UNIQUENESS OF STEEL

Among the common structural materials for fire-resistant construction—masonry, concrete, and steel—steel alone has useful tensile strength, which, along with compressive strength, it possesses in great abundance (Figure 9.84). A relatively small amount of steel can do a structural job that would take a much greater amount of another material. Thus steel, the most dense of structural materials, is also the one that produces the lightest structures and those that span the greatest distances.

The infrastructure needed to bring steel shapes to a building site—the mines, the mills, the fabricators—is vast and complex. An elaborate sequence of advance planning and preparation activities is required for the making of a steel building frame. Once on the site, however, a steel frame goes together quickly and with relatively few tools, in an erection process that is rivaled for speed and all-weather reliability only by certain precast concrete systems. With proper design and planning, steel can frame almost any shape building, including irregular angles and curves. Ultimately, of course, steel produces only a frame. Unlike masonry or concrete, it does not lend itself easily to forming a total building enclosure except in certain industrial applications. But this is of little consequence, because steel mates easily with glass, masonry, and panel systems of enclosure, and because steel does its own job, that of carrying loads high and wide with apparent ease, so very well.

Material	Allowable Tensile Strength	Allowable Compressive Strength	Density	Modulus of Elasticity
Wood (average)	700 psi (4.83 MN/m²)	1,100 psi (7.58 MN/m²)	30 pcf (480 kg/m³)	1,200,000 psi (8,275 MN/m²)
Brick masonry (average)	0	250 psi (1.72 MN/m²)	120 pcf (1,920 kg/m³)	1,200,000 psi (8,275 MN/m²)
Steel (ASTM A36)	22,000 psi (151.69 MN/m²)	22,000 psi (151.69 MN/m²)	490 pcf (7,850 kg/m³)	29,000,000 psi (200,000 MN/m²)
Concrete (average)	0	1,350 psi (9.31 MN/m²)	145 pcf (2,320 kg/m³)	3,150,000 psi (21,720 MN/m²)

FIGURE 9.84
Steel is many times stronger and stiffer than other structural materials.

FIGURE 9.86
Span-to-depth relationships for preliminary selection of open-web steel joists. Each heavy horizontal line corresponds to a standard depth of joist. For lightly loaded joists, read at or near the right-hand end of the heavy lines. For heavier loads, read further to the left. This chart is for preliminary design use only and does not take into account joist spacing, joist weight, concentrated loadings, and many other factors.

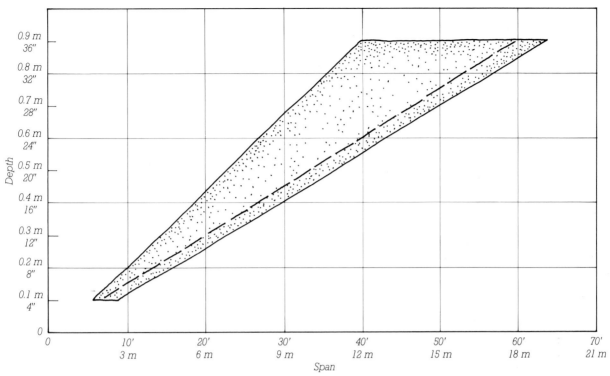

FIGURE 9.85

Span-to-depth relationships for preliminary selection of steel beams and girders. For floor beams uniformly loaded to capacity, read upward from the desired span to the dashed line, then to the left to read the required depth. For girders, read to the left from above the dashed line within the shaded area. For roof beams, read from below the dashed line. This chart is for preliminary design use only and does not take into account beam weight, concentrated loadings, wind loadings, lateral support conditions, and many other factors.

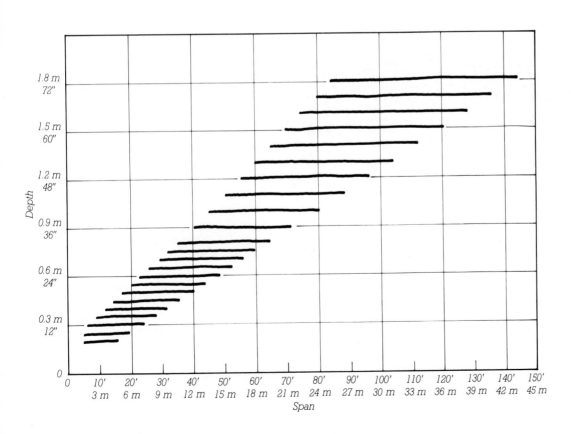

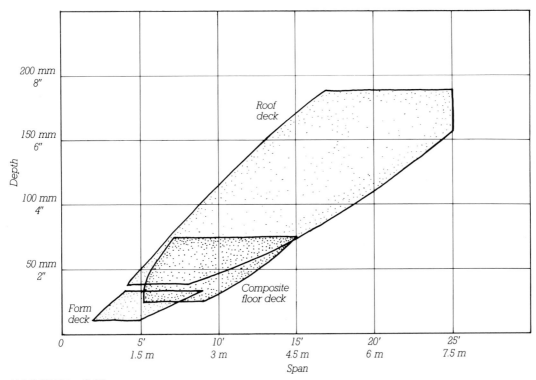

FIGURE 9.87

Span-to-depth relationships for preliminary selection of corrugated steel floor and roof decking. This chart is for preliminary design use only and does not take into account the number of continuous spans of decking, the thickness and corrugation pattern of the decking, the thickness of the concrete fill, and many other factors.

FIGURE 9.88

Architect Philip Johnson's famous house of steel and glass was built in 1949 in New Canaan, Connecticut. (Courtesy of John Burgee Architect with Philip Johnson)

FIGURE 9.89
Two views of the Chicago Police Training Center: (a) *Under construction.* (b) *Completed.* (Architect: Jerome R. Butler, Jr. Engineer: Louis Koncza. Photo courtesy of American Institute of Steel Construction)

FIGURE 9.90
A college art center spans a ravine with a steel truss structure. (Architect: Craig Ellwood Associates. Engineer: Norman J. Epstein. Photo courtesy of American Institute of Steel Construction)

FIGURE 9.91
Open-web steel joists and joist girders form the finish ceiling of a high-school library. (Architect: Rowe Holmes Associates. Structural Engineer: Rast Associates. Photo by Gordon Schenck, Jr. Courtesy of American Institute of Steel Construction)

FIGURE 9.92
Exposed steel framing in the lobby of an office building. (Architect and engineer: Ellerbe Associates. Photo courtesy of American Institute of Steel Construction)

FIGURE 9.93
Chicago is famous for its role in the development of the steel frame skyscraper (see Figure 9.4). The tallest of them is the Sears Tower, seen in the foreground of this photograph. (Architect and engineer: Skidmore, Owings and Merrill. Photo by Chicago Convention and Tourism Bureau, Inc. Courtesy of American Institute of Steel Construction)

C.S.I./C.S.C. Masterformat Section Numbers for Steel Frame Construction	
05100	STRUCTURAL METAL FRAMING
05120	Structural Steel
05150	Wire Rope
05160	Framing Systems Space Frames
05200	METAL JOISTS
05210	Steel Joists
05300	METAL DECKING
05310	Steel Deck
05320	Structural-Electrified Floor System
05400	COLD FORMED METAL FRAMING

S E L E C T E D   R E F E R E N C E S

1. American Iron and Steel Institute. *Fire-Resistant Steel Frame Construction*, 2nd Edition. Washington, D. C., 1974.

The problem of fireproofing steel building elements is discussed, and a range of fireproofing details is illustrated in this clear, concise booklet. (Address for ordering: 1000 16th Street N.W., Washington, D.C. 20036.)

2. American Institute of Steel Construction, Inc. *Manual of Steel Construction*, 8th Edition. Chicago, 1980.

This is the bible of the steel construction industry in the United States. It contains detailed tables on the dimensions and properties of all the standard rolled steel sections, data on standard connections, and specification and code information. (Address for ordering: 400 North Michigan Avenue, Chicago, IL 60611.)

3. Parker, Harry, and Harold D. Hauf. *Simplified Design of Structural Steel*, 4th Edition. New York, John Wiley and Sons, Inc., 1974.

This is an excellent introduction to the calculation of steel beams, columns, and connections.

4. Rapp, William G. *Construction of Structural Steel Building Frames*, 2nd Edition. New York, John Wiley and Sons, Inc., 1980.

Rapp covers in detail the process of erecting a steel building frame. Unfortunately, there does not seem to be a similarly clear and complete book that deals with fabrication of steel components.

5. Steel Joist Institute. *Standard Specifications, Load Tables, and Weight Tables for Steel Joists and Joist Girders*. Richmond, Virginia (updated frequently).

Load tables, sizes, and specifications for open-web joists are given in this booklet. (Address for ordering: 1703 Parham Road, Suite 204, Richmond, VA 23229.)

6. American Institute of Steel Construction. *Detailing for Steel Construction*. Chicago, 1983.

This is the definitive guide to structural steel drafting practices, written for the novice drafter. (Address for ordering: see reference 2.)

K E Y   T E R M S   A N D   C O N C E P T S

iron
cast iron
wrought iron
Bessemer process
steel
mild steel

basic oxygen process
ingot
bloom
American Standard (I-beam) shapes
steel section
wide-flange

$W21 \times 93$
angle
gusset plate
channel
tee
plate

bar
open-web steel joist
rivet
carbon steel bolt
high-strength bolt
friction or slip-critical connection
impact wrench
turn-of-nut method
load indicator washer
tension control bolt
arc welding
electrode
backup bar
framed connection
seated connection
end plate connection
diagonal bracing

shear panels
shear connection
moment connection
drift
diaphragm action
cope
fabricator
erector
shop drawings
camber
ironworker
tier
baseplate
leveling plate
plumbing up
guy derrick
climbing crane

drift pin
topping out
cellular decking
composite construction
composite metal decking
subpurlins
girts
fireproofing
intumescent mastic
castellated beam
plate girder
truss
chord
space truss
arch
anticlastic
stay

REVIEW QUESTIONS

1. What is the difference between iron and steel?

2. How are steel structural shapes produced? How are the weights and thicknesses of a shape changed?

3. How does the work of the fabricator differ from that of the erector?

4. Explain the designation W21 × 68.

5. How can you tell a shear connection from a moment connection? What is the role of each?

6. Give two reasons why a beam might be coped.

7. What is the advantage of composite construction?

8. Explain the advantages and disadvantages of a steel building structure with respect to fire.

9. List three different structural systems in steel that might be suitable for the roof of an athletic fieldhouse.

EXERCISES

1. For a simple multistory office building of your design:
   a. Draw a steel framing plan for a typical floor.
   b. Draw an elevation or section showing a suitable method of giving lateral stability to the building.
   c. Make a preliminary determination of the approximate sizes of the decking, beams, and girders, using the information in Figures 9.85 through 9.87.
   d. Sketch details of the typical connections in the frame, using actual dimensions from the *Manual of Steel Construction* (reference 2) for the member sizes you have determined, and working to scale.

2. Select a method of fireproofing, and sketch typical column and beam fireproofing details for the building in Exercise 1.

3. What fireresistance ratings in hours are required for the following elements of a four-story department store with 17,500 square feet of area per floor? (The necessary information is found in Figures 1.1 and 1.2.)
   a. Lower-floor columns
   b. Floor beams
   c. Roof beams
   d. Walls around elevator shafts and stairways

4. Find a steel building frame under construction. Observe the connections carefully and figure out why each is detailed as it is. If possible, arrange to talk with the structural engineer of the building to discuss the design of the frame.

CONCRETE CONSTRUCTION

A physical sciences center at Dartmouth College, built in a highly irregular space bounded by three existing buildings, typifies the potential of reinforced concrete to make expressive, highly individual buildings. *(Architects: Shepley Bulfinch Richardson and Abbott. Photo by Ezra Stoller, ©ESTO)*

Concrete is the universal material of construction. The raw materials for its manufacture are readily available in every part of the globe, and concrete can be made into buildings with tools ranging from a primitive shovel to a computerized precasting plant. It does not rot or burn; it is relatively low in cost; and it can be used for every building purpose, from lowly pavings to sturdy structural frames to handsome exterior claddings and interior finishes. But it has no form of its own and no useful tensile strength, so before its limitless architectural potential can be realized, the designer must learn to combine it skillfully with steel to bring out the best characteristics of each material and to mold and shape it to forms appropriate to its qualities.

History

The ancient Romans, in quarrying limestone for mortar, accidentally discovered a silica- and alumina-bearing mineral on the slopes of Mount Vesuvius that, when mixed with limestone and burned, produced a cement that exhibited the unique property of hardening underwater as well as in the air. This cement was also harder, stronger, and much more adhesive than the ordinary lime mortar to which they were accustomed. In time, this mortar not only became the preferred type for use in all their building projects but began also to alter the character of Roman construction. Masonry of stone or brick was used to build only the surfaces of masonry piers, walls, and vaults, and the hollow interiors were filled entirely with large volumes of the new type of mortar (Figure 10.2). We now know that this mortar contained the essential ingredients of modern portland cement and that the Romans were the inventors of concrete construction.

Knowledge of concrete construction was lost with the fall of the Roman empire, not to be regained until the latter part of the eighteenth century when a number of English inventors began experimenting with both natural and artificially produced cements. Joseph Aspdin, in 1824, patented an artificial cement that he named *portland cement*, after the gray English limestone whose color it resembled. His cement was soon in great demand, and the name portland remains in use to the present day.

Reinforced concrete was developed simultaneously in the 1850s by several people, including J. L. Lambot, who built several reinforced concrete boats in Paris in 1854, and an American, Thaddeus Hyatt, who made and tested a number of reinforced concrete beams. But the combination of steel and concrete did not come into widespread use until a French gardener, Joseph Monier, obtained a patent for reinforced concrete flower pots in 1867 and went on to build concrete water tanks

FIGURE   10.1

At the time concrete is placed, it has no form of its own. This bucket of fresh concrete was filled on the ground by a transit-mix truck and hoisted to the top of the building by a crane. The worker at the right has opened the valve in the bottom of the bucket to discharge the concrete into the formwork. Four bundles of posttensioning tendons can be seen coming toward the camera at the left side of the photograph. (Courtesy of Portland Cement Association, Skokie, Illinois) and bridges of the new material. By the end of the nineteenth century, engineering design methods had been developed for structures of reinforced concrete, and a number of major structures had been built. By this time, the earliest experiments in *prestressing* had also been carried out, although it remained for Freyssinet in the 1920s to establish a scientific basis for the design of *prestressed concrete* structures.

CEMENT AND CONCRETE

Concrete is a rocklike material produced by combining coarse and fine *aggregates*, portland cement, and water, and allowing the mixture to harden. Coarse aggregate is normally gravel or crushed stone, and fine aggregate is sand. Portland cement, hereafter referred to simply as *cement*,

is a fine gray powder. During the hardening of concrete, considerable heat (called *heat of hydration*) is given off as the cement combines chemically with water to form strong crystals that bind the aggregates together. During this *curing* process, and the drying of the excess water from the concrete after curing is complete, concrete shrinks slightly.

In properly formulated concrete, the majority of the volume consists of coarse and fine aggregate, proportioned and graded so the fine particles completely fill the spaces between the coarse ones (Figure 10.3). Each particle is completely coated with a paste of cement and water to join it fully to the surrounding particles.

Cement

Portland cement may be manufactured from any of a number of raw

FIGURE 10.2

Hadrian's Villa, a large palace built near Rome between 125 and 135 A.D., used unreinforced concrete extensively for structures such as this dome. (Photo by the author)

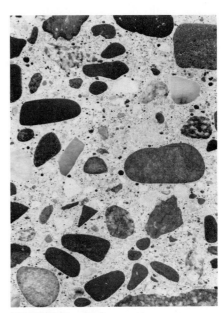

FIGURE 10.3

Photograph of a polished cross section of hardened concrete, showing the close packing of coarse and fine aggregates and the complete coating of every particle with cement paste. (Courtesy of Portland Cement Association, Skokie, Illinois)

materials, providing they are combined to yield the necessary amounts of lime, iron, silica, and alumina. Lime is commonly furnished by limestone, marble, marl, or seashells. Iron, silica, and alumina may be obtained in the form of clay or shale. The exact ingredients used depend on what is readily available, and the recipe varies widely from one geographic region to another, often including slag or flue dust from iron furnaces, chalk, sand, ore washings, bauxite, and other minerals. The selected constituents are crushed, ground, proportioned and blended, then conducted through a rotating kiln at temperatures of 2600 to 3000 degrees Fahrenheit (1400 to 1650° C) to produce *clinker* (Figures 10.4, 10.5). After cooling, the clinker is pulverized (along with a small amount of gypsum to retard the curing process) to a powder finer than flour. This powder, portland cement, is either packaged in bags or shipped in bulk. In the United States, a standard bag of cement contains one cubic foot of volume and weighs 94 pounds (43 kg).

The quality of cement is established by ASTM-C150, which identifies eight different types of portland cement:

Type I   normal

Type IA   normal, air-entraining

Type II   moderate resistance to sulfate attack

Type IIA   moderate resistance, air-entraining

Type III   high early strength

Type IIIA   high early strength, air-entraining

Type IV   low heat of hydration

Type V   high resistance to sulfate attack

Type I cement is used for most purposes in construction. Types II and V are used where the concrete will be

STEPS IN THE MANUFACTURE OF PORTLAND CEMENT

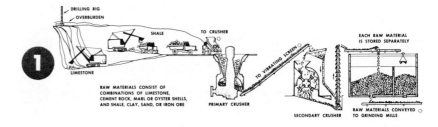

STONE IS FIRST REDUCED TO 5-IN. SIZE, THEN 3/4-IN., AND STORED

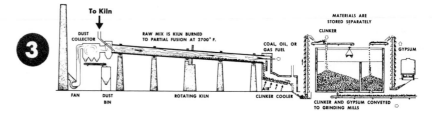

BURNING CHANGES RAW MIX CHEMICALLY INTO CEMENT CLINKER

FIGURE 10.4
A rotary kiln manufacturing cement clinker. (Courtesy of Portland Cement Association, Skokie, Illinois)

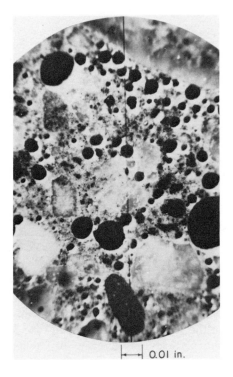

├──┤ 0.01 in.

FIGURE 10.6
A photomicrograph of a small section of air-entrained concrete shows the bubbles of entrained air (0.01 inch equals 0.25 mm). (Courtesy of Portland Cement Association, Skokie, Illinois)

in contact with water with a high sulfate concentration. Type III is employed in situations where a reduced curing period is desired (as may be the case in cold weather), in the precasting of concrete structural elements, or when the construction schedule must be accelerated. Type IV is used in massive structures such as dams where the heat emitted by the curing concrete may accumulate and raise the temperature of the concrete to damaging levels.

Air-entraining cements contain small amounts of additives that cause microscopic air bubbles to form in the concrete during mixing (Figure 10.6). These bubbles, which usually comprise 2 to 8 percent of the volume of the finished concrete, give improved workability during placement of the concrete and, more importantly, greatly increase the resistance of the cured concrete to damage caused by repeated cycles of freezing and thawing. Air-entrained concrete is commonly used for pavings and exposed architectural concrete in cold climates. With appropriate adjustments in the formulation of the mix, air-entrained concrete can achieve the same structural strength as normal concrete.

FIGURE 10.5
Steps in the manufacture of portland cement. (Courtesy of Portland Cement Association, Skokie, Illinois)

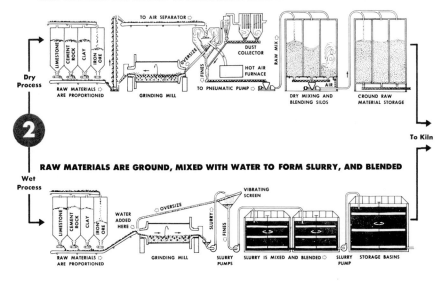

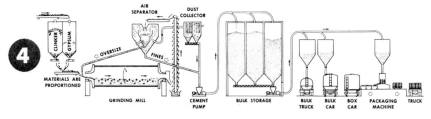

Aggregates and Water

Because aggregates make up roughly three-quarters of the volume of concrete, the structural strength of a concrete is heavily dependent on the quality of its aggregates. Aggregates for concrete must be strong, clean, resistant to freeze-thaw deterioration, chemically stable, and properly graded for size. An aggregate that is dusty or muddy will contaminate the cement paste with inert particles that weaken it, and an aggregate containing any of a number of chemicals from sea salt to organic compounds can cause problems ranging from corrosion of reinforcing steel to retardation of the curing process and ultimate weakening of the concrete. A number of standard ASTM laboratory tests are used to assess the various qualities of aggregates.

Size of aggregate is important because a range of sizes must be included in each concrete mix to achieve close packing of the aggregate particles. A concrete aggregate is graded for size by passing a sample of it through a standard assortment of sieves with diminishing mesh spacings, then weighing the percentage of material that passes each sieve. This test makes it possible to compare the grading of an actual aggregate with the ideal grading for a particular concrete mixture. Size of aggregate is also significant because the largest particle in a concrete mix must be small enough to pass easily between the most closely spaced reinforcing bars and to fit easily into the formwork. In general, the maximum aggregate size should not be greater than three-fourths of the clear spacing between bars or one-third the depth of a slab. For very thin slabs and toppings, a $\frac{3}{8}$ inch (9 mm) maximum aggregate diameter is often specified. A $\frac{3}{4}$ inch or $1\frac{1}{2}$ inches maximum (19 mm or 38 mm) is common for most slab and structural work, but aggregate diameters up to 6 inches (150 mm) are used in dams and other massive structures. Producers of concrete aggregates sort their product for size using screens and can furnish aggregates graded to order.

Lightweight aggregates are used instead of sand and crushed stone for various special types of concrete. *Structural lightweight aggregates* are made from minerals such as shale: The shale is crushed to the desired particle sizes, then heated in an oven to a tem-

FIGURE 10.7
Taking a sample of coarse aggregate from a crusher yard for testing. (Courtesy of Portland Cement Association, Skokie, Illinois)

perature at which the shale becomes plastic in consistency and the small amounts of water that occur naturally in the shale turn to steam and "pop" the particles of aggregate like popcorn. Concrete made from this aggregate has a density about 80 percent of that of normal concrete, while retaining most of the strength of normal concrete. This reduces the dead weight of the components produced from structural lightweight concrete. Nonstructural lightweight concretes are made as insulating roof toppings in densities only one-fourth to one-sixth that of normal concrete. The aggregates in these concretes are usually expanded mica (*vermiculite*) or expanded volcanic glass (*perlite*), and the density of the concretes is further reduced by admixtures that entrain large amounts of air during mixing.

Mixing water for concrete must be free of harmful substances, especially organic material, clay, and salts such as chlorides and sulfates. Water suitable for drinking is generally suitable for concrete.

Admixtures

Ingredients other than cement, aggregates, and water are often added to concrete to alter its properties in various ways.

• *Air-entraining admixtures* may be put in the mix, if they are not already in the cement, to increase workability of the wet concrete, reduce freeze-thaw damage, or in larger amounts, to create very lightweight nonstructural concretes with thermal insulating properties.

• *Water-reducing admixtures* allow a reduction in the amount of mixing water while retaining the same workability, which results in a higher-strength concrete.

• *Accelerating admixtures* cause the concrete to cure more rapidly, and *retarding admixtures* slow its curing to allow more time for working with the wet concrete.

• *Pozzolans* are various natural or artificial materials that react with the calcium hydroxide in wet concrete to form cementing compounds; they are used for purposes such as reducing the internal temperatures of curing concrete, reducing the reactivity of concrete with aggregates containing sulfates, or improving the workability of the concrete.

• *Workability agents* make the wet concrete easier to place in forms and finish by improving its plasticity. They include two admixtures already mentioned, pozzolans and air-entraining admixtures, along with certain fly ashes and organic compounds.

• *Superplasticizers* are organic compounds that transform a stiff concrete mix into a free-flowing liquid. They are used either to facilitate placement of concrete under difficult circumstances or to reduce the water content of a concrete mix to increase its strength.

• *Coloring agents* are dyes and pigments used to alter and control the color of concrete for building components whose appearance is important.

MAKING AND PLACING CONCRETE

Proportioning Concrete Mixes

The quality of cured concrete is measured by any of several criteria, depending on the end use of the concrete. For structural columns, beams, and slabs, compressive strength is important. For paving and floor slabs, surface smoothness and abrasion resistance are also important. For pavings and exterior concrete walls, a high degree of weather resistance is required. Watertightness is important in concrete tanks and dams, and for concrete building walls. But regardless of the criterion to which one is working, the rules for making high-quality concrete are much the same: Use clean, sound ingredients; mix them in the correct proportions; handle the wet concrete properly to avoid segregating its ingredients; and cure the concrete carefully under controlled conditions.

The design of concrete mixtures is a science that can be described here only in its broad outlines. The starting point of any mix design is to establish the desired workability characteristics of the wet concrete, the desired physical properties of the cured concrete, and the acceptable cost of the concrete, keeping in mind that there is no need to spend money to make concrete better than it needs to be for a given application. Concretes with ultimate compressive strengths as low as 2000 pounds per square inch (13.8 MN/m²) are satisfactory for some foundation elements, and strengths of 9000 pounds per square inch (62 MN/m²) are sometimes required for lower-floor columns in tall buildings. Acceptable workability is achievable at any of these strength levels.

Given a proper gradation of satisfactory aggregates, the strength of cured concrete is primarily dependent on the amount of cement in the mix, and the *water-cement ratio*. While water is required as a reactant in the curing of

concrete, much larger amounts of water must be added to a concrete mix than are needed for the hydration of the cement in order to give the wet concrete the necessary fluidity and plasticity for placing and finishing. The extra water eventually evaporates from the concrete, leaving microscopic voids that impair the strength and surface qualities of the concrete (Figure 10.8). Absolute water-cement ratios by weight should be kept below 0.60 for most applications, meaning that the weight of the water in the mix should not be more than 60 percent of the weight of the cement. Higher water-cement ratios than this are often favored by concrete workers because they produce a fluid concrete that is easy to place in the forms, but the resulting concrete is likely to be deficient in strength and surface qualities. Lower water-cement ratios will make concrete that is dense and strong, but unless air-entraining or workability admixtures are included in the mix to substitute for the lost water, the concrete will not flow easily into the forms and will leave large voids. It is therefore important that concrete be formulated with the right quantity of water for each situation, enough to assure workability, but not enough to adversely affect the final properties of the concrete.

Most concrete in North America is proportioned at central batch plants,

using the best of laboratory equipment and engineering knowledge to produce concrete with the specified properties. The concrete is *transit-mixed* en route in a rotating drum on the back of a truck so it is ready to pour by the time it reaches the job site (Figures 10.9, 10.10). For very small jobs, concrete is mixed at the job site, either in a small power-driven mixing drum, or on a flat surface with shovels. For these small jobs, where the quality of the finished concrete does not need to be precisely controlled, proportioning is usually done by rule of thumb. Typically, the dry ingredients are measured volumetrically, using a shovel as a measuring device, in proportions such as one shovel of cement to two of sand to three of gravel, with enough water to make a wet concrete that is neither soupy nor stiff.

Each load of transit-mixed concrete is delivered with a certificate from the batch plant that lists its ingredients and their proportions. As a further check on quality, a *slump test* may be performed at the time of pouring to determine if the desired degree of workability has been achieved without making the concrete too wet (Figures 10.11, 10.12). For structural concrete, standard test cylinders are also poured from each truckload and cured for a specified period alongside the concrete in the forms. The cylinders are

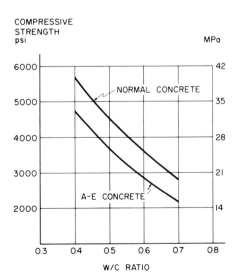

FIGURE   10.8
The effect of water-cement ratio on the strength of concrete. A-E concrete is air-entrained concrete. (Reprinted with permission of the Portland Cement Association from *Design and Control of Concrete Mixtures*)

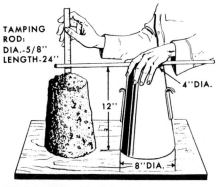

FIGURE 10.11
A slump test for concrete consistency. The hollow metal cone is filled with concrete and tamped with the rod according to a standard procedure. The cone is carefully lifted off, allowing the wet concrete to slump under its own weight. The slump in inches is measured in the manner shown. (From the U.S. Department of Army, Concrete, Masonry, and Brickwork)

FIGURE 10.9
Charging a transit-mix truck with measured quantities of cement, aggregates, and water at a central *batch plant. (Courtesy of Portland Cement Association, Skokie, Illinois)*

FIGURE 10.10
A transit-mix truck discharges its concrete, which was mixed en route in the rotating drum, into a truck mounted concrete pump, which forces it through *a hose to the point in the building at which it is being poured. (Courtesy of Portland Cement Association, Skokie, Illinois)*

FIGURE 10.12
A slump test taking place on a construction site. (Courtesy of Portland Cement Association, Skokie, Illinois)

FIGURE 10.13
Inserting a standard concrete test cylinder in a structural testing machine, where it will be crushed to determine its strength. (Courtesy of Portland Cement Association, Skokie, Illinois)

then tested for compressive strength in a laboratory (Figure 10.13). If the laboratory results are not up to the required standard, test cores are drilled from the actual members poured from the questionable batch of concrete. If the strength of these core samples is also deficient, the contractor will be required to cut out the defective concrete and replace it.

Handling and Placing Concrete

Wet concrete is not a liquid, but a slurry, an unstable mixture of solids and liquid. If wet concrete is vibrated excessively, dropped from very much of a height, or moved horizontally any distance in formwork, it is likely to *segregate*. The coarse aggregate works its way to the bottom of the form, and the water and cement paste rise to the top. The result is concrete of nonuniform and generally unsatisfactory properties. Segregation is prevented

by depositing the concrete, fresh from the mixer, as close to its final resting place as possible. If concrete must be dropped a distance of more than 3 or 4 feet (a meter or so), it should be deposited through *dropchutes* that break the fall of the concrete. If concrete must be moved a large horizontal distance to reach inaccessible areas of the formwork, it should be pumped through hoses (Figure 10.14) or conveyed in buckets or buggies, rather than pushed across or through the formwork.

Concrete must be compacted in the forms to eliminate trapped air and to

FIGURE 10.14
Concrete being placed in a basement floor slab with the aid of a concrete pump. Concrete can be pumped for long horizontal distances, and many stories into the air. Note the soldier beams, lagging, and rakers for bracing the wall of the excavation. (Courtesy of Portland Cement Association, Skokie, Illinois)

fill completely around the reinforcing bars and into all the corners of the formwork. This may be done by repeatedly thrusting a rod, spade, or immersion-type vibrator into the concrete at intervals. Excessive agitation of the concrete must be avoided, however, or segregation will occur.

Curing Concrete

Because concrete cures by hydration, it is essential that it be kept moist until its required strength is achieved. The curing reaction takes place over a very long period of time, but concrete is commonly designed to be used at the strength it reaches after 28 days (4 weeks) of curing. If it is allowed to dry at any point during this time period, the strength of the cured concrete will be reduced, and its surface qualities will be adversely affected (Figure 10.15). Concrete elements cast in formwork are largely protected from dehydration by the formwork, but the top surfaces must be kept moist by repeated spraying or flooding with water, by covering with moisture-resistant sheets of paper or film, or by the spraying on of a *curing compound* that seals the surface of the concrete against loss of moisture. These measures are particularly important for finished concrete slabs, whose large surface areas make them especially susceptible to drying. Premature drying is a particular danger when slabs are poured in hot or windy weather. Temporary windbreaks may have to be erected, shade may have to be provided, and frequent fogging of the surface of the slab with a fine spray of water is required until the slab is hard enough to be covered or sprayed with curing compound.

At low temperatures, the curing reaction in concrete proceeds at a much reduced rate. If concrete is exposed to subfreezing temperatures while curing, the curing reaction stops completely until the temperature of the concrete rises above the freezing mark. It is important that concrete be pro-

tected from low temperatures and especially from freezing until it is fully cured. If freshly poured concrete is covered or insulated, its heat of hydration is often sufficient to maintain an adequate temperature in the concrete. Under more severe winter conditions, the ingredients of the concrete may have to be heated before mixing, and both a temporary enclosure and a temporary source of heat may have to be provided.

In very hot weather, the hydration reaction is greatly accelerated, and concrete may begin curing before it can be placed and finished. This tendency can be controlled by using cool ingredients and, under extreme conditions, by replacing some of the mixing water with an equal quantity of crushed ice, making sure that the ice has melted fully and the concrete has been thoroughly mixed before placing.

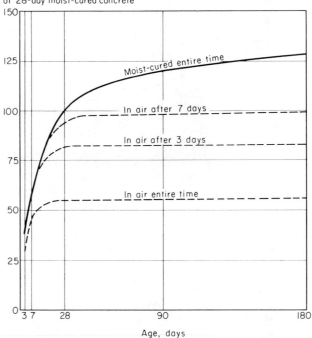

Compressive strength, percent
of 28-day moist-cured concrete

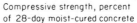

FIGURE   10.15

The growth of compressive strength in concrete over time. Moist-cured concrete is still gaining in strength after 6 months, while air-dried concrete

FORMWORK

Because concrete is put in place as a shapeless slurry with no physical strength, it must be shaped and supported by *formwork* until it has cured sufficiently to support itself. Formwork is usually made of wood, metal, or plastic. It is constructed as a negative of the shape intended for the concrete. Formwork for a beam or slab serves as a temporary working surface during the construction process and as the temporary means of support for reinforcing bars. Formwork must be strong enough to support the considerable weight and fluid pressure of wet concrete without excessive deflection, which often requires temporary supports that are major structures in themselves. During curing, the formwork helps to retain the necessary water of hydration in the concrete. When

virtually stops curing altogether. (Courtesy of Portland Cement Association, Skokie, Illinois)

curing is complete, the formwork must drop away cleanly from the concrete surfaces without damage either to the concrete or to the formwork, which is usually used repeatedly as a construction project progresses. This means that the formwork should have no reentrant corners that will trap or be trapped by the concrete and that the formwork must be tapered where an element of it, such as a joist pan, must be withdrawn directly from a location in which it is fully surrounded by concrete. All formwork surfaces in contact with concrete must be coated with a *parting compound*, which is an oil, wax, or plastic that prevents adhesion of the concrete to the form.

The quality of concrete surfaces can be no better than the quality of the forms in which they are cast, and the requirements for surface quality and structural strength of formwork are rigorous. Top-grade wooden boards and plastic-overlaid plywoods are frequently used to achieve high-quality surfaces. The ties and temporary framing members that support the boards or plywood are spaced closely to avoid bulging of the forms under the extreme pressure of the wet concrete.

In a sense, formwork constitutes an entire temporary building that must be erected and demolished in order to produce a second, permanent building of concrete. The cost of formwork is therefore a major component of the overall cost of a concrete building frame. This cost is one of the forces that has led to the development of *precasting*, a process in which concrete is cast into permanent, reusable forms at an industrial plant, then transported as fully cured structural units to the job site, where these units are hoisted into place and connected much as if they were structural steel shapes. The opposite of precasting is *sitecasting*, or *cast-in-place construction*, in which concrete is poured into forms that are erected in place on the job site. In the two chapters that follow, formwork is shown for both sitecast and precast concrete.

Reinforcing

The Concept of Reinforcing

Concrete has no useful tensile strength (Figure 10.16) and was limited in its structural use until the concept of steel *reinforcing* was developed. The compatibility of steel and concrete is a fortuitous accident. If they had grossly different coefficients of thermal expansion, a reinforced concrete structure would tear itself apart during cycles of temperature variation. If they were chemically incompatible, the steel would corrode or the concrete would be degraded. If concrete did not adhere to steel, a very different and more expensive configuration of reinforcing would be necessary. But concrete and steel change dimension at nearly the same rate in response to temperature changes; steel is protected from corrosion by the alkaline chemistry of concrete; and concrete bonds strongly to steel, providing a convenient means of adapting brittle concrete to structural elements that must resist tension, shear, and bending, as well as compression.

The basic theory of concrete reinforcing is extremely simple: Put the steel where there is tension in a structural member and let the concrete resist the compression. This accounts fairly precisely for the location of most of the reinforcing steel that one sees in a concrete structure. But there are some important exceptions: Steel is used to resist some of the compression in concrete columns and in beams whose height must be reduced for architectural reasons. It is also used to resist cracking that might otherwise be caused by curing shrinkage, and by thermal expansion and contraction in slabs and walls.

Steel Bars for Concrete Reinforcement

Reinforcing bars for concrete construction are hot-rolled in much the

Material	Allowable Tensile Strength	Allowable Compressive Strength	Density	Modulus of Elasticity
Wood (average)	700 psi (4.83 MN/m²)	1,100 psi (7.58 MN/m²)	30 pcf (480 kg/m³)	1,200,000 psi (8,275 MN/m²)
Brick Masonry (average)	0	250 psi (1.72 MN/m²)	120 pcf (1,920 kg/m³)	1,200,000 psi (8,275 MN/m²)
Steel (ASTM A36)	22,000 psi (151.69 MN/m²)	22,000 psi (151.69 MN/m²)	490 pcf (7,850 kg/m³)	29,000,000 psi (200,000 MN/m²)
Concrete (average)	0	1,350 psi (9.31 MN/m²)	145 pcf (2,320 kg/m³)	3,150,000 psi (21,720 MN/m²)

FIGURE 10.16
Concrete, like masonry, has no useful tensile strength, but its compressive strength is considerable, and when combined with steel reinforcing, concrete can be used for every type of structure.

same fashion as structural shapes, with surface ribs for better bonding to concrete formed as the bars reach their final dimensions (Figures 10.17, 10.18). The bars are cut to a standard length [commonly 60 feet (18.3 m) in the United States], bundled, and shipped to local fabricating shops.

Reinforcing bars are rolled in eleven standard diameters, as shown in Figure 10.19. Bars are specified by a simple numbering system in which the number corresponds to the number of eighths of an inch (3.2 mm) of bar diameter. A number 6 reinforcing bar is $\frac{6}{8}$ or $\frac{3}{4}$ inch (19 mm) in diameter, and a number 8 is 1 inch (25.4 mm). Bars larger than number 8 vary slightly from these nominal diameters in order to correspond to convenient cross-sectional areas of steel. The structural engineer, in selecting reinforcing bars for a given beam or column, knows from calculations the required cross-sectional area of bars. This area may be achieved with a larger number of smaller bars, or a smaller number of larger bars, in any of a number of combinations. The final bar arrangement is selected based on the physical space available in the concrete member, the required cover dimensions, the clear spacing required between bars to allow passage of the concrete aggregate, and the sizes and numbers of bars that will be most convenient to fabricate and install.

Reinforcing bars are manufactured according to ASTM standards A615,

FIGURE 10.17
Glowing strands of steel are reduced to reinforcing bars as they snake their way through a rolling mill. (Courtesy of Bethlehem Steel Company)

FIGURE 10.18
The deformations rolled into the surface of a reinforcing bar help it to bond tightly to concrete. (Photo by the author)

A616, A617, and A706. They are available in grades 40, 50, and 60, corresponding to steel with yield strengths of 40,000, 50,000, and 60,000 pounds per square inch (275, 345, and 415 MN/m²) as shown in Figure 10.20. Grade 60 is the most readily available of the three. The higher-strength bars are useful where there is a restricted amount of space available for bars in a concrete member, and they have proved to be the most economical for vertical bars in columns.

Reinforcing steel is also produced in sheets or rolls of *welded wire fabric,* a grid of wires or round bars spaced 2 to 12 inches (50 to 300 mm) apart (Figure 10.21). The lighter styles of welded wire fabric resemble cattle fencing and are used to reinforce concrete slabs on grade and certain precast concrete elements. The heavier styles find use in concrete walls and structural slabs. The principal advantage of welded wire fabric is economy of labor in placing the reinforcing, especially where a large number of small bars can be replaced by a single sheet of material. The size and spacing of the wires or bars for a particular application are specified by the structural engineer or architect of the building.

The fabrication of reinforcing steel for a concrete construction project is analogous to the fabrication of steel shapes for a Steel Frame building (Chapter 9). The fabricator receives the engineering drawings for the building from the contractor and prepares shop

ASTM STANDARD REINFORCING BARS

Bar size	U.S. customary			Metric		
	Nominal diameter, in.	Nominal area, sq in.	Nominal weight, lb per ft	Nominal diameter, mm	Nominal area, mm²	Nominal mass, kg/m
#3	0.375	0.11	0.376	9.52	71	0.560
4	0.500	0.20	0.668	12.70	129	0.994
5	0.625	0.31	1.043	15.88	200	1.552
6	0.750	0.44	1.502	19.05	284	2.235
7	0.875	0.60	2.044	22.22	387	3.042
8	1.000	0.79	2.670	25.40	510	3.973
9	1.128	1.00	3.400	28.65	645	5.060
10	1.270	1.27	4.303	32.26	819	6.404
11	1.410	1.56	5.313	35.81	1006	7.907
14	1.693	2.25	7.650	43.00	1452	11.38
18	2.257	4.00	13.600	57.33	2581	20.24

FIGURE 10.19

Standard sizes of reinforcing bars. Notice that the diameters of the bars correspond very closely to a rule-of-thumb value of ⅛ inch (3.2 mm) per bar size number. (Courtesy of American Concrete Institute)

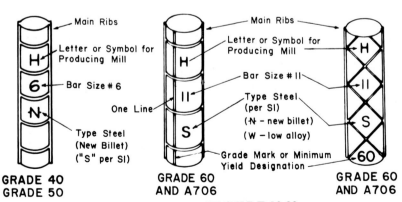

GRADE 40
GRADE 50

GRADE 60
AND A706

GRADE 60
AND A706

FIGURE 10.20

Examples of the identification markings rolled onto the surface of reinforcing bars. (Courtesy of Concrete Reinforcing Steel Institute)

drawings for the reinforcing bars. After the shop drawings have been checked by the engineer or architect, the fabricator sets to work cutting the reinforcing bar stock to length, making the necessary bends (Figure 10.22) and tying the fabricated bars into bundles that are tagged to indicate their destination in the building. The bundles are stored in organized piles, then shipped as needed to the building site. There they are broken down, lifted by hand or hoisted by crane into place, and wired together awaiting pouring of the concrete. The wire has a temporary function only, which is to hold the reinforcement in place until the concrete has set. Any transfer of load from one reinforcing bar to another in the completed building is done by the concrete. Where two bars must be spliced, they are overlapped a specified number of bar diameters, and the loads are transferred from one to the other by the surrounding concrete. The one common exception to this occurs in bar splices in columns where the reinforcing is heavy and space within the column is at a premium; here the bars are spliced end-to-end rather than

SECTIONAL AREA AND WEIGHT OF WELDED WIRE FABRIC

Wire Size Number[a]		Nominal Diameter Inches	Nominal Weight Lbs./Lin. Ft.	Area In Sq. In. Per Ft. Of Width For Various Spacings						
				Center-To-Center Spacing						
Smooth	Deformed			2″	3″	4″	6″	8″	10″	12″
W20	D20	0.505	.680	1.20	.80	.60	.40	.30	.24	.20
W18	D18	0.479	.612	1.08	.72	.54	.36	.27	.216	.18
W16	D16	0.451	.544	.96	.64	.48	.32	.24	.192	.16
W14	D14	0.422	.476	.84	.56	.42	.28	.21	.168	.14
W12	D12	0.391	.408	.72	.48	.36	.24	.18	.144	.12
W11	D11	0.374	.374	.66	.44	.33	.22	.165	.132	.11
W10.5		0.366	.357	.63	.42	.315	.21	.157	.126	.105
W10	D10	0.357	.340	.60	.40	.30	.20	.15	.12	.10
W9.5		0.348	.323	.57	.38	.285	.19	.142	.114	.095
W9	D9	0.338	.306	.54	.36	.27	.18	.135	.108	.09
W8.5		0.329	.289	.51	.34	.255	.17	.127	.102	.085
W8	D8	0.319	.272	.48	.32	.24	.16	.12	.096	.08
W7.5		0.309	.255	.45	.30	.225	.15	.112	.09	.075
W7	D7	0.299	.238	.42	.28	.21	.14	.105	.084	.07
W6.5		0.288	.221	.39	.26	.195	.13	.097	.078	.065
W6	D6	0.276	.204	.36	.24	.18	.12	.09	.072	.06
W5.5		0.265	.187	.33	.22	.165	.11	.082	.066	.055
W5	D5	0.252	.170	.30	.20	.15	.10	.075	.06	.05
W4.5		0.239	.153	.27	.18	.135	.09	.067	.054	.045
W4	D4	0.226	.136	.24	.16	.12	.08	.06	.048	.04
W3.5		0.211	.119	.21	.14	.105	.07	.052	.042	.035
W3		0.195	.102	.18	.12	.09	.06	.045	.036	.03
W2.9		0.192	.099	.174	.116	.087	.058	.043	.035	.029
W2.5		0.178	.085	.15	.10	.075	.05	.037	.03	.025
W2.1		0.162	.070	.126	.084	.063	.042	.031	.025	.021
W2		0.160	.068	.12	.08	.06	.04	.03	.024	.02
W1.5		0.138	.051	.09	.06	.045	.03	.022	.018	.015
W1.4		0.134	.048	.084	.056	.042	.028	.021	.017	.014

FIGURE 10.21
Standard configurations of welded wire fabric. The heaviest "wires" are more than ½ inch (13 mm) in diameter, making them suitable for structural slab reinforcing. (Courtesy of Concrete Reinforcing Steel Institute)

STANDARD HOOKS

All specific sizes recommended by CRSI below meet minimum requirements of ACI 318-77

RECOMMENDED END HOOKS
All Grades

D=Finished bend diameter

Bar Size	180° HOOKS			90° HOOKS
	D	A or G	J	A or G
# 3	2¼	5	3	6
# 4	3	6	4	8
# 5	3¾	7	5	10
# 6	4½	8	6	1-0
# 7	5¼	10	7	1-2
# 8	6	11	8	1-4
# 9	9½	1-3	11¾	1-7
#10	10¾	1-5	1-1¼	1-10
#11	12	1-7	1-2¾	2-0
#14	18¼	2-3	1-9¾	2-7
#18	24	3-0	2-4½	3-5

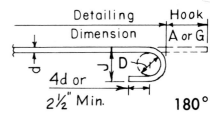

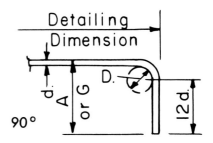

STIRRUP AND TIE HOOKS

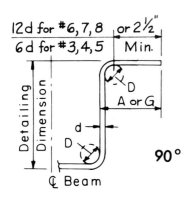

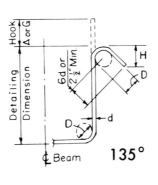

135° SEISMIC STIRRUP/TIE HOOKS

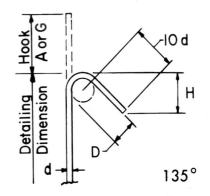

STIRRUPS
(TIES SIMILAR)

STIRRUP AND TIE HOOK DIMENSIONS
Grades 40-50-60 ksi

Bar Size	D (in.)	90° Hook	135° Hook	
		Hook A or G	Hook A or G	H Approx.
#3	1½	4	4	2½
#4	2	4½	4½	3
#5	2½	6	5½	3¾
#6	4½	1-0	7¾	4½
#7	5¼	1-2	9	5¼
#8	6	1-4	10¼	6

135° SEISMIC STIRRUP/TIE
HOOK DIMENSIONS
Grades 40-50-60 ksi

Bar Size	D (in.)	135° Hook	
		Hook A or G	H Approx.
#3	1½	5	3½
#4	2	6½	4½
#5	2½	8	5½
#6	4½	10¾	6½
#7	5¼	1-0½	7¾
#8	6	1-2¼	9

FIGURE 10.22

The bending of reinforcing bars is done according to precise standards in a fabricator's shop. (Courtesy of Concrete Reinforcing Steel Institute)

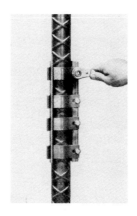

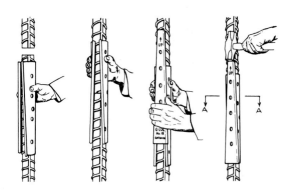

FIGURE 10.23
Two types of mechanical splicing devices for use on vertical bars in concrete columns. (Courtesy of Concrete Reinforcing Steel Institute)

FIGURE 10.24
A. *The directions of force in a simply supported beam under a uniform loading. The solid lines represent compression, and the broken lines represent tension. Near the ends of the beam, the lines of strongest tensile force move upward diagonally through the beam.* **B.** *Steel reinforcing for a simply supported beam under a uniform loading.* **C.** *A three-dimensional view of the same reinforcing.*

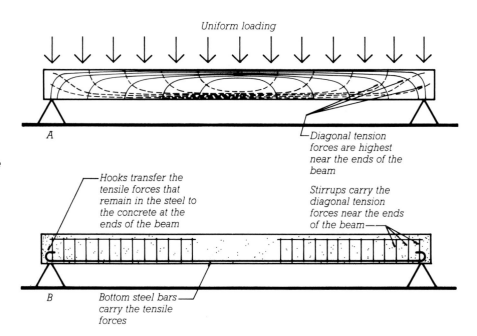

Uniform loading

A

Diagonal tension forces are highest near the ends of the beam

Hooks transfer the tensile forces that remain in the steel to the concrete at the ends of the beam

Stirrups carry the diagonal tension forces near the ends of the beam

B

Bottom steel bars carry the tensile forces

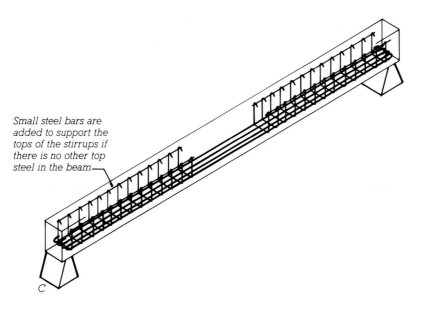

Small steel bars are added to support the tops of the stirrups if there is no other top steel in the beam

C

overlapped, and loads are transferred through welds or sleevelike mechanical splicing devices (Figure 10.23).

The composite action of concrete and steel in reinforced concrete structural elements is such that the reinforcing steel is always loaded axially in tension or compression, and occasionally in shear, but never in bending. The bending stiffness of the reinforcing bars themselves is of no consequence in imparting strength to the concrete.

Reinforcing a Simple Concrete Beam

In an ideal, simply supported beam under a uniform loading, compressive forces follow a set of archlike curves that indicate a maximum of compressive stress in the top of the beam at midspan, with progressively lesser compressive stresses toward either end. A mirrored set of curves shows the lines of tensile force, with stresses again reaching a maximum at the middle of the span (Figure 10.24). In an ideally reinforced concrete beam, steel reinforcing bars would be bent to follow these lines of tension, and the bunching of the bars at midspan would serve to carry the higher stresses at that point. But it is difficult to bend bars into these curves and to support the curved bars adequately in the formwork, so a simpler, rectilinear arrangement of reinforcing steel is substituted.

This is done with a set of *bottom bars* and *stirrups*. The bottom bars are placed near the bottom of the beam, leaving a specified amount of concrete below and to the sides of the rods as *cover* (Figure 10.25). The concrete cover provides a full embedment for the reinforcing bars and protects them against fire and corrosion. The bars are most heavily stressed at the midpoint of the beam span, with progressively smaller amounts of stress toward each of the supports. The differences in stress are dissipated from the bars into the concrete by means of the *bond* forces, the adhesive forces between the concrete and the steel, aided by the ribs

The country... near Taliesin my home and workshop, is the bed of an ancient glacier drift. Vast busy gravel pits abound there, exposing heaps of yellow aggregate once and still everywhere near, sleeping beneath the green fields. Great heaps, clean and golden, are always waiting there in the sun. And I never pass...without an emotion, a vision of the long dust-whitened stretches of the cement mills grinding to impalpable fineness the magic powder that would "set" my vision all to shape; I wish both mill and gravel endlessly subject to my will... Materials! What a resource.

Frank Lloyd Wright, in *Architectural Record*, October 1928.

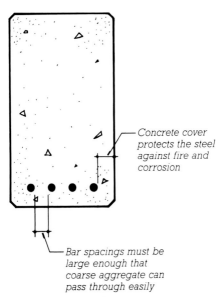

FIGURE 10.25
A cross section of a rectangular concrete beam showing cover and bar spacing.

on the surface of the bars. At the ends of the beam, some stress remains in the steel, but there is no further length of concrete into which the stress can be dissipated. This is solved by bending the ends of the bars into *hooks*, which are bends of standard dimension that are provided for this purpose.

The bottom steel does the heavy tensile work in the beam, but some lesser tensile forces remain in a diagonal orientation near the ends of the beam. These are resisted by a series of *stirrups*. The stirrups may be *U-stirrups* or *closed stirrup-ties*. U-stirrups are less expensive to make and install and are sufficient for many situations, but stirrup-ties are required in beams that will be subjected to torsional (twisting) forces or to high compressive forces in the top or bottom bars. In either case, the stirrups furnish vertical tensile reinforcing to resist the cracking forces that run diagonally across them. Obviously, diagonal stirrups would be better, but economy of installation dictates that stirrups should be oriented vertically.

When this simple beam is formed, the bottom steel is supported at the correct cover height by steel *chairs*, or in a broad beam or slab, by long chairs called *bolsters* (Figure 10.26). These remain in the concrete after pouring, because although their work is then finished, there is no way to get them out. In outdoor concrete work, the feet of the chairs and bolsters sometimes rust where they come in contact with the face of the beam, unless plastic-capped chairs are used. Where reinforced concrete is poured in direct contact with the ground, concrete bricks or small pieces of concrete are used instead of chairs to prevent rust from forming under the feet of the chairs and spreading up into the reinforcing bars.

The stirrups in this simple beam are supported by wiring them to the bottom steel and by tying their tops to #3 bars that have no function in the beam other than to support the stirrups until the concrete has been poured.

SYMBOL	BAR SUPPORT ILLUSTRATION	BAR SUPPORT ILLUSTRATION PLASTIC CAPPED OR DIPPED	TYPE OF SUPPORT	SIZES
SB	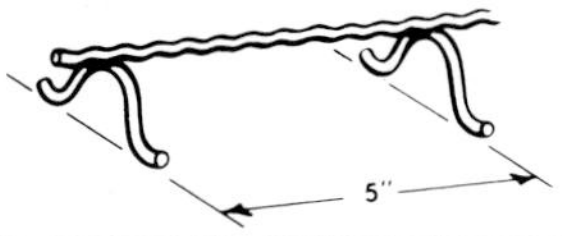5″	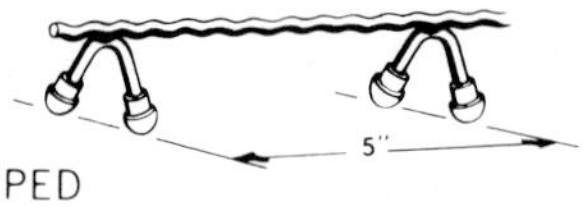5″ CAPPED	Slab Bolster	¾, 1, 1½, and 2 inch heights in 5 ft. and 10 ft. lengths
SBU	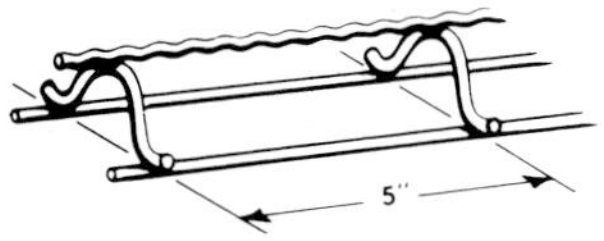5″		Slab Bolster Upper	Same as SB
BB	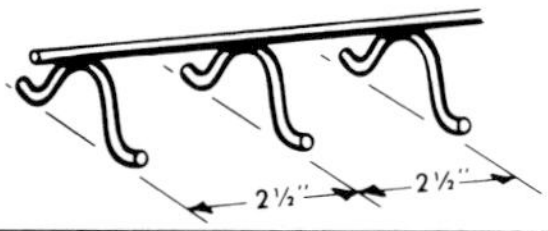2½″ 2½″	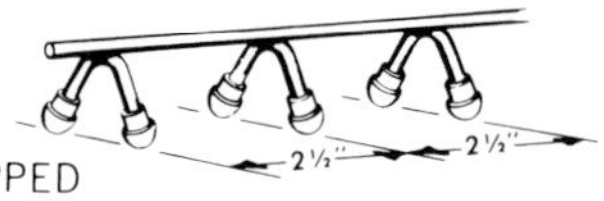2½″ 2½″ CAPPED	Beam Bolster	1, 1½, 2, over 2″ to 5″ heights in increments of ¼″ in lengths of 5 ft.
BBU	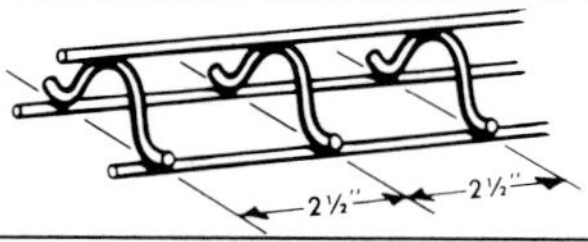2½″ 2½″		Beam Bolster Upper	Same as BB
BC	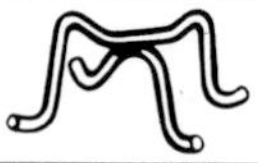	DIPPED 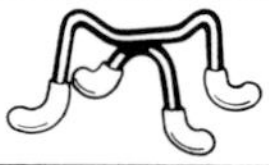	Individual Bar Chair	¾, 1, 1½, and 1¾″ heights
JC	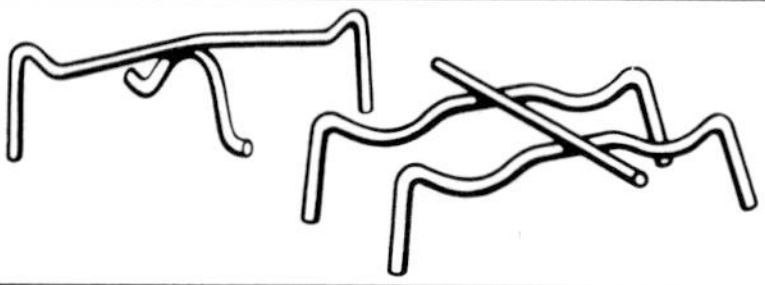	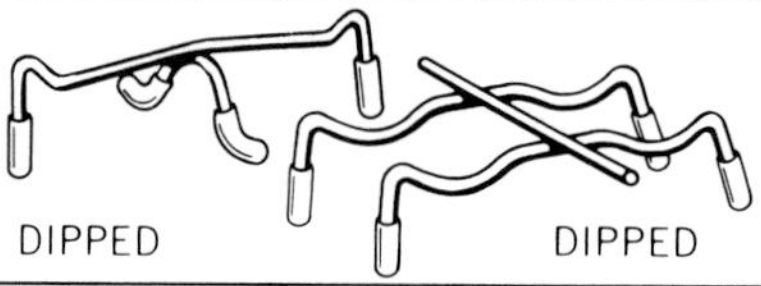 DIPPED DIPPED	Joist Chair	4, 5, and 6 inch widths and ¾, 1 and 1½ inch heights
HC	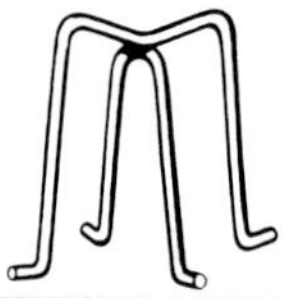	CAPPED 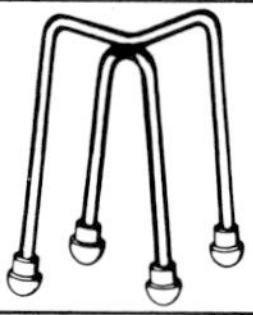	Individual High Chair	2 to 15 inch heights in increments of ¼ inch
HCM	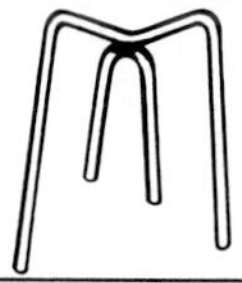		High Chair for Metal Deck	2 to 15 inch heights in increments of ¼ in.
CHC	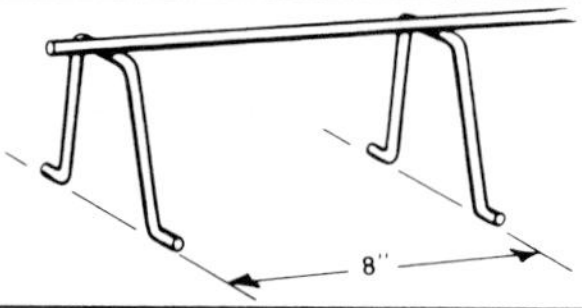8″	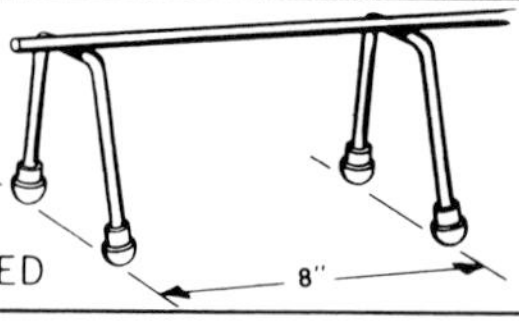8″ CAPPED	Continuous High Chair	Same as HC in 5 foot and 10 foot lengths
CHCU	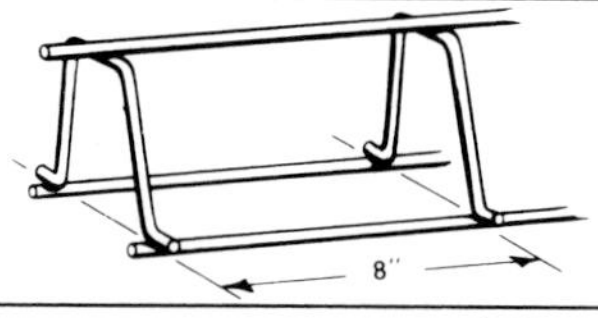8″		Continuous High Chair Upper	Same as CHC
CHCM	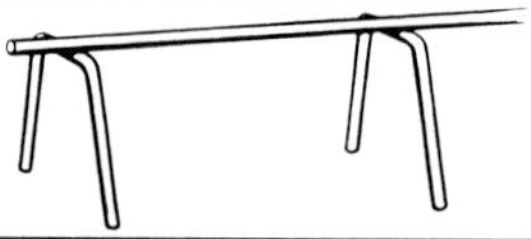		Continuous High Chair for Metal Deck	Up to 5 inch heights in increments of ¼ in.
JCU	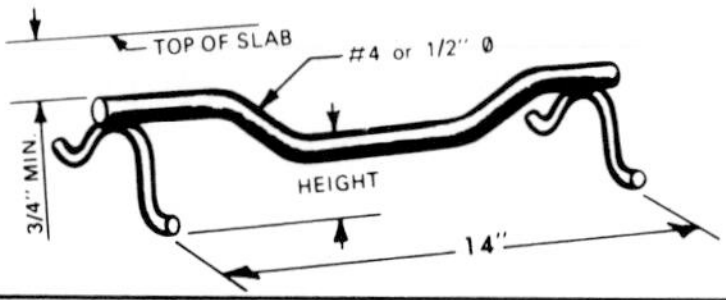TOP OF SLAB #4 or 1/2″ Ø 3/4″ MIN. HEIGHT 14″	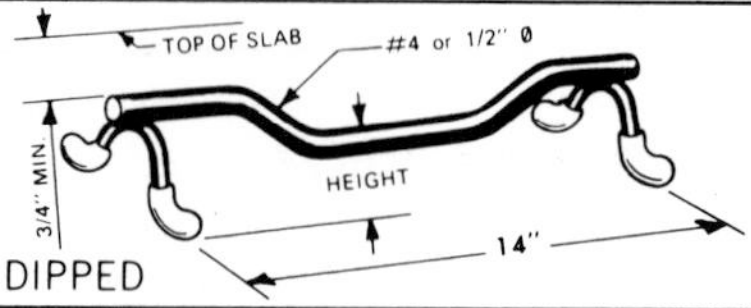TOP OF SLAB #4 or 1/2″ Ø 3/4″ MIN. HEIGHT 14″ DIPPED	Joist Chair Upper	14″ Span. Heights −1″ thru +3½″ vary in ¼″ increments

Reinforcing a Continuous Concrete Beam

Most sitecast concrete beams are not of this simple type because concrete lends itself most easily to one-piece structural frames with a high degree of structural continuity from one beam span to the next. In a continuous structure the bottom of the beam is in tension at midspan, and the top of the beam in tension at supporting columns or walls. This means that top bars must now be provided over the supports, and bottom bars in midspan, along with the usual stirrups, as illustrated in Figure 10.27. Until recently, it was com- mon practice to bend some of the horizontal bars up or down at the points of bending reversal in concrete beams so the same bars could serve both as bottom steel at midspan and top steel over the columns, but this has largely been abandoned in favor of the simpler practice of using straight bars only.

Reinforcing Structural Concrete Slabs

The reinforcing pattern for concrete structural slabs, which may be considered as very broad beams, is similar to the reinforcing pattern in beams,

FIGURE 10.26

Chairs and bolsters for supporting reinforcing bars in beams and slabs. Bolsters and continuous chairs are made in long lengths for use in slabs, while chairs support only one or two bars each. (Courtesy of Concrete Reinforcing Steel Institute)

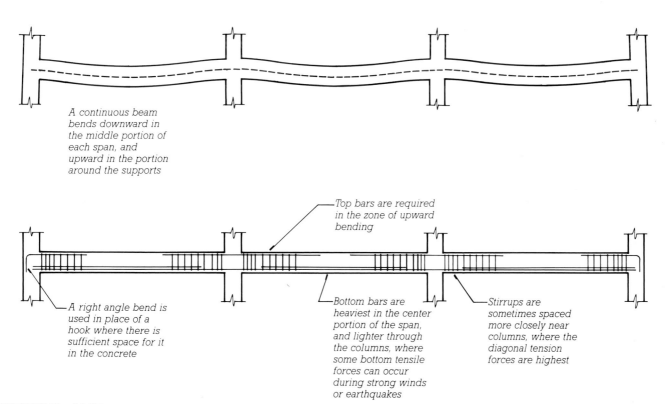

A continuous beam bends downward in the middle portion of each span, and upward in the portion around the supports

Top bars are required in the zone of upward bending

A right angle bend is used in place of a hook where there is sufficient space for it in the concrete

Bottom bars are heaviest in the center portion of the span, and lighter through the columns, where some bottom tensile forces can occur during strong winds or earthquakes

Stirrups are sometimes spaced more closely near columns, where the diagonal tension forces are highest

FIGURE 10.27

Reinforcing for a concrete beam that is continuous across several spans. The upper diagram shows in exaggerated form the shape taken by a continuous beam under a uniform loading; the broken line is the center line of the beam. The lower diagram shows the arrangement of bottom steel, top steel, and stirrups conventionally used in this beam. The bottom bars are usually placed at the same level, but they are shown on two levels in this diagram to demonstrate the way in which some of the bottom steel is discontinued in the zones near the columns. There is a simple rule of thumb for determining *where the bending steel must be placed in a beam: Draw an exaggerated diagram of the beam bending under load, as in the top drawing in this illustration, and put the heaviest bars as close as possible to the convex edges.*

except that the wide slab can usually resist the relatively weak diagonal tension forces near its supports without the aid of stirrups. Slabs are also provided with *shrinkage-temperature steel,* a set of light reinforcing bars set at right angles to, and on top of, the primary reinforcing in the slab in order to prevent cracking parallel to the primary reinforcing from concrete shrinkage, temperature-induced stresses, and miscellaneous forces that may occur in the building (Figure 10.28).

Two-Way Slab Action

A structural economy unique to concrete frames is easily realized through the use of *two-way action* in floor and roof structures. Two-way structures, which work well only for column spacing dimensions that are square or within about 20 percent of being square, are reinforced equally in both directions and share the bending load equally between the two directions. This allows two-way structures to be somewhat thinner than one-way structures, to use less reinforcing, and therefore to cost less. Figure 10.29 illustrates the concept of two-way slab action. Several different two-way concrete framing systems will be shown in detail in the next chapter.

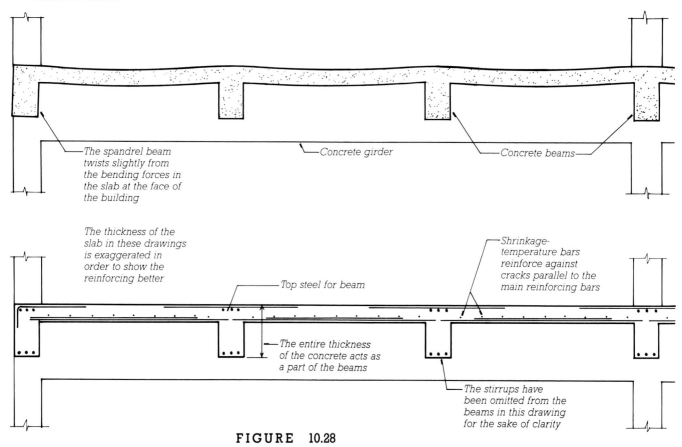

A concrete slab supported by a number of beams bends in the same pattern as a concrete beam supported by a number of columns

The spandrel beam twists slightly from the bending forces in the slab at the face of the building

Concrete girder

Concrete beams

The thickness of the slab in these drawings is exaggerated in order to show the reinforcing better

Top steel for beam

Shrinkage-temperature bars reinforce against cracks parallel to the main reinforcing bars

The entire thickness of the concrete acts as a part of the beams

The stirrups have been omitted from the beams in this drawing for the sake of clarity

FIGURE 10.28
Reinforcing for a one-way concrete slab. The reinforcing is similar to that for a continuous beam, except that stirrups are not usually required in the slab, and shrinkage-temperature bars must be added in the perpendicular direction. The slab does not sit on the beams; rather, concrete around the top of a beam is part of both the beam and the slab. A concrete beam in this situation is considered to be a T-shaped member, with a portion of the slab acting together with the stem of the beam, resulting in a greater structural efficiency and reduced beam depth.

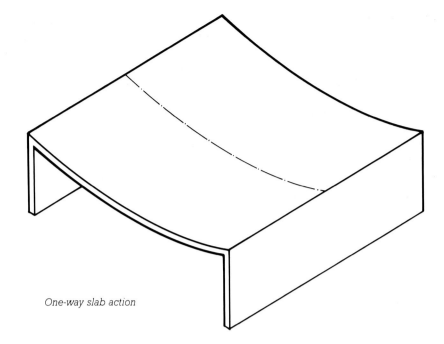

One-way slab action

FIGURE 10.29

One-way and two-way slab action. In practice, two-way action is usually continuous over many spans. The beams are then omitted and the structure becomes simply a slab and a number of supporting columns.

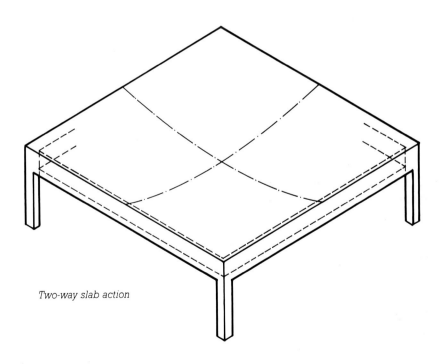

Two-way slab action

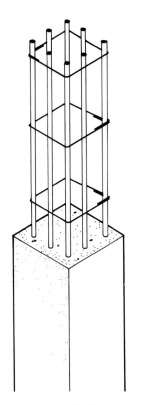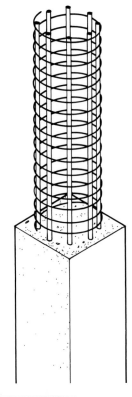

FIGURE 10.30

Reinforcing for concrete columns. To the left, a column with a rectangular arrangement of vertical bars and column ties. To the right, a circular arrangement of vertical bars with a column spiral. Either arrangement may be used in either a round or a square column.

FIGURE 10.31

Column spirals. Each double circle of vertical bars will be poured into a single rectangular column. The vertical bars are spliced with mechanical sleeves. (Courtesy of Concrete Reinforcing Steel Institute)

Reinforcing Concrete Columns

Columns contain two types of reinforcing: *Vertical bars* work with the concrete to carry the compressive loads and to resist the tensile stresses that occur in columns when a building frame is subjected to wind or earthquake loads. *Ties* of smaller steel bars wrapped around the vertical bars help to prevent them from buckling under load—inward buckling is prevented by the concrete core of the column, and outward buckling by the ties (Figure 10.30). The vertical bars may be arranged either in a circle or in rectangular patterns. The ties may be either of two types: *column ties* or *column spirals*. Spirals are shipped to the construction site as tight coils of rod that are then expanded accordian-fashion to the required spacing and wired to the vertical bars. They are generally used only for square or circular arrangements of vertical bars. For rectangular arrangements of vertical bars, discrete column ties must be wired on one by one. Each corner bar and alternate interior bars must be contained inside a bend of rod at each column tie, so two or more column ties are often attached together at each level (Figure 10.32). Column ties are generally more economical than spirals, so even columns with circular bar arrangements are often tied with discrete circular ties rather than spirals. The sizes and spacings of column ties and spirals are determined by the structural engineer. To minimize labor costs on the job site, the ties and vertical bars are usually wired together in a horizontal position at ground level, and the finished *column cage* is lifted into its final position with a crane.

Prestressing

When a beam carries a load, the compression side of the beam is squeezed slightly, and the tension side is stretched by a similar amount. In a reinforced concrete beam, the stretch-

FIGURE 10.32
Multiple column ties at each level are arranged so the four corner bars and alternate interior bars are contained in the corners of ties. (Courtesy of Concrete Reinforcing Steel Institute)

ing tendency is resisted by the reinforcing steel but not by the brittle concrete. When the steel elongates under tension, the concrete around it cracks from the edge of the beam to the horizontal plane in the beam where compressive forces take over. This cracking is visible to the unaided eye in reinforced concrete beams that are loaded to (or beyond) their full load-carrying capacity. In effect, over half the concrete in the beam is doing no useful work except to hold the steel in position and protect it from fire and corrosion (Figure 10.33).

If the reinforcing bars could be tightened to a high tension before the beam is loaded, and then released

against the concrete that surrounds them, they would place the concrete in the vicinity of the bars in compression. If a load were subsequently put on the beam, the tension in the steel would increase further, and the compression in the concrete surrounding the steel would diminish. But if the initial tension in the steel bars were of sufficient magnitude, the surrounding concrete would never go into tension, and no cracking would occur. Furthermore, mathematical analysis would show that the beam is capable of carrying a greater load with the same amounts of concrete and steel than if it were merely reinforced in the conventional manner. This is the rationale

behind *prestressing*. Prestressed members are lighter than reinforced members of equivalent strength and thus, in general, less expensive. The lighter weight also pays off by making precast, prestressed concrete members easier and cheaper to transport.

Pretensioning

Prestressing can be accomplished in either of two ways. In *pretensioning*, steel strands are stretched tightly between abutments in a precasting plant, and the concrete member (or more commonly, a series of concrete members laid end-to-end) is cast around the

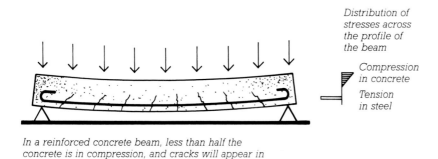

Distribution of stresses across the profile of the beam

Compression in concrete

Tension in steel

In a reinforced concrete beam, less than half the concrete is in compression, and cracks will appear in the bottom of the beam under full load

FIGURE 10.33
The rationale for prestressing concrete. In addition to the lack of cracks in the prestressed beam, the structural action is more efficient than that of a reinforced beam. The prestressed beam therefore uses less material. The small diagrams to the right indicate the distribution of stresses across the center vertical cross section of each of the beams.

Compression in concrete

When a concrete beam is prestressed, all the concrete acts in compression. The off-center location of the prestressing steel causes a camber in the beam

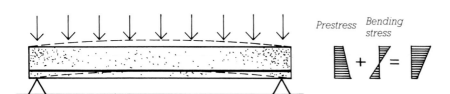

Prestress   Bending stress

Under loading, the prestressed beam becomes flatter, but all the concrete still acts in compression, and no cracks appear

steel. The steel bonds to the curing concrete along its entire length. After the concrete is fully cured, the steel is cut off at either end of the member. As the external tension on the steel is released it recoils slightly, pulling the concrete of the member into compression. If, as is usually the case, the steel is placed as closely as possible to the tension side of the member, the member takes on a decided *camber* at the time the steel strands are cut (Figure 10.34). Much or all of this camber disappears later when the member is put under a bending load. Because the strong abutments needed to hold the tensioned steel prior to the pouring of concrete are very difficult and expensive to create except in a single, fixed location where many concrete members can be created within the same set of abutments, pretensioning is useful only for precast members.

Posttensioning

Posttensioning is done almost exclusively on the building site. The steel strands are covered with grease or a steel tube to prevent bonding with the concrete and are not tensioned until the concrete has cured. Then one end of the steel strands is anchored to the end of the beam or slab, and a hydraulic jack is inserted between the other end of the strands and the other end of the member. The jack applies a large posttensioning force to the concrete and steel, and the steel is anchored to the second end of the member before the jack is removed (Figures 10.35, 10.36). (For very long members, the strands are jacked from both ends to be sure that frictional losses in the tubes do not prevent uniform tensioning). The net effect of posttensioning is identical to that of pretensioning. The difference is that abutments are not needed because the member itself provides the abutting force needed to tension the steel. When the posttensioning process is complete, the steel strands may be left unbonded, or

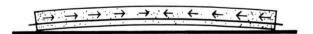

1. The first step in pretensioning is to stretch the steel prestressing strands tightly across the casting bed

2. Concrete is cast around the stretched strands and cured. The concrete bonds to the strands

3. When the strands are cut the concrete goes into compression and the beam takes on a camber

FIGURE 10.34
Pretensioning. Photographs of pretensioned steel strands for a beam are shown in Chapter 12.

if in a steel tube, they may be bonded by injecting grout to fill the space between the strands and the tube.

Even higher structural efficiencies are possible in prestressed construction if the steel strands are made to follow more closely the lines of tensile force. In a posttensioned beam or slab, this is easily done by keeping the two ends of the strand near the top of the member and allowing the strand to drape into a curve under the force of gravity, as seen in Figure 10.35. For pretensioned members, the pretensioning forces are far too high for this method to be followed, but it is possible to pull the strands up and down in the formwork to make a downward-pointing vee or flattened vee shape in each member (Figure 10.37).

Because of the high prestressing loads, there is a tendency for the concrete in a prestressed member to shorten progressively over an ex-

FIGURE 10.35

Posttensioning, using a draped tendon to more nearly approximate the flow of tensile forces in the beam.

1. In posttensioning, the concrete is not allowed to bond to the steel strands during curing

2. After the concrete has cured, the strands are tensioned with a hydraulic jack and anchored to the ends of the beam. If the strands are draped, as shown here, higher structural efficiency is possible than with straight strands

FIGURE 10.36

Posttensioning a large concrete beam with a hydraulic jack. Each tendon consists of a number of individual high-strength steel strands. The bent bars projecting from the top of the beam will be embedded in the concrete slab the beam will support, to allow them to act together as a composite structure. (Courtesy of Portland Cement Association, Skokie, Illinois)

tended period of time in a process known as *creep*, and for the steel strands to relax slightly. Prestressing forces must be increased slightly above the theoretically correct values to accommodate these long-term movements, along with the slight curing shrinkage present in all concrete, and small, short-term movements caused by elastic shortening of the concrete and frictional losses and anchorage set in post-tensioned members. If steel of ordinary structural strength were used in prestressed construction, it would be unable to overcome these movements; therefore, very high-strength, cold-drawn steel wires are used in prestressed members.

Succeeding chapters will discuss prestressed concrete, both pretensioned and posttensioned, in greater detail, showing its application to various standard precast and cast-in-place systems of construction.

FIGURE 10.37

Shaping pretensioning strands to improve structural efficiency. Examples of depressed and harped strands are shown in Chapter 12.

Straight pretensioning strands

Depressed pretensioning strands

Harped pretensioning strands

C.S.I./C.S.C. Masterformat Section Numbers for Concrete Construction	
03100	CONCRETE FORMWORK
03200	CONCRETE REINFORCEMENT
03210	Reinforcing Steel
03220	Welded Wire Fabric
03230	Stressing Tendons
03300	CAST-IN-PLACE CONCRETE
03400	PRECAST CONCRETE

SELECTED REFERENCES

1. Portland Cement Association. *Design and Control of Concrete Mixtures*, 12th Edition. Skokie, Illinois, 1979.

The fifteen chapters of this booklet summarize clearly and succinctly, with many explanatory photographs and tables, the state of current practice in making, placing, finishing, and curing concrete. (Address for ordering: 5420 Old Orchard Road, Skokie, IL 60077.)

KEY TERMS

cement
concrete
portland cement
reinforced concrete
aggregate
lightweight aggregate
structural lightweight aggregate
admixtures
air-entraining
water-cement ratio
curing
heat of hydration
28-day strength
curing compound

slump
segregation
formwork
parting compound
precast
sitecast
reinforcing
welded wire fabric
bottom bars
top bars
bond
hook
U-stirrups
stirrup-ties

chair
bolster
cover
shrinkage-temperature steel
one-way
two-way
vertical bars
column tie
column spiral
prestress
camber
pretension
posttension
creep

REVIEW QUESTIONS

1. What is the difference between cement and concrete?

2. List the conditions that must be met to make a satisfactory concrete mix.

3. List the precautions that should be taken to cure concrete properly. How do these change in very hot or very cold weather?

4. What problems are likely to occur if concrete has too low a slump? Too high a slump? How can the slump be increased without increasing the water content of the concrete mixture?

5. Explain how steel reinforcing works in concrete.

6. Explain the role of stirrups in beams.

7. Explain the role of ties in columns.

8. What does shrinkage-temperature steel do?

9. Why is there no shrinkage-temperature steel in a two-way slab?

10. Explain the difference between reinforcing and prestressing, and the relative advantages and disadvantages of each.

11. Under what circumstances is posttensioning used instead of pretensioning? Why?

EXERCISES

1. Sketch from memory the pattern of reinforcing for a continuous concrete beam. Add notes to explain the role of each feature of the reinforcing.

2. Design, form, reinforce, pour, cure, and load-test a small concrete beam, perhaps 6 to 12 feet (2 to 4 m) long. Get help from a teacher or professional, if necessary, in designing the beam.

SITECAST CONCRETE FRAMING SYSTEMS

*B*oston City Hall makes bold use of sitecast
concrete in its structure, facades, and interiors.
*(Architects: Kallmann, McKinnell, and Wood.
Photo by Ezra Stoller ©ESTO)*

Concrete cast into forms on the building site offers almost unlimited possibilities to the designer. Any shape that can be formed can be cast, with any of a limitless selection of surface textures, and the pages of books on modern architecture are filled with graphic examples of the realization of this extravagant promise. Certain types of concrete elements cannot be precast, but can only be cast on the site—foundation caissons and spread footings, slabs on grade, structural elements too large or too heavy to transport from a precasting plant, elements so irregular or special in form as to rule out precasting, slab toppings over precast floor and roof elements, and many types of structures with two-way slab action or full structural continuity from one member to another. In many cases where sitecast concrete could be replaced with precast, sitecast remains the method of choice simply because of its more massive, monolithic architectural character.

Sitecast concrete structures tend to be heavier than most other types of structures, a consideration that can lead to the selection of a precast concrete or structural steel frame instead if foundation loadings are critical. Sitecast buildings are also relatively slow to construct because each level must be formed, reinforced, poured, cured, and stripped of formwork before the building can be built significantly taller. In effect, each element of a sitecast concrete building is manufactured in place, often by relatively inefficient methods and under variable weather conditions, while the majority of the work on steel or precast concrete buildings is done in plants and shops where worker access, tooling, materials handling equipment, and environmental conditions are generally superior to those on the job site. But the technology of sitecasting has evolved rapidly in response to its own inherent limitations and the evolution of competing systems of construction, with streamlined methods of materials handling, systems of reusable formwork that can be erected and taken down almost instantaneously, extensive prefabrication of reinforcing elements, and mechanization of finishing operations, at a pace that has kept it among the construction techniques most favored by building owners, architects, and engineers.

FIGURE 11.1

Unity Temple in Oak Park, Illinois was constructed by architect Frank Lloyd Wright in 1906. Its structure and exterior surfaces were cast in concrete, making it one of the earliest buildings in the United States to be built primarily of this material. (Photo by John McCarthy. Courtesy of Chicago Historical Society ICHi-18291)

CASTING A CONCRETE SLAB ON GRADE

A concrete *slab on grade* is a level surface of concrete supported directly on the ground for a road, a sidewalk, an airport runway, or the ground floor of a building without a basement or crawlspace. It carries little or no structural stress except a direct transmission of compression between its superimposed loads and the ground beneath, so it furnishes a simple example of the operations involved in the sitecasting of concrete (Figure 11.2).

To prepare for the placement of a slab on grade, topsoil is scraped away to expose the subsoil beneath. A layer of $\frac{3}{4}$ inch (19 mm) diameter crushed stone about 4 inches (100 mm) deep is compacted over the subsoil as a drainage layer to keep water away from the slab. A simple edge form is constructed of wood or metal around the perimeter of the area to be poured and is coated with a form-release compound (ordinary motor oil is often used) to prevent the concrete from sticking (Figure 11.3). The top edge of the form is carefully leveled at a height above the surface of the crushed stone that is equal to the desired thickness of the slab. The thickness may range from 3 inches (75 mm) for a sidewalk to 4 inches (100 mm) for a residential floor, to a foot or more (300 mm) for an airport runway, which must carry very large concentrated loads. If the slab is to be the floor of a building, a moisture barrier (usually a heavy sheet of polyethylene plastic) is laid over the crushed stone. A reinforcing mesh of welded wire fabric, cut to a size just a bit smaller than the dimensions of the slab, is laid over the moisture barrier. The fabric most commonly used for lightly loaded slabs, such as those in houses, is 6×6-W1.4 $\times$ W1.4, which has a wire spacing of 6 inches (152 mm) in each direction and a wire diameter of 0.135 inches (3.43 mm); see Figure 10.21 for more information on welded wire fabric. For slabs in factories, warehouses, and airports, a fabric made of heavier wires or a grid of reinforcing bars may

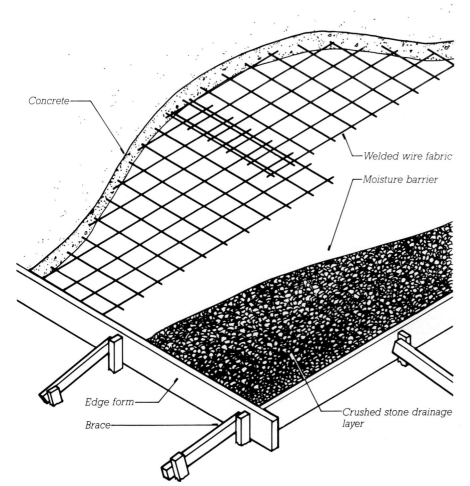

Concrete

Welded wire fabric

Moisture barrier

Edge form

Brace

Crushed stone drainage layer

FIGURE   11.2

The construction of a simple concrete slab on grade. Notice how the wire reinforcing fabric is overlapped where two sheets of fabric join.

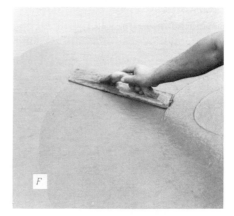

surface. (e) *A bull float can be used for preliminary smoothing of the surface immediately after straightedging.* (f) *Hand floating brings cement paste to the surface and produces a plane surface.* (g) *Floating can be done by machine instead of by hand.* (h) *Steel troweling after floating produces a dense, hard, smooth surface.* (i) *A section of concrete slab on grade finished and ready for curing. The dowels inserted through the screed joint will connect to the sections of slab that will be poured next.* (j) *One method for damp-curing a slab is to cover it with polyethylene sheeting to retain the moisture inside the concrete.* (Photos a, b, c, and i courtesy of Vulcan Metal Products, Inc., Birmingham, Alabama; photos d, e, f, g, h, and j courtesy of Portland Cement Association, Skokie, Illinois)

FIGURE 11.3

Constructing and finishing a concrete slab on grade: (a) *Attaching a proprietary slab edge form (screed joint) to a supporting stake. The profile of the screed joint forms a key to interlock two adjacent pours, and the hole knockouts allow for the placement of steel dowels to tie the pours together.*

(b) *To the right, a crushed stone drainage layer for a slab on grade, and to the left, a slab section ready to pour, with moisture barrier, welded-wire fabric reinforcing, and edge forms in place.* (c) *This asphalt-impregnated fiber board forms a control joint when cast into the slab. The plastic cap is removed immediately after slab finishing to create a clean slot for the later insertion of an elastomeric joint sealant.* (d) *Striking off the surface of a concrete slab on grade just after pouring, using a motorized straightedging device. The motor vibrates the straightedge end-to-end to work the wet concrete into a level*

be used instead. The reinforcing protects the slab against cracking that might be caused by concrete shrinkage, temperature stresses, concentrated loads, frost heaving, or settlement of the ground beneath. If the slab is to be a large one, *control joints* are provided at intervals. A control joint is a straight, intentional crack formed with a fiberboard strip (Figure 11.3*c*), or tooled into the surface of the slab before the concrete has hardened. Its function is to provide a place where the forces that cause cracking can be relieved without disfiguring the slab. The reinforcing mesh is discontinued at each control joint as a further inducement for cracking to occur in this location.

Pouring and Finishing the Slab On Grade

Pouring of the slab commences with the placing of concrete into the formwork. This may be done directly from the chute of a transit-mix truck, or with wheelbarrows, concrete buggies, a large crane-mounted concrete bucket, a conveyor belt, or a concrete pump and hoses; the method selected will depend on the scale of the job and the accessibility of the slab to the truck delivering the concrete. The concrete is distributed by hand with shovels or rakes until the form is full, and the same tools are used to agitate the concrete slightly, especially around the edges, to eliminate air pockets. Next, using hand hooks, the masons reach into the concrete and raise the welded wire fabric to the midheight of the slab, so it will be able to resist tensile forces caused by loads acting either upward or downward.

The first operation in finishing the slab is to *strike off* or *straightedge* the concrete by drawing a stiff plank of wood or metal across the edges of the formwork to achieve a level surface (Figure 11.3*d*). This is done with an end-to-end sawing motion that avoids tearing the projecting pieces of coarse aggregate from the surface of the wet concrete. A bulge of concrete is main-

tained in front of the straightedge as it progresses across the slab, so when a low point is encountered, concrete from the bulge will flow down to fill it. When straightedging is completed, the top of the slab is level but rather rough. If a concrete topping will later be poured over the slab, or if a floor finish of terrazzo, stone, brick, or quarry tile will be applied, the slab may be left to cure without further finishing.

If a smoother surface is desired, the slab is next *floated* (Figure 11.3*e*, *f*, and *g*). The masons wait until the watery sheen has evaporated from the surface, then smooth the concrete with a flat tool called a *float*. Floats may be large or small; for large slabs, rotary power floats may be used. The working surfaces of floats are made of wood, or of metal with a slightly rough surface. As the float is drawn across the surface, its scraping action vibrates the concrete gently and brings cement paste to the surface, where it is smoothed over the coarse aggregate and into low spots. If too much floating is done, however, an excess of paste and free water rises to the surface, and it is almost impossible to get a good finish. Experience on the part of the mason is essential to floating, as it is to all slab finishing operations, to know when to begin each operation, and when to stop.

After the floating operation, specially shaped hand tools are used to form neatly rounded edges and control joints. The floated slab has a lightly textured surface that is appropriate for outdoor walks and pavings without further finishing.

For a completely smooth, dense surface, the slab must be *troweled*. This is done either by hand with a smooth rectangular steel trowel (Figure 11.3*h*), or with a rotary power trowel. Troweling is done several hours after pouring, when the slab is becoming quite firm. Its purpose is to smooth and compact the surface of the concrete into a dense, hard, almost polished plane. If the concrete mason cannot reach all areas of the slab from around the edges, *knee boards* are placed on

the surface of the concrete to distribute the mason's weight sufficiently that he or she can kneel on the surface without making indentations. Any marks left by the knee boards are removed by the trowel as the finisher works backward across the surface from one edge to the other.

If a nonslip surface is required, a stiff-bristled janitor's broom is drawn across the surface of the slab after troweling to produce a striated texture called a *broom finish*.

When the finishing operations are complete, the slab should be cured under damp conditions for at least a week; otherwise its surface may crack or become dusty from premature drying. Damp curing may be accomplished by covering the slab with an absorbent material such as sawdust, earth, sand, straw, or burlap, and maintaining the material in a damp condition for the required length of time. Alternatively, an impervious sheet of plastic or waterproof paper can be drawn over the slab soon after troweling to prevent the escape of moisture from the concrete (Figure 11.3*j*). The same effect may be obtained by spraying the concrete surface with one or more applications of a liquid *curing compound*, which forms an invisible moisture barrier membrane over the slab.

Casting a Concrete Wall

A reinforced concrete wall at ground level usually rests on a poured concrete strip footing (Figures 11.4, 11.5). The footing is formed and poured much like a concrete slab on grade. Its cross-sectional dimensions, and reinforcing, if any, are determined by the structural engineer. A *key* is sometimes formed in the top of the footing with strips of wood which are temporarily embedded in the wet concrete. The key is a groove that forms a mechanical connection to the wall. *Dowels* of steel reinforcing bars are usually installed in the footing before pouring; these will later be overlapped with the vertical bars in the walls to form a strong structural connection. After pouring, the top of the footing is straightedged, and the footing is left to cure for at least a day before the wall forms are erected.

Wall forms can be custom-built of lumber and plywood for each job, but it is more usual for standard prefabricated formwork panels to be employed. The panels for one side of the form are set on the footing, aligned carefully, braced, and coated with a form release compound. The wall reinforcing, in one or two layers as specified by the engineer, is next installed, with the bars wired to one an-

other at the intersections. The vertical bars are overlapped with the corresponding dowels projecting from the footing, and the horizontal bars are installed as L-shaped bends at the corners to maintain full structural continuity between the two walls. If the wall will connect to a concrete floor or another wall at its top, rods are left projecting from the top of the formwork to form a continuous connection. The *form ties*, which are small-diameter steel rods specially shaped to hold the formwork together under the pressure of the wet concrete, are inserted through holes provided in the formwork panels and secured to the back of the form with the devices supplied with the form ties; both ties and fasteners vary in detail from one manu-

facturer to another (Figures 11.6, 11.7). The ties go straight through the concrete wall from one side to another and remain in the wall after it is poured. This may seem like an odd way to go about holding wall forms together, but the pressures on the forms are so large that there is no other economical way of dealing with them.

When the ties are in place and the reinforcing has been inspected, the formwork for the second side of the wall is erected, the walers and braces are added, and the forms are inspected to be sure they are straight, plumb, correctly aligned, and adequately tied and braced. A surveyor's transit is used to establish the exact height to which the concrete will be poured, and this height is marked all around the inside

FIGURE 11.4

Formwork and reinforcing for a concrete wall. No key between the footing and the wall is shown in this example.

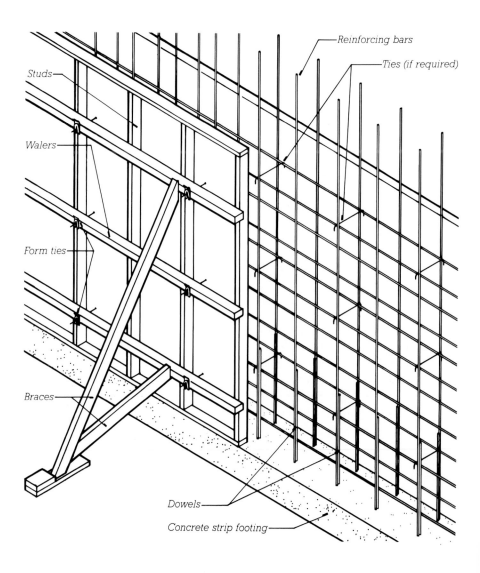

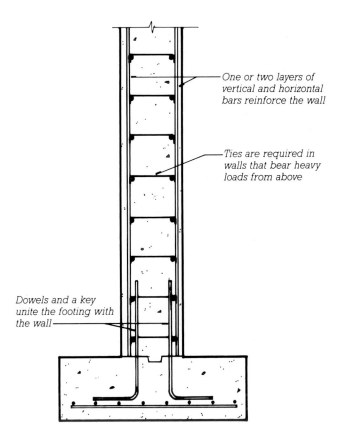

One or two layers of
vertical and horizontal
bars reinforce the wall

Ties are required in
walls that bear heavy
loads from above

Dowels and a key
unite the footing with
the wall

FIGURE   11.5
*Section through a reinforced concrete
wall.*

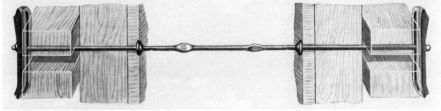

FIGURE   11.6
*Detail of a form tie assembly. Two
washers just inside the faces of the form
maintain the correct wall thickness.
Tapered, slotted wedges at the ends
transmit force from the tie to the walers.
The notches just inside the washers are
the points at which the tie will be
snapped off after the forms are stripped.*
(Courtesy of Richmond Screw Anchor
Co., Inc., 7214 Burns St., Fort Worth, TX
76118)

FIGURE   11.7
*Detail of a heavy-duty form tie. This
assembly is tightened with special
screws that engage a helix of heavy
wire welded into the tie. The wire
components remain in the concrete, but
the screws and the plastic cone on the
right are removed and reused after
stripping. The purpose of the cone is to
give a neatly finished hole in the
exposed surface of the concrete.*
(Courtesy of Richmond Screw Anchor
Co., Inc., 7214 Burns St., Fort Worth, TX
76118)

of the forms. Pouring may then proceed. Concrete is brought to the site, test cylinders are made, and a slump test is performed to check for the proper pouring consistency. Concrete is then transported to the top of the wall by bucket or by hose. Workers standing on planks at the top of the forms deposit the concrete in the forms, compacting it frequently with a vibrator (Figure 11.8). When the form is filled and compacted to the line inside the formwork, hand floats are used to smooth and level the top of the wall. The top of the form is then covered with a plastic sheet or canvas, and the wall is left to cure.

After a few days of curing, the bracing and walers are taken down, the connectors are removed from the ends of the form ties, and the formwork is *stripped* from the wall (Figure 11.9). This leaves the wall bristling with projecting ends of form ties. These are twisted off by workers with heavy pliers, and the holes they leave in the surfaces of the wall are carefully filled with mortar. If required, major defects in the wall surface caused by defects in the formwork or inadequate filling of the forms with concrete can be repaired at this time. The wall is now complete.

Casting a Concrete Column

A column is formed and cast much like a wall, with a few important differences. The footing is usually an isolated column footing, a pile cap, or a caisson, rather than a strip footing (Figure 11.10). The dowels are sized and spaced in the footing to match those in the column. The cage of column reinforcing is assembled with wire ties and hoisted into place over the dowels. Usually the vertical bars in the column simply overlap the dowels, but if space is very tight, the bars are spliced end to end with welds or mechanical connectors. The column form can be a square box of plywood panels, a cy-

FIGURE 11.8
Compacting wet concrete after pouring, using a mechanical vibrator immersed in the concrete. (Courtesy of Portland Cement Association, Skokie, Illinois)

FIGURE 11.9
Three stages in the construction of a reinforced concrete wall on a strip footing: In the foreground, the reinforcing bars have been wired to the dowels projecting from the footing, ready for erection of the formwork. In the center, a section of wall has been poured in steel formwork tied with small steel straps secured by wedges. In the background, the forms have been stripped, and some of the ties have been snapped off. (Courtesy of Portland Cement Association, Skokie, Illinois)

FIGURE 11.10
(a) *A column footing almost ready for pouring, but lacking dowels. The reinforcing bars are supported on pieces of concrete brick.* (b) *Column footings poured, with projecting dowels to connect to both round and rectangular columns.* (Photos by the author)

FIGURE 11.11
In the foreground, a square column form tied with pairs of L-shaped steel brackets. In the background, a worker braces a round column form made of steel sheet. (Courtesy of The Ceco Corporation)

lindrical steel or plastic tube bolted together in halves so it can later be removed, or a waxed cardboard tube that is stripped after curing by unwinding the layers of paper that make up the tube (Figures 11.11, 11.12). Unless the column is broad and wall-like, form ties through the concrete are not required. At the top of the column, the vertical bars are left projecting, either for splicing to the bars in the column for the story above, or bent over at right angles to splice into the roof slab.

ONE-WAY FLOOR AND ROOF FRAMING SYSTEMS

The One-Way Solid Slab System

A one-way solid slab (Figure 11.13, 11.14, 11.15) spans across parallel lines of support furnished by walls and/or beams. The walls and columns are poured prior to erecting the formwork

for a one-way slab, but the girders and beams are nearly always formed and poured at the same time as the slab.

The girder and beam forms are erected first, then the slab forms are erected. A form-release compound is applied to all surfaces that will be in contact with concrete. The forms are supported on temporary joists and beams of metal or wood, and the temporary beams are supported on temporary *shores* (adjustable-length columns). The weight of uncured concrete that must be supported is enormous, and the temporary beams and shoring must be both numerous and strong. Formwork is, in fact, designed by a concrete contractor's structural engineers as if it were a permanent building, because structural failures in formwork can cause intolerable risk to workers and damage to property.

Formwork is usually made so as to eliminate sharp edges on the concrete. Sharp edges often break during stripping to leave a ragged edge that is almost impossible to patch. In service, sharp edges are easily damaged by, and are potentially damaging to, people, furniture, and vehicles. Edges are beveled or rounded by inserting shaped strips of wood or plastic into the corners of the formwork to produce the desired profile.

In accordance with reinforcing diagrams and schedules prepared by the structural engineer, the girder and beam reinforcing—bottom bars, top bars, and stirrups—is installed in the forms, supported on chairs and bolsters. Next the slab reinforcing—bottom bars, top bars, and shrinkage-temperature reinforcing—is placed on bolsters. After the reinforcing and formwork have been inspected, the girders, beams, and slab are poured in a single operation, with the usual sample cylinders made for later testing. Slab depths are typically 4 to 10

FIGURE 11.12
Round columns may also be formed with single-use cardboard tubes. (Courtesy of Sonoco Products Company)

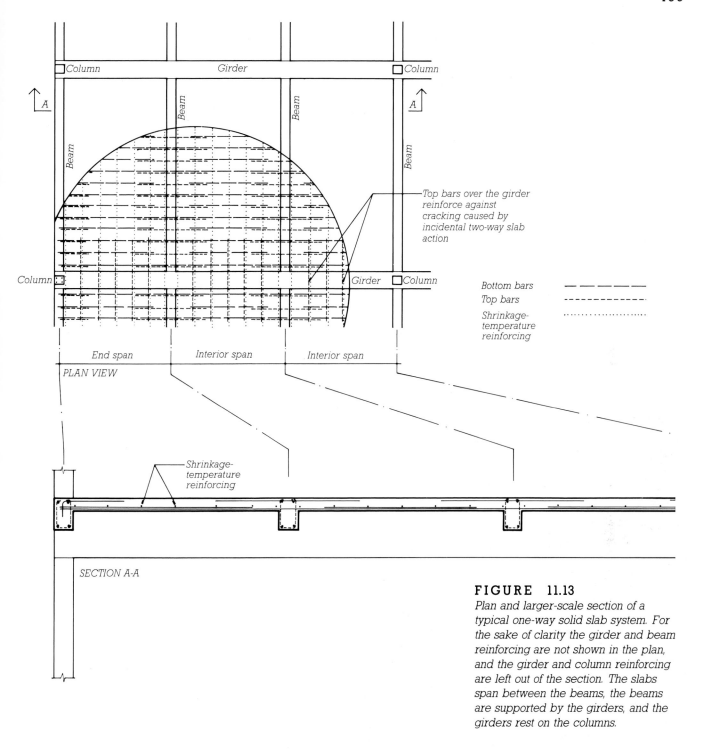

Top bars over the girder reinforce against cracking caused by incidental two-way slab action

Bottom bars	— — — — —
Top bars	- - - - - - -
Shrinkage-temperature reinforcing	··············

End span   Interior span   Interior span

PLAN VIEW

Shrinkage-temperature reinforcing

SECTION A-A

FIGURE 11.13

Plan and larger-scale section of a typical one-way solid slab system. For the sake of clarity the girder and beam reinforcing are not shown in the plan, and the girder and column reinforcing are left out of the section. The slabs span between the beams, the beams are supported by the girders, and the girders rest on the columns.

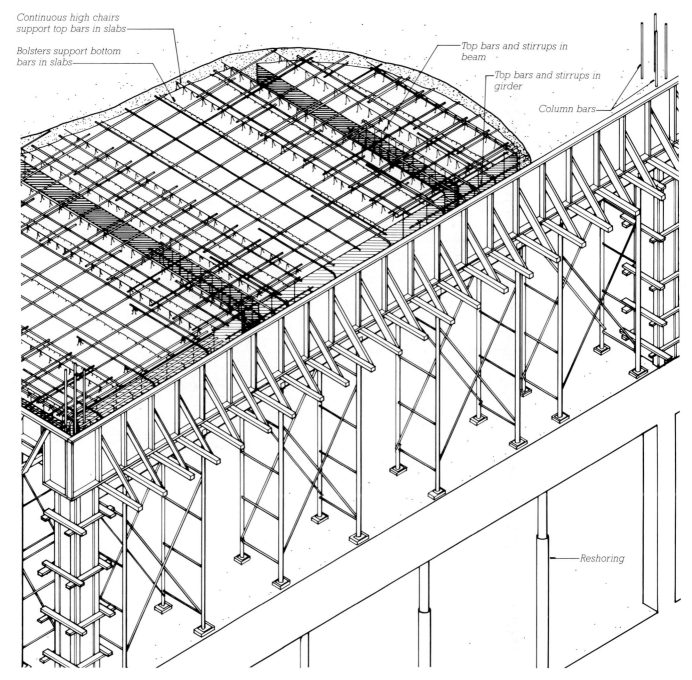

Continuous high chairs
support top bars in slabs

Bolsters support bottom
bars in slabs

Top bars and stirrups in
beam

Top bars and stirrups in
girder

Column bars

Reshoring

FIGURE   11.14
*Isometric view of a one-way solid slab
system under construction. The slab,
beams, and girders are created in a
single pour.*

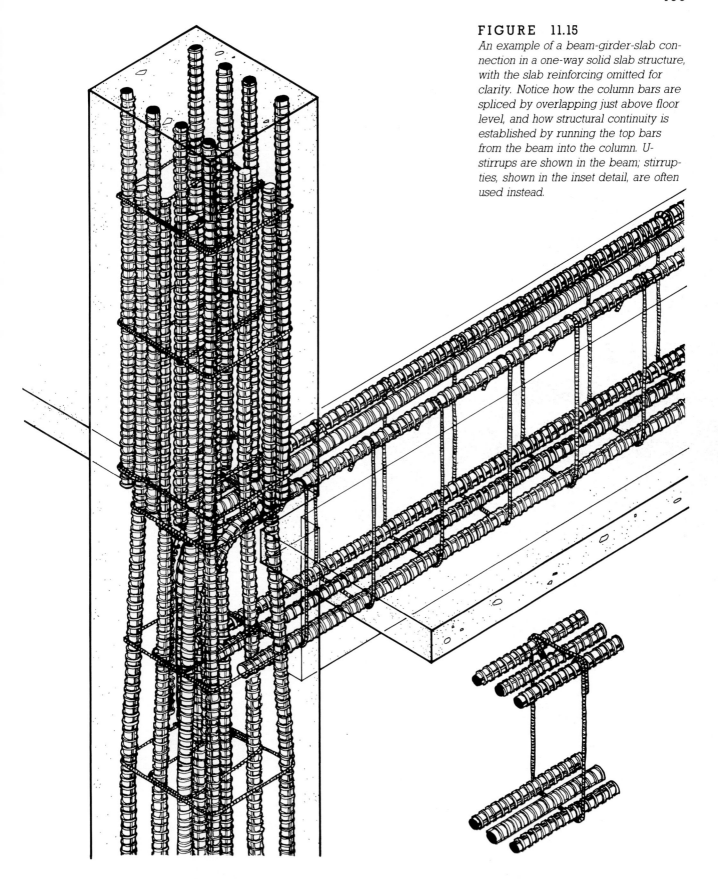

FIGURE 11.15

An example of a beam-girder-slab connection in a one-way solid slab structure, with the slab reinforcing omitted for clarity. Notice how the column bars are spliced by overlapping just above floor level, and how structural continuity is established by running the top bars from the beam into the column. U-stirrups are shown in the beam; stirrup-ties, shown in the inset detail, are often used instead.

inches (100 to 250 mm). The top of the slab is finished in the same manner as a slab on grade, usually to a steel-trowel finish, and the slab is sealed or covered for damp curing. The only construction left projecting above the slab surface is the offset column bars, which are now ready to splice to the column bars for the level above.

When the slab and beams have attained enough strength to support themselves safely, the formwork is stripped and the slabs and beams are *reshored* with vertical props to relieve them of loads until they have reached full strength, which will take several more weeks. Meanwhile, the formwork and the remainder of the shoring are cleaned and moved up a level above the slab and beams just poured, where the cycle of forming, reinforcing, pouring, and stripping is repeated (Figure 11.16).

FIGURE 11.16
A one-way solid slab under construction. (Courtesy of Portland Cement Association, Skokie, Illinois)

FIGURE 11.17
This helical ramp is a special application of one-way solid slab construction. The formwork here is made of overlaid plywood for a smooth surface finish. (Courtesy of American Plywood Association)

The One-Way Concrete Joist System (Ribbed Slab)

The one-way solid slab system is economical for structures in which the slab does not span very far between beams. For longer one-way spans, a progressively thicker slab is required, and the weight of the slab itself becomes an excessive burden, unless a substantial portion of the nonworking concrete in the lower part of the slab can be eliminated to lighten the load. This is the rationale for the *one-way concrete joist system* (Figure 11.18). The bottom steel is concentrated in spaced ribs or joists, and the thin slab that spans across the top of the joists is reinforced only by shrinkage-temperature bars. There is little concrete in this system that is not working, with the result that it can span considerably longer distances than a one-way solid slab. Each individual joist is reinforced as a small beam, except that stirrups are not usually used in concrete joists because of the restricted space in the joist and the large number of stirrups that would be re-

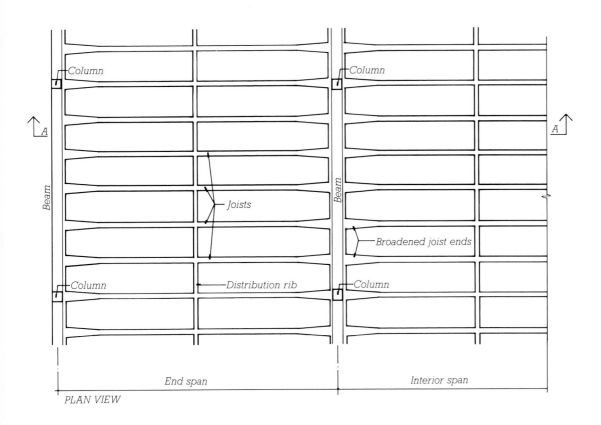

Column

A

Beam

Joists

Column

Distribution rib

Column

Column

Broadened joist ends

Beam

A

End span

Interior span

PLAN VIEW

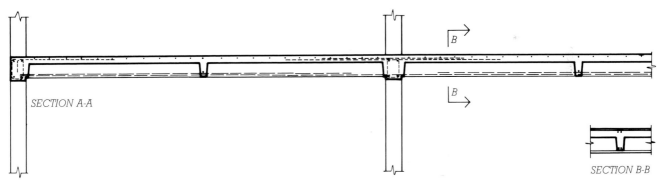

SECTION A-A

B

B

SECTION B-B

FIGURE 11.18

*Plan and larger-scale section of a
typical one-way concrete joist system.
For the sake of clarity no reinforcing is
shown in the plan, and the column
reinforcing is not shown on the section.
All bottom and top reinforcing occurs in
the ribs, and all shrinkage-temperature
bars are placed in the slab.*

quired. Instead, the ends of the joists are broadened sufficiently that the concrete itself can resist the diagonal tension forces.

The joists are formed with formed metal or plastic *pans* supported on longitudinal strips of wood or on a plywood deck. Pans are available in two standard widths, 20 inches (508 mm) and 30 inches (762 mm), and in depths ranging up to 20 inches (508 mm), as shown in Figure 11.19. (Larger pans are also available for forming floors designed as one-way solid slab systems.)

A slope on the sides of the pans allows them to drop easily out of the hardened concrete during stripping. The joist width can be varied by placing the rows of pans closer together or farther apart, with the bottom of the joist formed by the wood deck or strip of wood. The broadening of the joist ends is accomplished with standard end pans whose width tapers. A *distribution rib* is formed across the joists at midspan to distribute concentrated loads to more than one joist. After application of a form-release compound, the beam steel and joist steel are placed, the shrinkage-temperature steel is laid crosswise over the pans, and the entire system is poured and finished (Figures 11.20, 11.21).

For greater economy of formwork, one-way concrete joists are often supported on *joist-bands*, which are broad beams that are only as deep as the joists. While a deeper beam would be more efficient structurally, a joist-band can be poured on the same plywood deck that supports the pans, which simplifies the formwork considerably.

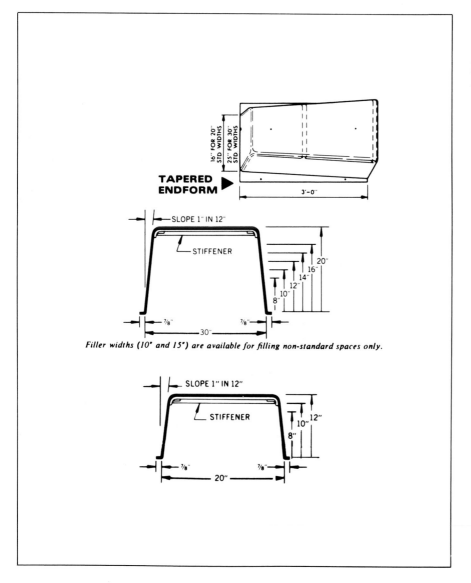

FIGURE 11.19
Standard steel form dimensions for one-way concrete joist construction.
(Courtesy of The Ceco Corporation)

FIGURE 11.20

Reinforcing being placed for a one-way concrete joist floor. Electrical conduits and boxes have been put in place, and welded wire fabric is being installed as shrinkage-temperature reinforcing. Both the tapered end pans and the square endcaps for the midspan distribution rib are clearly visible. (Courtesy of The Ceco Corporation)

FIGURE 11.21

A one-way concrete joist system after stripping of the formwork, showing broadened joist ends and a distribution rib. The dangling wires are hangers for a suspended finish ceiling (see Chapter 18). (Courtesy of Portland Cement Association, Skokie, Illinois)

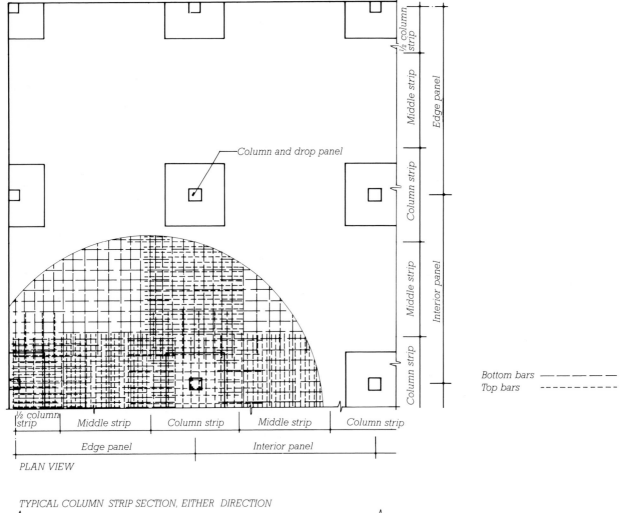

Column and drop panel

½ column strip

Middle strip — Edge panel

Column strip

Middle strip — Interior panel

Column strip

Bottom bars ————
Top bars --------

½ column strip | Middle strip | Column strip | Middle strip | Column strip

Edge panel | Interior panel

PLAN VIEW

TYPICAL COLUMN STRIP SECTION, EITHER DIRECTION

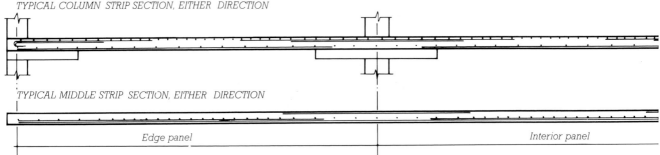

TYPICAL MIDDLE STRIP SECTION, EITHER DIRECTION

Edge panel | Interior panel

FIGURE 11.22

Plan and larger-scale section of a typical two-way flat slab system. The reinforcing pattern consists of column strips and middle strips, with each strip changing pattern slightly to accommodate the different bending forces that occur in the edge panels around the perimeter of the building. The system shown uses only drop panels, without mushroom capitals. The reinforcing in a two-way flat plate system is essentially identical to this example; the only difference is that the flat plate has no drop panels.

TWO-WAY FLOOR AND ROOF FRAMING SYSTEMS

The Two-Way Flat Slab and Two-Way Flat Plate Systems

Two-way concrete framing systems are generally more economical than one-way systems in buildings where the columns can be spaced in bays that are square or nearly square in proportion. A *two-way solid slab* is a system in which the slab is supported by a grid of beams running in both directions over the columns. This system is occasionally used for very heavily loaded industrial floors, but most two-way floor and roof framing systems, even for heavy loadings, are made without beams. The slab is reinforced in such a way that the varying stresses in the different zones of the slab are accommodated within a uniform thickness of concrete.

The *two-way flat slab* (Figure 11.22), a system suited to heavily loaded buildings such as storage and industrial buildings, illustrates this concept. The formwork is completely flat except for a thickening of the concrete to resist the high shear forces around the top of each column. Traditionally, this thickening was accomplished with both a *mushroom capital* and a *drop panel*, but today the capital is usually eliminated to achieve a greater economy of formwork cost, leaving a drop panel to do the work (Figure 11.23). Typical depths for the slab itself lie in the 6 to 12 inch (150 to 300 mm) range. The reinforcing is laid in both directions in half-bay-wide strips of two fundamental types: *Column strips* are designed to carry the higher bending forces encountered in the zones of the slab that cross the columns, and *middle strips* have a lighter reinforcing pattern. Shrinkage-temperature steel is not needed in two-way systems because the concrete is already reinforced in both directions. The drop panel and capital (if any) have no additional reinforcing beyond that provided by the column strip; the greater thickness of concrete furnishes the required shear resistance.

In more lightly loaded buildings, such as hotels, hospitals, dormitories, and apartment buildings, the slab need not be thickened at all over the columns. This makes the formwork extremely simple and even allows some columns to be moved off the grid a bit if it will facilitate a more efficient floor plan arrangement (Figure 11.24). The completely flat ceilings of this system allow room partitions to be placed anywhere with equal ease, and the story heights of the building to be kept to an absolute minimum, which saves on the cost of exterior cladding (Figure

Reinforced concrete made "pilotis" possible. The house is in the air, away from the ground; the garden runs under the house, and it is also above the house, on the roof...Reinforced concrete is the means which makes it possible to build all of one material...Reinforced concrete brings the free plan into the house! Floors no longer have to stand simply one on top of the other. They are free...Reinforced concrete revolutionizes the history of the window. Windows can run from one end of the facade to the other...

Le Corbusier and P. Jeanneret, *Oeuvre Complete 1910–1929, Zurich, 1956, p. 128.*

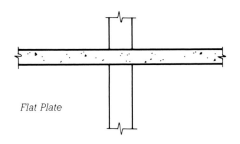

Flat Plate

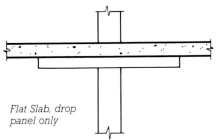

Flat Slab, drop panel only

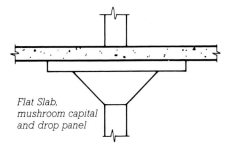

Flat Slab, mushroom capital and drop panel

FIGURE 11.23
Column capitals for two-way concrete framing systems. For slabs bearing heavy loads, shear stresses around the column are reduced by means of a mushroom capital and drop panel, or a drop panel alone. For lighter loads, no thickening of the slab is required.

FIGURE 11.24

A two-way flat plate floor of a high-rise apartment building is poured with the aid of a concrete pump at the base of the building. The concrete is delivered up a telescoping tower to a nozzle at the end of a telescoping boom. The columns in the center of the newly poured area demonstrate the ability of flat plate construction to adapt to irregular patterns and spacings of columns. (Courtesy of Schwing America, Inc.)

FIGURE 11.25

The underside of a two-way flat plate, with sprinkler pipes installed. (Courtesy of Portland Cement Association, Skokie, Illinois)

11.25). Typical slab depths for this *two-way flat plate system* range from 5 to 10 inches (125 to 250 mm).

The zones along the exterior edges of both the two-way flat slab system and the two-way flat plate system require special attention. To take full advantage of structural continuity, the slabs should be cantilevered beyond the last row of columns a distance equal to about thirty percent of the interior span. If columns are required at the faces of the building, additional reinforcing can be added to the slab edges to carry the higher stresses that will result.

The Two-Way Waffle Slab System

The *waffle slab*, or *two-way concrete joist system* (Figure 11.26), is the two-way equivalent of the one-way concrete joist system. Metal or plastic pans called *domes* are used as formwork to eliminate the nonworking concrete from a deeper slab, allowing a greater economy in longer spans. The standard domes form joists 6 inches (152 mm) wide on 36 inch (914 mm) centers, or 5 inches (127 mm) wide on 24 inch (610 mm) centers, in a variety of depths up to 20 inches (500 mm), as shown in Figures 11.27 through 11.30. Special domes are also available in larger sizes. Solid concrete *heads* are created around the tops of the columns by leaving the domes out of the formwork; these serve the same function as drop panels in the two-way flat slab system. If the slab cannot be cantilevered at the perimeter of the building, a perimeter beam must be provided. Stripping of the domes is facilitated in many cases by a compressed air fitting at the top of each. This allows the dome to be popped out of the concrete with the application of a puff of compressed air. Waffle slabs are often employed in a building not only for their economy, but also for their richly coffered undersides, which can be left exposed as ceilings (Figure 11.31).

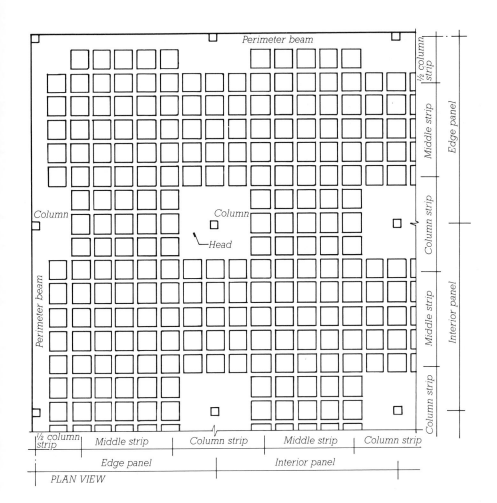

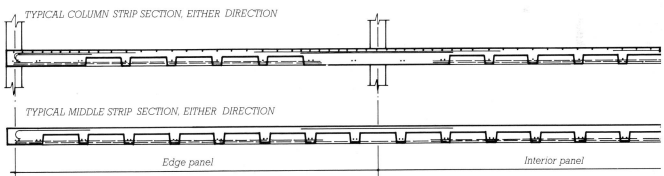

FIGURE 11.26

Plan and larger-scale section of a typical two-way concrete joist system, also known as a waffle slab. No reinforcing is shown on the plan drawing for the sake of clarity, and the section does not show the welded wire fabric that is spread over the entire form before pouring.

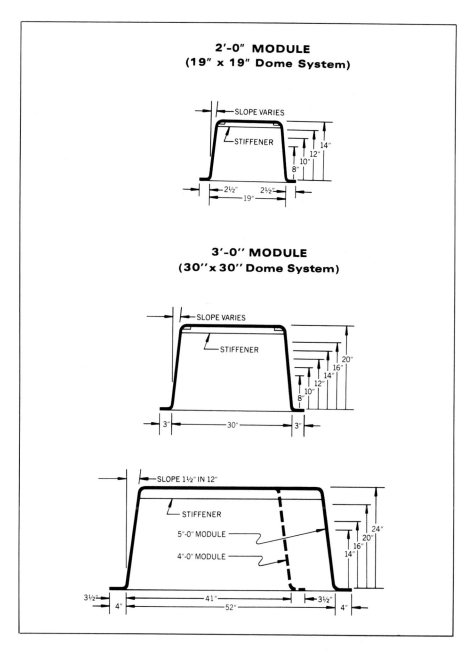

2'-0" MODULE
(19" x 19" Dome System)

SLOPE VARIES

STIFFENER

14"
12"
10"
8"

2½"   2½"
19"

3'-0" MODULE
(30" x 30" Dome System)

SLOPE VARIES

STIFFENER

20"
16"
14"
12"
10"
8"

3"   30"   3"

SLOPE 1½" IN 12"

STIFFENER

5"-0" MODULE

4"-0" MODULE

24"
20"
16"
14"

3½"   41"   3½"
4"   52"   4"

FIGURE 11.27
Standard steel dome forms for two-way concrete joist construction. (Courtesy of The Ceco Corporation)

FIGURE 11.28
Steel domes being placed on a temporary plywood deck to form a two way concrete joist floor. Pans are omitted around columns to form heads. (Courtesy of The Ceco Corporation)

FIGURE 11.29
Plastic dome formwork being assembled for a two-way concrete joist roof. Notice the electrical conduit and junction box in the foreground, ready to be poured into the slab. (Courtesy of Molded Fiber Glass Concrete Forms Company)

FIGURE 11.30
Stripping plastic domes after removal of the temporary plywood deck. (Courtesy of Molded Fiber Glass Concrete Forms Company)

FIGURE 11.31
The underside of a two-way concrete joist floor. (Courtesy of The Ceco Corporation)

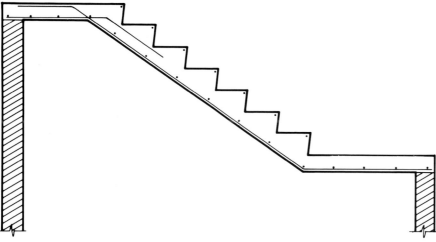

FIGURE 11.32
Concrete stair.

CONCRETE STAIRS

A concrete stair (Figure 11.32) is reinforced and poured as an inclined one-way solid slab with additional concrete added to make risers and treads. The underside of the form is planar, and the top is built with riser forms, usually inclined to give additional tread width. The concrete is poured in one operation, and the treads are tooled to a steel trowel finish. Nosings, as used in wood stair construction, are generally avoided in concrete stairs because they would have a tendency to break off. The inclined riser design is also preferable for handicapped access because it does not catch the toes of persons climbing the stairs on crutches.

SITECAST POSTTENSIONED FRAMING SYSTEMS

Posttensioning can be applied to any of the sitecast concrete framing systems. It is used in beams or girders to reduce member sizes and extend the spanning capability, as well as in slabs, both one-way and two-way.

Posttensioning may be either bonded or unbonded construction. In bonded construction, the tendons, cables of twelve to fifteen strands of high-strength steel, are not usually installed until after the concrete has cured. Instead, before the concrete is poured, a galvanized metal *duct* is placed in the formwork in the desired location for each tendon. If there is a danger that the duct may collapse during pouring operations, a flexible plastic pipe is inserted to add rigidity. After curing, the plastic pipes are removed and the duct is flushed clean with compressed air or a mixture of water and lime (calcium hydroxide). The tendon is slipped into the duct, tensioned with a hydraulic jack, and anchored to the ends of the beam with any of a number of patented systems of wedges that lock the strands of the tendon securely in place. The jack is then removed, the protruding ends of the strands are cut off, and grout is pumped into the duct under pressure to bond the tendon to the duct and thence to the concrete of the beam.

In unbonded construction, the tendons are typically individual strands of 0.5 inch or 0.6 inch (12.7 mm or 15.2 mm) diameter, each encased in a plastic sheath. These are wrapped in paper or tape as a precaution against bonding to the concrete. The wrapped tendons are installed in their final positions in the formwork, and the concrete is cast. After the concrete has cured, the tendons are tensioned in a manner similar to that used for bonded tendons, but are not grouted. Because of the possibility that water may find its way into the space around unbonded tendons, bonded tendons have less tendency to corrode in service and are therefore favored for many applications.

During installation, the conduits or unbonded tendons are usually draped into curves that follow the lines of maximum tension in the beams and slabs, dipping to a low point at the middle of each span and rising to a high point over each support, thus gaining additional structural efficiency. Tendons can be draped in both directions of a slab to allow for two-way action.

As with any prestressed concrete framing system, both short-term and long-term losses of prestressing force must be anticipated. The short-term losses in posttensioning are elastic shortening of the concrete, friction between the tendons and the concrete, and initial movement (set) in the anchorages. The long-term losses are concrete shrinkage, concrete creep, and steel relaxation. The structural engineer calculates the total of these expected losses and specifies an additional amount of initial posttensioning force to compensate for them.

SELECTING A SITECAST CONCRETE FRAMING SYSTEM

In the absence of other factors, two-way concrete framing systems are pre-

If we were to train ourselves to draw as we build, from the bottom up...stopping our pencil to make a mark at the joints of pouring or erecting, ornament would grow out of our love for the expression of method.

Louis I. Kahn, quoted in Vincent Scully, Jr., *Louis I. Kahn*, New York, George Braziller, 1962, p. 27.

ferred to one-way systems for their greater economy. Within either two-way or one-way systems, the choice is governed largely by the desired span and the expected loading. The overlap between spanning capabilities of the various systems leaves considerable discretion to the designer (Figures 11.33, 11.34).

There are many reasons beside structural efficiency to choose one system over another. One-way solid slab systems can be proportioned to suit the designer, with the girders and beams forming a pattern that relates strongly to the overall form and module of a building. Waffle slabs lose much of their visual impact if the space below is partitioned into small rooms, and the location of partitions is regu-

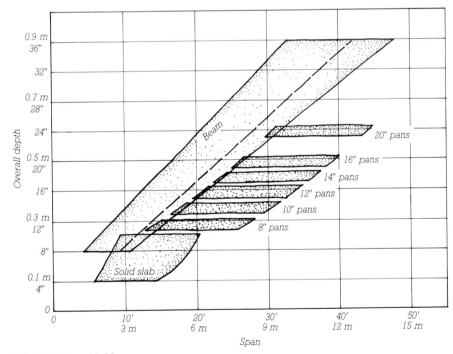

FIGURE 11.33

Span-to-depth relationships for preliminary sizing of one-way concrete framing systems. Floor and roof beams may be sized by reading upward from the desired span to the area marked "beam." For ordinary floor beams, read to the left from the intersection of the span line and the broken line to determine a preliminary overall depth (including cover) for the beam. For girders, read to the left from a point in the shaded area above the dashed line, and for lightly loaded beams, read from below the dashed line. One-way solid slabs and one-way concrete joist systems may be sized approximately from the other shaded areas of the diagram. This chart is for preliminary design use only and does not take into account many factors that must be considered in the final design of a structural system.

lated by the grid of the pans. Two-way flat plate ceilings allow complete freedom in partition location but add no visual character to the interior of a building. If a finish ceiling is to be suspended under the slab, of course, the choice of a concrete framing system is of little visual consequence.

Innovations in Sitecast Concrete Construction

The development of sitecast concrete construction continues along several lines. The basic materials, concrete and steel, are constantly undergoing research and development, leading to higher allowable strengths and decreased weight of the structure itself. Lighter-weight structural concretes that utilize expanded mineral aggregates are being used more widely to reduce

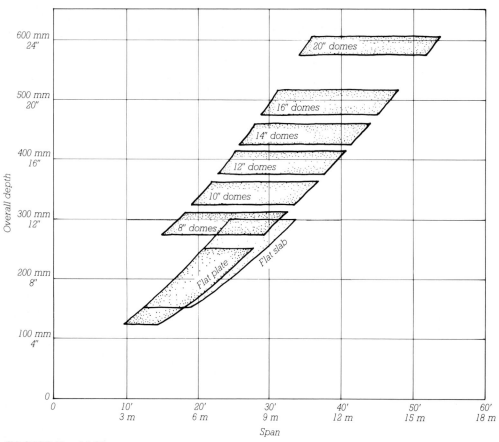

FIGURE 11.34

Span-to-depth relationships for preliminary sizing of two-way concrete framing systems. Read upward from the desired span to the shaded area indicating the desired type of structural system, then to the left to determine an approximate overall depth, including cover. For more heavily loaded structures, stay within the left-hand region of each shaded area. This chart is for preliminary design use only, and does not take into account many factors that must be considered in the final design of a structural system.

loads still further. Shrinkage-compensating cements have been developed for use in concrete structures that cannot be allowed to shrink during curing.

The high cost of formwork has led to many innovations. *Lift-slab* construction, used chiefly with two-way flat plate structures, almost eliminates formwork by casting the slabs of a building in a stack on the ground, then using hydraulic jacks to lift the slabs up the columns to their final positions, where they are welded in place using special cast-in-place steel slab collars (Figure 11.35). *Flying formwork* is fabricated in large sections supported on deep metal trusses; the sections are moved from one floor to the next by crane, eliminating much of the labor usually expended on stripping and re-erecting formwork (Figure 11.36). *Slip-forming* is useful for tall walled structures such as elevator shafts, stairwells, and storage silos. A ring of formwork is pulled steadily upward by jacks supported on the vertical reinforcing bars, while workers add concrete and reinforcing in a continuous process. In *tilt-up* construction, a floor slab is cast on the ground, and reinforced concrete wall panels are poured over it in a horizontal position, then tilted into position and grouted together, thereby eliminating most of the usual wall formwork. *Shotcrete* (pneumatically placed concrete) is sprayed into place from a hose by a stream of compressed air and can be deposited without formwork even on vertical surfaces; it is used primarily for repairing damaged concrete on the faces of beams and columns, and for the production of free-form structures such as swimming pools and playground structures. The ultimate saving in formwork, of course, is realized by casting the concrete into reusable molds in a precasting plant, which is the subject of the next chapter.

Advancements in reinforcing for sitecast concrete, other than the adoption of posttensioning, include a move to higher strength steels, and a trend toward increased prefabrication of reinforcing bars prior to installation in

FIGURE 11.35

Lift-slab construction in progress. The paired steel rods silhouetted against the sky to the right are part of the lifting jacks seen at the tops of the columns. (Courtesy of Portland Cement Association, Skokie, Illinois)

FIGURE 11.36

Flying formwork for a one-way concrete joist system being moved from one floor to the next in preparation for pouring. Stiff metal trusses allow a large area of formwork to be handled by a crane as a single piece. (Courtesy of Molded Fiber Glass Forms Company)

FASTENING TO CONCRETE

As a concrete frame is finished, many things need to be fastened to it: exterior wall panels and facings; interior partitions; hangers for pipes, ducts, and conduits; suspended ceilings; stair railings; cabinets; machinery; and many more. Drawings A through H are examples of fastening systems that must be cast into the concrete. A is the familiar *anchor bolt.* B is a steel plate welded to a bent rod or strap anchor; this *weld plate* furnishes a surface to which steel components can be welded. The steel angle in C has a threaded stud welded to it so another component can be attached by bolting. D is an adjustable insert of malleable iron that is nailed to the formwork through the slots in the ears on either side. A special nut twists and locks into the slot to accept a bolt or threaded rod from below. A slightly different form of this type of insert is shown in the detail of masonry shelf angle attachment in Chapter 15. E and F are two different designs of threaded inserts cast into the concrete. The dovetail slot in G is used with special anchor straps as shown to tie masonry facings to a frame or wall. H is simply a dovetailed wood nailer strip cast into the concrete, a detail that is risky because the wood may swell and crack the concrete, or shrink and become loose. A through F are heavy-duty devices that can be selected in capacities sufficient to anchor heavy building components and machinery.

Details I through P depict fastening devices that are inserted in holes drilled into the cured concrete. Concrete is drilled fairly easily with carbide-tipped drill bits. I shows the steel post of a railing anchored into an oversize hole in a concrete slab using grout, poured lead, or epoxy; this system is also used to fasten bolts to concrete and can carry heavy loads if properly designed and installed. J is a plastic sleeve, K a wood or fiber plug, and L a lead sleeve; all three are inserted into drilled holes and expand to grip the sides of the hole when a screw is driven into them. M is a similar type of metal sleeve but has a special nail that expands the sleeve as it is driven. N is a special bolt with a steel sleeve over a tapered shank at the inner end. The sleeve catches against the concrete as the bolt is driven into the hole and is expanded by the taper as the bolt is tightened. O is a special screw and P is a special nail, both designed to grip tightly when inserted into drilled holes of the correct dimension. J through P are light- to medium-duty fasteners, with the exception of N, which can carry rather heavy loads.

Q and R depict driven anchors. Q is the familiar concrete nail or masonry nail, made of hardened steel. If driven through a strip of wood with a few blows of a very heavy hammer, or inserted with a nail gun, it will penetrate concrete just enough to provide some shear resistance for furring strips and sleepers, but it has a tendency to loosen, particularly if driven with too many blows. Shown in R are three examples of *powder-driven* or *powder-actuated* fasteners, which are driven into steel or concrete with the energy from an exploding cartridge of gunpowder. The first fastener is a simple pin used for attaching wood or sheet metal components to a wall or slab. The middle one is threaded to accept a nut. The eye on the fastener to the right allows a wire, such as a hanger wire for a suspended ceiling, to be attached. Powder-driven fasteners are rapid to install, economical, and have a moderately high load-carrying capacity. S typifies devices whose perforated metal plates adhere securely with a mastic-type adhesive to surfaces of concrete or masonry. The fastener shown here has a thin sheet metal spike, over which foam plastic insulation can be impaled. The spike is then bent across the face of the insulation panel to hold it in place. Roof edge systems employing perforated plates and adhesives to fasten them to the building are illustrated in Chapter 13.

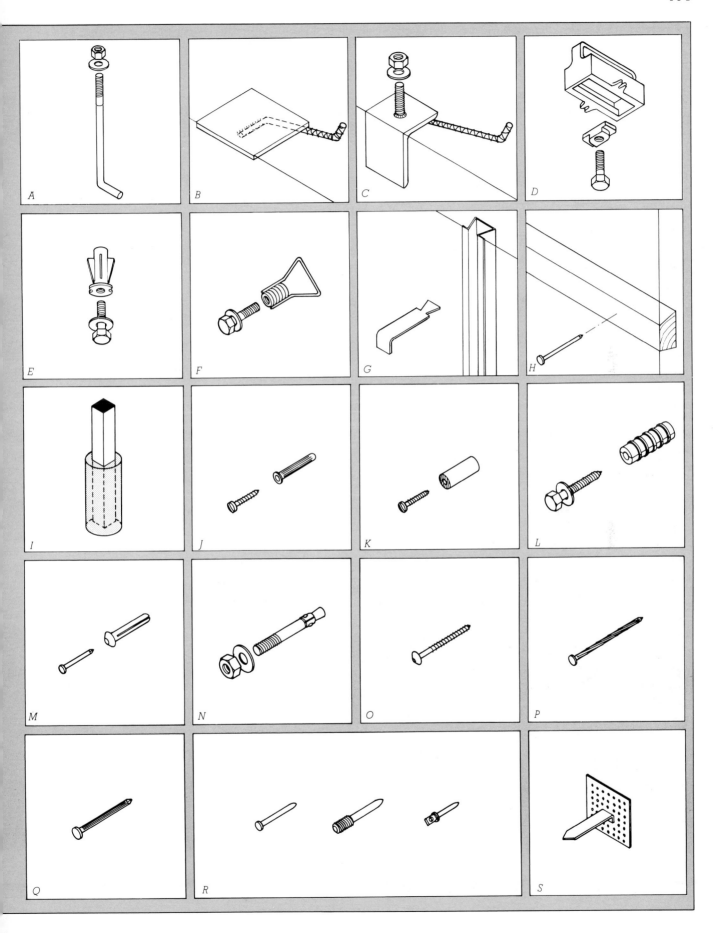

A

B

C

D

E

F

G

H

I

J

K

L

M

N

O

P

Q

R

S

the forms. With developments in welding and fabricating machinery, welded wire fabric is pushing beyond the familiar grid of heavy wire to include complete cages of column reinforcing and entire bays of slab reinforcing.

A long chain of development continues in the area of surface finishes for concrete (Figures 11.37 through 11.40). *Exposed aggregate* finishes have been popular for decades. These involve the scrubbing and hosing of concrete surfaces shortly after the initial set of the concrete to remove the cement paste from the surface and leave the rough texture of the aggregate. This process is often aided by chemicals that retard the set of the cement paste; these are either sprayed on the surface of a slab, or used as a coating inside the formwork. Because concrete can take on almost any texture that can be imparted to the surface of formwork, much work has gone into developing formwork surfaces that range from almost glassy smooth, to ribbed, veined, board-textured, and corrugated. After partial curing, other steps can be taken to change the texture of concrete, in-

FIGURE 11.37
Exposed wall surfaces of sitecast concrete. Narrow boards were used to form the walls, and form tie locations were carefully worked out in advance. (Architect: Eduardo Catalano. Photo by Erik Leigh Simmons. Courtesy of the architect)

cluding sandblasting, rubbing with abrasive stones, grinding smooth, and hammering with any of a number of types of flat, pointed, or toothed masonry hammers. Many types of pigments, dyes, paints, and sealers can be used to add color or gloss to concrete surfaces, and to give protection against weather, dirt, and wear.

Exposed wall surfaces of concrete need special attention from the designer and contractor (Figure 11.39). Chairs and bolsters need to be designed so they will not create rust spots on exterior concrete surfaces. Form tie locations in exposed concrete walls should be patterned to harmonize with the layout of the walls themselves, and the holes left in the concrete surfaces by snapped-off ties patched or plugged securely to prevent rusting through. Joints between pours can be concealed gracefully with recesses in the face of the concrete. The formulation of the concrete needs to be closely controlled for color consistency from one batch to the next. In cold climates, air entrainment is advisable to prevent freeze-thaw damage of exterior wall surfaces.

FIGURE 11.38
Exposed wall surfaces of concrete, sandblasted to expose the aggregate. (Architect: Eduardo Catalano. Photo by Gordon H. Schenck, Jr. Courtesy of the architect)

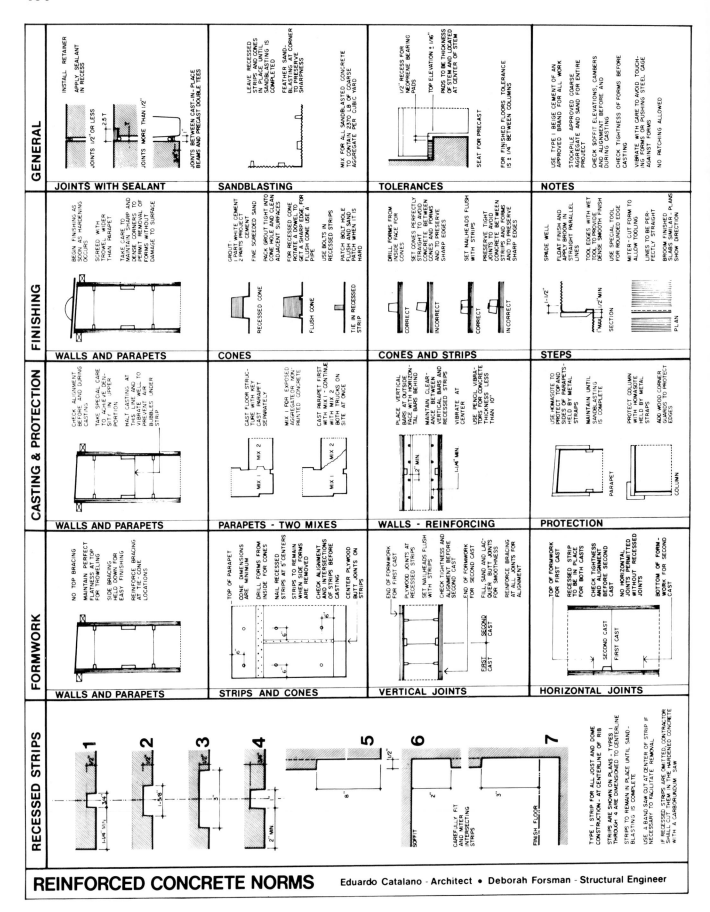

GENERAL
JOINTS WITH SEALANT   SANDBLASTING   TOLERANCES   NOTES

FINISHING
WALLS AND PARAPETS   CONES   CONES AND STRIPS   STEPS

CASTING & PROTECTION
WALLS AND PARAPETS   PARAPETS - TWO MIXES   WALLS - REINFORCING   PROTECTION

FORMWORK
WALLS AND PARAPETS   STRIPS AND CONES   VERTICAL JOINTS   HORIZONTAL JOINTS

RECESSED STRIPS

REINFORCED CONCRETE NORMS

Eduardo Catalano - Architect • Deborah Forsman - Structural Engineer

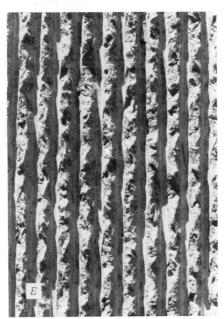

FIGURE 11.39

Precautions specified by the architect to assure satisfactory visual quality in the exposed concrete walls of the buildings illustrated in Figures 11.37 and 11.38. (Courtesy of Eduardo Catalano, Architect)

FIGURE 11.40

Close-up photographs of some surface textures for exposed concrete walls. (a) Ten-year-old concrete cast against overlaid plywood to obtain a very smooth surface shows a crazing pattern of hairline cracks. (b) The boat-shaped patches and rotary-sliced grain figure of A-veneered plywood formwork are mirrored faithfully in this surface. A neatly plugged form tie hole is seen at the upper left, and several lines of overspill from a higher pour have dribbled over the surface. (c) This exposed aggregate surface was obtained by coating the formwork with a curing retarder, and scrubbing the surface of the concrete with water and a stiff brush after stripping the formwork. (d) The bush hammered surface of this concrete column is framed by a smoothly formed edge. (e, f) Architect Paul Rudolph developed the technique of casting concrete walls against ribbed formwork, then bush hammering the ribs to produce a very heavily textured, deeply shadowed surface. In the example to the right, the ribbed wall surface is contrasted to a board-formed slab edge. (Photos by the author)

LONGER SPANS IN SITECAST CONCRETE

The ancient Romans built unreinforced concrete vaults and domes to roof temples, baths, palaces, and basilicas (Figure 10.2). Impressive spans were constructed, including a dome over the Pantheon in Rome, still standing, that approaches 150 feet (45 m) in diameter. Today, the arch, dome, and vault remain favorite devices for spanning long distances in concrete because of concrete's suitability to structural forms that work entirely in compression (Figures 11.41, 11.42). With folding or scalloping of vaulted forms, or with the use of warped geometries such as the hyperbolic paraboloid, the required resistance to buckling can be achieved with a surprisingly thin layer of concrete, often proportionally thinner than the shell of an egg (Figures 11.43*a*, 11.43*b*).

Improved beams and trusses are possible in concrete, including posttensioned beams and girders, and reinforced deep girders analogous to steel plate girders and rigid frames.

Concrete trusses and space frames are not common but are built from time to time; by definition a truss includes strong tensile forces as well as compressive forces and is heavily dependent on steel reinforcing or prestressing.

Barrel shells and *folded plates* (Figures 11.42, 11.43*c*, 11.44) derive their stiffness from the folding or scalloping of a thin concrete plate to increase its rigidity without adding material. Each of these forms depends on reinforcing or posttensioning to resist the tensile forces involved.

FIGURE 11.41
The same wooden centering was used four times to form this concrete arch bridge. (Courtesy of Gang-Nail Systems, Inc.)

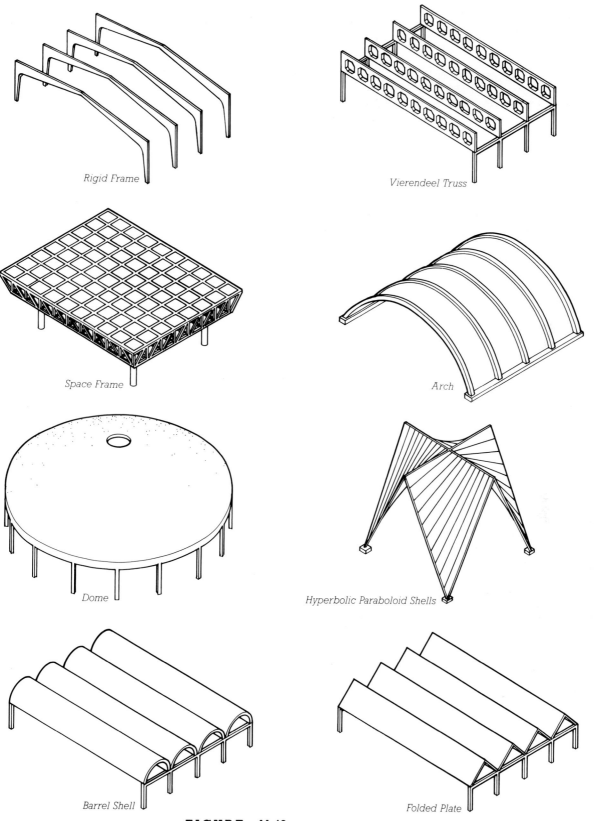

Rigid Frame

Vierendeel Truss

Space Frame

Arch

Dome

Hyperbolic Paraboloid Shells

Barrel Shell

Folded Plate

FIGURE 11.42

Examples of eight types of longer-span structures in concrete. Each is a special case of an infinite variety of forms. All can be sitecast, but the rigid frame, space frame, and Vierendeel truss are more likely to be precast for most applications.

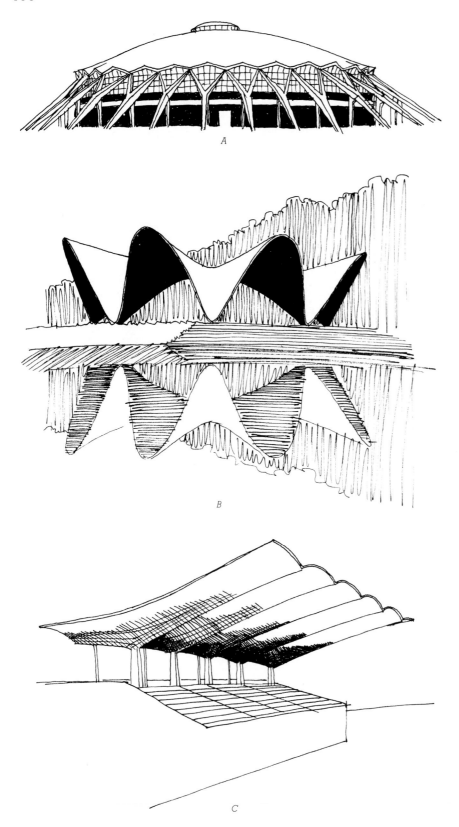

A

B

C

FIGURE 11.43
Three concrete shell structures by twentieth-century masters of concrete engineering: (a) A domed sports arena by Pier Luigi Nervi. (b) A lakeside restaurant of hyperbolic paraboloid shells by Felix Candela. (c) A racetrack grandstand roofed with cantilevered concrete barrel shells, by Eduardo Torroja. (Drawings by the author)

FIGURE 11.44
*Flying formwork is removed from a bay
of a folded plate concrete roof for an air
terminal.* (Architects: Thorshov and
Cerny. Photo courtesy of American
Plywood Association)

Basically there are two approaches to the problem of producing a good surface finish on concrete. One is to remove the cement that is the cause of the blemishes and expose the aggregate. The other is to superimpose a pattern or profile that draws attention from the blemishes.

Henry Cowan, *Science and Building: Structural and Environmental Design in the Nineteenth and Twentieth Centuries,* New York, John Wiley and Sons, 1978, p. 283.

Sitecast concrete and the building codes

Concrete structures are inherently fire resistant. When fire attacks concrete, the water of hydration is gradually driven out and the concrete loses its strength, but this deterioration is slow because considerable heat is needed to raise the temperature of the mass of concrete to the point where dehy-dration begins, and a large additional quantity of heat is required to vaporize the water. The steel reinforcing bars or prestressing strands are buried beneath a cover of concrete that affords them protection for an extended period of time. Except under unusual circumstances, such as a prolonged fire fueled by stored petroleum products, concrete structures usually survive fires with only cosmetic damage and are repaired with relative ease. Thus concrete structures with adequate cover over reinforcing are classified as Type 1 buildings in the tables in Figures 1.1 and 1.2. This gives the designer freedom to build in concrete with little thought to height and area restrictions, although the usual directives on exitways and other life safety features still apply. And many concrete buildings are provided with sprinkler systems, not so much to protect the building as to protect against the loss of its contents and its inhabitants in case of fire.

FIGURE 11.45
An early example of high-quality sitecast concrete surfaces: Kips Bay Plaza Apartments, New York, 1962.
(Architects: I. M. Pei and Partners. Photo by George Cserna)

Sitecast concrete buildings have rigid joints and in many cases need no additional structural elements to achieve the necessary resistance to wind and seismic loads. More restrictive seismic design provisions in the building codes have, however, increased the amount of attention paid by structural engineers to column ties and beam stirrups, particularly in the zones where beams and columns meet, to be sure that vertical bars in columns and horizontal bars in beams are adequately restrained against the unusually strong forces that can occur in these zones under seismic loadings.

THE UNIQUENESS OF SITECAST CONCRETE

Concrete is a shapeless material to be given form by the designer. For economy, the designer can adopt a standard system of concrete framing. For excitement, one can invent new shapes and textures, a route taken by many of the leading practitioners of architecture in this century. Some have pursued its sculptural possibilities; others its surface patterns and textures, still others its structural logic. From each of these routes have come masterpieces—LeCorbusier's chapel at Ronchamp (Figure 11.48), Wright's Unity Temple (Figure 11.1), and the elegant structures of Maillart, Torroja, Candela, and Nervi, examples of which

FIGURE 11.46
The plastered surfaces of Frank Lloyd Wright's Guggenheim Museum (1943–1956) cover a helical ramp of cast-in-place concrete. (Photo by Wayne Andrews)

are sketched in Figure 11.43. Many of these masterpieces, especially from the latter group of designers, were also constructed with impressive economy. Sitecast concrete can do almost anything, be almost anything, at almost any scale, and in any type of building. It is the most potent of architectural materials, and therefore the material both of spectacular architectural achievements and of dismal architectural failures. A material so malleable demands skill and restraint from those who would build with it, and a material so commonplace requires imagination if it is to rise above the mundane.

Designers have learned to express the internal composition of concrete by exposing its aggregates at the surface, or to show the beauties of the

FIGURE   11.47
A sitecast concrete house in Lincoln, Massachusetts. (Architects: Mary Otis Stevens and Thomas F. McNulty)

FIGURE   11.48
LeCorbusier's most sculptural building in his favorite material, concrete: The chapel of Notre Dame du Haut at Ronchamp, France (1950–1955). (Drawing by the author)

formwork in which it was cast, by leaving the marks of the ties and the textures of the form boards. But they have yet to discover how to reveal in the finished structure the lovely and complex geometries of the steel bars which are half the structural equation that makes concrete buildings stand up.

C.S.I./C.S.C. Masterformat Section Numbers for Sitecast Concrete Framing Systems	
03300	CAST-IN-PLACE CONCRETE
03310	Structural Concrete
03345	Concrete Finishing
03350	Special Concrete Finishes
03365	Post-Tensioned Concrete
03370	Concrete Curing
03430	Structural Precast Concrete (Site Cast) Lift-Slab Concrete
03470	Tilt-Up Precast Concrete (Site Cast)

FIGURE 11.49
The TWA Terminal at John F. Kennedy Airport, New York, 1956–1962.
(Architect: Eero Saarinen. Photo by Wayne Andrews)

SELECTED REFERENCES

1. American Concrete Institute. *Building Code Requirements for Reinforced Concrete* (ACI Standard 318-77). Detroit, 1977. (Address for ordering: Box 19150, Redford Station, Detroit, MI 48219.)

This booklet establishes the basis for the engineering design and construction of reinforced concrete structures in the United States.

2. *CRSI Handbook*, 5th Edition. Schaumburg, Illinois, Concrete Reinforcing Steel Institute, 1982.

Structural engineers working in concrete use this handbook, which is based on the ACI Code (reference 1) as their major reference. It contains examples of engineering calculation methods, and hundreds of pages of tables of standard designs for reinforced concrete structural elements. (Address for ordering: 933 North Plum Grove Road, Schaumburg, IL 60195.)

3. Parker, Harry, and Harold D. Hauf. *Simplified Design of Reinforced Concrete*, 4th Edition. New York, John Wiley and Sons, 1976.

This is an excellent introduction to the calculation of reinforced concrete columns, beams, and slabs, with a brief summary of calculation procedures for prestressed beams.

4. Concrete Reinforcing Steel Institute. *Placing Reinforcing Bars: Recommended Practices*, 4th Edition. Schaumburg, Illinois, 1983.

Written as a handbook for those engaged in the business of fabricating and placing reinforcing steel, this small volume is clearly written and beautifully illustrated with diagrams and photographs of reinforcing for all the common concrete framing systems. (Address for ordering: see reference 2.)

KEY TERMS

slab on grade
welded wire fabric
control joints
strike off or straightedge
floating
float
troweling
steel trowel
rotary power trowel
broom finish
curing compound
key
dowels
form ties
one-way solid slab

shoring
shores
reshoring
one-way concrete joist system (ribbed slab)
pans
distribution rib
joist-band
two-way flat slab system
mushroom capital
drop panel
column strip
middle strip
two-way flat plate system
waffle slab

two-way concrete joist system
domes
heads
tendons
set
lift-slab
flying formwork
slip form
tilt-up
shotcrete
exposed aggregate
barrel shell
folded plate

REVIEW QUESTIONS

1. Distinguish one-way concrete framing systems from two-way systems. Are steel and wood framing systems one-way or two-way? Is one-way more efficient structurally than two-way?

2. List the common one-way and two-way concrete framing systems and indicate the possibilities and limitations of each.

3. List from memory the steps in finishing a concrete slab. Why can't the surface be finished in one operation, instead of waiting for hours before final finishing?

4. Why posttension a concrete structure rather than merely reinforce it?

E X E R C I S E S

1. Propose a suitable reinforced concrete framing system for each of the following buildings and determine an approximate thickness for each:

    a. An apartment building with a column spacing of about 16 feet (5 m) in each direction.

    b. A newsprint warehouse, column spacing 20' × 22' (6 m × 6.6 m).

    c. An elementary school, column spacing 24' × 32' (7.3 m × 9.75 m).

    d. A museum, column spacing 36' × 36' (11 m × 11 m).

    e. A hotel where overall building height must be minimized.

2. Look at several sitecast concrete buildings. Determine the type of framing system used in each and explain why you think it was selected.

3. Observe a concrete building under construction. What is its framing system? Why? What types of forms are used for its columns, beams, and slabs? How is the concrete mixed? How is it raised into position? How is it compacted in the forms? How is it cured? Are samples taken for testing? How soon after pouring are the forms stripped? Are the forms reused? Keep a diary of your observations over a period of a month or more.

PRECAST CONCRETE FRAMING SYSTEMS

A precast concrete hollow-core slab is lifted from the casting bed where it was manufactured, using suction devices. *(Courtesy of The Flexicore Co., Inc.)*

Structural *precast concrete* elements—slabs, beams, girders, columns, and wall panels—are cast and cured in industrial plants, then transported to the construction job site and erected as rigid components. Precasting offers many potential advantages over sitecasting of concrete: The production of precast elements is carried out conveniently at ground level. The mixing and pouring operations are often highly mechanized and frequently, in difficult climates, they are carried out under shelter. Control of the quality of materials and workmanship is generally better than on the construction job site. The concrete is cast in *beds*, permanent forms of steel, concrete, glass-fiber reinforced plastic, or varnished wood, whose excellent surface properties are mirrored in the high-quality surfaces of the finished precast elements they produce. The beds are reused hundreds or thousands of times before they have to be renewed, so formwork costs per unit of finished concrete are low. The beds are equipped to pretension the precast elements for greater structural efficiency, which translates into longer spans, lesser depths, and lower weights than for comparable reinforced concrete elements. The wet concrete is vibrated mechanically in the beds to achieve maximum density and highest surface quality. Concrete and steel of superior strength are used in precast elements, typically 5000 psi (35 MN/m^2) concrete and 270,000 psi (1860 MN/m^2) prestressing steel.

Precast concrete elements are usually *steam cured*. Steam furnishes heat to accelerate the curing of the concrete, and moisture for full hydration. Steam curing, coupled with use of Type III (High Early Strength) cement, enables a precasting plant to produce a fully cured structural element, from the laying of the prestressing strands to the removal of the finished element from the bed, in 24 hours.

When the elements produced by this expeditious technique are delivered to the construction job site, further advantages are realized: The erection process is similar to that of structural steel, but it is often faster because most precast concrete systems produce a deck as an integral part of the major spanning elements, without the need for placing additional joist or decking components (Figures 12.1, 12.2). Erection is much faster than that of sitecast concrete because there is no formwork to be erected and stripped, and little or no waiting for concrete to cure. And erection of precast structures can take place under some types of adverse weather conditions, such as extremely high or low temperatures, that would not permit the sitecasting of concrete.

When choosing between precast and sitecast concrete, the designer must weigh these potential advantages of precasting against some potential disadvantages. The precast structural elements, although light compared to similar elements of sitecast concrete, are nevertheless heavy and bulky to transport over the roads and hoist into place. This restricts somewhat the size and proportions of most precast elements: They can be rather long, but only as wide as the maximum legal vehicle width of 12 to 14 feet (3.66 to 4.27 m). This restricted width usually precludes utilization of the efficiencies of two-way structural action in precast slabs. The standardized, modular nature of precast components sometimes makes them unsuitable for use on odd-shaped building sites and in specially shaped buildings. Furthermore, the fully three-dimensional sculptural possibilities of sitecast concrete are largely absent in precast.

Although whole walls or rooms are sometimes precast as single units in concrete, this chapter will focus on the standard precast elements commonly mass-produced as structural components. Precast exterior curtain wall panels will be covered separately in Chapter 15.

FIGURE 12.1
A fully precast building frame under construction. A poured concrete topping will cover the hollow-core slabs and the beams to create a smooth floor and tie the precast elements together. (Courtesy of The Flexicore Co., Inc.)

FIGURE 12.2
Workers guide two hollow-core slabs, lowered by a crane in wire rope slings, onto the precast concrete beams that will support them. (Courtesy of The Flexicore Co., Inc.)

PRECAST, PRESTRESSED CONCRETE STRUCTURAL ELEMENTS

Precast Concrete Slabs

The most fully standardized precast elements are those used for making floor and roof slabs (Figure 12.3). These may be supported by bearing walls of precast concrete or masonry, or by frames of steel, sitecast concrete, or precast concrete. Four kinds of precast slab elements are commonly produced: For short spans and minimum slab depth, *solid slabs* are appropriate. For longer spans, deeper elements must be used and precast solid slabs, like their sitecast counterparts, become inefficient because they contain too much deadweight of nonworking concrete. In *hollow-core* slabs, precast elements suitable for intermediate spans, internal longitudinal voids replace much of the nonworking concrete. For the longest spans, still deeper elements are required, and *double tees* and *single tees* eliminate still more nonworking concrete.

For most applications, precast slab elements of any of the four types are produced with a rough top surface. After the elements have been erected, a concrete *topping* is poured over them and finished to a smooth surface. The topping, usually 2 inches (50 mm) in thickness, bonds during curing to the rough top of the precast elements and becomes a working part of their structural action. The topping also helps the precast elements to act as a structural unit rather than as individual planks in resisting concentrated loads and diaphragm loads, and conceals the slight differences in camber that are likely to be found in prestressed components. Structural continuity across a number of spans can be achieved by casting reinforcing bars into the topping over the supporting beams or walls. Underfloor electrical conduits can also be embedded in the topping. (Smooth-surface precast slabs are sometimes used without topping, as discussed later in this chapter.)

Either normal-density concrete or structural lightweight concrete may be selected for any of the precast slab elements. The lightweight concrete, approximately 20 percent less dense than normal concrete, reduces the load on the frame and foundations of a building, but is more expensive than normal-density concrete.

As can be seen from Figure 12.4, there is considerable overlapping of the spanning capabilities of the different kinds of precast slab elements, allowing the designer some latitude in choosing which to use in a particular situation. Solid slabs and hollow-core slabs save on overall building height in multistory structures, and their smooth undersides can be painted and used as finish ceilings in many applications. For longer spans, double tees are generally preferred to the older single-tee design because they do not need to be supported against tipping during erection.

Precast Concrete Beams, Girders, and Columns

Precast concrete beams and girders are made in several standard shapes (Figure 12.5). The projecting ledgers on L-shaped beams and inverted tees provide direct support for precast slab elements to conserve headroom in a building as compared to rectangular beams with slab elements resting on top. AASHTO (American Association of State Highway and Transportation Officials) girders were designed originally as efficient shapes for bridge structures, but they are used in buildings as well. Precast columns are usually square or rectangular in section and may be prestressed or simply reinforced.

Precast Concrete Wall Panels

Precast solid slabs are commonly used as loadbearing wall panels in many types of lowrise and highrise buildings. The prestressing strands are located symmetrically in wall panels to

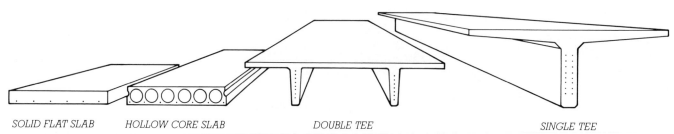

SOLID FLAT SLAB   HOLLOW CORE SLAB   DOUBLE TEE   SINGLE TEE

FIGURE 12.3
The four major types of precast concrete slab elements. Hollow-core slabs are produced by different companies in a variety of cross-sectional patterns, using several different processes. Single tees are less commonly used than double tees because they need temporary support against tipping until they are permanently fastened in place.

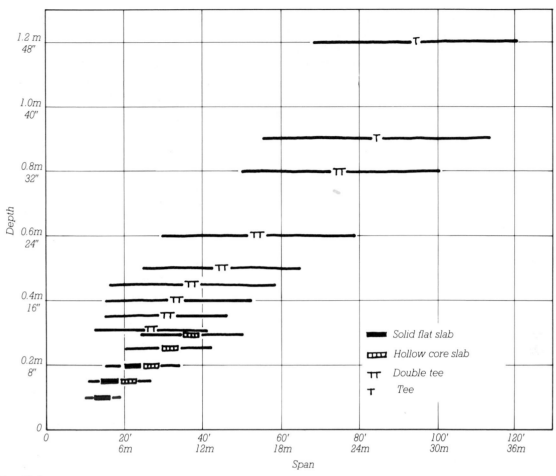

FIGURE 12.4

Feasible span ranges for various precast, prestressed concrete slab elements. Standard depths can be read from the vertical axis; some precasting plants also produce a 60-inch (1524-mm) tee that can span up to 153 feet (46.6 m). If floor or roof loadings are light, read upward from the required span to the right-hand region of the span-range line for a given element, and if loadings are heavy, read to the left-hand region. For example, given a 55-foot (16.5-m) span and a light loading, a 20- or 24-inch (508- or 610-mm) double tee might be appropriate. For a heavily loaded structure, a better choice might be a 32-inch (813-mm) double tee. This table is valid for preliminary design purposes only; the final depth, number, and type of strands and strand arrangement must be determined through detailed structural analysis.

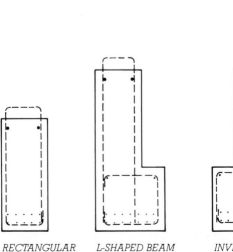
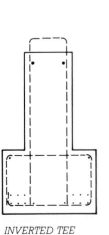
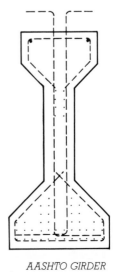

RECTANGULAR BEAM   L-SHAPED BEAM   INVERTED TEE BEAM   AASHTO GIRDER

FIGURE 12.5

Standard precast concrete beam and girder shapes. The larger dots represent mild steel reinforcing, and the smaller dots represent prestressing strands. The broken lines show mild steel stirrups. Stirrups usually project above the top of the beam, as shown, to bond to the topping for composite structural action.

strengthen the panels against buckling and to eliminate camber. Rigid foam insulation can be cast into wall panels for thermal insulation, with suitable wire shear ties between the inner and outer wythes of concrete (Figures 12.13, 12.14).

Assembly Concepts for Precast Concrete Buildings

Figure 12.6 shows a building whose precast slab elements (double tees in this example) are supported on a skeleton frame of L-shaped precast girders and precast columns. The slab elements in Figure 12.7 are supported on precast loadbearing wall panels. Figure 12.8 illustrates a building whose slabs are supported on a combination of wall panels and girders. These three

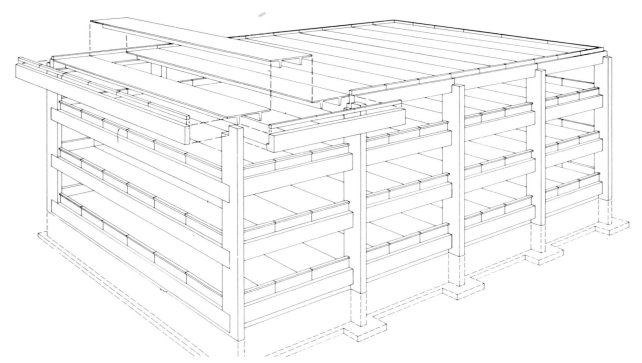

FIGURE 12.6
Double-tee slab elements supported on a frame of precast columns and L-shaped girders. (Courtesy of Prestressed Concrete Institute)

fundamental ways of supporting precast slabs—on a precast concrete skeleton, on precast loadbearing wall panels, and on a combination of the two—occur in endless variations in buildings. The skeleton may be one bay deep or many bays; the loadbearing walls are often of reinforced masonry, or of any of a variety of configurations of precast concrete; the slab elements may be solid, hollow-core, or double tee, topped or untopped.

One of the principal virtues of precast concrete as a structural material is that it is locally manufactured to order and is easily customized to an individual building design, usually at minimal additional cost.

FIGURE 12.7
Hollow-core slab elements supported on precast concrete loadbearing wall panels. (Courtesy of Prestressed Concrete Institute)

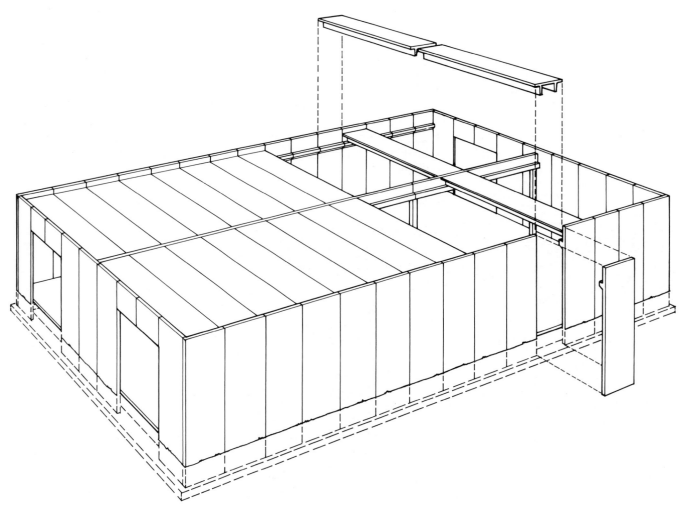

FIGURE 12.8
*Double-tee slab elements supported on
a perimeter of precast concrete
loadbearing wall panels, and an interior
structure of precast columns and
girders.* (Courtesy of Prestressed
Concrete Institute)

THE MANUFACTURE OF PRECAST CONCRETE STRUCTURAL ELEMENTS

Casting Beds

The casting beds for precast concrete elements average 400 feet (125 m) in length, but extend 800 feet (250 m) or more in some plants (Figure 12.9). A cycle of precasting usually begins in the morning, as soon as the elements cast the previous day have been lifted from the beds. High-strength steel reinforcing strands are strung between the abutments at the extreme ends of the bed. After the strands are pretensioned with hydraulic jacks, transverse bulkhead separators may be placed along the bed at the required intervals to divide the individual components from one another. (For solid slabs, cored slabs, and wall panels, the bulkheads are often omitted; the cured slab is simply sawed into the required lengths before it is removed from the bed.)

When pretensioning has been completed, the concrete is placed in the bed, vibrated to eliminate voids, and struck off level. If required, the top surface is finished further with floats and trowels. Live steam or radiant heat is then applied to the concrete to accelerate its curing. Ten to twelve hours after pouring, the concrete has reached a compressive strength of 3500 to 4000 psi (24 to 28 MN/m^2) and has bonded to the steel strands. The next morning, after the strength of the concrete has been verified by breaking test cylinders in the laboratory, the exposed ends of the strands are cut, prestressing the concrete. Asymmetrically prestressed slab and beam elements immediately arch up from the casting bed as the prestressing force is released into them. When the elements have been separated from one another, they are hoisted off the bed and stockpiled ready for shipment, and a new cycle of casting begins.

A

FIGURE 12.9
Manufacturing double-tee slabs. (a) *A worker inserts weld plates in the casting bed prior to pouring the concrete. The prestressing strands and wire fabric shear reinforcement have been installed in the stems of the double tees, and a portion of the welded wire fabric for the top slab can be seen in the foreground.* (b) *The top surface of the concrete is finished by machine.* (c) *The next morning, following steam curing, the prestressing strands are cut between bulkheads with an oxyacetylene cutting torch. The welded wire fabric is exposed at the ends of the elements because they will be used as untopped slabs (see Figure 12.23).* (d) *Using lifting loops cast into the ends, the slabs are lifted from the bed and stockpiled outside. The dapped stems will rest on inverted tee and L-shaped girders; the notched corner will fit around a column.* (e) *Loading the double tees for trucking.* (Photos by Alvin Ericson)

...a project should be of a certain minimum size in order to spread out certain fixed costs such as plant set-up, mold costs, and erection mobilization costs over a sufficiently broad base to make the use of precast concrete economically viable. Minimum project sizes are those that would generate 10,000 square feet (1000 m²) of architectural precast concrete panels, 15,000 square feet (1500 m²) of prestressed concrete deck members, or 1000 lineal feet (300 m) of standard prestressed or precast concrete components such as girders, columns, or piling.

William R. Phillips and David A. Sheppard: Prestressed Concrete Institute. *Plant Cast Precast and Prestressed Concrete: A Design Guide,* 2nd Edition. Chicago, 1982, p. 3.

FIGURE 12.9b

FIGURE 12.9c

FIGURE 12.9d

FIGURE 12.9e

Prestressing and Reinforcing Steel

Solid slabs, hollow-core slabs, and wall panels are cast around horizontal strands. Tees, double tees, beams, and girders are often cast around depressed or harped strands to give a greater efficiency of structural action (Figures 10.37, 12.12, 12.24).

Ordinary steel reinforcing is also cast into prestressed concrete elements for various purposes. Beams or slabs intended to cantilever beyond their supports are given top reinforcing bars over the cantilever points. Welded wire fabric is used to reinforce the flanges of tees and double tees and for general reinforcing of wall panels. Where stirrups are required in the stems of tees and double tees, either mild steel reinforcing bars or welded wire fabric is used. Additional reinforcing may be installed to strengthen around openings in panels or slabs for pipes, ducts, columns, and hatchways. Weld plates and other metal connecting devices are cast into the elements as required. Projecting steel loops are cast into many types of elements as crane attachments for lifting.

Hollow-Core Slab Production

The longitudinal voids in hollow-core slabs can be formed by any of a number of patented processes. In some processes, extrusion devices squeeze a stiff concrete mix through an extrusion die to produce the voided shape directly. This method has the disadvantage that vertical openings and weld plates cannot easily be cast in; where openings are required in extruded slabs, they must be cut out of the stiff but still wet concrete just after extrusion, or sawed after curing. Another process deposits a bottom layer of wet concrete in the casting bed, then a second layer of concrete with dry crushed stone or lightweight aggregate carefully introduced to form the voids.

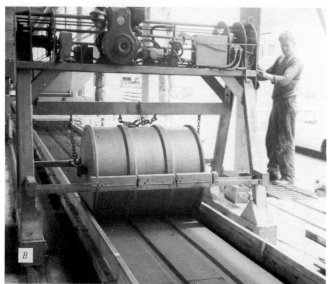

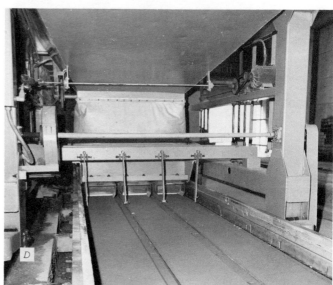

FIGURE 12.10
Steps in the manufacture of Span-Deck, a proprietary hollow-core slab. (a) Depositing a thin bottom slab of rather dry concrete into the casting bed with a traveling hopper. (b) A ridged roller compacts the bottom slab and makes indentations for the keying of the concrete webs. (c) Four prestressing strands are pulled onto the bed; these will be pretensioned with jacks. (d) After pretensioning, an extrusion device travels the length of the bed, forming the webs and top of the slab and filling the cores with dry lightweight aggregate that serves as temporary support for the fresh concrete. (e) The next day, after curing of the slab has been completed, a saw cuts it into the required lengths. Each length is then lifted from the bed and transported by an overhead crane to the aggregate recovery area, where the dry aggregate is poured out and saved for reuse. (f) Finished Span-Deck slabs stockpiled, ready for transportation to the construction site. The wads of paper in the cores are to prevent concrete from filling the cores during the pouring of the topping. (Courtesy of Blakeslee Prestress Inc.)

Special forms may be placed in the bed to make openings as required, and weld plates may be cast in by this process. After curing, the slabs are conveyed to an aggregate recovery area where the dry stone is poured out of the voids and saved for reuse (Figure 12.10). In a third type of process, air-inflated tubes are used to form the voids. In some plants, hollow-core slabs are cast atop one another in stacks rather than in a single layer and are wet cured for seven days rather than steam cured overnight.

Column Production

Precast columns are reinforced or pretensioned as directed by the architect or structural engineer. Columns are often made and shipped in multistory sections with *corbels* to support beams or slabs. Columns with corbels on one side or on two opposing sides are easily cast in flat beds. If corbels are required on three sides, box forms are set atop the upper side of the column as it lies in the bed. For corbels on the fourth side, steel plate inserts are cast into

FIGURE 12.11
Workers install side forms for inverted tee beams in an outdoor casting bed. Mild steel reinforcing is used extensively for stirrups. (Courtesy of Blakeslee Prestress Inc.)

the bottom of the column in the bed, to which reinforcing bars are welded after the column is removed from the bed. The corbels on the fourth side are then cast around the reinforcing bars in a separate operation.

FIGURE 12.12

A precasting bed being readied for the pouring of an AASHTO girder. Side forms for the mold can be seen in the background to the right. The depressed strands are held down in the center of the beam by steel pulleys that will be left in the concrete after pouring. The bed is long enough that several girders are being cast end-to-end, with the depressed strands pulled up and down as required. Mild steel deformed bars are used for stirrups. The projecting tops of the stirrups will bond to the sitecast topping. Vertical twists of prestressing strand near the end of the girder will serve as lifting loops. (Courtesy of Blakeslee Prestress Inc.)

FIGURE 12.13

A tilting table being used for the casting of a foam-insulated concrete wall panel. To the left, a concrete panel face with welded wire fabric reinforcing is being cast. Sheets of rigid foam insulation with wire ties can be seen in the center, and the welded wire fabric for the second panel face is seen at the right, with a pair of vertical bars and a pocket at the bottom being cast in as part of a system of connections. Notice the pipes at the left edge of the table for heating the mold to accelerate curing. (Courtesy of Blakeslee Prestress Inc.)

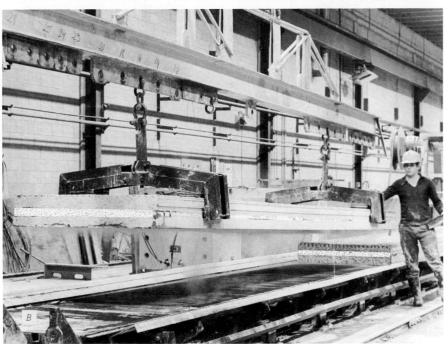

FIGURE 12.14
Manufacturing Corewall, a proprietary system of foam-core precast concrete panels. (a) At the conclusion of the casting process, rollers apply a ribbed texture to the exterior face of the panels. (b) Completed slabs are lifted from the casting bed. (Courtesy of Butler Manufacturing Co.)

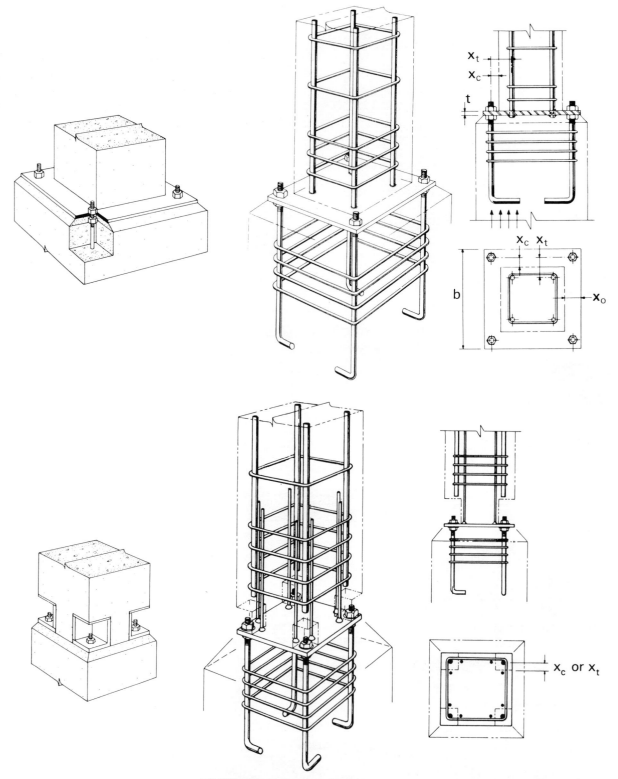

FIGURE 12.15

Some typical base details for precast concrete columns. The open corners of the lower detail are dry packed with grout after the column is aligned and *bolted; this protects the metal parts of the connection from fire and corrosion.* (Courtesy of Prestressed Concrete Institute)

JOINING PRECAST CONCRETE ELEMENTS

Figures 12.15 through 12.29 show some frequently used connection details for precast concrete construction. Bolting, welding, and grouting are all commonly employed in these connections. Connections can be posttensioned to produce continuous beam action at points of support (Figure 12.19). Exposed metal connectors not covered by topping are usually *dry packed* with stiff grout after being joined, to protect them from fire and corrosion.

The simplest joints in precast concrete construction are those that rely upon gravity by placing one element atop another, as is done where slab elements rest on a bearing wall or beam, or where a beam rests on the corbel of a column. *Bearing pads* are usually inserted between the concrete members at bearing points to avoid the con-

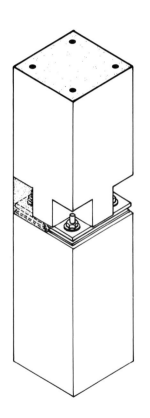

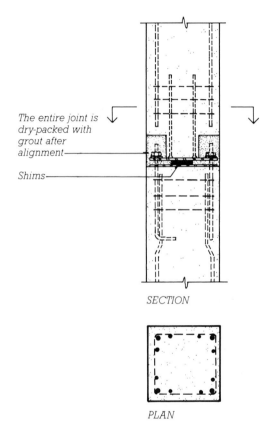

The entire joint is dry-packed with grout after alignment

Shims

SECTION

PLAN

FIGURE 12.16
A bolted column-to-column connection detail.

FIGURE 12.17
Section through a column-to-column connection using a proprietary sleeve that is cast into the column. Projecting reinforcing bars from the lower column section are inserted in the sleeves, where injected grout makes the connection.

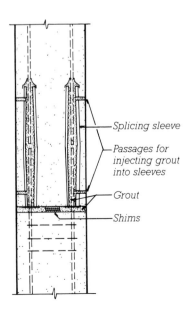

Splicing sleeve

Passages for injecting grout into sleeves

Grout

Shims

crete-to-concrete contact that might create points of high stress and to allow for expansion and contraction in the members. For hollow-core slabs these pads are strips of high-density plastic. Under elements with higher point loadings such as tees and beams, pads of synthetic rubber are used. For seismic and wind loadings, the members in these simple joints must be tied together laterally. Slab elements are often joined over supports by reinforcing bars cast into the topping or the grout keys. Double tees and tees may additionally be connected by the welding together of steel plate inserts that have been precast into the tops of the elements in the plant. The stems of these mem-

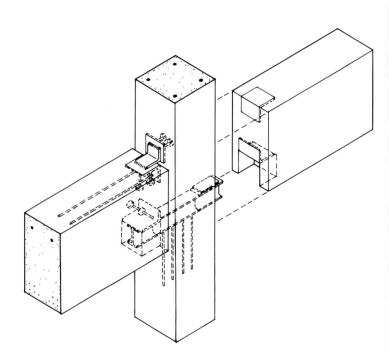

bers are never welded to the supports but are left free to move on their bearing pads (Figures 12.22, 12.23).

Precast slab elements, especially solid slabs, may require temporary shoring at midspan to help support the weight of the topping until it has cured. For construction economy, smooth-topped precast slab elements are sometimes used without topping, especially for roofs, where any unevenness between elements is bridged by the rigid thermal insulation, and also for floors that will be finished with a pad and carpet, and for parking garages. Untopped slabs require special connection details (Figures 12.20, 12.23).

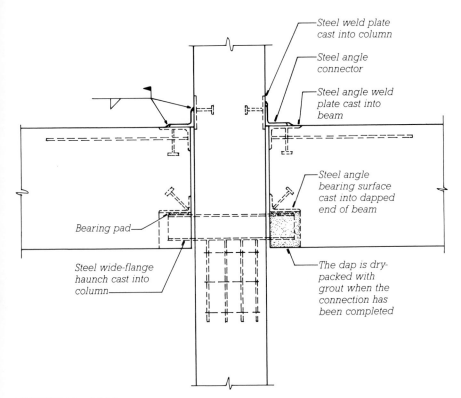

FIGURE   12.18
A beam-column connection in which the dapped beam ends rest on a steel wide-flange haunch.

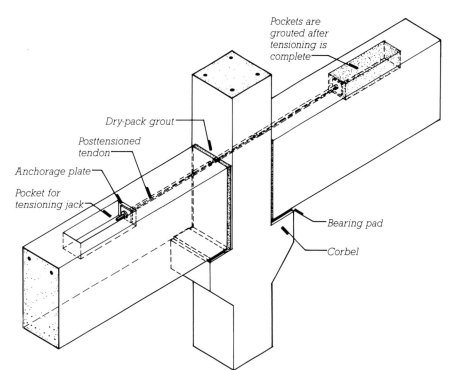

FIGURE   12.19
A posttensioned, structurally continuous beam-column connection.

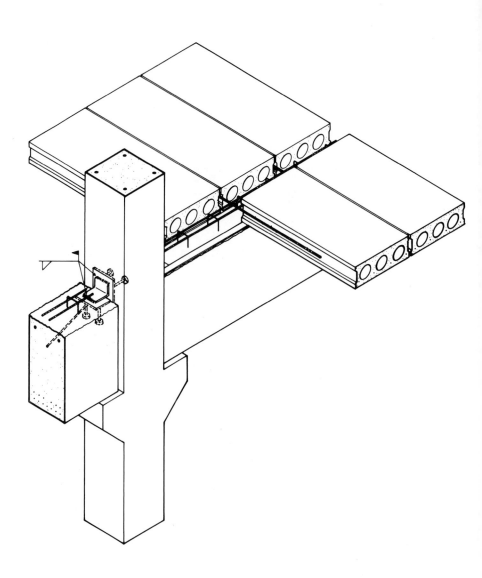

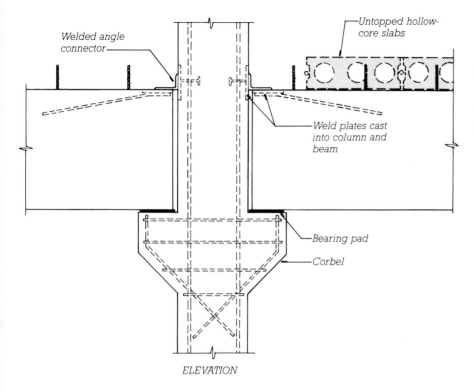

Welded angle connector

Untopped hollow-core slabs

Weld plates cast into column and beam

Bearing pad

Corbel

ELEVATION

FIGURE 12.20
The beams in this system of framing rest on concrete corbels that are cast as part of the column. The hollow-core slabs are detailed for use without topping.

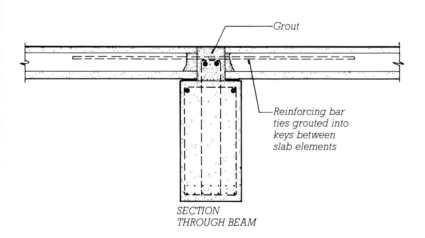

Grout

Reinforcing bar ties grouted into keys between slab elements

SECTION THROUGH BEAM

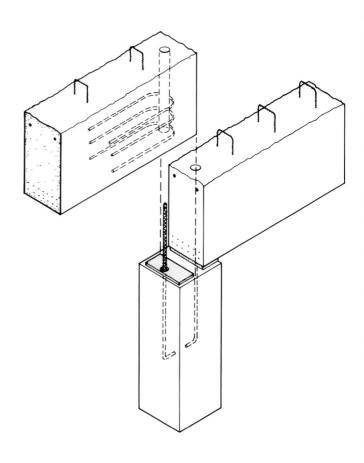

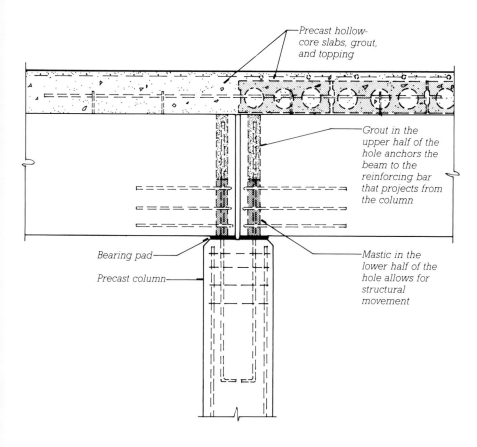

Precast hollow-core slabs, grout, and topping

Grout in the upper half of the hole anchors the beam to the reinforcing bar that projects from the column

Bearing pad

Precast column

Mastic in the lower half of the hole allows for structural movement

FIGURE 12.21
Topped hollow-core roof slabs supported on beams joined to a column with vertical rods. The same beam-column connection can be used for floor beams resting on corbels.

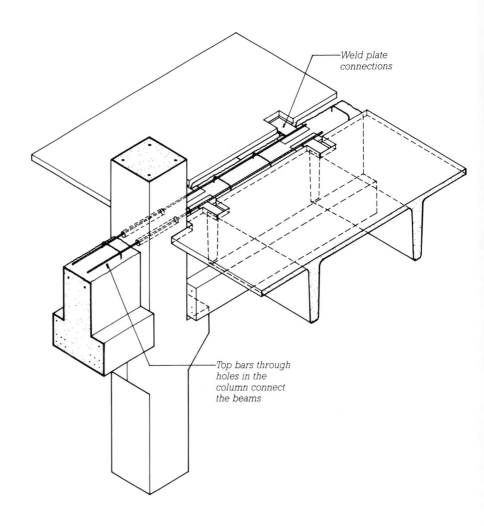

Weld plate connections

Top bars through holes in the column connect the beams

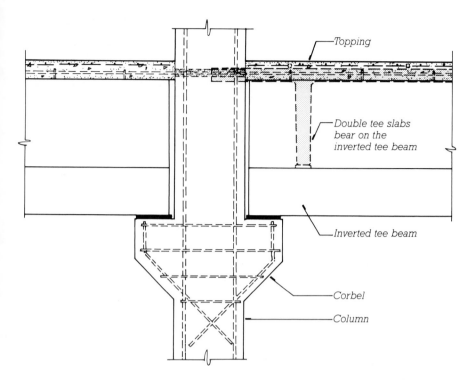

FIGURE 12.22
Topped double-tee floor slabs resting on inverted tee beams.

Topping

Double tee slabs bear on the inverted tee beam

Inverted tee beam

Corbel

Column

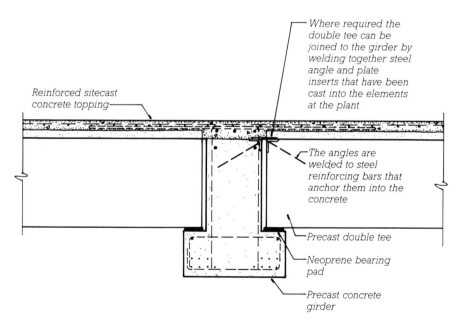

Where required the double tee can be joined to the girder by welding together steel angle and plate inserts that have been cast into the elements at the plant

Reinforced sitecast concrete topping

The angles are welded to steel reinforcing bars that anchor them into the concrete

Precast double tee

Neoprene bearing pad

Precast concrete girder

FIGURE 12.23

A minimum headroom, minimum cost floor system for parking garages using untopped double tees. Refer to Figure 12.9 for photographs of how the ends of the tees are detailed for use in this system.

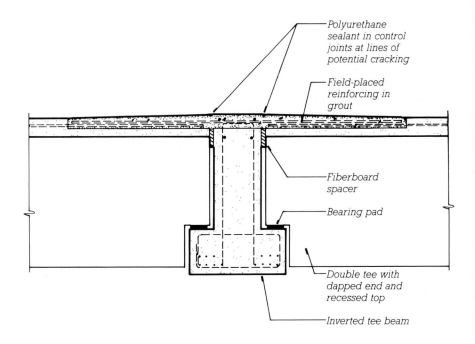

Polyurethane sealant in control joints at lines of potential cracking

Field-placed reinforcing in grout

Fiberboard spacer

Bearing pad

Double tee with dapped end and recessed top

Inverted tee beam

FIGURE 12.24

A cross section through a topped double-tee slab. The prestressing strands are harped in the stems of the double tee.

A sitecast concrete topping with welded wire reinforcing fabric bonds to the rough top of the double tees to form a composite structural unit

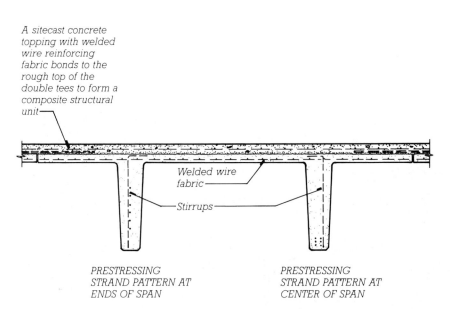

Welded wire fabric

Stirrups

PRESTRESSING STRAND PATTERN AT ENDS OF SPAN

PRESTRESSING STRAND PATTERN AT CENTER OF SPAN

FIGURE 12.25

A cross section through a topped hollow-core slab.

A sitecast concrete topping with reinforcing fabric bonds to the rough tops of the precast slabs to form a composite structural unit

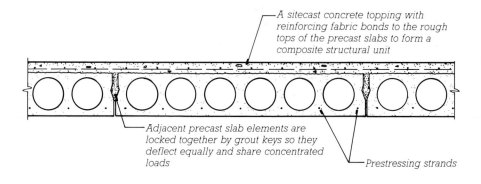

Adjacent precast slab elements are locked together by grout keys so they deflect equally and share concentrated loads

Prestressing strands

To save expense and dead weight in roof structures, hollow-core slabs are sometimes installed a distance apart, with the intervening space spanned by corrugated steel decking supported on the edges of the slabs. A concrete topping reinforced with welded wire fabric ties the entire construction together (Figure 12.26).

Posttensioning can be used for combining large precast elements into even larger ones on the site, as in the case of very long, deep girders made up of precast sections (Figure 12.27), or precast shear walls in multistory buildings. In either case, ducts are placed accurately in the sections before casting, so they will mate perfectly when the sections are assembled on the site. After assembly, tendons are run through the ducts, horizontally in the case of girders or vertically in the case of shear walls, and tensioned with portable hydraulic jacks, then grouted if required.

Joint design is the area in which precast concrete technology is developing most rapidly. Many new joining systems are patented each year, and as grouts and adhesives develop further there will be further simplifications and improvements of many kinds in precast concrete framing details.

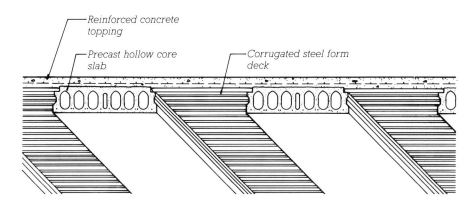

FIGURE 12.26
A low-cost roof system using steel form deck in combination with hollow-core slabs.

FIGURE 12.27
The Linn Cove Viaduct in Linnville, North Carolina, was built of precast sections that were posttensioned together into a continuous box girder deck as they were placed. A section is being placed at the extreme right; notice the absence of shoring as the girder grows toward its next supporting pier, which allows construction to proceed with almost no disruption of the natural environment below. The maximum clear span is 180 feet (55 m). (Engineer: Figg and Muller Engineers, Inc. Photo courtesy of Prestressed Concrete Institute)

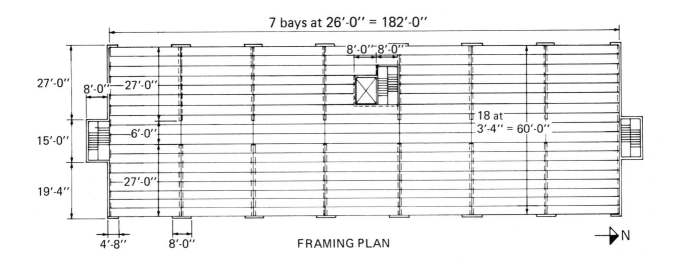

7 bays at 26'-0'' = 182'-0''

27'-0''

8'-0''   27'-0''

8'-0'' 8'-0''

18 at
3'-4'' = 60'-0''

15'-0''   6'-0''

19'-4''   27'-0''

4'-8''   8'-0''

FRAMING PLAN

N

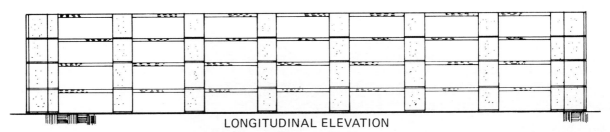

LONGITUDINAL ELEVATION

FIGURE 12.28

*A framing plan and elevation of a
simple four-story building made of
loadbearing precast concrete wall
panels and hollow-core slab elements.*
(Courtesy of Prestressed Concrete
Institute)

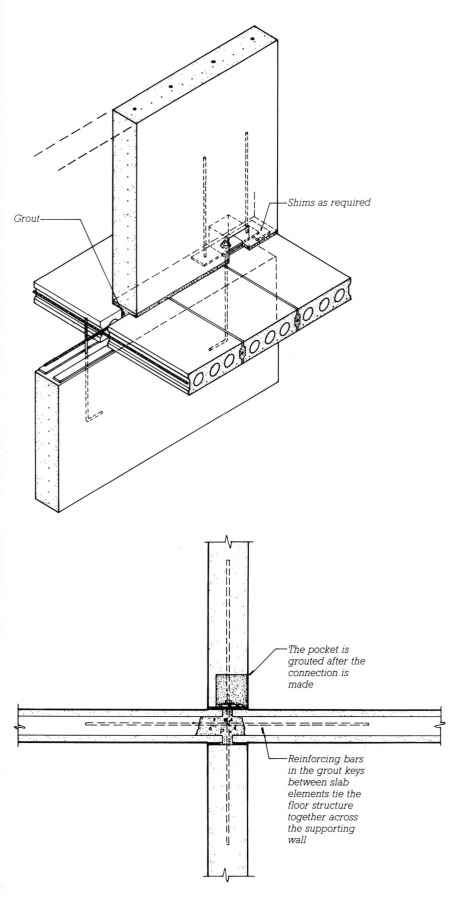

Grout

Shims as required

FIGURE 12.29

Typical details for slab-wall junctions in the structure shown in Figure 12.28. The reinforcing in the wall panels and the prestressing steel in the slabs is omitted for the sake of clarity.

The pocket is grouted after the connection is made

Reinforcing bars in the grout keys between slab elements tie the floor structure together across the supporting wall

FIGURE 12.30
Erecting a building of precast loadbearing wall panels. (Courtesy of Blakeslee Prestress Inc.)

FIGURE 12.31
A closer view of another building using precast loadbearing wall panels. The rectangular pockets at the lower edges of the wall panels are for bolted connections of the type shown in Figure 12.29. Steel pipe braces are used to support the panels until all connections have been made. (Courtesy of Blakeslee Prestress Inc.)

THE CONSTRUCTION PROCESS

The construction process for precast concrete framing is directly parallel to that for steel framing. The architect or structural engineer forwards the structural drawings for the building to the precasting plant, where engineers and drafters prepare shop drawings that cover the details of the individual elements and how they are to be connected. These drawings are reviewed by the engineer and architect for conformance with their design intentions and corrected as necessary. Then the production of the precast components proceeds, beginning with construction of any special molds that are required and fabrication of reinforcing cages, then continuing through cycles of casting, curing, and stockpiling as previously described. The finished elements, marked to designate their final positions in the building, are transported to the construction site as needed and put in place by crane in accordance with erection drawings prepared by the precasting plant.

FIGURE 12.33
A crane hoists a one-piece precast stair for a highrise building of loadbearing precast concrete wall panels. (Courtesy of Blakeslee Prestress Inc.)

FIGURE 12.32
Exterior loadbearing wall panels are often made of specially colored concrete, with textured surfaces cast in.
(Courtesy of Blakeslee Prestress Inc.)

FIGURE 12.34
A precast column section is lifted from the ground by a crane. (Courtesy of Blakeslee Prestress Inc.)

FIGURE 12.35
Erecting an upper-story column, using a connection of the general type illustrated in Figure 12.16. (Courtesy of Blakeslee Prestress Inc.)

FIGURE 12.36
Placing a hollow-core slab deck on a precast frame. (Courtesy of Blakeslee Prestress Inc.)

Precast concrete and the building codes

Precast concrete building frames and bearing wall panels can be designed to achieve any desired degree of fire resistance, depending on the amount of cover and topping specified. The typical building code table shown in Figure 1.2 indicates the fireresistance ratings in hours required for components of Type 1 and 2 construction. When the building type has been determined by the architect or engineer, the precaster can assist in determining how the necessary degrees of fire resistance can be achieved in each component of the building. Slab elements are readily available in 1 and 2 hour fireresistance ratings, and beams and columns in ratings ranging from 1 to 4 hours.

The uniqueness of precast concrete

Precast, prestressed concrete structural elements are crisp, slender in relation to span, precise, repetitive, and highly finished. They combine the rapid all-weather erection of structural steel framing with the self-fireproofing of sitecast concrete framing to offer singular functional qualities. Aesthetically, architects have so far been most successful with precast concrete in longer-span building types, especially parking structures, warehouses, and industrial plants, where its awesome structural potential and efficient serial production of identical elements are fully utilized. Solid and hollow-core precast slab components are functional and economical at shorter spans, as in schools, apartment buildings, hotels, and hospitals, but have little direct effect on the appearance of the building. It is well to remember, however, that precast concrete did not come into widespread use in the United States until after World War II, which makes

FIGURE 12.37

The erection of an all-precast parking garage in New Haven, Connecticut. (a) Corbelled concrete columns are installed with the aid of two crawler-mounted cranes. (b) A specially designed column tree is brought to the site by truck. (c, d, e) The column tree is lifted from the truck and guided carefully to its location in the building. (f) Single tees, dapped at the outside end, are used as slab elements. (g) The finished garage shows the results that can be achieved by a close working relationship among the architect, the structural engineer, and the precaster. The architect for this project is Carlton Granbury Associates. The precaster is Blakeslee Prestress Inc. (Photos by Charles J. Miller, Jr. Courtesy of Blakeslee Prestress Inc.)

(continued)

it the newest and least developed of the major framing materials for buildings. It is reasonable to expect that as its rapid evolution continues, precast concrete will lead the designers of buildings to create structural forms that are fresh and uniquely expressive of this sleek, sinewy new material of construction.

C.S.I./C.S.C. Masterformat Section Numbers for Precast Concrete Framing Systems	
03400	**PRECAST CONCRETE**
03410	Structural Precast Concrete (Plant Cast)
	Precast Concrete Hollow-Core Planks
	Precast Concrete Slabs
	Structural Precast Pretensioned Concrete

G

FIGURE 12.38
An office building in Texas shows a carefully detailed frame of precast concrete. (Architects: Omniplan Architects. Photo courtesy of Prestressed Concrete Institute)

FIGURE 12.39
Precast concrete girders span the roof of a paper mill in British Columbia. (Engineer: Swan Wooster Engineering Co., Ltd. Photo courtesy of Prestressed Concrete Institute)

FIGURE 12.40
The precast walls and slabs of these condominium apartments were erected during the winter in the New Mexico mountains. (Architect: Antoine Predock. Photo courtesy of Prestressed Concrete Institute)

FIGURE 12.41
Highly customized precast concrete framing in a courthouse building by architect Eduardo Catalano. (Photo by Gordon H. Schenck, Jr. Courtesy of the architect)

S E L E C T E D   R E F E R E N C E S

1.   Prestressed Concrete Institute. *PCI Design Handbook*, 2nd Edition. Chicago, 1980.

This is the major reference handbook for those engaged in designing precast, prestressed concrete buildings. It includes basic building assembly concepts, load tables for standard precast elements, engineering design methods, and a few suggested connection details. (Address for ordering: 201 North Wells Street, Chicago, IL 60606.)

2.   Martin, L. D., and W. J. Korkosz: Prestressed Concrete Institute. *Connections for Precast Prestressed Concrete Buildings, Including Earthquake Resistance*. Chicago, 1982.

Much of this book is concerned with engineering calculation of connections in precast concrete buildings. The fourth and last section of the book contains 109 pages of beautifully clear illustrations of precast connections that will be of use to anyone designing a building in this material. (Address for ordering: see reference 1.)

3.   Phillips, William R., and David A. Sheppard: Prestressed Concrete Institute. *Plant Cast Precast and Prestressed Concrete: A Design Guide*, 2nd Edition. Chicago, 1982.

A comprehensive text on precast, prestressed concrete, this book covers production, transportation, erection, and engineering design procedures. (Address for ordering: see reference 1.)

K E Y   T E R M S

precast concrete
casting bed
steam curing
solid slab
hollow-core slab
double tee

(single) tee
topping
corbel
L-shaped beam
inverted tee
AASHTO girder

weld plate
bearing pad
grout
dry pack grout
dap

R E V I E W   Q U E S T I O N S

1.   In what circumstances might a designer choose a precast concrete framing system over a sitecast system? In what circumstances might a sitecast system be favored?

2.   Why are precast concrete structural elements usually cured with steam?

3.   Explain several methods of producing hollow-core slabs.

4.   Diagram from memory several different ways of connecting precast concrete beams to columns.

5.   Diagram from memory a method of connecting a pair of untopped double-tee slabs to an inverted tee beam. Then work out a similar way of connecting a double tee to an L-shaped beam, as would occur around the perimeter of the same building.

E X E R C I S E S

1.   Design a simple two-story rectangular warehouse, 90' × 180' (27 m × 54 m), using precast concrete for at least the floor and roof structure, and perhaps for the walls as well. Use the preliminary structural information given in Figure 12.4 to help determine the column spacing, the types of elements to use, and the depths of the elements. Draw a framing plan and typical connections for the building.

2.   Locate a concrete precasting plant in your area and arrange a visit to view the production process. If possible, arrive at the plant early in the morning when the strands are being cut and the elements are being lifted from the molds.

3.   Learn from the management of the precasting plant where there is a precast building being erected, and then visit the building site. Try to trace a typical precast concrete structural element from raw materials through precasting, transporting, and erecting. Are there ways in which this process could be made more efficient? Sketch a few of the typical connections being used on the project.

## CHAPTER THIRTEEN

ROOFING

Flat Roofs

Roof Decks

Thermal Insulation and Vapor Retarder

The Flat Roof Membrane

Edge and Drainage Details for Flat Roofs

Pitched Roofs

Shingles

Sheet Metal Roofing

Roofing and the Building Codes

A copper finial crowns this conical slate roof.
(Photo by the author)

A building's roof is its first line of defense against the weather. The roof protects the interior of the building from rain, snow, and sun. The roof helps to insulate the building from extremes of heat and cold and to control the accompanying problems with condensation of water vapor. And like any front-line defender, it must itself take the brunt of the attack: A roof is subject to the most intense solar radiation of any part of a building. At midday the sun broils a roof with radiated heat and ultraviolet light. On clear nights a roof radiates heat to the blackness of space and becomes colder than the surrounding air. From noon to midnight the surface temperature of a roof can vary from near boiling to below freezing. In cold climates, snow and ice cover a roof after winter storms, and cycles of freezing and thawing gnaw at the materials of the roof. A roof is vital to the sheltering function of a building and is singularly vulnerable to the destructive forces of nature.

Roofs can be covered with many different materials. These can be organized conveniently into two groups: those that work on sloping (*pitched*) roofs and those that work on flat or nearly flat roofs. The distinction is important: A pitched roof drains itself quickly of water, giving wind and gravity little opportunity to push or pull water through the roofing material. Pitched roofs can, therefore, be covered with roofing materials that are fabricated and applied in small, overlapping units—*shingles* of wood, slate, or artificial composition; tiles of fired clay or concrete; or even bundles of reeds, leaves, or grasses (Figures 13.1, 13.2). There are several advantages to these materials: Many of them are inexpensive. The small, individual units are easy to handle and install. Repair of localized damage to the roof is easy. The effects of thermal expansion and contraction, and of movements in the structure that supports the roof, are minimized by the ability of the small roofing units to move with respect to one another. Water vapor vents itself easily from the interior of the building through the loose joints in the roofing material. And a pitched roof of well-chosen materials skillfully installed can be a delight to the eye.

Flat roofs have none of these advantages. Water drains relatively slowly from their surfaces and small errors in design or construction can cause them to trap puddles of standing water. Slight structural movements can tear the single-piece membrane that keeps the water out of the building. Water vapor pressure from within the building can blister and rupture the membrane. But flat roofs also have overriding advantages: A flat roof can cover a building of any horizontal dimension, while a pitched roof becomes uneconomically tall when used on a very broad building. A flat-roofed building has a much simpler, and often cheaper, geometry. And flat roofs, when appropriately detailed, can serve as balconies, decks, patios, and even landscaped parks.

FLAT ROOFS

A flat roof is a complex, highly interactive assembly of several components. The *deck* is the structural surface that supports the roof. *Thermal insulation* is usually installed as a part of the roof assembly to slow the passage of heat into and out of the building. A *vapor retarder* is essential in colder climates and on buildings with high interior humidity to prevent moisture from accumulating within the insulation. The *membrane* is the impervious sheet of material that keeps water out of the building. *Drainage* components remove the water that runs off the membrane. Around the membrane's edges and wherever it is penetrated by pipes, vents, expansion joints, electrical conduits, or roof hatches, special *flashings* and details must be designed to prevent water penetration.

Roof Decks

Previous chapters of this book have presented the types of structural decks ordinarily used under flat roofs: plywood over wood joists, solid wood decking over heavy timber framing, corrugated steel decking, panels of wood fiber bonded together with portland cement, poured gypsum over insulating formboard, sitecast concrete, and precast concrete. For a durable

FIGURE 13.1
A pitched roof can be made waterproof with any of a variety of materials. This thatched roof is being constructed by fastening bundles of reeds to the roof structure in overlapping layers, in such a way that only the butts of the reeds are left exposed to the weather.

FIGURE 13.2
The finished thatch roof has gently rounded contours and a pleasing surface texture. The decorative pattern of the ridge cap is the unique signature of the thatcher who made the roof. (Photos of thatch roofs courtesy of Warwick Cottage Enterprises, 2944 Greenhedge, Anaheim, California)

flat roof installation, it is important that the deck be adequately stiff under expected roof loadings and fully resistant to wind uplift forces. The deck should slope toward drainage points at an angle sufficient to drain reliably despite the effects of structural deflections. Minimum slopes of $\frac{1}{8}$ to $\frac{1}{4}$ inch per foot of run (1:100 to 1:50) are usually recommended. To produce these slopes, the beams that support the deck are often sloped by shortening some of the columns. Alternatively, a tapered fill of lightweight insulating concrete or lightweight aggregate with an asphaltic binder is poured over a dead-level structural deck to create the slopes, or a system of tapered boards of rigid insulation is laid over the deck. If a roof is insufficiently sloped, puddles of water will stand for extended periods of time in the low spots that are inevitably present, leading to premature deterioration of the roofing materials in those areas. If water accumulates in low spots caused by structural deflections, progressive structural collapse becomes a possibility, with deepening puddles attracting more and more water during rainstorms and becoming heavier and heavier, until the beams or joists become loaded to the point of failure (Figure 13.3).

If large in extent, the deck should be provided with enough expansion joints to control the effects of expansion and contraction on the roof membrane. If the expansion joints in the structure of the building are too far apart to satisfy the requirements of the membrane, *area dividers*, which are much like expansion joints but do not extend below the surface of the roof deck, may be installed (Figure 13.26).

The roof membrane must be laid over a smooth surface. If it is to be laid

directly on the deck, a wood deck should have no large gaps or knotholes. A concrete deck should be troweled smooth, and a precast plank deck, if not topped with a concrete fill, must be smoothed over with mortar at junctions between planks to fill the cracks and form smooth transitions between planks with varying degrees of camber. A corrugated steel deck must be covered with rigid boards, usually insulating boards, to bridge the flutes in the deck and create a smooth surface.

It is extremely important that the deck be dry at the time roofing operations commence, to avoid later problems with water vapor trapped under the membrane. A deck should not be roofed when rain, snow, or frost is present in or on the deck material. Concrete decks and insulating fills must be fully cured and thoroughly air dried.

Thermal Insulation and Vapor Retarder

Thermal insulation can be installed in any of three positions: below the structural deck of a flat roof, between the deck and the membrane, and above the membrane (Figure 13.4).

Insulation Below the Deck

Below the deck, batt insulation of mineral fiber or glass fiber is installed above a vapor retarder, either between wood joists or on top of a suspended ceiling assembly. A ventilated airspace should be provided between the insulation and the deck to dissipate stray water vapor. Insulation in this position is relatively economical and trouble-free, but it leaves the deck and the mem-

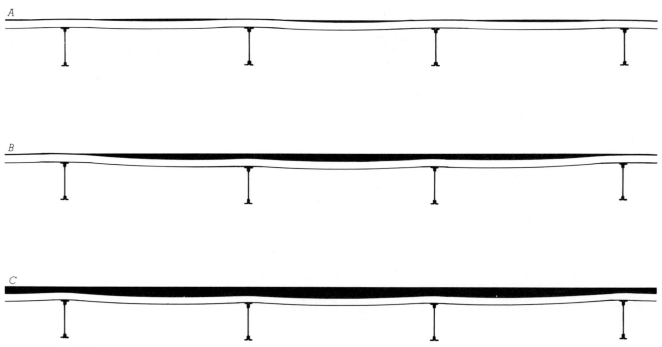

FIGURE 13.3

*A flat roof with insufficient pitch to drain is subject to structural failure through progressive collapse, as demonstrated in these three cross sections: **A.** Water stands on the roof in puddles, causing slight deflections of the roof deck*

*between supporting beams or joists. **B.** If heavy rainfall continues, the puddles grow and join, and the accumulating weight of the water begins to cause serious deflections in the supporting structural elements. The deflections*

*encourage water from a broader area of the roof to run into the puddle. **C.** As structural deflections increase, the breadth and depth of the puddle increase more and more rapidly, until the overloaded structure collapses.*

brane exposed to the full range of outdoor temperature fluctuation.

Insulation Between the Deck and the Membrane

The traditional position for flat roof insulation is between the deck and the membrane. Insulation in this position is in the form of low-density rigid panels or lightweight concrete because it must support the loads on the roof membrane without allowing puncture of the membrane. It protects the deck from temperature extremes and is itself protected from the weather by the membrane. But the roof membrane in this type of installation is subjected to extreme temperature variations, and any water vapor in the insulation is trapped beneath the membrane, which can lead to decay of the insulation and roof deck, and blistering and eventual rupture of the membrane from vapor pressure. (See the discussion on pages 517–519 for an explanation of insulation and vapor problems.)

Two precautions are required in cold climates for insulation located between the deck and the membrane. A vapor retarder must be installed below the insulation, and the insulation must be ventilated to allow the escape of any moisture that may accumulate there. Ventilation is accomplished through the installation of *topside vents*, one per thousand square feet (100 m²) or so, that direct escaping moisture upward through the membrane (Figures 13.5, 13.6). Topside vents are most effective with a loose-laid membrane,

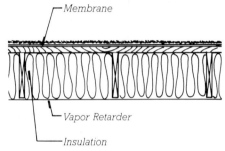 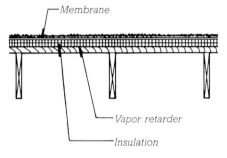 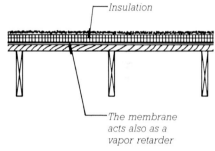

FIGURE 13.4

Flat roofs with thermal insulation in three different positions, shown here on a wood joisted roof deck. At left, insulation below the deck, with a vapor retarder on the warm side of the insulation. (This detail would be better if a ventilated airspace were provided between the insulation and the deck sheathing.) In the center, insulation between the deck and the membrane, with a vapor retarder on the warm side of the insulation. At right, a protected membrane roof, in which the insulation is above the membrane.

FIGURE 13.5

Topside roof vents are installed to release vapor pressure that may build up beneath a roof membrane. (Courtesy of Manville Corporation)

which allows trapped moisture to work its way toward the vents from any part of the insulating layer.

Insulation Above the Membrane: The Protected Membrane Roof

Insulation above the roof membrane is a relatively new concept. It offers two major advantages: The membrane is protected from extremes of heat and cold, and the membrane is on the warm side of the insulation where it is virtually immune to vapor blistering problems. Because the insulation itself is exposed to moisture when placed above the membrane, the insulating material must be one that retains its insulating value when wet and does not decay or disintegrate. Extruded polystyrene foam board has all these qualities and is the insulating material most used in this type of roof construction (Figure 13.7). The insulating board is either embedded in a coat of hot asphalt to adhere it to the membrane below, or laid loose. It is held down and protected from sunlight (which disintegrates polystyrene) by a layer of stone *ballast* or a thin concrete layer factory laminated to the upper surface

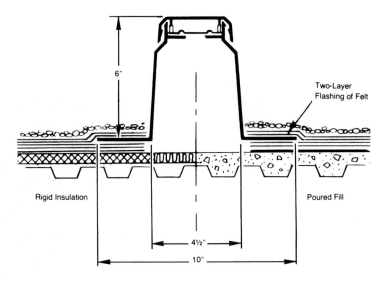

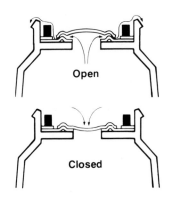

FIGURE 13.6
This proprietary topside roof vent, made of molded plastic with a synthetic rubber valve, allows moisture vapor to escape from beneath the membrane, but closes to prevent water or air from entering. (Courtesy of Manville Corporation)

of the insulating board. Critics of this roofing system (called variously the *inverted roof*, *upside-down roof*, or *protected membrane roof (PMR)*) originally predicted that the membrane would disintegrate quickly because of its continual exposure to dampness trapped under and around the insulating boards, but experience of more than 20 years has shown that the membrane ages little when thus protected from sunlight and temperature extremes, despite the presence of moisture.

Rigid Insulating Materials for Flat Roofs

An insulating material for flat roofs should have a high resistance to heat flow, adequate resistance to denting and gouging, moisture decay, and fire, and the ability to contact hot asphalt without melting or dissolving. No single material has all these virtues. Some rigid insulating materials commonly used on flat roofs in North America are listed in Figure 13.8, along with the advantages and disadvantages of each. There is an increasing preference for composite insulating materials, which approach the ideal material by com-

FIGURE 13.7
Installing expanded polystyrene foam insulation over a roof membrane to create a protected membrane roof. (Courtesy of Dow Chemical Company)

	Composition	Advantages	Disadvantages
Cellulose fiber board	A rigid, low-density board of wood or sugar cane fibers and a binder	Economical	Lower insulating efficiency than plastic foams; susceptible to absorption of moisture
Glass fiber board	A rigid, low-density board of glass fibers and a binder	Inert; fire resistant; non-decaying; dimensionally stable; ventilates moisture freely	Lower insulating efficiency than plastic foams
Polystyrene foam board	A closed-cell rigid foam of polystyrene plastic	High insulating efficiency; resistant to moisture	Flammable; high coefficient of thermal expansion; recommended for use only in protected membrane roofs
Polyurethane foam board	A closed-cell rigid foam of polyurethane, sometimes with saturated felt facings	Very high insulating efficiency	Flammable; high coefficient of thermal expansion; best when combined with other materials to increase its resistance to fire and hot bitumens
Polyisocyanurate foam board	A closed-cell rigid foam of polyisocyanurate, sometimes with glass fiber reinforcing and saturated felt facings	Very high insulating efficiency	Best when combined with other materials to increase its resistance to fire and hot bitumens
Cellular glass board	A closed-cell rigid foam of glass	Inert; fire resistant; moisture resistant; dimensionally stable; compatible with hot bitumens	Much lower insulating efficiency than plastic foams
Perlitic board	Granules of expanded volcanic glass and a binder pressed into a rigid board	Inert; fire resistant; compatible with hot bitumens, dimensionally stable	Much lower insulating efficiency than plastic foams
Lightweight concrete fill, Lightweight gypsum fill	Concretes made from very lightweight mineral aggregates (perlite or vermiculite), a cementing agent (portland cement or gypsum), and a high volume of entrained air	Can easily produce a tapered insulation layer for positive roof drainage	Much lower insulating efficiency than plastic foams; residual moisture from mixing water can cause blistering of membrane
Lightweight fill with asphaltic binder	Lightweight mineral aggregate with asphaltic binder	Can easily produce a tapered insulation layer for positive roof drainage; does not introduce mixing water under membrane	Much lower insulating efficiency than plastic foams
Composite insulating boards	Sandwich layers of foam plastic and other materials such as perlite board, glass fiber board, and saturated felt	Combine the high insulating efficiency and moisture resistance of foam plastics with the fire resistance, structural rigidity, and/or bitumen compatibility of other materials	

FIGURE 13.8
A comparative summary of some rigid insulating materials for flat roofs.

THERMAL INSULATION AND VAPOR RETARDER

Thermal Insulation

Thermal insulation is any material added to a building assembly for the purpose of slowing the conduction of heat through that assembly. Insulation is almost always installed in new roof and wall assemblies in North America, and often in floors and around foundations and concrete slabs-on-grade, anywhere that heated or cooled interior space comes in contact with unheated space or the outdoors.

The effectiveness of a building assembly in resisting the conduction of heat is expressed in terms of its *thermal resistance*, abbreviated as *R*. R is expressed either in English units of square foot-hour-degree Fahrenheit per Btu, or in metric units of square meter-degree Celsius per watt. The higher the R-value, the higher the insulating value.

Every component of a building assembly contributes in some measure to its overall thermal resistance. The amount of the contribution depends on the amount and type of material. Metals have very low R-values, and concrete and masonry materials are only slightly better. Wood has a sub-

stantially higher thermal resistance, but not nearly as high as that of an equal thickness of any of the common insulating materials. Most of the thermal resistance of any insulated building assembly is attributable to the insulating material.

In wintertime, it is warm inside a building and cold outside. The inside surface of a wall or roof assembly is warm, and the outside surface is cold. Between the two surfaces, the temperature varies from the inside temperature to the outside temperature according to the thermal resistances of the various layers of the assembly. The largest temperature difference within the assembly is between the two surfaces of the insulation (Figure A).

Water Vapor and Condensation

Water exists in three different physical states, depending on its temperature and pressure: solid (ice), liquid, and vapor. Water vapor is an invisible gas. The air always contains water vapor. The higher the temperature of the air, the more water vapor it is capable of containing. At a given temperature, the amount of water vapor the air *actually* contains, divided by the maximum amount

FIGURE A

Temperature variation and dew point location in a roof assembly and a wall assembly under conditions of 0 degrees Fahrenheit outdoors ($-18°$ C), 70 degrees F (21° C) and 50 percent relative humidity indoors. The dark solid line is a plot of the temperature at each point in the assembly. The broken line indicates the dew point temperature for the indoor air mass, and the line of dots represents the plane at which condensation takes place. Notice in both examples that condensation will occur within the layer of insulation unless an effective vapor retarder is placed on the inside of the insulation to prevent water vapor from migrating into the insulation.

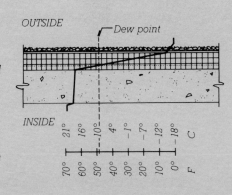

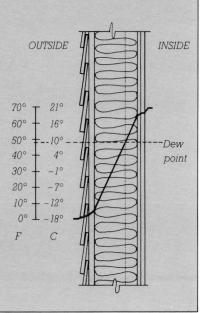

of water vapor it *could* contain, is the *relative humidity* of the air. Air at 50 percent relative humidity contains half as much water vapor as it could contain at the given temperature.

If a mass of air at 50 percent relative humidity is cooled, its relative humidity rises. The amount of water vapor in the air mass has not changed, but the ability of the air mass to contain water vapor has diminished because the air has become cooler. If the cooling of the air mass continues, a temperature will be reached at which the humidity is 100 percent. This temperature is known as the *dew point.* The dew point is different for every air mass. A roomful of very humid air has a high dew point, which is another way of saying that the air in the room would not have to be cooled very much before it would reach 100 percent humidity. A roomful of dry air has a low dew point; it can undergo considerable cooling before it reaches saturation.

When a mass of air is cooled below its dew point, it can no longer retain all its water vapor. Some of the vapor is converted to liquid water,

usually in the form of fog. The further the air mass is cooled below its dew point, the more fog will be formed. The process of converting water vapor to liquid by cooling is called *condensation.*

Condensation takes place in buildings in many different ways. In winter, room air circulating against a cold pane of glass is cooled to below its dew point, and a fog of water droplets forms on the glass. If the air is very humid, the droplets will grow in size, then run down the glass to accumulate in puddles on the window sill. If the glass is very cold, the condensate will freeze as it forms and create patterns of ice crystals. On a hot, humid summer day, the moisture in the air in the vicinity of a cold water pipe or a cool basement wall will condense in a similar fashion.

In an insulated wall or roof assembly, condensation can become a serious problem under wintertime heating conditions. The air inside the building is at a higher temperature than the air outside and usually contains much more water vapor, especially in densely populated areas of the building or where cooking, wet industrial proc-

FIGURE B

Perm ratings of some common building materials.

	U.S. Perms	CSI Perms
Aluminum foil		
1 mil (0.025 mm)	0.0	0.0
Built-up roofing	0.0	0.0
Polyethylene		
10 mil (0.25 mm)	0.03	0.0005
4 mil (0.10 mm)	0.08	0.0014
2 mil (0.05 mm)	0.16	0.0028
PVC, plasticized		
4 mil (0.10 mm)	0.8–1.4	0.014–0.024
Interior primer plus one coat flat oil paint on plaster	1.6–3.0	0.028–0.052
Exterior oil paint, three coats on wood	0.3–1.0	0.005–0.017
Hot melt asphalt		
2 oz/ft² (0.6 kg/m²)	0.5	0.009
3.5 oz/ft² (1.1 kg/m²)	0.1	0.002
Brick masonry		
4″ (100 mm) thick	0.8	0.014
Plaster on metal lath		
¾″ (19 mm) thick	15	0.26
Gypsum wallboard		
⅜″ (9.5 mm) thick	50	0.87
Plywood, exterior glue		
¼″ (6 mm) thick	0.7	0.012

esses, bathing, or washing take place. Indoor air leaking through the assembly toward the outside becomes progressively cooler and reaches its dew point somewhere inside the assembly, almost always within the thermal insulation. Where air itself does not leak through the assembly, water vapor still migrates from indoors to outdoors, driven by the difference in vapor pressure between the moist indoor air and the drier air outdoors, and again reaches its dew point somewhere within the insulation. In both cases the result is that the insulation becomes wet, and portions of it may become frozen with ice. The insulating value is lost. The materials of the roof or wall become wet and subject to rust or decay. Water may accumulate to such an extent that it runs or drips out of the assembly and spoils finishes or contents of the building. And when the outside of the assembly is heated by warmer outdoor air or bright sunlight, the water begins to vaporize and move toward the outdoors. As it does so, its vapor pressure can raise blisters in paint films or roof membranes, blisters that can rupture like an overinflated balloon if the pressure is not relieved in time. (Most cases of peeling paint on wooden buildings are caused not by a poor painting job, but by lack of a vapor retarder in the wall behind the paint.)

Under summertime cooling conditions in hot, humid weather, the flow of water vapor through building assemblies can be reversed, as moisture moves from the warm, damp outside air toward the cooler, drier air within. This condition is not usually as severe as the winter condition because the differences in temperature and humidity between indoors and outdoors are not as great. And in most areas of North America the cooling season is short compared to the heating season, allowing the designer to neglect the summer vapor problem and concentrate on the winter problem. But in areas of the American South, the summer problem is more severe than the winter problem, and must be solved first.

The Vapor Retarder

To prevent condensation inside building assemblies, a *vapor retarder* (often called, inexactly, a *vapor barrier*), is installed on the warmer side of the insulation layer. This is a continuous sheet, as nearly seamless as possible, of plastic sheeting,

aluminum foil, kraft paper laminated with asphalt, roofing felt laminated with asphalt, troweled mastic, or some other material that is highly resistant to the passage of water vapor. The effect of the vapor retarder is to diminish the flow of air and vapor through the building assembly, preventing the moisture from reaching the point in the assembly where it would condense. For buildings in most parts of this continent the vapor retarder should be placed on the inside of the insulation. In areas where warm weather cooling is the predominant problem, these positions should be reversed, or in some extremely mild climates, the vapor retarder may not be required at all.

The part of the building assembly toward the cooler side of the vapor retarder should be allowed to "breathe," to ventilate freely by means of attic ventilation, topside roof vents, or vapor-permeable exterior materials. This helps prevent stray moisture from becoming trapped between the vapor retarder and another near-impermeable surface.

The performance of a vapor retarder material is measured, in the United States, in *perms.* A perm is defined by ASTM Standard E96 as the passage of one grain of water vapor per hour through one square foot of material at a pressure differential of one inch of mercury between the two sides of the material. In Canada, the CSI perm is measured in terms of one nanogram per second per square meter per pascal of pressure difference. One CSI perm is equal to 57.38 U.S. perms (Figure B). For vapor retarders below flat roofs, the National Roofing Contractors Association of the United States recommends a perm rating approaching zero; a well constructed vapor retarder of asphalt and felt, the best type in general use, has a rating of about 0.005 U.S. perms, and single-ply vapor retarders of foil or kraft paper range up to about 0.30, which is satisfactory for most purposes. For sloping roofs and walls, polyethylene films have U.S. perm ratings of 0.16 for a 0.002 inch thickness, and 0.08 for a 0.004 inch thickness. (For metric and C.S.I units, see Figure B.)

For a more complete discussion of thermal insulation and vapor retarders in buildings, consult Stein, Benjamin, John S. Reynolds, and William J. McGuinness, *Mechanical and Electrical Equipment for Buildings,* 7th Edition. New York, John Wiley and Sons, 1986.

bining two or more materials to exploit the best qualities of each.

If rigid insulating boards are located below the roof membrane, they may be fastened to the deck mechanically with nails, screws, or any of a broad selection of fasteners made especially for the purpose. Alternatively, they may be adhered to the deck with hot asphalt, but mechanical fasteners are favored by insurance companies as better protection against wind uplift (Figures 13.9, 13.10).

Poured deck fill insulations are economical. They may be applied directly to corrugated steel decking and rough concrete decks and can easily be tapered during installation to slope toward points of roof drainage. Thermal resistances per inch are not as high as for plastic foams, so greater thicknesses are required to achieve similar insulating values. Portland-cement-based or gypsum-based fill insulations contain large amounts of free water. They should be cured and dried as thoroughly as possible before application of the membrane, and topside vents should be installed to allow the escape of moisture vapor from the insulation during the life of the roof.

Vapor Retarders for Flat Roofs

The membrane in a protected membrane roof serves also as the vapor retarder. In other flat roof constructions, a separate vapor retarder is advisable except in warm, humid climates where wintertime condensation is not a problem and summertime air conditioning can cause a reversal of vapor migration through the roof.

The most common type of vapor retarder for a flat roof consists of two layers of asphalt-saturated roofing felt bonded together and to the roof deck with hot asphalt. Proprietary vapor retarder sheets of various materials can be equally satisfactory and more economical. Polyethylene sheeting, which makes an excellent vapor retarder in many other types of construction, is not used in flat roofs because it melts at the application temperature of the

FIGURE 13.9
Workers bed rigid insulation boards in strips of hot asphalt over a corrugated metal roof deck. (Courtesy of GAF Corporation)

FIGURE 13.10
For more secure attachment of insulation over metal decks, screws and large washers are used. (Courtesy of GAF Corporation)

hot asphalt used in most roof membranes.

The vapor retarder must be located at such a point in the roof assembly that it will always be warmer than the dew point of the air under any conceivable condition of use. Usually this means putting the vapor retarder below the insulation, but this is not always possible. A vapor retarder installed directly over a corrugated steel deck, for example, would have to bridge across the corrugations and would be unusually vulnerable to damage until it was covered by insulation. In this case, a thin layer of insulation is laid over the deck, then the vapor retarder, followed by a thicker layer of insulation board. The designer must carefully calculate the dew point location in this assembly to be sure the vapor retarder lies below it, or condensation of moisture can occur within the lower layer of insulation.

The Flat Roof Membrane

The membranes used for flat roofing fall into three general categories: the *built-up roof membrane (BUR)*, the *elastomeric/plastomeric roof membrane*, and the *fluid-applied roof membrane*.

The Built-Up Roof Membrane

A built-up roof membrane (BUR) is assembled in place from multiple plies of asphalt-impregnated *felt* bedded in *bitumen* (Figures 13.11 through 13.13). The felt may consist of cellulose fibers or glass fibers. It is saturated with asphalt at the factory and delivered to the site in rolls. The bitumen is usually asphalt derived from the distillation of petroleum, but for dead-level or very low-pitch roofs, coal-tar bitumen or coal-tar pitch is used instead, because of its greater resistance to standing water. (Coal-tar bitumen is distinguished from coal-tar pitch by having fewer volatile components.) Both asphalt and coal tar are applied hot in order to merge with the saturant bitumens in the felt and form a single-

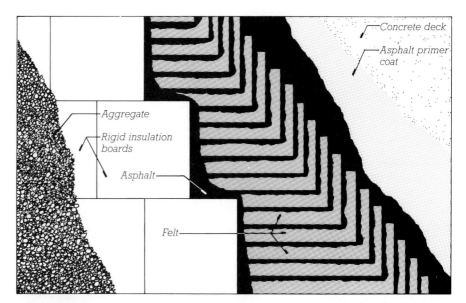

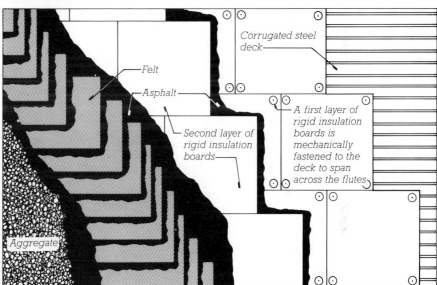

FIGURE 13.11

Two typical built-up roof constructions, as seen from above. The top diagram is a cutaway view of a protected membrane roof over a poured concrete roof deck. The membrane is made from plies of felt overlapped in such a way that it is never less than four plies thick. Rigid foam insulation boards are bedded in hot asphalt over the membrane and ballasted with stone aggregate to keep them in place and *protect them from sunlight. The bottom diagram shows how rigid insulation boards are attached to a corrugated steel roof deck in two staggered layers to provide a firm, smooth base for application of the membrane. A three-ply membrane is shown. In cold climates, a vapor retarder should be installed between the layers of insulation.*

piece membrane. The felt is laminated in overlapping layers to form a membrane that is two to four plies thick. The more plies used, the more durable the roof. To protect the membrane from sunlight and physical wear, a layer of aggregate (crushed stone or other mineral granules) is embedded in the surface. (Figure 13.14)

Cold-applied mastics can be used in lieu of hot bitumen in built-up roof membranes. A roofing mastic is compounded of asphalt and other substances to bond to felts or to synthetic fabric reinforcing mats at ordinary temperatures. The mastic may be sprayed or brushed on and hardens by the evaporation of solvents.

FIGURE 13.12
A base sheet of asphalt-saturated felt is installed over rigid insulation, using a machine that unrolls the felt and rolls it into a layer of hot asphalt. (Courtesy of The Celotex Corporation)

FIGURE 13.13
Overlapping layers of roofing felt are hot-mopped with asphalt to create a four-ply roof membrane. (Courtesy of Manville Corporation)

FIGURE 13.14
Roofers embed stone ballast in hot asphalt to hold down and protect the panels of rigid insulation in a protected membrane roof. The area of the membrane behind the wheelbarrow has not yet received its insulation. (Courtesy of The Celotex Corporation)

FIGURE 13.15
A cutaway detail of a proprietary type of protected membrane roof shows, from bottom to top, the roof deck, the membrane, polystyrene foam insulation, a polymeric fabric that separates the ballast from the insulation, and the ballast. (Courtesy of Dow Chemical Company)

Elastomeric/Plastomeric Membranes: The Single-Ply Roof

Elastomeric/plastomeric membranes are a diverse and rapidly growing group of sheet materials that are applied to the roof in a single layer (Figures 13.16 through 13.21). As compared to built-up membranes, they require less on-site labor, and are usually more elastic and therefore less prone to cracking and tearing. They are affixed to the roof by any of several means: with adhesives; by the weight of a gravel ballast spread over them; by fasteners concealed in the seams between sheets; or if they are sufficiently flexible, with ingenious mechanical fasteners that do not penetrate the membrane (Figures 13.17 through 13.19).

The materials presently used for elastomeric/plastomeric membranes include the following:

• *Neoprene* (polychloroprene) is a high-performance synthetic rubber compound. It is applied in sheets ranging from 0.030 to 0.120 inches (0.75 to 3.0 mm) in thickness and is joined at the seams with an adhesive. Because it is vulnerable to attack by ultraviolet light, it is usually coated with a protective layer of chlorosulfonated polyethylene. It may be adhered to the roof deck, or laid loose and ballasted with aggregate to prevent wind uplift. It can be used in a protected membrane roof.

• *EPDM* (ethylene propylene diene monomer) is the most widely used material for single-ply roof membranes. It is relatively low in cost. It is a synthetic rubber manufactured in sheets from 0.030 to 0.060 inches in thickness (0.75 to 1.5 mm). It is joined with adhesive, and may be laid loose, adhered, or used in a protected membrane roof.

• *PVC* (polyvinyl chloride) is a thermoplastic compound commonly known as *vinyl*. It is relatively low in cost. PVC sheet for roofing is 0.032 to 0.060 inches thick (0.81 to 1.5 mm). Its seams are sealed either by solvent welding

FIGURE 13.16
Workers unfold a large single-ply roof membrane. (Courtesy of Carlisle SynTec Systems)

1. Roll membrane over knobbed base plate

2. Roll and snap on white retainer clip

3. Snap and screw on threaded black cap

FIGURES 13.17, 13.18
A proprietary nonpenetrating attachment system for a single-ply roof membrane. (Courtesy of Carlisle SynTec Systems)

FIGURE 13.19
Another proprietary nonpenetrating attachment system folds the membrane into a continuous slot, where it is held by a synthetic rubber spline inserted with a wheeled tool. (Courtesy of Firestone)

FIGURE 13.20
Roofers bond a composite single-ply membrane to a concrete deck with a cold-applied adhesive. The seams will be heat-fused together. (Courtesy of Koppers Company, Inc.)

or hot air welding. It may be laid loose, adhered, or used as a protected membrane.

• *Chlorinated polyethylene* and *chlorosulfonated polyethylene* sheets are highly resistant to ultraviolet deterioration and can be manufactured in light, heat-reflective colors. They are used on roofs where aggregate ballasting is unacceptable for reasons of appearance or excessive slope.

• *Polymer-modified bitumens* are formed into composite sheets with various other materials to give them the necessary physical properties to be handled as sheet material. Some are intended to be laid loose, others to be adhered to the roof deck or insulation (Figures 13.20, 13.21).

Elastomeric/plastomeric roofing technology is developing rapidly, and the latest manufacturers' literature should be consulted for current information.

Fluid-Applied Membranes

Fluid-applied membranes are used primarily for domes, vaults, and other complex shapes that are difficult to roof by conventional means. Such shapes are too flat on top for shingles, but too steep on the sides for built-up roof membranes, and if doubly curved are difficult to fit with single-ply membranes. Fluid-applied membranes are applied with a roller or spray gun, usually in several coats, and cure to form

a rubbery membrane. Materials applied by this method include neoprene (with a weathering coat of chlorosulfonated polyethylene), silicone, polyurethane, butyl rubber, and asphalt emulsion.

Traffic Decks

Traffic decks are installed over flat roof membranes for walks, roof terraces, and sometimes driveways or parking surfaces. Two different details are used: In one, low blocks of plastic or concrete are set on top of the roof membrane to support the corners of heavy square paving stones or slabs with open joints. In the other, a drainage layer of gravel or no-fines concrete (a very

FIGURE 13.21

A roofer heat-fuses a seam between two sheets of an aluminum-faced single-ply membrane. The membrane in this example consists of a flexible plastic sheet laminated between layers of bitumen, with protective layers of polyethylene below and embossed aluminum foil above. The bitumen layers provide the bonding elements for heat fusion. The aluminum foil protects the inner plies of the membrane from the sun, and reflects solar heat from the roof. (Courtesy of Koppers Company, Inc.)

porous concrete whose aggregate consists solely of a single size of coarse stone) is leveled over the membrane, and open-jointed paving blocks are installed on top. In either detail, water falls through the joints in the paving and is caught and drained away by the membrane below. Notice that the membrane is not pierced in either detail.

Edge and Drainage Details for Flat Roofs

Some typical details of flat roofs are presented in Figures 13.22 through 13.31. All are shown with built-up roof membranes, but details for elastomeric/plastomeric membranes are similar in principle.

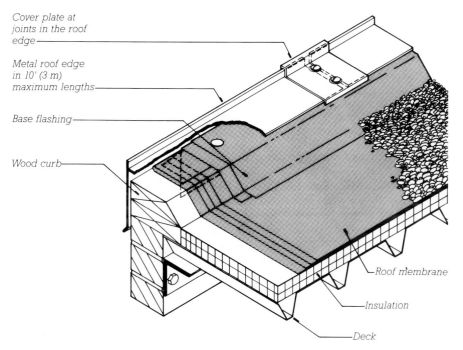

Cover plate at joints in the roof edge

Metal roof edge in 10' (3 m) maximum lengths

Base flashing

Wood curb

Roof membrane

Insulation

Deck

FIGURE 13.22

A roof edge with conventional built-up roof. The membrane consists of four plies of felt bedded in asphalt with a gravel ballast. The base flashing is composed of two additional plies of felt that seal the edge of the membrane and reinforce it where it bends over the curb. The curb directs water toward interior drains or scuppers rather than allowing it to spill over the edge. The exposed vertical face of the metal roof edge is called a fascia.

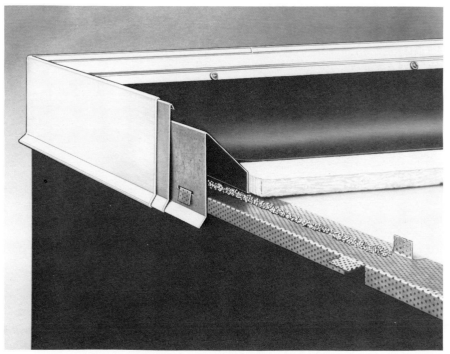

FIGURE 13.23

A proprietary roof edge system for flat roofs. The perforated metal strip is fastened to the roof with a mastic adhesive that pushes through the perforations for a tighter bond. When the adhesive has hardened, a galvanized steel curb is fastened in place with the tabs of perforated metal, and an aluminum roof edge is hooked on, with a backup piece at end joints as shown to prevent leakage. Lastly, the roof edge and the membrane are locked in place simultaneously by installing a clamping strip that engages the hook on the top of the aluminum roof edge. The clamping strip is held in place by screws that pass through the edge of the membrane into the galvanized curb, as seen at the top of the photograph. (Product of W. P. Hickman Company, Asheville, North Carolina)

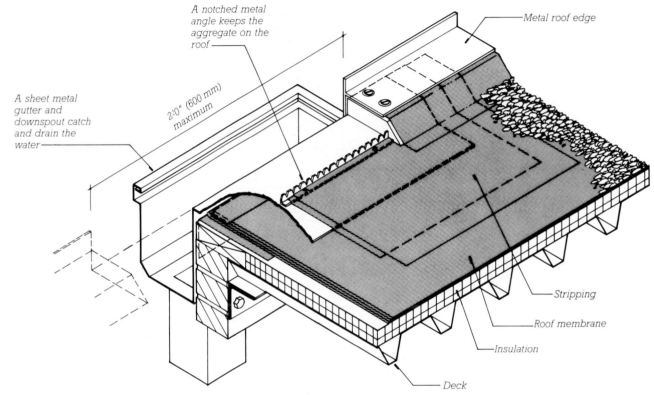

A notched metal angle keeps the aggregate on the roof

A sheet metal gutter and downspout catch and drain the water

2'-0" (600 mm) maximum

Metal roof edge

Stripping

Roof membrane

Insulation

Deck

FIGURE 13.24

Detail of a scupper. The curb is discontinued to allow water to spill off the roof into a gutter and downspout. Additional layers of felt, called stripping, seal around the sheet metal components. Most roofs use interior drains (Figure 13.29) rather than scuppers.

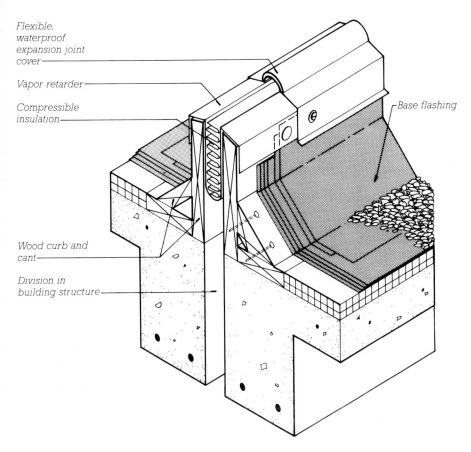

Flexible, waterproof expansion joint cover

Vapor retarder

Compressible insulation

Wood curb and cant

Division in building structure

Base flashing

FIGURE 13.25
An expansion joint in a membrane roof. Large differential movements between the adjoining parts of the structure can be tolerated with this type of joint because of the ability of the flexible joint cover to adjust to movement without tearing. A two-ply base flashing seals the edge of the membrane.

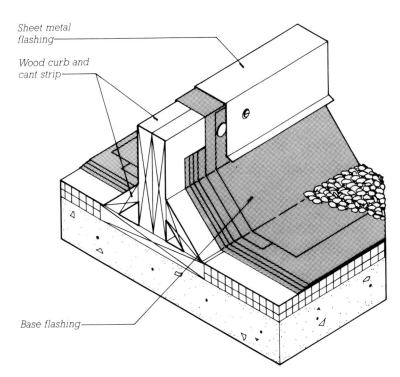

Sheet metal flashing

Wood curb and cant strip

Base flashing

FIGURE 13.26
An area divider is similar to an expansion joint but is designed only to allow for some movement in the membrane itself, not the entire structure.

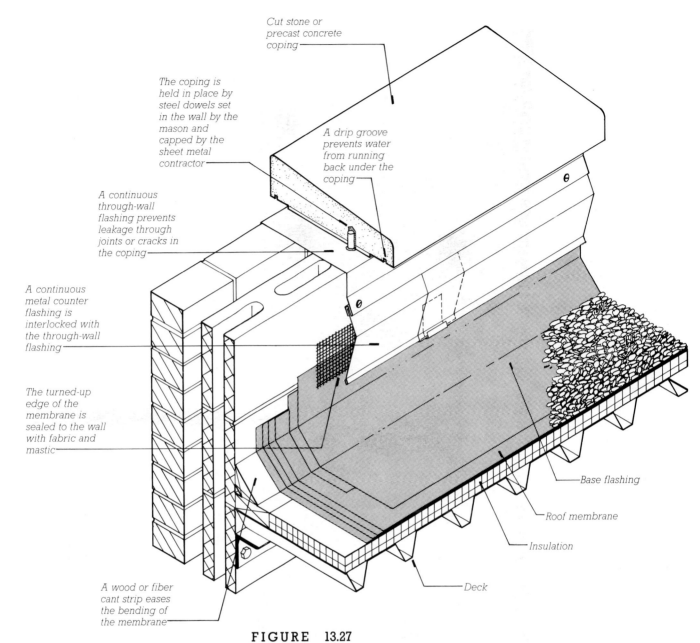

Cut stone or
precast concrete
coping

The coping is
held in place by
steel dowels set
in the wall by the
mason and
capped by the
sheet metal
contractor

A drip groove
prevents water
from running
back under the
coping

A continuous
through-wall
flashing prevents
leakage through
joints or cracks in
the coping

A continuous
metal counter
flashing is
interlocked with
the through-wall
flashing

The turned-up
edge of the
membrane is
sealed to the wall
with fabric and
mastic

A wood or fiber
cant strip eases
the bending of
the membrane

Base flashing

Roof membrane

Insulation

Deck

FIGURE   13.27

*A conventional parapet design. The
coping attachment and flashing are very
labor intensive, and therefore costly.*

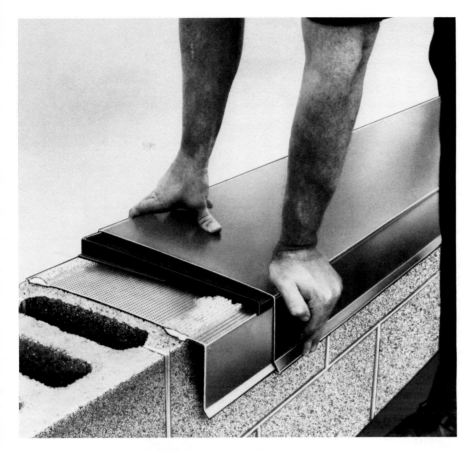

FIGURE 13.28
A proprietary parapet coping system. The perforated metal channel is fastened to the masonry with a mastic adhesive. Sections of metal coping are snapped on over the channel, with a special pan element beneath the joints to drain leakage. (Product of W. P. Hickman Company, Asheville, North Carolina)

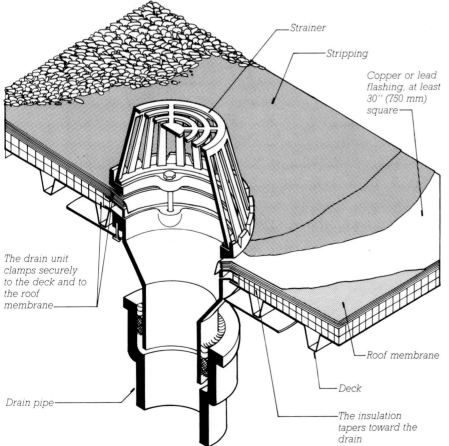

Strainer

Stripping

Copper or lead flashing, at least 30" (750 mm) square

The drain unit clamps securely to the deck and to the roof membrane

Roof membrane

Drain pipe

Deck

The insulation tapers toward the drain

FIGURE 13.29
A conventional cast-iron interior roof drain. Two plies of felt stripping seal around the sheet metal flashing.

FIGURE 13.30
A proprietary single-piece roof drain made of molded plastic. (Product of W. P. Hickman Company, Asheville, North Carolina)

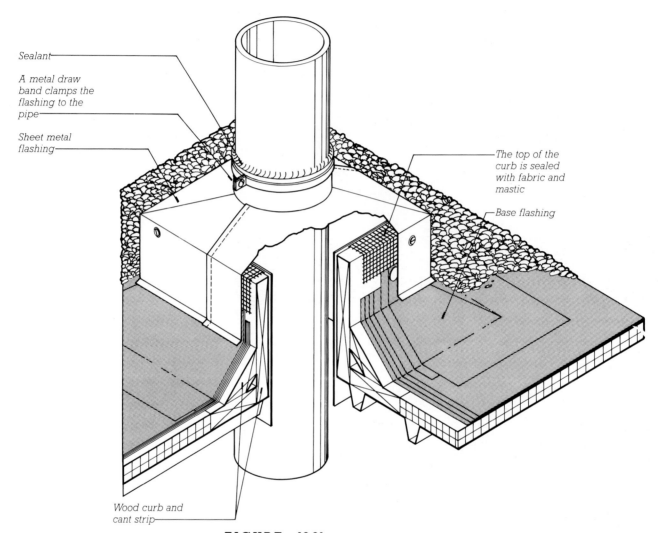

Sealant

A metal draw band clamps the flashing to the pipe

Sheet metal flashing

The top of the curb is sealed with fabric and mastic

Base flashing

Wood curb and cant strip

FIGURE 13.31

A roof penetration for a plumbing vent stack. Notice how this and all the previous edge and penetration details for a flat roof use the curb, cant strip, and stripping to keep standing water away from the edge of the membrane.

PITCHED ROOFS

Roof coverings for pitched roofs fall into three general categories: *thatch*, *shingles*, and *sheet metal*. Thatch, an attractive and effective roofing consisting of bundles of reeds, grasses, or leaves (Figures 13.1 and 13.2), is highly labor intensive and is used today primarily for historic restorations and highly specialized buildings. Shingle and sheet metal roofs of many types are common to every type of building, and range in price from the most economical of roof coverings to the most expensive.

The insulation and vapor retarder in most pitched roofs are installed below the roof sheathing or deck; typical details of this practice are shown in Chapters 6 and 7. Where the underside of the deck is to be left exposed as a finish surface, a vapor retarder and rigid insulation are applied above the deck, just below the roofing. A layer of plywood is then nailed over the insulation panels as a *nail base* for fastening the shingles or sheet metal, or special composite insulation panels with an integral nail base layer can be used.

Shingles

The word "shingles" is used here in a generic sense to include wood shingles and shakes, asphalt shingles, slates, clay tiles, and concrete tiles. What these materials share in common is their small unit size and their application to the roof in overlapping layers with staggered vertical joints.

Wood shingles are thin, tapered slabs of wood sawed from short pieces of tree trunk with the grain of the wood running approximately parallel to the face of the shingle (Figures 13.32, 13.50). *Shakes* are split from the wood, rather than sawn, and exhibit a much rougher face texture than wood shingles (Figures 13.33, 13.34). Most wood shingles and shakes in North America are made from Red cedar, White cedar, or Redwood because of the natural decay resistance of these woods. Wood roof coverings are not highly re-

FIGURE 13.32

Applying Red cedar shingles, in this example as reroofing over asphalt shingles. Small corrosion-resistant nails are driven near each edge at the mid-height of the shingle. Each succeeding course covers the cracks and nails in the course below. (Courtesy of Red Cedar Shingle and Handsplit Shake Bureau)

sistant to fire unless they have been pressure-treated with fire-retardant chemicals.

Asphalt shingles are die-cut from heavy sheets of asphalt-impregnated felt faced with mineral granules that act as a wearing layer and decorative finish. The most common type of asphalt shingle, which covers probably 90 percent of the single-family houses in North America, is 12″ × 36″ (305 mm × 914 mm) in size. (A metric size shingle 337 mm × 1000 mm is also widely marketed). Each shingle is slotted twice to produce a roof that looks as though it were made of smaller shingles (Figures 13.35 through 13.39). An assortment of other shingle styles are also available. Asphalt shingles are inexpensive to buy, quick to install, moderately fire-resistant, and have an expected lifetime of 10 to 20 years, depending on their exact composition.

The same sheet material from which asphalt shingles are cut is also sold in rolls 3 feet (900 mm) wide as *asphalt roll roofing*. Roll roofing is very inexpensive and is used primarily on storage and agricultural buildings. Its chief drawbacks are that thermal ex-

FIGURE 13.33

Application of handsplit Red cedar shakes, again over an existing roof of asphalt shingles. Compare the shape, thickness, and surface of these shakes with those of the cedar shingles in the preceding illustration. Each course is interleaved with strips of heavy asphalt-saturated felt 18 inches (460 mm) wide as extra security against the passage of wind and water between the highly irregular and therefore loosely mated shakes. The nail stripper hung around the roofer's neck speeds his work by holding the nails and aligning them with points down, ready for driving.
(Courtesy of Red Cedar Shingle and Handsplit Shake Bureau)

FIGURE 13.34

Shake application over a new roof deck using air-driven, heavy-duty staplers for greater speed. The strips of asphalt-saturated felt have all been placed in advance. Each course of shakes is laid out, then quickly fastened by roofers walking across the roof and inserting staples as fast as they can pull the trigger. (Courtesy of Senco Products, Inc.)

FIGURE 13.35
Installing asphalt shingles. The slots make each shingle appear as if it were three smaller shingles when the roof is finished. (Photo by the author)

FIGURE 13.36
Reroofing an asphalt shingle roof using air-driven, wide-crown staples for higher efficiency and greater holding power. When the second layer of shingles becomes worn, both layers will have to be removed before reroofing. (Courtesy of Senco Products, Inc.)

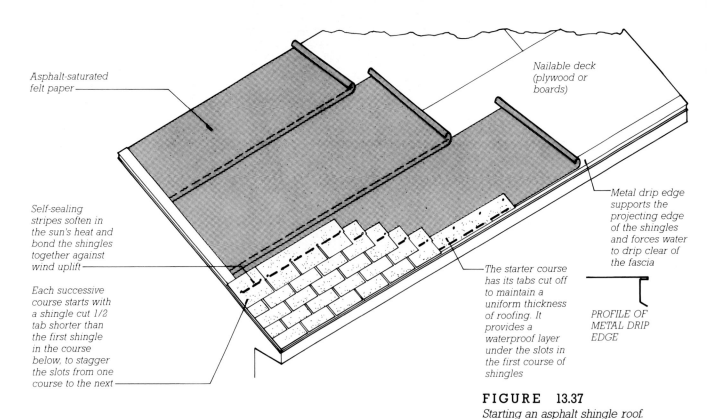

Asphalt-saturated felt paper

Nailable deck (plywood or boards)

Self-sealing stripes soften in the sun's heat and bond the shingles together against wind uplift

Each successive course starts with a shingle cut 1/2 tab shorter than the first shingle in the course below, to stagger the slots from one course to the next

The starter course has its tabs cut off to maintain a uniform thickness of roofing. It provides a waterproof layer under the slots in the first course of shingles

Metal drip edge supports the projecting edge of the shingles and forces water to drip clear of the fascia

PROFILE OF METAL DRIP EDGE

FIGURE 13.37
Starting an asphalt shingle roof.

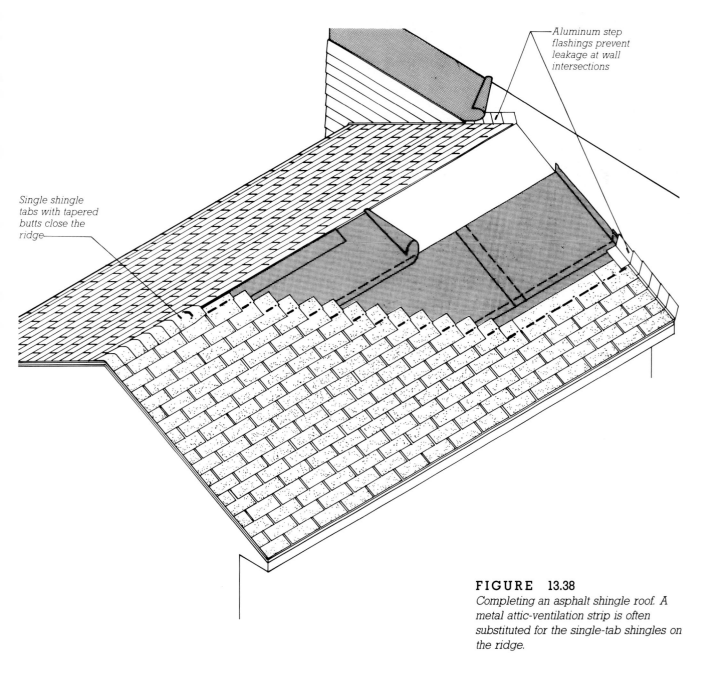

Aluminum step flashings prevent leakage at wall intersections

Single shingle tabs with tapered butts close the ridge

FIGURE 13.38
Completing an asphalt shingle roof. A metal attic-ventilation strip is often substituted for the single-tab shingles on the ridge.

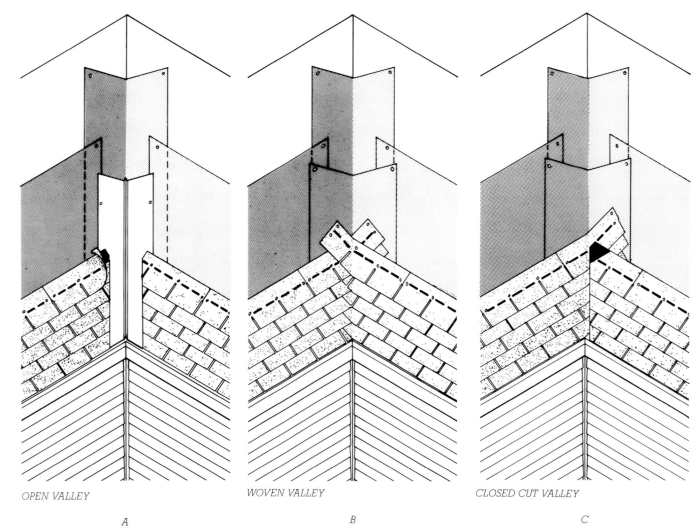

OPEN VALLEY

WOVEN VALLEY

CLOSED CUT VALLEY

A

B

C

FIGURE 13.39

Three alternative methods of making a valley in an asphalt shingle roof. **A.** *The open valley uses a sheet metal flashing; the ridge in the middle of the flashing helps prevent water coming off one slope from washing up under the shingles on the opposite slope. The woven valley* **B.** *and cut valley* **C.** *are favorites of roofing contractors because they require no sheet metal. The solid black areas on shingles in the open and cut valleys indicate areas to which asphaltic roofing cement is applied to adhere shingles to each other.*

pansion of the roofing or shrinkage of the wood deck can cause unsightly ridges to form in the roofing and that thermal contraction can tear it.

Slate for roofing is delivered to the site split, trimmed to size, and punched or drilled for nailing (Figures 13.40,

13.41). It forms a fire-resistant, long-lasting roof that is suitable for buildings of the finest materials. It is relatively costly.

Clay tiles have been used on roofs for thousands of years. It is said that the tapered cylindrical roofing tiles

traditional to the Mediterranean region (similar to the mission tiles in Figure 13.42) were originally formed on the thigh of the tilemaker. Many other patterns of clay tiles are now available, both glazed and unglazed. Concrete tiles are generally less expensive than

FIGURE   13.40
Splitting slate for roofing. The thin slates in the background will next be trimmed square and to dimension, after which nail holes will be punched in them.
(Photo by Flournoy, courtesy of Buckingham-Virginia Slate Corporation)

FIGURE   13.41
A slate roof during installation.
(Courtesy of Buckingham-Virginia Slate Corporation)

clay tiles and are available in some of the same patterns. Tile roofs in general are heavy, durable, highly resistant to fire, and relatively expensive in first cost.

Each type of shingle, slate, or tile must be laid on a roof deck that slopes sufficiently to assure leakproof performance. Minimum slopes for each material are specified by the manufacturer, and often by the building code. Slopes greater than the minimum should be used in locations where water is likely to be driven up the roof surface by heavy storm winds.

Sheet Metal Roofing

Sheets of lead and copper have been used for roofing since ancient times. Both metals are self-protecting and last for many decades. They are installed in small sheets using ingenious systems of joining and fastening to maintain watertightness at the seams (Figures 13.43 through 13.47). The seams, especially standing and batten seams, create a strong visual pattern that can be manipulated by the designer to emphasize the qualities of the roof shape.

Lead roofs oxidize in time to a white color. Copper turns a beautiful blue-green in clean air, and a dignified black in an industrial atmosphere; various treatments can be used to obtain and preserve the desired color. *Terne* roofing is made of stainless steel sheet or steel sheet coated with an alloy of lead and tin. Terne-coated stainless steel is self-protecting and weathers to a soft gray color. A terne coating over plain steel must be painted to protect it against pinhole corrosion. Aluminum, zinc alloys, and stainless steel can also be used for roofing in much the same

FIGURE 13.42
Two styles of clay tile roofs. The mission tile has very ancient origins.

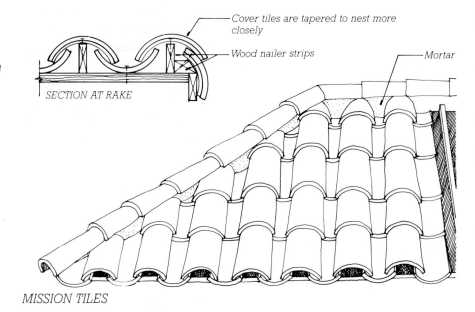

Cover tiles are tapered to nest more closely

Wood nailer strips

Mortar

SECTION AT RAKE

MISSION TILES

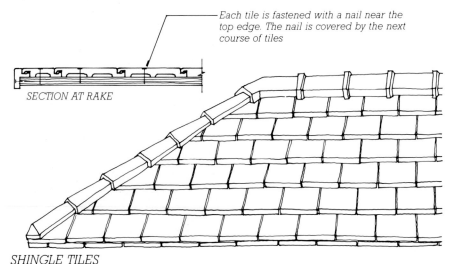

Each tile is fastened with a nail near the top edge. The nail is covered by the next course of tiles

SECTION AT RAKE

SHINGLE TILES

FIGURE 13.43
Standing seam copper roofs. (Designer: Emil Hanslin. Courtesy of Copper Development Association, Inc.)

FIGURE 13.44
An automatic roll seamer, moving under its own power, locks standing seams in a copper roof. A cleat is visible at the lower right. (Courtesy of Copper Development Association, Inc.)

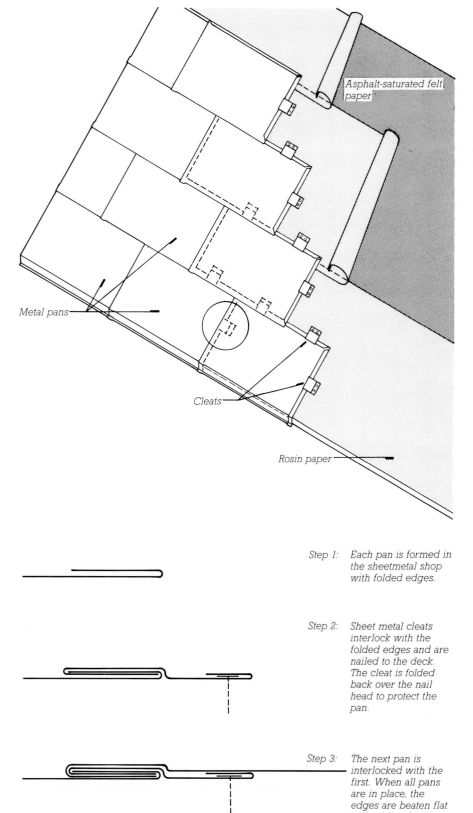

Metal pans

Asphalt-saturated felt paper

Cleats

Rosin paper

FIGURE 13.45

Installing a flat seam metal roof. The three diagrams at the bottom of the illustration show the three steps in creating the seam, viewed in cross section. The cleats, which fasten the roofing to the deck, are completely concealed when the roof is finished.

Step 1: *Each pan is formed in the sheetmetal shop with folded edges.*

Step 2: *Sheet metal cleats interlock with the folded edges and are nailed to the deck. The cleat is folded back over the nail head to protect the pan.*

Step 3: *The next pan is interlocked with the first. When all pans are in place, the edges are beaten flat and soldered or sealed.*

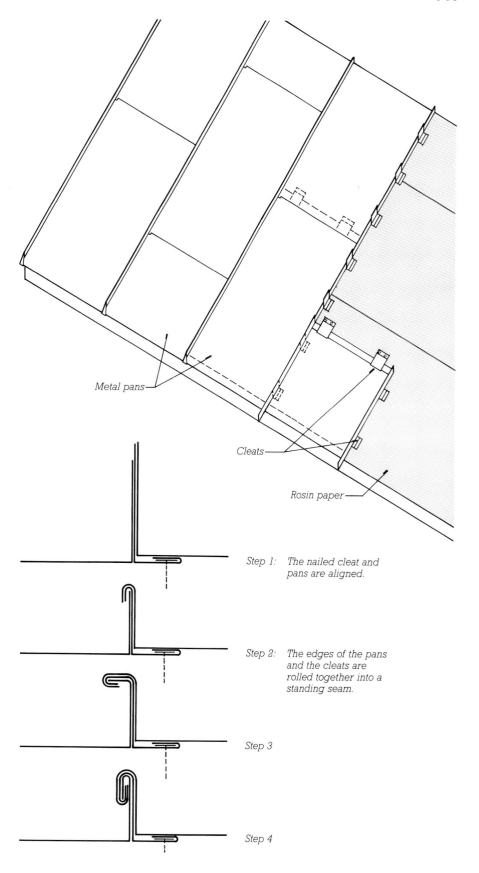

FIGURE 13.46
Installing a standing seam metal roof.

Metal pans

Cleats

Rosin paper

Step 1: The nailed cleat and pans are aligned.

Step 2: The edges of the pans and the cleats are rolled together into a standing seam.

Step 3

Step 4

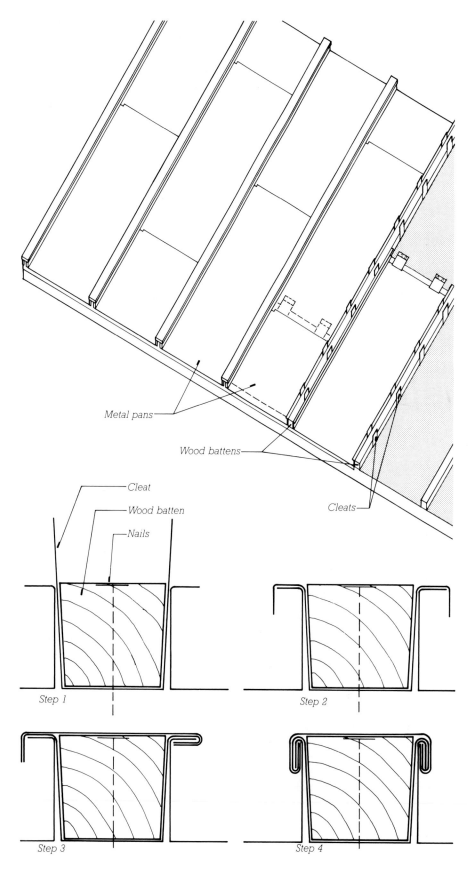

Metal pans

Wood battens

Cleat

Wood batten

Nails

Cleats

Step 1

Step 2

Step 3

Step 4

FIGURE 13.47
Installing a batten seam metal roof. The battens are tapered in cross section to allow for expansion of the roofing metal.

manner as copper, lead, and terne. Sheet metal roofs of any of these materials are relatively high in first cost but can be expected to last for many decades.

Large corrugated sheets of aluminum or steel are used extensively for roof coverings, especially in industrial and agricultural buildings. Many different fluting patterns and surface coatings are available. Most corrugated roofing sheets have sufficient strength and stiffness that sheathing or decking is not required if the purlins of the roof are appropriately spaced. A number of manufacturers produce proprietary systems of metal roofing whose corrugations resemble the standing seams of conventional sheet metal roofing; these are designed to be installed at pitches as low as $\frac{1}{4}$ inch per foot (1:48), enabling them to be used to cover flat-roofed structures. Corrugated metal roofing is generally low in cost and very long lasting. It can be attractive on buildings of any type if thoughtfully selected and detailed (Figure 13.48).

The same metal should be used for every component of a sheet metal roof, including the fasteners and flashings. If this is impossible, metals of similar galvanic activity should be used. Where strongly dissimilar metals touch in the presence of rainwater, which is generally acidic, galvanic action causes rapid corrosion. Figure 13.49 will be helpful to the designer in avoiding destructive combinations of metals.

FIGURE 13.48
Townhouses with a proprietary wide batten aluminum roofing system. The aluminum is coated with a blue fluorocarbon coating, one of a number of available colors, for weather protection. (Architect: The Design Advocates, Inc., Gerald G. Curts, A.I.A., principal architect. Photo courtesy of Howmet/Alumax Building Specialties Division)

| Aluminum |
| Zinc and galvanized steel |
| Chromium |
| Steel |
| Stainless steel |
| Cadmium |
| Nickel |
| Tin |
| Lead |
| Brass |
| Bronze |
| Copper |

FIGURE 13.49
A galvanic series for metals used in buildings. Each metal is corroded by all those that follow it in the list. The wider the separation of two metals on the list, the more severe the corrosion is likely to be. Some families of alloys, such as stainless steels, are difficult to place with certainty because some alloys within the family behave differently than others. To be certain, the designer should consult with the manufacturer of any metal product before installing it in contact with dissimilar metals in an outdoor environment.

ROOFING AND THE BUILDING CODES

Manufacturing standards and installation procedures for roofing materials are specified by many building codes. Building codes also regulate the type of roofing that may be used on a building, based on a required level of fire resistance as measure by ASTM procedure E108. Roofing materials are grouped into four classes:

• Class A roof coverings are effective against severe fire exposure. They include slate, concrete tiles, clay tiles, and other materials certified as Class A by approved testing agencies. They may be used on any building in any type of construction.

• Class B roof coverings are effective

The roof plays a primal role in our lives. The most primitive buildings are nothing but a roof. If the roof is hidden, if its presence cannot be felt around the building, or if it cannot be used, then people will lack a fundamental sense of shelter.

Christopher Alexander, *et al. A Pattern Language.* New York, Oxford University Press, 1977, p. 570.

against moderate fire exposure, and include sheet metal roofings and some composition shingles. These are the minimum class that may be used on buildings of Type 1 construction as defined in the table in Figure 1.1.

• Class C roof coverings are effective against light fire exposure. They include most asphalt shingles and fire-retardant treated wood shingles and shakes. These are the minimum class that may be used on Types 2, 3, and 4A construction.

• Nonclassified roof coverings such as untreated wood shingles may be used on Type 4B construction and on some agricultural, accessory, and storage buildings.

The required class of roofing for a particular building may also be affected by an urban fire zone in which the building is located, and by the proximity of the building to its neighbors.

C.S.I./C.S.C. Masterformat Section Numbers for Roofing	
07190	VAPOR AND AIR RETARDERS
07200	INSULATION
07210	Building Insulation
07220	Roof and Deck Insulation
07300	SHINGLES AND ROOFING TILES
07310	Shingles
07320	Roofing Tiles
07500	MEMBRANE ROOFING
07510	Built-Up Bituminous Roofing
07515	Cold-Applied Bituminous Roofing
07520	Prepared Roll Roofing
07530	Elastomeric/Plastomeric Sheet Roofing
07535	Modified Bitumen Sheet Roofing
07540	Fluid Applied Roofing
07550	Protected Membrane Roofing
07570	TRAFFIC TOPPING
07600	FLASHING AND SHEET METAL
07610	Sheet Metal Roofing
07620	Sheet Metal Flashing and Trim
07630	Sheet Metal Roofing Specialties
07650	Flexible Flashings
07700	ROOF SPECIALTIES AND ACCESSORIES
07800	SKYLIGHTS

FIGURE 13.50
A house both roofed and sided with red cedar shingles. (Architect: William Isley. Photo by Paul Harper. Courtesy of Red Cedar Shingle and Handsplit Shake Bureau)

FIGURE 13.51
A standing seam metal roof with beautifully detailed overhangs, designed by architects Kallmann and McKinnell. (Photo by Steve Rosenthal)

SELECTED REFERENCES

1. Sheet Metal and Air Conditioning Contractors National Association, Inc. *Architectural Sheet Metal Manual*, 3rd Edition. Vienna, Virginia, 1980.

Sheet metal roofs are detailed in this excellent reference, along with every conceivable flashing, fascia, gravel stop, and gutter for flat and shingled roofs. (Address for ordering: 8224 Old Court House Road, Tysons Corner, Vienna, VA 22180.)

2. Baker, Maxwell C. *Roofs:Design, Application and Maintenance*. Montreal, Multiscience Publications Limited, 1980.

Only flat roofs are covered in this book, but the coverage is detailed and the writing and illustrations are clear. Both the novice and the expert will find this a useful volume. (This book can be obtained from the NRCA—for the address, see reference 3.)

3. National Roofing Contractors Association. *The NRCA Roofing and Waterproofing Manual*. Chicago, 1983.

This thick looseleaf volume is a comprehensive guide to current U.S. practice for both flat and sloping roofs. The treatment is exhaustive and both the diagrams and the text are excellent. Fluid-applied membrane roofing and sheet metal roofing are not covered. (Address for ordering: 8600 Bryn Mawr Avenue, Chicago, IL 60631.)

KEY TERMS

pitched roof
flat roof
shingle
tile
roof membrane
deck
vapor retarder
area divider
topside vent
ballast
protected membrane roof
deck fill insulation
built-up roof membrane
elastomeric/plastomeric roof membrane
liquid-applied roof membrane
coal tar

bitumen
felt
ply
cold-applied mastic
neoprene
EPDM
PVC
chlorinated polyethylene
chlorosulfonated polyethylene
polymer-modified bitumens
parapet
fascia
scupper
cant
coping
flashing
counter flashing

traffic deck
thatch
shake
asphalt shingle
asphalt roll roofing
terne
standing seam
batten seam
flat seam
Class A, B, C roof coverings
R value
water vapor
dew point
condensation
perm

REVIEW QUESTIONS

1. What are the major differences between a flat roof and a pitched roof? What are the advantages and disadvantages of each type?

2. Discuss the three positions in which thermal insulation can be installed in a flat roof, and the advantages and disadvantages of each.

3. Explain in precise terms the function of a vapor retarder in a flat roof.

4. Compare a built-up roof membrane to an elastomeric/plastomeric roof membrane.

5. What is the difference between cedar shingles and cedar shakes?

6. What metals are used for sheet metal roofing? What are the strengths and drawbacks of each?

7. What are the ASTM fire-resistance classifications for roofing materials? List one or two roofing materials that fall into each classification.

E X E R C I S E S

1. For a flat-roofed university classroom building with a masonry bearing wall, steel interior frame, corrugated steel roof deck, and parapet:

   a. Show two ways of achieving a 1:50 roof slope on structural bays 36 feet (11 m) square.

   b. Sketch a set of details of the parapet edge, expansion joint, area divider, and roof drain for a flat roof system of your choice. Show insulation, vapor retarder (if any), roof membrane, and flashings.

2. Sketch a fascia detail for a flat roof system of your choice, assuming the wall below is made of precast concrete panels and the roof deck of precast concrete slab elements.

3. Find a flat roof system being installed and take notes on the process until the roof is completed. Ask questions of the roofers, the architect, or your instructor about anything you don't understand.

4. Examine a number of existing flat roofs around your campus or neighborhood, looking for problems such as cracking, blistering, tearing, and leaking. Explain the reasons for each problem you discover.

GLASS AND GLAZING

———

———

*G*laziers install panels of reflective glass weighing up to 125 pounds (57 kg) each in one of four 80-foot-high (24 m) spires atop a new office building. *(Designers: John Burgee Architects with Philip Johnson. Photo courtesy of PPG Industries)*

Glass plays many roles and takes many forms in buildings—Gothic church windows made of thousands of jewel-like pieces of colored glass; breathtaking expanses of smooth, uninterrupted glass that fill whole walls of today's buildings; Elizabethan casement windows with tiny diamond panes set in lead; skyscrapers that shimmer in facets of reflective glass mirroring the sky; cozy windows; comfortable windows; windows that bring soft, natural light; windows that frame spectacular views; windows that welcome winter sunlight to warm a room. But glass can also form windows that make privacy impossible; windows that admit a harsh, glaring light; winter-cold surfaces that chill the body and tax the heating system; windows that broil a room in summer-afternoon sunlight. Glass skillfully used in buildings contributes strongly to our enjoyment of architecture, but glass thoughtlessly used can make a building unattractive and uncomfortable to inhabit.

HISTORY

The origins of glass are lost in prehistory. Initially a material for colored beads and small bottles, glass was first used in windows in Roman times. The largest known piece of Roman glass, a crudely cast sheet used for a window in a public bath at Pompeii, was nearly 3 feet by 4 feet (800 mm × 1100 mm) in size.

By the tenth century A.D., the Venetian island of Murano had become the major center of glassmaking, producing *crown glass* and *cylinder glass* for windows. Both the crown and cylinder processes were begun by blowing a large glass sphere. In the crown process, the heated glass sphere was adhered to an iron rod called a *punty*, opposite the blowpipe. The blowpipe was then removed, leaving a hole opposite the punty, and the

FIGURE 14.1
The entire facade of an office building for an English insurance brokerage firm consists of suspended glazing ½ inch (12 mm) thick. (Architects: Foster Associates. Photo courtesy of Pilkington Brothers Limited)

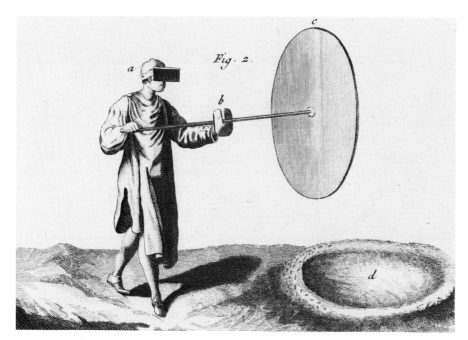

FIGURE 14.2

The glassworker in this old engraving wore a face shield (a) *and hand shield* (b) *to protect against the heat of the large glass crown* (c) *which he had just spun on the end of a punty. After cooling, the crown was cut into small lights of window glass.* (Courtesy of the Corning Museum of Glass, Corning, New York)

FIGURE 14.3

Making cylinder glass in the nineteenth century, Pittsburgh, Pennsylvania. Elongated glass bottles were blown by swinging the blowpipe back and forth in the pit in front of the furnace (center). As each bottle solidified (left), it was brought to another area where the ends were cut off to produce cylinders (right). The cylinders were then reheated and flattened into sheets from which window glass was cut. (Courtesy of the Corning Museum of Glass, Corning, New York)

sphere was reheated, whereupon the glassworker would spin the punty rapidly, using centrifugal force to open the sphere into a large disk, or *crown*, 30 inches (750 mm) or more in diameter (Figure 14.2). When the crown was cut into panes, one always contained the "bullseye" where the punty was attached before being cracked off. In the cylinder process, the sphere, heated to a plastic state, was swung back and forth in a pendulum fashion on the end of the blowpipe to elongate it into a cylinder. The hemispherical ends were cut off and the remaining cylinder slit lengthwise, reheated, and flattened into a rectangular sheet of glass that was later cut into panes of any desired size (Figure 14.3). Prior to the introduction of modern glassmaking techniques, crown glass was favored over cylinder glass for its surface finish, which was smooth and brilliant because it was formed without contacting another material. Cylinder glass, though more economical to produce, was limited in surface quality by the texture and cleanliness of the surface on which it was flattened.

Neither crown glass nor cylinder glass was of sufficient optical quality for the fine mirrors desired by the seventeenth-century nobility; for this reason, *plate glass* was first produced, in France, in the late seventeenth century. Molten glass was cast into frames, spread into sheets by rollers, cooled, then ground flat and polished with abrasives, first on one side and then the other. The result was a costly glass of near-perfect optical quality, in sheets of unprecedented size. Before long, mechanization of the grinding and polishing operations brought down the price of plate glass to a level that allowed it to be used for storefronts in both Europe and America.

In the nineteenth century, the cylinder process evolved into a method of drawing cylinders of molten glass vertically from a crucible, enabling the routine and economical production of cylinders 40 to 50 feet (12 to 15 m) long. In 1851 the Crystal Palace in London (Figure 9.2) was glazed with 900,000 square feet (84,000 m²) of cylinder glass supported on a cast-iron structure.

In the early years of the twentieth century, cylinder glass production was gradually replaced by processes that pulled flat sheets of *drawn glass* directly from the molten glass. Production lines for the grinding and polishing of plate glass were established, with rough glass sheets entering the line continuously at one end and finished sheets emerging at the other.

In 1959 the English firm of Pilkington Brothers Ltd. started production of a new kind of glass, *float glass*, which has since been licensed to other glassmakers and has become the worldwide standard, replacing both drawn glass and plate glass. In this process, a ribbon of molten glass is floated across a bath of molten tin, where it hardens before touching a solid surface (Figures 14.4 through 14.6). The resulting sheets of glass have parallel surfaces, high optical quality (virtually indistinguishable from plate glass), and a brilliant surface finish. Float glass has been produced in America since 1963, and now accounts for more than 90 percent of domestic flat glass production.

The terminology still associated with glass developed early in this long history. The term *glazing* as it applies to building refers to the installing of glass in an opening, or to the transparent material (usually glass) in a glazed opening. The installer of glass is known as a *glazier*. Individual pieces of glass are known as *lights*, or sometimes, to avoid confusion with visible light, *lites*.

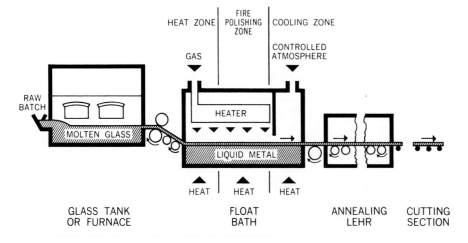

FIGURE 14.4

In the float glass process, molten glass from the furnace is floated on a bath of liquid tin to form a continuous sheet of glass. The annealing lehr cools the glass at a controlled rate to avoid internal stresses, after which it is cut into smaller sheets. (Courtesy of PPG Industries)

FIGURE 14.5
The superior flatness and bright surface finish of float glass are readily seen in the reflections on the glass ribbon emerging from the annealing lehr. (Courtesy of LOF Glass, a Libby-Owens-Ford Company)

FIGURE 14.6
Track-mounted cutting devices score the ribbon of cooled float glass as part of a computer-controlled cutting operation that automatically produces the glass sizes ordered by customers. (Courtesy of PPG Industries)

THE MATERIAL GLASS

The major ingredient of glass is sand (silicon dioxide), which is mixed with soda ash (sodium hydroxide or sodium carbonate), lime, and small amounts of alumina, potassium oxide, and elements to control color, then heated to form glass. The finished material, while seemingly crystalline and convincingly solid, is actually a supercooled liquid, for it has no fixed melting point, and an open, noncrystalline microstructure.

Ordinary, clear, single-strength glass transmits about 85 percent of the visible light incident upon it. When drawn into small fibers it is stronger than steel, but in larger pieces the microscopic imperfections that are an inherent characteristic of glass reduce its useful strength to significantly lower levels, particularly in tension. When a surface of a sheet of glass is placed in sufficient tension, as happens when an object strikes the glass, cracks propagate from an imperfection near the point of maximum tension, and the glass shatters.

Thicknesses of Glass

Glass is typically manufactured in a series of thicknesses ranging from approximately $\frac{3}{32}$ inch (2.5 mm), which is called *single-strength*, through $\frac{1}{8}$ inch (3 mm), or *double-strength*, to a maximum of $\frac{7}{8}$ inch (22 mm) and, on special order, 1 inch (25.4 mm). Glass thickness for a particular window is determined by the size of the sheet of glass, the expected maximum wind loads on the glass, and the predicted breakage rate that can be accepted. For low buildings with relatively small windows, single-strength and double-strength glass are usually sufficient. For larger windows and for windows in tall buildings, where high wind velocities are experienced at higher altitudes, thicker glass is generally required, along with increased attention to how the glass is supported in its frame. It has become standard practice for archi-

tects and structural engineers to order extensive wind tunnel testing of models of tall buildings during the design process to establish the expected maximum wind pressures and suctions on the windows. Because of unavoidable manufacturing defects in the glass and the probability of subsequent damage to the glass during installation and while it is in service, a certain amount of breakage must always be anticipated in a large building. Charts are furnished by glass manufacturers to help the designer strike a satisfactory balance between safety and economy (Figure 14.7).

During its manufacture, ordinary window glass is *annealed*, cooled slowly under controlled conditions, to avoid locked-in thermal stresses that might cause it to behave unpredictably in use. But several other types of glass have come into use for particular purposes in buildings.

Tempered Glass

Tempered glass is produced by cutting annealed glass to the required sizes for use, reheating it to approximately 1200 degrees Fahrenheit (650° C), and cooling both its surfaces rapidly with a blast of air while its core cools much more slowly. This process induces permanent compressive stresses in the edges and faces of the glass, and tensile stresses in the core. The resulting glass is about four times as strong in bending as annealed glass, and much more resistant to thermal stress and impact. When tempered glass does break, the sudden release of its internal stresses reduces it instantaneously to small, square-edged granules rather than long, sharp-edged shards. These properties make tempered glass useful for windows exposed to heavy wind pressures or intense heat or cold. From the point of view of safety, the breakage characteristics of tempered glass make it especially well suited for use in and around exterior doors, where people may accidentally bump against the glass, and for floor-to-ceiling sheets of

FIGURE 14.7

Thickness selection charts for annealed float glass supported on four sides by mullions. To use these charts, start by calculating the area of a light of glass in square feet, and its aspect ratio, which is the long dimension of the light divided by the short dimension. Determine the expected wind load on the glass in pounds per square foot (psf), either from the local building code or from wind tunnel test results. Select the chart for a trial thickness and find the intersection of the calculated aspect ratio and the given wind loading. If this intersection lies above the line for the calculated area of the light, try the chart for the next thicker glass, and repeat until a satisfactory glass thickness is found. Similar charts are published for annealed glass supported on two sides, and for other types of glass. These charts are based on an anticipated failure rate of 8 lights of glass per 1000 installed. (Courtesy of PPG Industries)

PPG ANNEALED FLOAT GLASS

4-SIDE

Four Sides Supported in Weathertight Rabbet

Architect's Specified Probability of Breakage = 8/1000 Lights for a 1-Minute Uniform Wind Load Duration

FOOTNOTES:

1. CONTACT YOUR PPG REPRESENTATIVE FOR:
   A. NON-LISTED PPG ARCHITECTURAL PRODUCTS.

2. IMPACT AND OTHER LOAD CONSIDERATIONS REQUIRE SPECIAL STUDY.

3. SUPPORT DEFLECTION SHALL NOT EXCEED 1/175 SPAN AT DESIGN LOAD.

4. CHECK SPECIFICATIONS AND CODE REQUIREMENTS.

A. DESIGN SPECIFICATIONS NOT SHOWN.

glass, which are often walked into by people who mistake them for openings in the wall. Tempered glass is also used for all-glass doors that have no frame at all (Figure 14.8), for whole walls of squash and handball courts, and for basketball backboards.

Tempered glass does have its drawbacks. It is substantially more costly than annealed glass. It usually has noticeable optical distortions created by the tempering process. And all cutting, drilling, and edging must be done before the glass is tempered because any such operations after tempering will release the stresses in the glass and cause it to disintegrate.

Heat-Strengthened Glass

For many applications, lower-cost *heat-strengthened glass* is preferable to tempered glass. The heat-strengthening process is similar to tempering, but the induced compressive stresses in the surfaces and edges are about one-third as high (typically 5000 pounds per square inch compared to 15,000 pounds per square inch for tempered glass, or 34 MN/m² versus 104 MN/m²). Heat-strengthened glass is about twice as strong in bending as annealed glass, and much more resistant to thermal stress. It usually has fewer distor-

FIGURE 14.8
Tempered glass is used for strength and breakage safety in both the doors and windows of this store in a downtown shopping mall. (Photo by the author)

tions than tempered glass. Its breakage behavior is more like that of annealed glass than tempered glass.

Laminated Glass

Laminated glass is made by sandwiching a transparent vinyl interlayer between sheets of glass and bonding the three layers together under heat and pressure. When laminated glass breaks, the soft vinyl holds the shards of glass in place rather than allowing them to fall out of the frame of the window. This makes laminated glass useful for skylights and overhead glazing, because it reduces the risk of injury to people below in case of breakage (Figure 14.9). Laminated glass is also a better barrier to the transmission of sound than solid glass. It is used to glaze windows of residences, class-rooms, hospital rooms, and other rooms that must be kept quiet in the midst of noisy environments. It is especially effective when installed in two or more layers with airspaces between. *Security glass*, used for drive-in banking windows and other facilities that need to be resistant to burglary, is made of multiple layers of glass and vinyl, and is available in a range of thicknesses to stop any desired caliber of bullet.

FIGURE 14.9
Laminated glass provides safety against falling glass in this overhead sloped glazing installation. (Courtesy of PPG Industries)

Wired Glass

Prior to the introduction of tempered and laminated glass, *wired glass*, produced by rolling a mesh of small wires into a sheet of hot glass, was the only available type of safety glazing material. When wired glass breaks from thermal stress, the wires hold the sheet of glass together. This makes wired glass valuable for windows in fire doors and fire walls, where it maintains its integrity as a fire barrier for an extended period of time during a fire.

Patterned Glass

Hot glass can be rolled into sheets with many different surface patterns for use where light transmission is desired but vision must be obscured for privacy (Figure 14.10).

Spandrel Glass

Special opaque glasses are produced for covering the spandrel areas (the bands of wall around the edges of floors) in glass curtain wall construction (Figure 14.11). Some spandrel glasses are made as similar as possible in exterior appearance to the glass that will be used for the windows on a specific project. It is, however, very difficult, even with reflective coated glass, to make the spandrels indistinguishable from the windows under all lighting conditions. Other spandrel glasses are given a colored backup coating or a heat-fused ceramic frit on the interior side of the glass to give them a different appearance from the windows of the building. Factory-applied insulation on the interior of the glass can be supplied by many glass companies, complete with vapor retarder. Spandrel glass is usually tempered or heat-strengthened to resist the thermal stresses that can be caused by accumulations of solar heat behind the spandrel.

FIGURE 14.10
Patterned glass is used here to divide space without significantly reducing the flow of light. (Courtesy of AFG Industries, Inc., Kingsport, Tennessee)

Imagine a city iridescent by day, luminous by night, imperishable! Buildings, shimmering fabrics, woven of rich glass; glass all clear or part opaque and part clear, patterned in color or stamped to harmonize with the metal tracery that is to hold it all together, the metal tracery to be, in itself, a thing of delicate beauty consistent with slender steel construction...

Frank Lloyd Wright,
in *Architectural Record*, April 1928.

Tinted and Reflective Coated Glass

Solar heat buildup can be problematic in the inhabited spaces of buildings with large areas of glass, especially during the warm part of the year. Fixed sun shading devices outside the windows are the best way of blocking unwanted sunlight, but glass manufacturers have developed tinted and reflective glasses that are designed to reduce glare and cut down on solar heat gain.

Tinted Glass

Tinted glass is made by adding small amounts of selected chemical elements to the molten glass mixture to produce the desired hue and intensity of color in grays, bronzes, blues, greens, and golds. The visible light transmitted by commercially available tinted glasses varies from 14 percent in a very dark gray glass to 75 percent in the lightest tints, as compared to about 85 percent for clear glass. The total amount of transmitted solar heat is significantly higher than these figures, however, because the heat absorbed by the glass must go somewhere, and a substantial portion of it is transmitted to the interior of the building (Figure 14.12). To simplify comparisons among various types of heat-reducing glasses, a quantity called the *shading coefficient* is used; it is the ratio of total solar heat transmission through a particular glass to total solar heat transmission through double-strength clear glass. Shading coefficients for tinted glasses range from about 0.50 to 0.75, indicating that they transmit from one-half to three-quarters of the incident solar

FIGURE 14.11
Lever House in New York, an early glass curtain wall building designed by architects Skidmore, Owings and Merrill, uses dark-green glass for the spandrels and lighter-green glass for the windows. (Courtesy of PPG Industries)

energy that would be transmitted by double-strength clear glass, whose shading coefficient is established as 1.0.

Reflective Coated Glass

To achieve lower shading coefficients, thin, durable films of metal or metal oxide can be deposited on a surface of either clear or tinted glass sheets under closely controlled conditions. Depending on the composition of the film, the film side of the glass may be turned either toward the inside of the building or toward the outside. The film is thin enough to see through, but turns away a substantial proportion of the incident solar energy before it can enter the building. Shading coefficients for reflective coated glasses vary from about 0.31 to 0.70, depending on the density of the metallic coating. Reflective coated glasses appear as mirrors from the outside on a bright day and are often chosen by architects for this property alone (Figure 14.13). At night, with lights on inside the building, they appear as dark but transparent glass. The sunlight reflected by a building of reflective coated glass can be helpful in some circumstances by lighting an otherwise dark urban street space, but it can also create problems in other circumstances by bouncing solar heat and glare into neighboring buildings and onto the street.

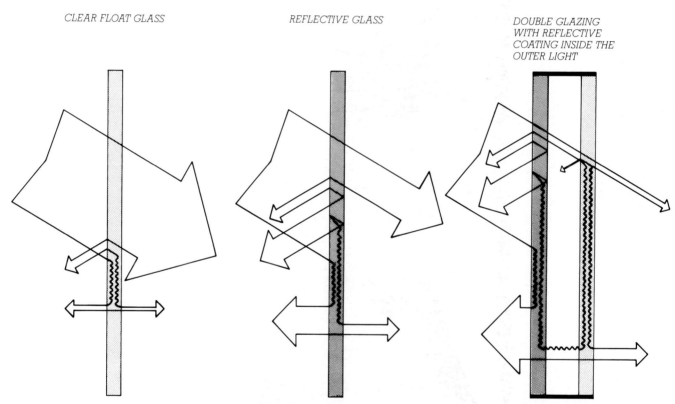

CLEAR FLOAT GLASS

REFLECTIVE GLASS

DOUBLE GLAZING WITH REFLECTIVE COATING INSIDE THE OUTER LIGHT

FIGURE 14.12

A schematic representation of the effect of three different glazing assemblies on incoming sunlight. Outdoors is to the left. The relative widths of the arrows indicate the relative percentages of the incoming light transmitted, reflected, and absorbed. In clear float glass, to the left, most of the light is transmitted, with small quantities reflected, absorbed, and reradiated as heat. Reflective coated glass, center, bounces a large proportion of the light back to the outdoors, and also absorbs and reradiates a significant amount. In double glazing many different combinations of types of glass are possible; the one shown to the right of this diagram utilizes glass with an inside reflective coating for the outer light.

FIGURE 14.13
Reflective coated windows with subtly different reflective coated glass spandrels. (Architects: Paul Rudolph and 3D International. Photo courtesy of PPG Industries)

Who when he first saw the sand and ashes...would have imagined that in this shapeless lump lay concealed so many conveniences in life...by some such fortuitous liquefaction was mankind taught to procure a body at once in a high degree solid and transparent; which might admit the light of the sun, and exclude the violence of the wind; which might extend the sight of the philosopher to new ranges of existence...

Dr. Samuel Johnson, writer and lexicographer, 1709–1784.

Insulating Glass

Window glass is an extremely poor thermal insulator. A single sheet of glass conducts heat about five times as fast as an inch (25 mm) of polystyrene foam insulation, and ten to twenty times as fast as a well-insulated wall. A second sheet of glass applied to a window with an airspace between the two sheets cuts this rate of heat loss in half, and a third sheet reduces the rate of heat loss to about a third of the rate through a single sheet. The practice of doubling or tripling layers of glass in windows has long been used in cold climates, but was frequently attended by problems of moisture condensation and frost inside the outer sheet of glass. In recent times the factory production of hermetically sealed *double-glazed* or *triple-glazed* window units has virtually eliminated these problems. Two kinds of edge seals are currently in use (Figure 14.14). For small lights of double glass, the edges of the two sheets can be fused together, with dry air or a dry, low-conductivity gas inserted in the space between. For lights of any size, and for any number of thicknesses of glass, a hollow metal spacer bar can be inserted between the edges of the sheets of glass, and the edges closed with an organic sealant compound. A small amount of a chemical drying agent is left inside the spacer bar to remove any residual moisture from the trapped air. The air or gas is always inserted at atmospheric pressure to avoid structural pressures on the glass. The thickness of the airspace is less critical to the insulating value of the glass than the mere presence of an airspace: From $\frac{3}{8}$ inch (9 mm) up to about an inch (25 mm) of thickness, the insulating value of the airspace increases somewhat, but above that thickness little additional benefit is gained. A standard overall thickness for large lights of double glazing is 1 inch (25.4 mm), which results in an airspace $\frac{1}{2}$ inch thick (13 mm) if quarter-inch (6 mm) glass is used.

The thermal performance of double glazing can be improved substantially by the use of a *low-emissivity* coating on one or both of the sheets of glass. A low-emissivity coating allows most sunlight to pass through the window but reflects back to the interior of the building most of the infrared radiation from warm surfaces indoors. Coated double glazing commercially available through many glass and window manufacturers equals or slightly exceeds the thermal performance of uncoated triple glazing. One proprietary system inserts a very thin sheet of coated polyester film in the middle of the airspace between two sheets of uncoated glass to produce an assembly about equal in insulating value to quadruple glazing.

Low-Iron Glass

Ordinary glass contains traces of iron that give it a greenish tinge when viewed from the edge of the sheet. This iron is responsible for absorbing

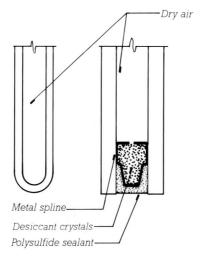

FIGURE   14.14

Two methods of sealing the edge of double glazing: fused glass edges, to the left, and a metal spacer bar and organic sealant, to the right. Desiccant crystals in the spline or spacer bar absorb any residual moisture in the airspace.

a small but significant amount of the total radiant energy in sunlight as it passes through the glass. For use in solar energy collectors, special low-iron glasses are produced by a number of manufacturers. The higher transparency of these glasses is especially useful in double-glazed collectors, where any reduction in energy transmission through a single sheet of glass is compounded mathematically by losses through the second sheet.

Plastic Glazing Sheets

Transparent plastic sheets are often used instead of glass for certain specialized glazing applications. The two most common types of plastic glazing materials are *acrylic* and *polycarbonate*. Both are more expensive than glass. Both have very high coefficients of thermal expansion, which cause them not merely to expand and contract with temperature changes, but also to bow visibly toward the warm side when subjected to high indoor-outdoor temperature differentials. This in turn requires that plastic sheet materials be installed in their frames with relatively expensive glazing details that allow for plenty of linear movement and rotation. Both polycarbonate and acrylic are soft and easily scratched, although more scratch-resistant formulations are available. Despite all these problems, plastic glazing sheets find use in many situations where glass is inappropriate: The plastics can be cut to shapes with inside corners (L-shapes and T-shapes, for example) that are likely to crack if cut from glass. They can be bent easily to fit in curved frames. They can be heat-formed into domed glazing for skylights. And the plastics, especially polycarbonate, which is literally impossible to break under ordinary conditions, are widely used for windows in buildings where vandalism is a problem.

Translucent but nontransparent plastic sheets reinforced with glass fibers are also used in building. Corrugated sheets are used for industrial skylights and residential patio roofs. Thin, flat sheets of a special formulation with a high translucency to solar energy are used for skylights and low-cost solar collector glazing.

There is a general problem with any type of plastic glazing sheet of discoloration with time, caused by ultraviolet degradation of the molecules of the plastic. The yellowing of acrylic and polycarbonate sheets has been largely brought under control by improved material formulations, but most of the glass-reinforced materials change appearance noticeably over a span of a few years and require periodic replacement. Plastic glazing sheets are also combustible, which restricts their use in some applications.

GLAZING

Glazing Small Lights

Small lights of glass are subjected neither to large wind force stresses nor to large amounts of thermal expansion and contraction. They may be glazed by very simple means (Figure 14.15). In traditional wood sash, the glass is first held in place by small metal *glazier's points*, and then sealed on the outside with *putty*, a simple compound of linseed oil and pigment which gradually hardens by oxidation of the oil. Putty must be protected from the weather by subsequent painting. For factory-glazed sash, improved putties or *glazing compounds* are employed. These are generally more adhesive and more elastic than traditional putty, and are formulated so as not to harden and become brittle with age.

Glazing Large Lights

Large lights of glass (those over 6 square feet or 0.6 m² in area) require more care in glazing. Wind load stresses on each light of glass are higher, and the glass must span farther between its supporting edges. Any irregularities in the frame of the window may result in distortion of the glass, highly concentrated pressures on small areas of the glass, or in metal frames, glass-to-metal contact, any of which can lead to abrasion or fracture of the glass. In large walls of glass, thermal expansion and contraction can cause a buildup of stresses in the glass.

The design objectives for a large-light glazing system are:

1. To support the weight of the glass in such a way that the glass is not subjected to intense or abnormal stress patterns.

2. To support the glass against wind pressure and suction.

3. To isolate the glass from the effects of structural deflections in the frame of the building and the smaller framework of *mullions* that supports the glass.

4. To allow for expansion and contraction of both glass and frame without damage to either.

5. To avoid contact of the glass with the frame of the window or with any other material that could abrade or stress the glass.

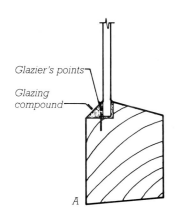

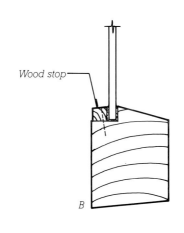

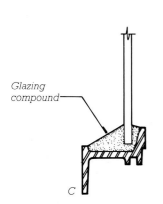

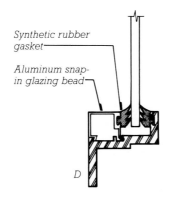

FIGURE 14.15
Alternative methods of single-glazing small lights of glass. Glass is traditionally mounted in wood sash using glazier's points and glazing compound (A), or a wood stop nailed to the sash (B). Metal sash is often glazed in the manner shown in (C), but most aluminum sash is now glazed with some type of snap-in beads and synthetic rubber gaskets, as exemplified in (D).

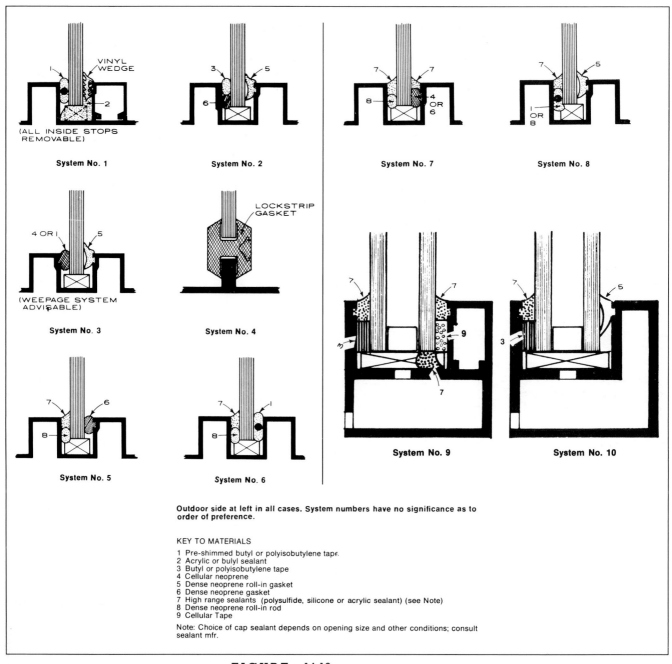

Outdoor side at left in all cases. System numbers have no significance as to order of preference.

KEY TO MATERIALS

1 Pre-shimmed butyl or polyisobutylene tape.
2 Acrylic or butyl sealant
3 Butyl or polyisobutylene tape
4 Cellular neoprene
5 Dense neoprene roll-in gasket
6 Dense neoprene gasket
7 High range sealants (polysulfide, silicone or acrylic sealant) (see Note)
8 Dense neoprene roll-in rod
9 Cellular Tape

Note: Choice of cap sealant depends on opening size and other conditions; consult sealant mfr.

FIGURE 14.16
An assortment of typical large-light glazing details, with the outdoor side to the left. (Reprinted with permission from AAMA Curtain Wall Design Guide Manual)

The weight of the glass is supported in the frame by synthetic rubber *setting blocks*, normally two per light, located at the quarter points of the bottom edge of the light. For support against wind loads, a specified amount of *bite* (depth of grip on the edge of the glass) is provided by the supporting mullions. If the bite is too little, the glass may pop out under wind loading; if too much, the glass may not be able to deflect enough under heavy wind loads without being stressed at the edge. The mullions, of course, must be stiff enough to transmit the wind loads from the glass to the frame of the building without deflecting enough to overstress the glass. The resilient glazing material used to seal the glass-to-mullion joint must be of sufficient dimension and elasticity to allow for any anticipated thermal movement and for possible irregularities in the mullions.

The glazing materials that are most commonly used between the mullions and the glass include:

• *Preformed solid tape sealant*, a thick ribbon of very sticky synthetic that is adhered by pressure to the glass and the mullions. Several different formulations (butyl, polyisobutylene) are usually lumped under the name *polybutene*. This material exerts an extremely strong hold on the glass and stays plastic indefinitely to allow for movement in the glazing system. Systems 1, 2, 3, and 6 in Figure 14.16 illustrate applications of preformed solid tape sealant to large-light glazing.

• *Liquid sealants* of various formulations are injected into the joint between the glass and the mullion with a sealant gun, and cure to form synthetic rubber. To maintain the required thickness of joint while the sealant is being injected and cured, synthetic rubber *centering shims* are inserted between glass and frame below the space to be occupied by the sealant. Liquid sealant materials are covered in more detail in Chapter 15; their application in glazing is shown in Systems 5 through 10 in Figure 14.16.

• *Compression gaskets* of rubber, synthetic rubber, or plastic are compressed between the glass and the mullion to form a watertight seal and cushion for the glass. These are shown in Systems 1, 2, 3, 5, 8, and 10 in Figure 14.16. System 4 is a special case of the compression gasket, a two-piece *lockstrip gasket* that is a completely self-contained glazing system. Figures 14.17 through 14.20 show some applications of lockstrip gasket glazing.

The properties that solid tape sealants, liquid sealants, and compression gaskets have in common are that they possess the required degree of resiliency, they can be installed in the thickness required to cushion the glass against all expected movements, and they form a watertight seal against both glass and frame. To guard against possible leakage and moisture condensation, however, *weep holes* should be provided to drain water from the horizontal mullions, as seen in Systems 9 and 10 in Figure 14.16.

FIGURE 14.17
Installing lockstrip gasket on an aluminum mullion. (Courtesy of Standard Products Company)

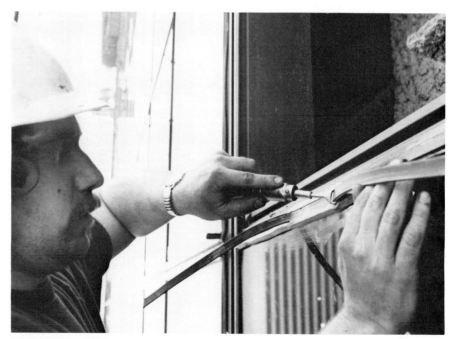

FIGURE 14.18
Inserting the lockstrip to expand the gasket and seal it against the glass.

(Courtesy of Standard Products Company)

FIGURE 14.19
A finished lockstrip gasket glazing installation. (Courtesy of Standard Products Company)

FIGURE 14.20
Lockstrip gasket glazing in precast double-tee concrete wall panels. (Courtesy of Standard Products Company)

Advanced Glazing Systems

In their search for ever more minimal buildings, architects have encouraged the development of several systems of glazing that seem, in varying degrees, to defy gravity. In *butt-joint glazing*, the head and sill of the glass sheets are supported conventionally in metal frames, but the vertical mullions are eliminated, the vertical joints between sheets of glass being made by the injection of a clear silicone sealant. This gives a strong effect of unbroken horizontal bands of glass wrapping continuously around the building (Figures 14.21 and 14.22). In *structural silicone flush glazing*, the metal mullions are entirely inside the glass, with the glass adhered to the mullions by silicone sealant. The outside skin of the building is therefore completely flush, unbroken by protruding mullions (Figures 14.23 through 14.25).

Perhaps most striking of all is the *glass mullion system* or *suspended glazing system*, used primarily for high

FIGURE   14.21
Mullionless butt-joint glazing uses only a bead of clear silicone sealant at the vertical joints in the glass. The glass in this example is ¾ inch (19 mm) thick. (Courtesy of LOF Glass, a Libby-Owens-Ford Company)

FIGURE   14.22
Another mullionless butt-joint glazing installation, as seen from the outside. (Architects: Neuhaus & Taylor. Photo courtesy of PPG Industries)

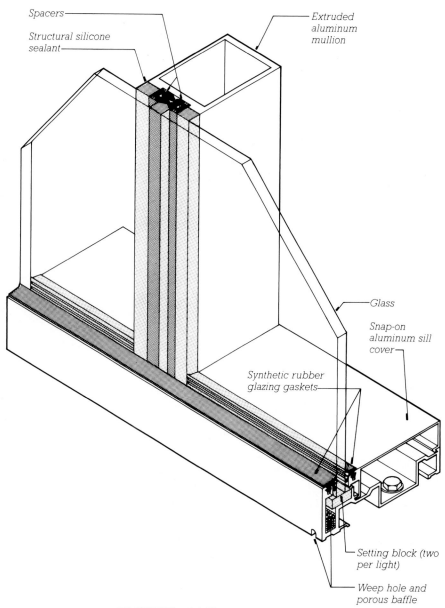

Spacers

Structural silicone
sealant

Extruded
aluminum
mullion

Glass

Snap-on
aluminum sill
cover

Synthetic rubber
glazing gaskets

Setting block (two
per light)

Weep hole and
porous baffle

FIGURE 14.23
*Horizontal strip windows that need to
appear mullionless only from the
exterior can be created by adhering the
glass to interior mullions with structural
silicone sealant. The sill and head are
conventionally glazed, using snap-on
aluminum covers to hold the interior
glazing gaskets. Either single glazing, as
shown, or double glazing can be used
with this type of system (Copied by
permission from PPG EFG System 401
details, courtesy of PPG Industries)*

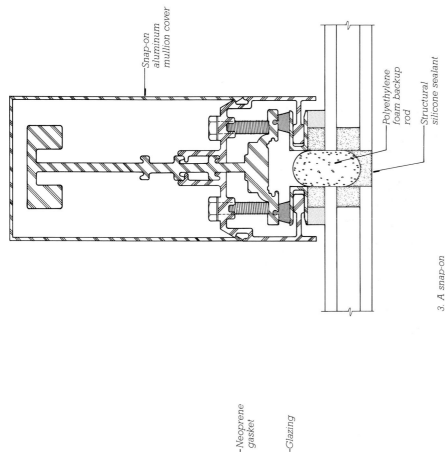

Snap-on aluminum mullion cover

Polyethylene foam backup rod

Structural silicone sealant

3. A snap-on mullion cover and an exterior sealant joint complete the assembly.

Neoprene gasket

Glazing

Extruded aluminum structural mullion

Structural silicone sealant

2. The machine screw and attachment strip clamp the glazing unit into place.

Machine screw

Extruded aluminum attachment strip

Glazing strip

Backup

1. Glazing is factory fabricated with metal glazing strips adhered to the glass using structural silicone sealant.

FIGURE 14.24

Steps in the assembly of a mullion for a four-side structural silicone exterior flush glazing system. This system is used to construct multistory glass walls with no metal exposed on the exterior of the building. The adhesive action of structural silicone sealant is the sole means by which the glass is held in place. This system is applicable either to double glazing, as shown, or to single glazing. Notice that the internal complexities of the aluminum components and sealant are completely concealed when the installation is finished—from the inside, one sees only a simple rectangular aluminum mullion, and from the outside, only glass and a thin bead of silicone sealant. (Copied by permission from PPG EFG System 712 details. Courtesy of PPG Industries)

FIGURE 14.25
A hotel clad with dark glass using a four-side structural silicone exterior flush glazing system. (Architects: Patrick and Associates. Photo courtesy of PPG Industries)

FIGURE 14.26
The vertical glass mullions allow for maximum unobstructed view of the racetrack in this suspended glazing installation. (Architects: Akel, Logan and Shafer. Photo courtesy of PPG Industries)

FIGURE 14.27
Another racetrack building with suspended glazing uses tapered glass stabilizers and metal patch plates.

(Architects: Century A. E. Photo courtesy of Pilkington Brothers Limited)

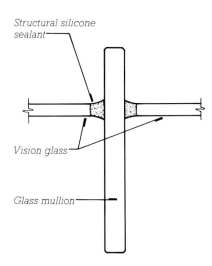

Structural silicone sealant

Vision glass

Glass mullion

FIGURE 14.28
A typical detail of a glass mullion in a suspended glazing assembly.

walls of glass around building lobbies and enclosed grandstands. The tempered glass sheets in this system are suspended from above on special clamps and are stabilized against wind pressure by perpendicular stiffeners, also of tempered glass. The components are joined only by polysulfide or structural silicone sealant, and sometimes by metal patch plates at corners of large sheets (Figures 14.26 through 14.28).

GLASS AND ENERGY

Glass is a two-way pipeline for the flow of both conducted and radiated heat. As noted previously, glass, even when doubled or tripled, conducts heat very rapidly into or out of a building and can collect and trap large amounts of solar heat inside a building.

In residential buildings, the conduction of heat through glass is generally to be minimized in the extremely hot or cold seasons of the year, so multiple glazings, low-emissivity coatings, and snug curtains or shutters are desirable features for windows. Warming sunlight is welcome in winter, but highly undesirable in summer, which leads to the orientation of major windows toward the south, with overhangs or sunshades above to protect them against the high summer sun. Large east or west windows lead to severe overheating in summer and should be avoided unless they are shaded by nearby trees.

In nonresidential buildings, heat generated within the building by lights, people, and machinery is often sufficient to maintain comfort throughout much of the winter. In warmer weather, this heat must be removed from the building by a cooling system, along with any solar heat that has entered the windows. In this situation, north-facing windows contribute the least to the cooling load on the building, and south-facing windows with horizontal sunshades above allow the entry of solar

FIGURE 14.29
Leaded diamond-pane windows of handmade rolled glass in an Elizabethan English bay window. (Photo by the author)

FIGURE 14.30
Discrete windows in an office building. (Architects: Skidmore, Owings and Merrill. Photo courtesy of PPG Industries)

FIGURE 14.31
Limestone and metal mullions for a Gothic church window. (Courtesy of Indiana Limestone Institute)

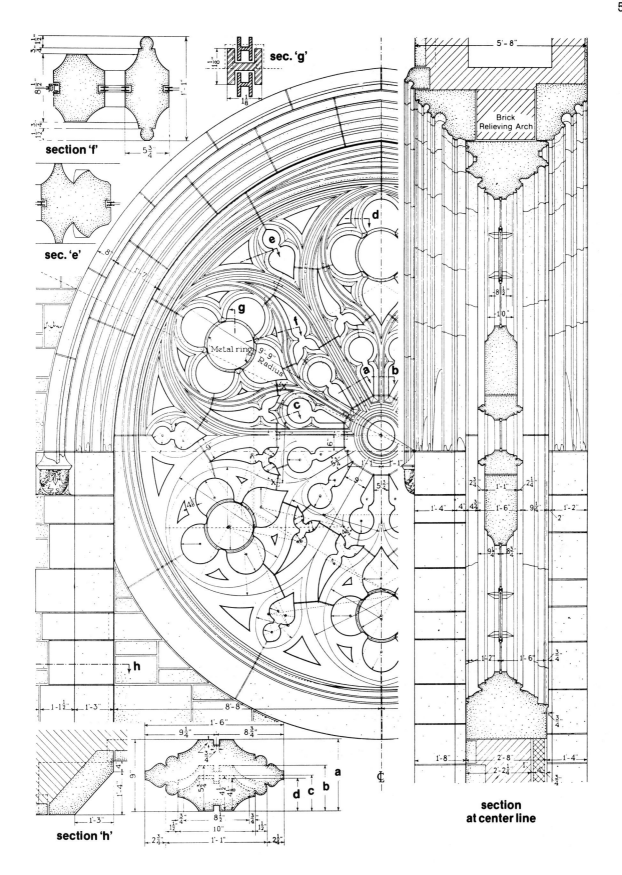

section 'f'

sec. 'g'

sec. 'e'

section 'h'

Brick Relieving Arch

Metal ring 9'-9" Radius

section at center line

heat only in the winter. East and west windows are undesirable, as they contribute strongly to summertime overheating and are very difficult to shade. Shades or blinds inside the glass are helpful in eliminating the glare from such windows, but they do little to keep out the heat because once sunlight strikes them, its heat is already inside the building and little of it will escape.

Tinted and reflective glasses are of obvious value in controlling the entry of solar heat into buildings, to the point that they might be perceived as encouraging the designer to pay little attention to window size and orientation. But there is a small, slowly growing body of buildings that point the way toward a future time when architectural style may spring less from an interest in abstract surface geometries and more from a thoughtful response to local climate. These buildings are characterized by different fenestration schemes for the different sides of the building, each designed to create an optimal flow of heat into and out of the building for that orientation, and each making creative use of the available types of glass to help accomplish this objective. The results, as measured in occupant comfort and energy savings, are generally impressive, and the aesthetic possibilities are intriguing.

This last statement could apply equally well to the role of glass in admitting light to a building. Electric lighting is often the major consumer of energy in a commercial building, especially when the heat generated by the lights must be removed from the building by a cooling system. Daylight shining through windows and skylights, distributed throughout a space by reflecting and diffusing surfaces, can reduce or eliminate the need for electric lights under many circumstances and is often more pleasant than artificial illumination. New design methods are making it progressively easier to predict the levels of daylighting that can be achieved with alternative designs, and more and more architects and engineers are becoming expert in this field.

FIGURE 14.32
Reflective coated glass and matching spandrel glass. (Architects: Odell Associates. Photo courtesy of PPG Industries)

FIGURE 14.33
Leaded stained glass in the Robie House, Chicago, 1906. (Architect: Frank Lloyd Wright. Photo by the author)

Winter dining rooms and bathrooms should have a southwestern exposure, for the reason that they need the evening light, and also because the setting sun, facing them in all its splendor but with abated heat, lends a gentler warmth to that quarter in the evening. Bedrooms and libraries ought to have an eastern exposure, because their purposes require the morning light...Dining rooms for Spring and Autumn to the east; for when the windows face that quarter, the sun, as he goes on his career from over against them to the west, leaves such rooms at the proper temperature at the time when it is customary to use them.

Marcus Vitruvius Pollio, Roman architect, 1st century B.C.

FIGURE 14.34
Reflective coated glass in a hotel complex. (Architect: Welton Becket Associates. Photo courtesy of LOF Glass, a Libby-Owens-Ford Company)

GLASS AND THE BUILDING CODES

Building codes concern themselves with several functional aspects of glass: its role in providing natural light in habitable rooms; its breakage safety; its safety in preventing the spread of fire through a building; and its role in determining the energy consumption in a building.

The BOCA Code, while permitting the use of artificial illumination alone in many types of buildings, encourages the use of natural light and suggests that each room have a glass area equal to at least 8 percent of its floor area. Minimum ceiling heights are established in the same section of the code, in part to assist in the propagation of natural light by successive reflections and diffusions within the room.

Breakage safety is regulated in skylights and overhead glazing, in glass located in or near doors, and in large sheets of glass that might be mistaken for clear openings in a wall. Laminated glass and plastic glazing sheets, because they will not drop out of the skylight if broken, are the only skylight glazings that are permitted without restriction under most codes. Annealed glass, heat-strengthened glass, wired glass, and tempered glass are permitted only in skylights equipped with wire screens below to catch the falling glass in case of breakage. Annealed glass skylights are further required to have a protective screen above the glass to catch falling objects. In and around doors, and in windows having a glazed area in excess of 9 square feet (0.9 m²) whose lowest edge is within 18 inches (450 mm) of the floor, tempered glass, laminated glass, or plastic glazing materials must be used to prevent the disfiguring and sometimes fatal accidents that could occur if annealed glass were used in such situations.

Wired glass must be used for openings in required fire doors and fire sep-aration walls. The areas of glazed openings in these locations are regulated by all building codes. Most codes also require that windows aligned above one another in certain types of buildings over three stories in height be separated vertically by fire-resistive spandrels of a specified minimum height. The intent of this provision is to restrict the spread of fire from one floor of a building to the floors above. If a glass spandrel is used under this type of code provision, it must be backed up inside with a material that offers the necessary fire resistance.

Code provisions relating to glass and energy consumption are often as simple as specifying a required thermal resistance for windows in certain types of buildings. Many codes require, in effect, that double glazing or storm windows, as a minimum, be installed in residential buildings. Future codes are likely to require that an analysis be performed by the designer to show that the overall performance of glass with respect to heat loss, heat gain, and daylighting, has been optimized.

THE FUTURE OF GLASS

The current pace of development in glass technology is such that it is difficult to predict what will happen in the field over the coming years. It is certain that many new developments will become available for improving the energy characteristics of glass: selectively reflective coatings, new types of multiple glazing, double glazing with an evacuated space between, glasses that vary their degree of transparency with the intensity of the light incident upon them, between-the-glazings louvers and movable insulating layers, and multiple glazing edge seals that are less conductive of heat. Another direction for future development is stronger glass, through new methods of heat or chemical treatment, surface coatings or laminated interlayers of stronger materials, or fiber reinforcing. A third direction is new glazing techniques that do away with more and more of the metal supporting structure and replace it with glass. But perhaps the most promising area for future development is in the growing ability of the designers of buildings, the architects and engineers, to design with glass not merely as a formal element, but also as a sophisticated tool for regulating the flow of light and heat through the skin of a building, and as a poetic means of expression through the selective opening of a building's walls to light and views.

C.S.I./C.S.C. Masterformat Section Numbers for Glass	
08800	GLAZING
08810	Glass
08840	Plastic Glazing
08850	Glazing Accessories

S E L E C T E D   R E F E R E N C E S

1. The most current information on glass will be found in manufacturers' catalogs. If you have access to *Sweet's Catalog File*, nearly every glass manufacturer's catalog is included in Section 8.26.

2. Flat Glass Marketing Association. *FGMA Glazing Manual*. Topeka, Kansas.

This handbook, updated frequently, summarizes current practice in the use of glass in buildings. (Address for ordering: 3310 Harrison, Topeka, KN 66611.)

K E Y   T E R M S

cylinder glass
crown glass
plate glass
drawn glass
float glass
glazing
glazier
lights (or lites) of glass
single-strength, double-strength
annealed glass
tempered glass
heat-strengthened glass

laminated glass
security glass
wired glass
patterned glass
spandrel glass
tinted glass
reflective coated glass
double glazing, triple glazing
acrylic
polycarbonate
putty
setting blocks

mullion
polybutene tape
solid tape sealant
liquid sealant
compression gasket
lockstrip gasket
butt-joint glazing
exterior flush glazing
suspended glazing
daylighting

R E V I E W   Q U E S T I O N S

1. What are the advantages of float glass over drawn glass? Over plate glass?

2. Name two situations in which you might use each of the following types of glass: a. tempered glass b. laminated glass c. wired glass d. patterned glass e. reflective glass f. polycarbonate plastic glazing sheet.

3. What are the design objectives for a large-light glazing system?

4. Discuss the role of glass facing each of the principle directions of the compass in adding solar heat to an air-conditioned office building in summer. How should windows be treated on each of these facades to minimize summertime solar heat gain?

5. In what ways does a typical building code regulate the use of glass, and why?

1. Examine the ways in which glass is mounted in several actual buildings and sketch a detail of each. Explain why each detail was used in its situation, and why you agree or disagree with the detail used.

2. Find a book on passive solar heating of houses, and in it, a table of solar heat gains for your area. For windows facing each of the four directions of the compass, plot a curve that represents the solar heat coming through a square foot of clear glass on an average day in each month of the year. Which window orientation maximizes wintertime heat and minimizes summertime heat? Is there an orientation that maximizes summertime heat and minimizes wintertime heat?

CLADDING

―

―

A Houston, Texas office building is clad in a sleek, reflective skin of structurally glazed glass (see Figure 14.24) and fluoropolymer-coated aluminum. *(Architect: Lloyd, Jones, Brewer. Photo by Wes Thompson. Courtesy of Cupples Products Division, H. H. Robertson Company)*

Cladding typifies a paradox of building: Those parts of a building that are seen are also those that must take the wear and weathering. The *cladding* (non-loadbearing exterior wall enclosure) is the most visible part of a building, one to which architects devote a great deal of time to achieve the desired visual effect. It is also the part of the building most subject to attack by natural forces that can literally tear the building apart, and its design is an intricate process that merges art, science, and craft to solve a very long and difficult list of problems.

THE DESIGN REQUIREMENTS FOR CLADDING

Primary Functions of Cladding

The major purpose of cladding is to separate the indoor environment of a building from the outdoors in such a way that the indoors can be brought to the desired set of environmental conditions for its intended use. This translates into a number of separate and diverse functional requirements.

Keeping Water Out

Cladding must prevent the entry of rain, snow, and ice into a building. This requirement is complicated by the fact that water on the face of a building is often driven by wind at high velocities and high air pressures, not just in a downward direction, but in every direction, even upward. Water problems are especially acute on tall buildings, which rise to altitudes where wind velocities are much higher than at ground level, and which present a large profile to the wind. Here enormous amounts of water must be drained from a windward building face during a heavy rainstorm, and the water, pushed by wind, tends to accumulate in crevices and against projecting mullions, where it will readily penetrate the smallest crack or hole and enter the building.

FIGURE 15.1
A closeup of the building shown in the opening photograph. (Architect: Lloyd, Jones, Brewer. Photo by Wes Thompson. Courtesy of Cupples Products Division, H. H. Robertson Company)

Preventing Air Leakage

The cladding of a building must prevent the unintended passage of air between indoors and outdoors. At a gross scale this is necessary to regulate air velocities within the building. Smaller air leaks are harmful because they waste conditioned air, carry water through the wall, allow moisture vapor to condense inside the wall, and allow noise to penetrate the building from outside.

Controlling Light

The cladding of a building must control the passage of light, especially sunlight. Sunlight is heat that may be welcome or unwelcome. Sunlight is visible light, useful for illumination but bothersome if it causes glare within a building. Sunlight includes destructive ultraviolet wavelengths that must be kept off human skin and away from interior materials that will fade or disintegrate.

FIGURE 15.2
A steel-framed Chicago office building during the installation of its aluminum, stainless steel, and glass curtain wall cladding. (Architects: Kohn Pederson Fox/Perkins & Will. Photo by Architectural Camera. Courtesy of American Institute of Steel Construction)

FIGURE 15.3
Chicago's Reliance Building, built in 1894–1895, has a curtain wall of glass and white terra cotta tiles. (Architect: Charles Atwood, of Daniel H. Burnham and Company. Photo by Wm. T. Barnum. Courtesy of Chicago Historical Society ICHi-18294)

Controlling the Radiation of Heat

Beyond its role in regulating the flow of radiant heat from the sun, the cladding of a building should also present to people inside the building interior surfaces that are at temperatures that will not cause radiant discomfort. A very cold interior surface will make nearby people feel chilly, even if the air in the building is warmed to a comfortable level, and a hot surface or direct sunlight in summer can cause overheating of the body despite the coolness of the interior air.

Controlling the Conduction of Heat

The cladding of a building must resist to the required degree the conduction of heat into and out of the building. This requires not merely a satisfactory overall resistance of the wall to the passage of heat, but the avoidance of *thermal bridges*, wall components such as metal framing members that are highly conductive of heat and therefore likely to cause localized condensation on interior surfaces.

Controlling Water Vapor

The cladding of a building must retard the passage of water vapor. Vapor moving through a wall assembly is likely to condense inside the assembly in cold weather and cause problems of staining, lost insulating value, corrosion, and freeze-thaw deterioration. The cladding must be constructed to prevent condensation of moisture insofar as possible and, where condensation is inevitable, to drain the condensate safely to the outdoors.

Controlling Sound

Cladding serves to isolate the interior of a building from noises outside, or vice versa. Noise isolation is best achieved by walls that are airtight, massive, and resilient. The required degree of noise isolation varies from one building to another, depending on the noise levels and noise tolerances of the inside and outside environments. Cladding for a hospital near a major airport requires a high level of noise isolation. Cladding for a commercial office in a suburban office park need not perform to as high a standard.

Secondary Functions of Cladding

The fulfillment of the primary functional requirements of cladding leads unavoidably to another set of requirements.

Resisting Wind Forces

The cladding of a building must be adequately strong and stiff to sustain the pressures and suctions that will be placed upon it by wind. For low buildings, which are exposed to relatively slow and predictable winds, this requirement is fairly easily met. The upper reaches of taller buildings are buffeted by much faster winds, and wind directions and velocities are often determined by aerodynamic effects from surrounding buildings. High suction forces can occur on some portions of the cladding, especially near corners of the building (Figure 15.4).

Adjusting to Movement

A building is never at rest. Several different kinds of forces are always at work throughout a building, tugging and pushing both the frame and the cladding: thermal expansion and contraction, moisture expansion and contraction, and structural movement. These forces must be anticipated and allowed for in designing a system of building cladding.

Thermal Expansion and Contraction The cladding of a building has to accommodate to thermal expansion and contraction at several levels: Indoor/outdoor temperature differences can cause warping of cladding

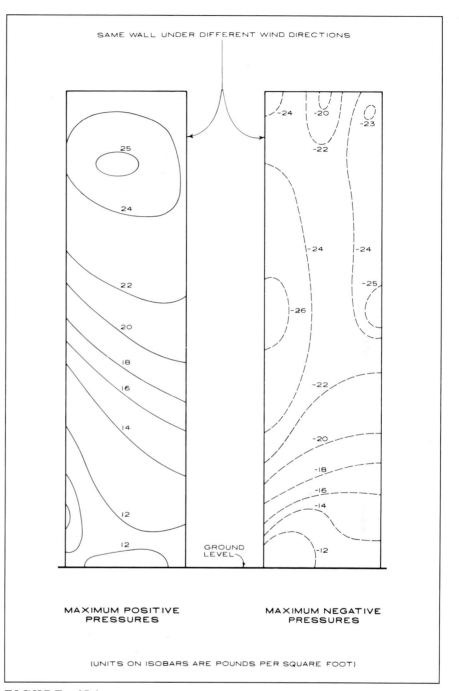

FIGURE 15.4

An example of expected positive and negative wind pressures on the cladding of a tall building, shown here in elevation, as predicted by wind tunnel testing. The building in this case is 64 stories tall and triangular in plan. Notice the high negative pressures (suctions) on the upper regions of the facade. The wind pressures on a building are dependent on many factors, including the shape of the building, its orientation, topography, wind direction, and surrounding buildings, so each building must be tested individually to determine the pressures it is expected to undergo. (Reprinted with permission from AAMA Curtain Wall Design Guide Manual)

panels due to differential expansion and contraction of their inside and outside faces (Figure 15.5). The cladding as a whole, exposed to outdoor temperature variations, grows and shrinks constantly with respect to the frame of the building, which is usually protected by the cladding from temperature extremes. And the building frame itself will expand and contract to some extent, especially between the time the cladding is installed and the time the interior space of the building is temperature controlled.

Moisture Expansion and Contraction
Masonry cladding materials must accommodate their own expansion and contraction caused by varying moisture content. Bricks and building stone expand slightly as they absorb moisture. Concrete block

shrinks slightly after installation in a building, as its curing is completed and excess moisture is given off. These movements are small but can accumulate to significant and potentially troublesome quantities in long or tall panels of masonry.

Structural Movements Cladding of every type must adjust to movements in the frame of the building. Building foundations may settle unevenly, causing distortions of the frame. Gravity forces shorten columns and deflect beams and girders to which cladding is attached. Wind and earthquake forces push laterally on building frames and wrack panels attached to the faces. Long-term creep causes significant shortening of concrete columns and sagging of concrete beams during the first year or two of a building's life.

If building movements from temperature differences, moisture differences, structural stresses, and creep are allowed to be transmitted between the frame and the cladding, unexpected things can happen. Cladding components can be subjected to forces for which they were not designed, which can result in broken glass, buckled cladding, sealant failures, and broken cladding attachments (Figure 15.6). In extreme cases, the building frame can end up supported by the cladding, rather than the other way around, or pieces of cladding can fall off the building.

Resisting Fire

The cladding of a building can interact in several ways with building fires. This has resulted in a number of building

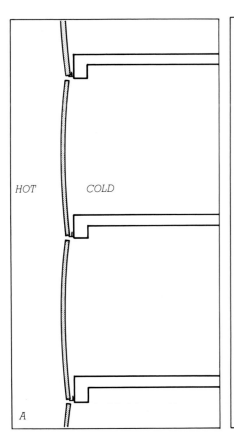

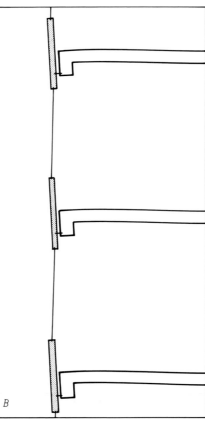

HOT   COLD

A

B

FIGURE 15.5
Distortions of curtain wall panels, illustrated in cross section: **A.** *Bowing due to greater thermal expansion of the outside skin of the panels under hot summertime conditions.* **B.** *Twisting of spandrel beams due to the weight of the curtain wall.*

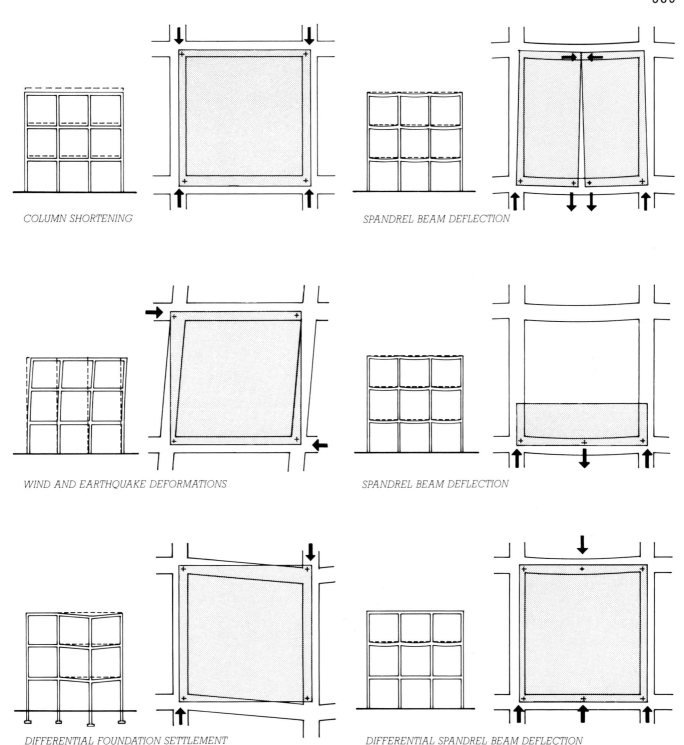

COLUMN SHORTENING

SPANDREL BEAM DEFLECTION

WIND AND EARTHQUAKE DEFORMATIONS

SPANDREL BEAM DEFLECTION

DIFFERENTIAL FOUNDATION SETTLEMENT

DIFFERENTIAL SPANDREL BEAM DEFLECTION

FIGURE 15.6

Forces on curtain wall panels caused by movements in the frame of the building, illustrated in elevation. In each of the six examples, the drawing to the left shows the movement in the overall frame of the building, and the larger-scale drawing to the right shows its consequences on the curtain wall panels (shaded in gray) covering one bay of the building. Points of attachment between the panels and the frame are shown as crosses. The black arrows indicate forces on the wall panels caused by the movement in the structure. The magnitude of the structural movements is exaggerated for clarity, and some inadvisable attachment schemes are shown to demonstrate their consequences. Forces such as these, if not taken into account in the design of the frame and cladding, can result in glass breakage, panel failures, and failure of the attachments between the panels and the frame.

code provisions relating to the construction of building cladding, as summarized in the last paragraph of this chapter.

Weathering Gracefully

To maintain the visual quality of a building, its cladding must weather gracefully. The inevitable dirt and grime should accumulate evenly, without streaking or splotching, and functional provisions must be made for maintenance operations such as glass and sealant replacement, and for periodic cleaning, including scaffolding supports and safety attachment points for window washers. The cladding must resist oxidation and ultraviolet degradation of organic materials, corrosion of metallic components, and freeze-thaw damage of stone, brick, concrete block, and tile.

Installation Requirements for Cladding

Cladding should be easy to install. There should be secure places for the installers to stand, preferably on the floors of the building rather than on scaffolding outside. There must be built-in adjustment mechanisms in all the fastenings to allow for the inaccuracies that are normally present in the structural frame of the building and the cladding components themselves. There must be dimensional clearances provided to allow the cladding components to be inserted without binding against adjacent components. And most importantly, there must be forgiving features that allow for a lifetime of trouble-free cladding function despite all the lapses in workmanship that inevitably occur—features such as backup air barriers and drainage channels to get rid of moisture that has leaked through a faulty sealant joint, or generous edge clearances that keep a sheet of glass from contacting the hard material of the frame even though the glass is installed slightly crooked.

BASIC CONCEPTS OF BUILDING CLADDING

The Curtain Wall

Until the advent of the steel building frame a century ago, exterior walls of buildings were invariably loadbearing. The first steel-framed skyscraper, the Home Insurance Building in Chicago (1883), introduced the concept of the *curtain wall*, an exterior cladding (made in this case of masonry) supported at each story by the steel frame, rather than bearing its own load to the foundations (Figure 9.4). The principal advantage of the curtain wall is that because it bears no vertical load, it can be thin and lightweight regardless of the height of the building, as compared to a masonry bearing wall, which becomes prohibitively thick and weighty at the base of a very tall building. The name "curtain wall" derives from the idea that the wall is thin and "hangs" like a curtain on the structural frame. (Most curtain wall panels do

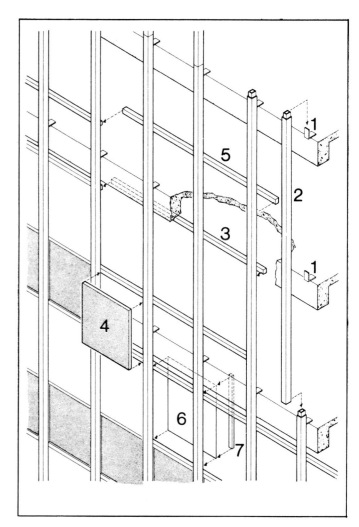

STICK SYSTEM—Schematic of typical version

1: Anchors. 2: Mullion. 3: Horizontal rail (gutter section at window head). 4: Spandrel panel (may be installed from inside building). 5: Horizontal rail (window sill section). 6: Vision glass (installed from inside building). 7: Interior mullion trim.

Other variations: Mullion and rail sections may be longer or shorter than shown. Vision glass may be set directly in recesses in framing members, may be set with applied stops, may be set in sub-frame, or may include operable sash.

A

not actually hang in tension from the frame, but are supported from the bottom at each floor level.) Curtain wall cladding today is made variously from brick, stone, or concrete masonry, slabs of cut stone, panels of precast concrete, a profusion of different metal-glass combinations, preassembled components of light cold-formed steel framing with skins of many different materials, panels molded from glass fiber reinforced polyester resin or glass-fiber-reinforced concrete, and so on. There are almost as many curtain wall systems as there are curtain wall buildings.

Curtain Wall Installation Methods

Curtain walls can be classified in groups according to their degree of assembly at the time of installation on the building (Figure 15.7).

Many metal-and-glass curtain walls are made up of a so-called *stick system* whose principle components are metal

FIGURE 15.7

*Schematic drawings of five different systems of curtain wall installation: **A.** Stick system. **B.** Unit system. **C.** Unit-and-mullion system. (overleaf) **D.** Panel system. **E.** Column-cover-and-spandrel system.* (Reprinted by permission from AAMA Aluminum Curtain Wall Design Guide Manual)

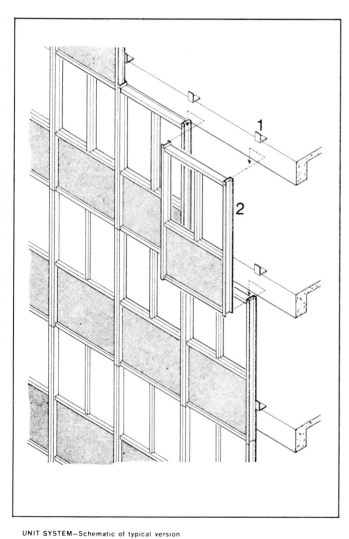

UNIT SYSTEM—Schematic of typical version

1: Anchor. 2: Pre-assembled framed unit.

Other variations: Mullion sections may be interlocking "split" type or may be channel shapes with applied inside and outside joint covers. Units may be unglazed when installed or may be pre-glazed. Spandrel panel may be either at top or bottom of unit.

B

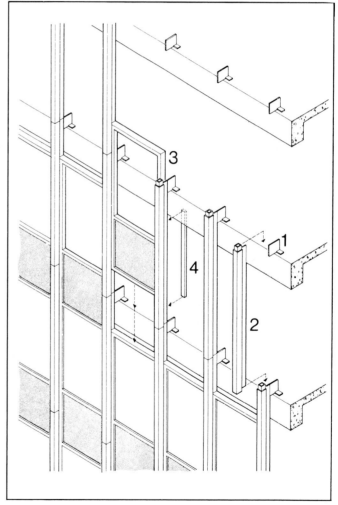

UNIT-AND-MULLION SYSTEM—Schematic of typical version

1: Anchors. 2: Mullion (either one- or two-story lengths). 3: Pre-assembled unit —lowered into place behind mullion from floor above. 4: Interior mullion trim.

Other variations: Framed units may be full-story height (as shown), either unglazed or pre-glazed, or may be separate spandrel cover units and vision glass units. Horizontal rail sections are sometimes used between units.

C

mullions and rectangular panels of glass or metal that are assembled in place (Figure 15.7a). Stick systems have the advantages of low shipping bulk and a high degree of ability to adjust to unforeseen site conditions, but they must be assembled on-site, under highly variable conditions, rather than in a factory with its ideal tooling, controlled environmental conditions, and generally lower wage rates.

The *unit system* of curtain wall installation takes full advantage of factory assembly and minimizes on-site labor, but the units require more space during shipping than stick components and more protection from damage (Figure 15.7b). The *unit-and-mullion system* (Figure 15.7c) offers a middle ground between the stick system and the unit system.

The *panel system* (Figure 15.7d) is made up of homogeneous units that may be formed from metal, precast from concrete, cut from stone, or molded of glass-fiber-reinforced polyester or concrete. Its advantages and disadvantages are similar to those of the unit system, but it involves the higher tooling costs of a custom-made mold, which makes it advantageous only for buildings requiring a large number of identical panels.

The *column-cover-and-spandrel system* (Figure 15.7e) emphasizes the structural module of the building rather

FIGURE 15.7 D, E

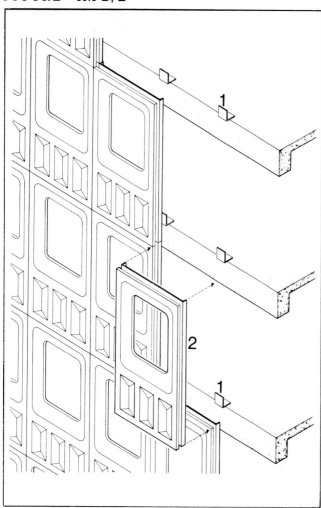

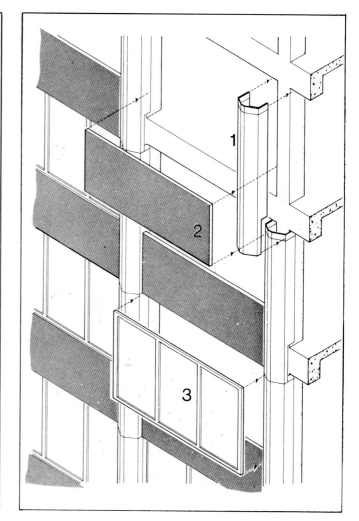

PANEL SYSTEM—Schematic of typical version

1: Anchor. 2: Panel.

Other variations: Panels may be formed sheet or castings, may be full story height (as shown) or smaller units, and may be either pre-glazed or glazed after installation.

COLUMN COVER AND SPANDREL SYSTEM—Schematic of typical version

1: Column cover section. 2: Spandrel panel. 3: Glazing infill.

Other variations: Column covers may be one piece or an assembly, may be of any cross-sectional profile, and either one or two stories in height. Spandrel panel may be plain, textured or patterned. Glazing infill may be a pre-assembly, either glazed or unglazed, or be assembled in place.

than creating its own applied grid on the facade, as do the previously described systems. A custom design must be created for each project because there is no standard column or floor spacing for buildings. Special care is required to ensure that the spandrel panels do not deflect downward when loads are applied to the spandrel beams of the building frame; otherwise the window strips could be subjected to unforeseen loadings that could deform the mullions and crack the glass.

In addition to the installation methods shown in Figure 15.7, curtain walls are also assembled entirely in place on the frame of the building in the case of masonry curtain walls, as will be described later in this chapter.

CONCEPTUAL APPROACHES TO WATERTIGHTNESS IN CLADDING

In order for water to penetrate a wall, three simple conditions must be satisfied simultaneously:

1. There must be water present at the outer face of the wall.

2. There must be an opening through which the water can move.

3. There must be a force to move the water through the opening.

If any one of these conditions is not satisfied, the wall will not leak. This suggests that there are three conceptual approaches to making a wall watertight:

1. One can try to keep water completely away from the wall. This is impossible, however, on any but the smallest of buildings. A very broad overhang can keep a single-story wall dry under most conditions, but on a taller building it must be assumed that the wall will get wet.

2. One can try to eliminate the openings from a wall. This is the traditional approach to making a wall watertight, especially in the United States. One simply builds very carefully, sealing every seam in the wall with sealant or gaskets, attempting to eliminate every hole and crack.

This approach works fairly well if done well, but it has inherent problems. Even a solid brick wall is porous and will allow the entry of some water if wetted for a prolonged period. In a wall made up of sealant-jointed components, the joints are unlikely to be perfect. If an edge is a bit damp, dirty, or oily, sealant may not stick to it. If the worker applying the sealant is insufficiently skilled or has to reach a bit too far to finish a joint, he or she may fail to fill the joint completely. Even if the joints are all made perfectly, building movements can tear the sealant or pull it loose. Because the sealant is on the outside of the building, it is exposed to the full destructive forces of sun, wind, water, and ice, and may fail prematurely from weathering. And whatever the cause of sealant failure, because the sealant joint is on the outside face of the wall, it is difficult to reach for inspection and repair.

In response to these problems, many cladding designers have adopted a theory of *internal drainage* or *secondary defense*, which accepts the uncertainties of external sealant joints, and attempts to deal with them by providing internal drainage channels within the cladding to carry away any leakage, and backup sealant joints to the inside of the drainage channels.

3. One can try to eliminate or neutralize all the forces which can move water through the wall. These forces are five in number: gravity, momentum, surface tension, capillary action, and air currents (Figure 15.8).

Gravity is a factor in pulling water through a wall only if the wall contains an inclined plane that slopes into, rather than out of, the building. It is usually

a simple matter to detail cladding so that no such inclined planes exist, though sometimes a loose gasket or an errant bead of sealant can create one despite the best efforts of the designer.

Momentum is the horizontal component of the energy of a raindrop falling at an angle toward the face of a building. It can drive water through a wall only if there is a suitably oriented slot or hole that goes completely through the wall. Momentum is easily neutralized by applying a cover to each joint in the wall, or by designing each joint as a simple *labyrinth*.

Surface tension of water that causes it to adhere to the underside of a cladding component can allow water to be drawn into the building. The provision of a simple *drip* on any underside surface to which water might adhere will eliminate the problem.

Capillary action is the surface-tension effect that pulls water through any opening that can be bridged by a water drop. It is the primary force that transports water through the pores of a masonry wall. It can be eliminated as a factor in the entry of water through a wall by making each of the openings in a wall wider than a drop of water can bridge, or if this is not feasible or desirable, by providing a concealed *capillary break* somewhere inside the opening. In porous materials such as brick, capillary action can be counteracted by applying an invisible coating of a silicone-based water repellant, which destroys the adhesive force between water and the walls of the pores in the brick.

This leaves *wind currents* remaining as the major force that must be neutralized if water is to be kept from penetrating through an opening in a wall. This is, indeed, the force most difficult to deal with in designing a wall for watertightness.

The Rainscreen Principle

The generic solution to the wind current problem is to let wind pressure

FIGURE 15.8

Five forces that can move water through an opening in a wall, illustrated in cross section with outdoors to the left. Each pair of drawings shows first a horizontal joint between curtain wall panels in which a force is causing water leakage through the wall, then an alternative design for the joint which neutralizes this force. Leakage caused by gravity is avoided by sloping internal surfaces of joints toward the outside. Momentum leakage can be prevented with a simple labyrinth as shown. A drip and a capillary break are shown here as means for stopping leakage from surface tension and capillary action, which are closely related forces. Air pressure differences between the outside and inside of the joint will result in air currents that can transport water through the joint. This is prevented by closing the area behind the joint with a pressure equalization chamber (PEC) as shown. When wind strikes the face of the building, a slight movement of air through the joint raises the pressure in the PEC until it is equal to the pressure outside the wall, after which all air movement ceases. Each joint in a curtain wall must be designed to neutralize all five of these forces.

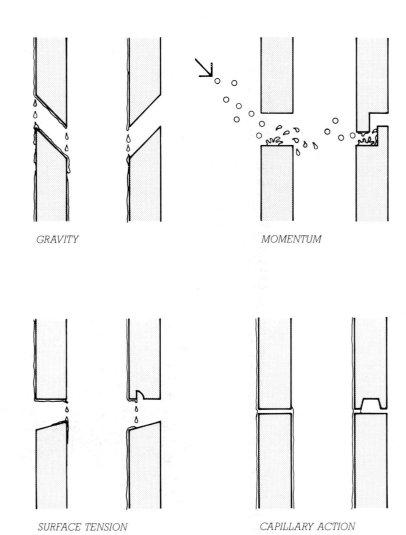

GRAVITY

MOMENTUM

SURFACE TENSION

CAPILLARY ACTION

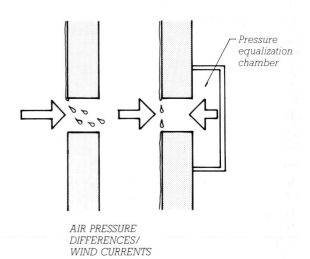

Pressure equalization chamber

AIR PRESSURE DIFFERENCES/ WIND CURRENTS

differences between the outside and inside of the cladding neutralize themselves through implementation of a concept known as the *rainscreen principle*. The implementation of this principle in *pressure-equalized wall design* involves the creation of a more-or-less airtight barrier at the internal side of the cladding, protected from direct exposure to the outdoors by a loose-fitting, labyrinth-jointed layer known as the *rainscreen*. As wind pressures on the cladding build up and fluctuate, extremely small currents of air pass back and forth through the unsealed joints, not enough to carry water, but just enough to equalize the pressure inside and outside the joint (Figure 15.8). A small flaw in the air barrier, such as a sealant bead that has pulled away from one side of the joint, is unlikely to cause a water leak because the volume of air that can pass through the flaw is still relatively small and probably insufficient to carry water. Such small amounts of water as may penetrate partway through the unsealed joints in the rainscreen are drained away to the face of the building by channels built into the joints. By contrast, any flaw, no matter how small, in an external sealant joint without an air barrier behind will cause a water leak, because the sealant joint itself is wetted (Figure 15.9).

Because wind pressures across the face of a building vary considerably at any given moment between one area of the face and another, the air chamber behind a rainscreen should be divided into compartments not more than two stories high and a bay or two wide. If this is not done, air may rush through the joints in higher-pressure areas of the face and flow across the air chamber to lower-pressure areas, carrying water with it as it goes.

The rainscreen principle, which we have come to understand fully only in recent years, is virtually foolproof in theory. In practice, as applied to the major joints in a curtain wall, it is sometimes difficult to implement, primarily because many of the joints in the cladding that need to be sealed to

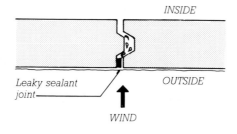

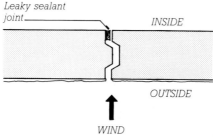

FIGURE 15.9
Leakage through a defective sealant joint between curtain wall panels, shown in plan view. In the upper example, the sealant joint is at the outside face of the panels, where it is wetted during a storm. Air passing through the defective joint carries water with it. In the lower example, with the defective sealant installed on the inside of the panels, air leakage through the joint is insufficient to transport water through the joint, and no water penetrates.

create an air barrier tend to fall outside columns and spandrel beams where they cannot be reached from inside the building for sealing, inspection, or repair. There are at present relatively few constructed buildings that rely completely upon the concept of pressure equalization for watertightness. But the rainscreen principle is steadily being adopted in more and more cladding designs and will undoubtedly become the governing principle in the design of future cladding systems as we learn how to use it to its full potential.

SEALANT JOINTS IN CLADDING

Sealant Materials

Gunnable Sealant Materials

Gunnable sealant materials are viscous, sticky liquids that are injected into the joints of a building with a *sealant gun* (Figure 15.10). They cure in the joint to become rubberlike materials that adhere to the surrounding surfaces and seal the joint against the passage of air and water. Gunnable sealants can be grouped conveniently in three categories according to the amount of change in joint size that each can withstand safely after curing:

Low-range sealants or *caulks* are materials with very limited elongation capabilities, up to plus or minus five percent. They are used mainly for filling minor cracks in small buildings. They include *oil-based caulks*, which harden by oxidation, and *latex caulks*, which cure by evaporation of mixing water. Most caulks shrink during curing, which further limits their usefulness. None is used for sealing of joints in curtain walls.

Medium-range sealants are materials such as butyl rubber or acrylic that have safe elongations in the plus or minus 5 to 10 percent range. They are used in building cladding for sealing of nonworking joints (joints that are fastened together mechanically as well as being filled with sealant, as shown in Figure 15.11). Because these sealants cure by the evaporation of water or an organic solvent, they undergo some shrinkage during curing.

High-range sealants can safely sustain elongations up to plus or minus 25 percent. They include various *polysulfides* and *polymercaptans*, which are usually site-mixed from two components to effect a chemical cure; *polyurethanes*, which may also cure from a two-component reaction, or else from

reacting with moisture vapor from the air, depending on the formulation; and *silicones*, which cure by reacting with moisture vapor from the air. None of these sealants shrinks upon curing because none relies upon the evaporation of water or a solvent to effect a cure. All adhere tenaciously to the sides

FIGURE 15.10
Applying polysulfide, a high-range gunnable sealant, to a joint between exposed-aggregate precast concrete curtain wall panels, using a sealant gun. The operator moves the gun slowly so a bulge of sealant is maintained at the nozzle to exert enough pressure on the sealant that it fully penetrates the joint. Following application, the operator will return to smooth and compress the wet sealant into the joint with a convex tool, much as a mason tools the mortar joints between masonry units. (Courtesy of Morton Thiokol, Inc., Morton Chemical Division)

of properly prepared joints. All are highly resilient, rubberlike materials that return to their original size and shape after being stretched or compressed, and all are durable for twenty years or more if properly formulated and installed. Sealants for the working joints in curtain wall systems are selected from among this group.

Solid Sealant Materials

In addition to the gunnable sealants, several types of solid materials are used for sealing seams in building cladding:

Gaskets are strips of various fully cured elastomeric (rubberlike) materials, manufactured in a number of different configurations and sizes for different purposes (Figure 15.11). They are either compressed into a joint to seal tightly against the surfaces on either side or inserted in the joint loose, then expanded with a *lockstrip* insert as illustrated in Figures 14.17 through 14.20.

Preformed cellular tape sealants are relatively new to the North American market. They are strips of polyurethane sponge material impregnated with a chemical sealant and delivered to the site compressed to one-fifth or one-sixth of their original volume. When the strip is inserted, it expands to fill the joint and the sealant material cures to form a watertight seam.

Preformed solid tape sealants are used only in lap joints, as in the mounting of glass in a metal frame, or the overlapping of two thin sheets of metal at a cladding seam. They are thick, sticky ribbons of polybutene or polyisobutylene that adhere to both sides of the joint and seal and cushion the junction.

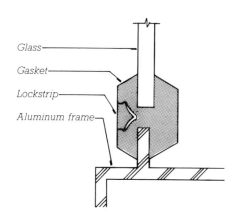

Glass
Gasket
Lockstrip
Aluminum frame

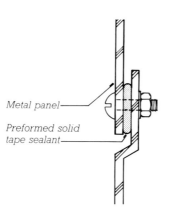

Metal panel
Preformed solid tape sealant

FIGURE 15.11
Some solid sealant materials. At the left, two examples of lockstrip gaskets. At the right, preformed solid tape sealants. The upper right example shows a nonworking joint, in which two sheets of aluminum are fastened with bolts so little movement can occur between them.

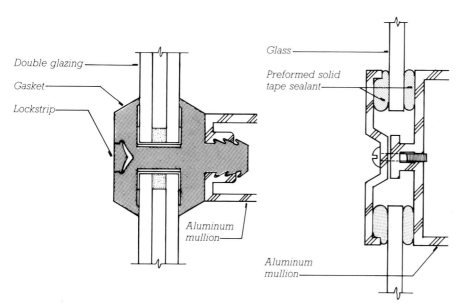

Double glazing
Gasket
Lockstrip
Aluminum mullion

Glass
Preformed solid tape sealant
Aluminum mullion

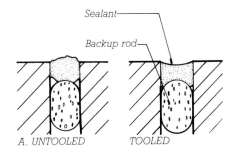

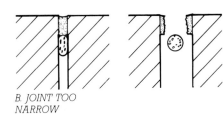

A. UNTOOLED *TOOLED*

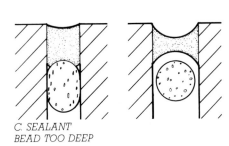

B. JOINT TOO NARROW

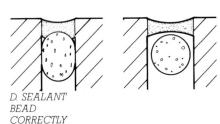

C. SEALANT BEAD TOO DEEP

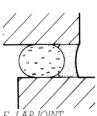

D. SEALANT BEAD CORRECTLY PROPORTIONED

E. LAP JOINT CORRECTLY PROPORTIONED

FIGURE 15.12

Good and bad examples of sealant joint design. **A.** *This properly proportioned joint is shown both untooled and tooled. The untooled sealant fails to penetrate completely around the backup rod and does not adhere fully to the sides of the joint.* **B.** *A narrow joint may cause the sealant to elongate beyond its capacity when the panels on either side contract, as shown to the right.* **C.** *If the sealant bead is too deep, sealant is wasted, and the four edges of the sealant bead are stressed excessively when the joint enlarges.* **D.** *A correctly proportioned sealant bead. The backup rod, made of a spongy material that does not stick to the sealant, is inserted into the joint to maintain the desired depth. The width is calculated so the expected elongation will not exceed the safe range of the sealant, and the depth is between ⅛ and ¾ inch (3 and 9.5 mm).* **E.** *A correctly proportioned lap joint. The thickness of the joint (the distance between the panels) should be twice the depth of the sealant bead, and also twice the expected movement in the joint.*

Sealant Joint Design

Figures 15.12 through 15.14 show the major principles that need to be kept in mind while designing a gunnable sealant joint. When sealing a joint between materials with high coefficients of expansion, the time of year when the sealant is to be installed must be taken into account when specifying the size of the joint and the type of sealant; sealant installed in cold weather will

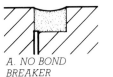 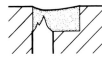

A. NO BOND BREAKER

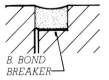 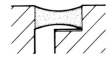

B. BOND BREAKER

FIGURE 15.13

In three-sided joints, tearing of the sealant is likely to occur unless a nonadhering plastic bond breaker is placed in the joint before the sealant.

have to stretch very little during its lifetime but will have to compress a great deal in summer, as the materials around it expand and crowd together. Sealant installed in hot weather will have to compress very little but will be greatly stretched in winter. Sealant joint design cannot be left to chance: The design of the joints themselves is the primary factor in the successful functioning of building cladding.

Installation procedures are also critical to the success of gunnable sealant joints in a cladding system. Each joint must be carefully cleaned of oil, dirt, oxide, moisture, or concrete formwork coating. Then the backup rod (and bond breaker, if any) are inserted, and any necessary *priming* is carried out. (Priming is the process of applying a coating to the joint surfaces of certain cladding materials to which sealant does not readily adhere in order to increase the adhesion.) The sealant is extruded into the joint from the nozzle of a sealant gun in such a way that enough pressure is created to fill the joint completely. Lastly, the sealant is mechanically tooled, much the same as a masonry mortar joint is tooled, to compress the sealant material firmly against the sides of the joint and give the desired surface profile.

Gasket sealants have generally proved to be less sensitive to installation problems than gunnable sealants. For this reason they are widely used in commercially available curtain wall systems.

FIGURE 15.14
Sealants are best applied at temperatures that are neither excessively hot nor excessively cold. If cold- or hot-weather application of sealants is anticipated, the joints should be proportioned to minimize overstretching or overcompression. Row A shows the behavior of a sealant bead applied at a middle temperature. Rows B and C show beads applied at summer and winter temperatures, respectively.

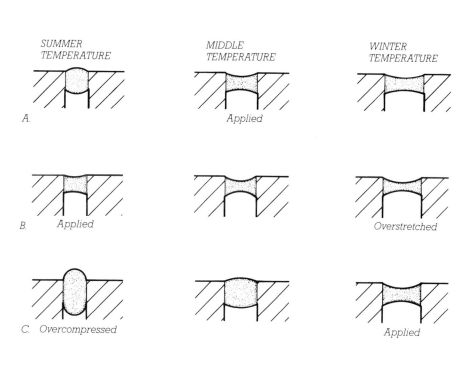

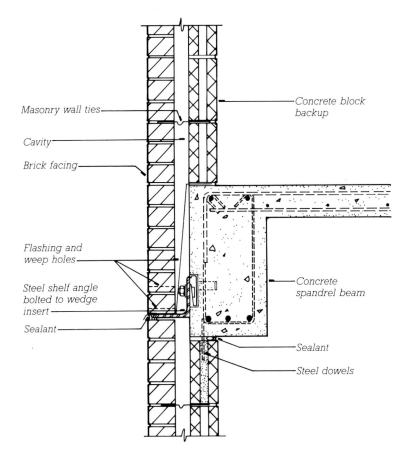

Masonry wall ties

Cavity

Brick facing

Flashing and weep holes

Steel shelf angle bolted to wedge insert

Sealant

Concrete block backup

Concrete spandrel beam

Sealant

Steel dowels

FIGURE 15.15

A simple brick curtain wall supported by a reinforced concrete frame. It is crucial that the horizontal brick joint below the shelf angle be filled with sealant rather than mortar, and that it be sufficiently thick to allow for any expected expansions and contractions in the masonry and the frame. This detail, used on thousands of buildings, effectively employs the rainscreen principle: The cavity between the brick and the backup wall acts as a pressure equalization chamber to neutralize air pressure differentials between the two sides of the masonry.

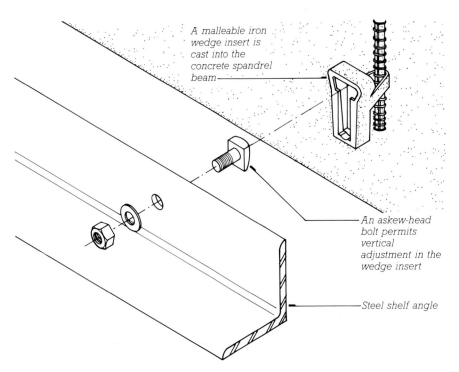

A malleable iron wedge insert is cast into the concrete spandrel beam

An askew-head bolt permits vertical adjustment in the wedge insert

Steel shelf angle

FIGURE 15.16

An example of a cast-in-place anchorage system for attaching a steel shelf angle to a concrete spandrel beam, as manufactured by the Richmond Screw Anchor Company, Inc.

CURTAIN WALL DESIGN: CASE STUDIES

Figures 15.15 through 15.48 illustrate examples of several types of curtain wall cladding commonly used in North America. The idea of the masonry curtain wall goes back a century or so, while the other types developed in the years following the Second World War. These examples are not in any way universal but are given here to demonstrate some of the issues involved in the detailing of each type of wall.

Masonry Curtain Walls

Most masonry curtain walls (Figures 15.15 through 15.21) are built in place brick by brick or stone by stone, starting from a steel *shelf angle* attached to the structural frame at each floor. Their construction process is essentially identical to that of a masonry facing applied to a single-story building at ground level. Masonry curtain walls must be divided both horizontally and vertically by expansion joints to allow the frame and the masonry cladding to move independently of one another; otherwise the thin masonry wythe could be subjected to forces it is incapable of sustaining. Horizontal expansion joints are made at each shelf angle (Figure 15.15), and vertical expansion joints should be made at intervals appropriate to the size of the building and the amount and character of its expected movements.

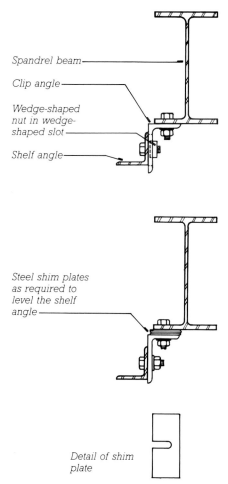

FIGURE 15.17
Two methods for attaching shelf angles to steel spandrel beams. The traditional method, below, uses steel clip angles with shim plates as needed to make up for dimensional inaccuracies in the components. A proprietary system, above, is based on a bolt with a wedge-shaped nut engaging a wedge-shaped slot punched in the clip angle.

Spandrel beam

Clip angle

Wedge-shaped nut in wedge-shaped slot

Shelf angle

Steel shim plates as required to level the shelf angle

Detail of shim plate

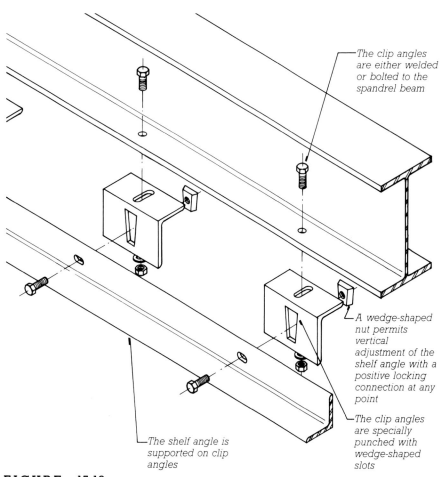

The clip angles are either welded or bolted to the spandrel beam

A wedge-shaped nut permits vertical adjustment of the shelf angle with a positive locking connection at any point

The clip angles are specially punched with wedge-shaped slots

The shelf angle is supported on clip angles

FIGURE 15.18
A three-dimensional view of shelf angle attachment using a proprietary system of wedge-shaped nuts and slots. The advantages of these fasteners are speed of connection, ease of adjustment, and secure long-term load-carrying capacity without welding. (Courtesy of SlotLock Fastener Systems, Hamden, Connecticut)

Stone and Precast Concrete Curtain Walls

Thin panels of cut stone are often used for curtain walls by bolting them directly to the face of the building and injecting sealant into the joints between panels (Figures 15.22 through 15.24). Even thinner sheets of stone can be mounted in aluminum mullions as if they were sheets of spandrel glass, or fastened to concealed steel angle frames as a means of combining them into larger panels and attaching them to the structure of the building (Figure 15.25). Stone is also used as a cladding in two other forms: unit stonemasonry supported on shelf angles, similar to the brick curtain walls shown in Figures 15.15 through 15.21; and self-supporting facing wythes of stonemasonry built from the foundations up, as illustrated in Chapter 8.

Curtain wall panels of precast concrete with conventional reinforcing or prestressing (Figures 15.26, 15.28) are simple in concept, but such panels require close attention to problems of surface finish, mold design, thermal insulation, and attachment to the building frame. Glass fiber reinforced concrete (GFRC), which requires no steel reinforcing, can be used to produce much thinner, lighter panels than conventionally reinforced concrete and is just coming into use in North America at the time of this writing.

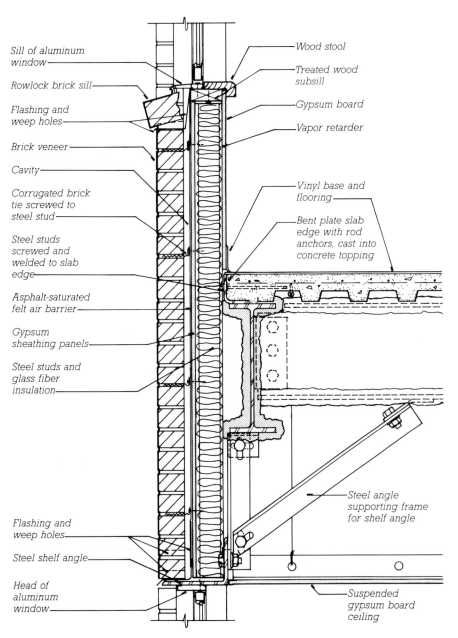

Sill of aluminum window

Rowlock brick sill

Flashing and weep holes

Brick veneer

Cavity

Corrugated brick tie screwed to steel stud

Steel studs screwed and welded to slab edge

Asphalt-saturated felt air barrier

Gypsum sheathing panels

Steel studs and glass fiber insulation

Flashing and weep holes

Steel shelf angle

Head of aluminum window

Wood stool

Treated wood subsill

Gypsum board

Vapor retarder

Vinyl base and flooring

Bent plate slab edge with rod anchors, cast into concrete topping

Steel angle supporting frame for shelf angle

Suspended gypsum board ceiling

FIGURE 15.19
An example of a brick curtain wall carried below the level of the spandrel beam on a steel supporting frame. The supporting frame becomes necessary when horizontal bands of windows are to be installed between brick spandrels. All the connections in the supporting frame are made with bolts in slotted holes, to allow for exact alignment of the shelf angle. Before the masonry work begins, the connections are welded to prevent slippage. Shelf angle constructions for masonry curtain walls require careful engineering to meet expected loads and structural deflections. Notice that this detail, like the detail in Figure 15.15, employs the Rainscreen Principle.

FIGURE 15.20
A brick curtain wall with horizontal bands of windows. (Photo by the author)

FIGURE 15.21
Constructing the ground floor of a windowless brick curtain wall over a backup of steel studs and insulating sheathing. The brick wythe is anchored with corrugated metal ties screwed through the sheathing to the studs. (Courtesy of Dow Chemical Company)

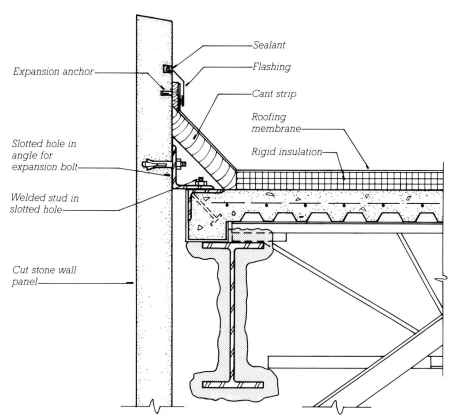

Sealant

Flashing

Expansion anchor

Cant strip

Roofing membrane

Rigid insulation

Slotted hole in angle for expansion bolt

Welded stud in slotted hole

Cut stone wall panel

A

FIGURE 15.22
A. *Parapet and* **B.** *Spandrel details for a stone panel curtain wall, appropriate to limestone, marble, or granite. The broken lines indicate the outline of the interior finish and thermal insulation components, which are not shown. Each support plate holds edges of two adjacent wall panels, which are pocketed as shown to rest on the plate. The vertical joints between panels are made weathertight with sealant.*

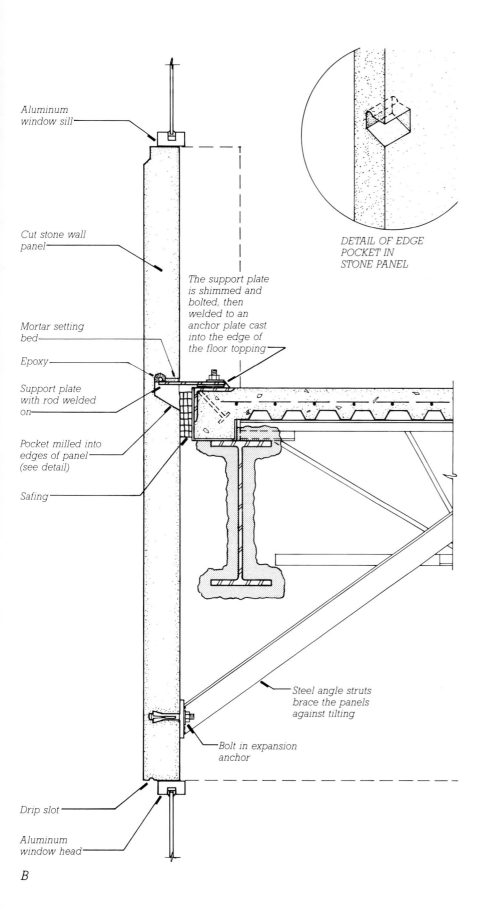

Aluminum
window sill

Cut stone wall
panel

The support plate
is shimmed and
bolted, then
welded to an
anchor plate cast
into the edge of
the floor topping

Mortar setting
bed

Epoxy

Support plate
with rod welded
on

Pocket milled into
edges of panel
(see detail)

Safing

DETAIL OF EDGE
POCKET IN
STONE PANEL

Steel angle struts
brace the panels
against tilting

Bolt in expansion
anchor

Drip slot

Aluminum
window head

B

FIGURE 15.23
Safing is a high-temperature, highly fire-resistant mineral batt material, which is inserted between a curtain wall panel and the edge of the floor slab to block the passage of fire from one floor to the next. It is seen here behind a metal-and-glass curtain wall with insulated spandrel panels. The safing is held in place by metal clips such as the one seen in the foreground. (Courtesy of United States Gypsum Company)

FIGURE 15.24
A granite panel curtain wall of the general type illustrated in Figure 15.22 wraps around the corner of a Boston office building. The upper-floor windows have not yet been installed, but the frames have been mounted in two of the middle floors, and the lower floors are glazed. (Architect: Hugh Stubbins and Associates. Photo by the author)

FIGURE 15.25
A curtain wall made of thin slabs of
polished red granite mounted on
aluminum frames. (Architects: John
Burgee with Philip Johnson. Photo
courtesy of Cupples Products, Division
of H. H. Robertson Company)

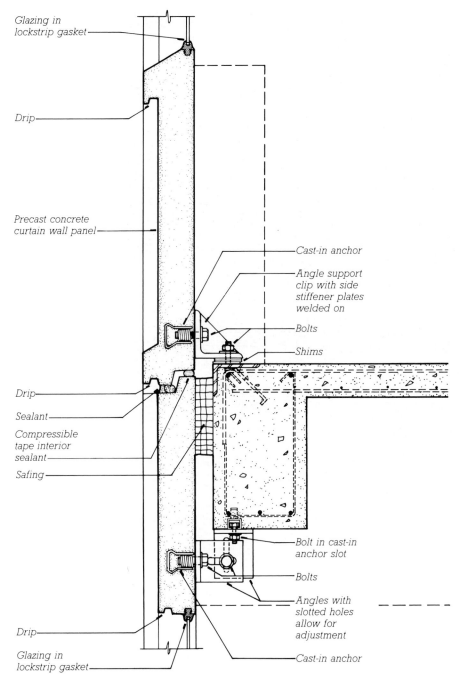

Glazing in
lockstrip gasket

Drip

Precast concrete
curtain wall panel

Cast-in anchor

Angle support
clip with side
stiffener plates
welded on

Bolts

Shims

Drip

Sealant

Compressible
tape interior
sealant

Safing

Bolt in cast-in
anchor slot

Bolts

Angles with
slotted holes
allow for
adjustment

Drip

Glazing in
lockstrip gasket

Cast-in anchor

FIGURE 15.26
*An example of a precast concrete
curtain wall on a sitecast concrete
frame. Panels are a full story high, each
containing a fixed window. The
reinforcing has been omitted from the
panel for the sake of clarity, and the
outline of the thermal insulation and
interior finishes, which are not shown, is
indicated by the broken lines.*

FIGURE 15.27
A precast concrete curtain wall. (Photo by the author)

FIGURE 15.28
Workers install a precast concrete curtain wall panel. (Photo by the author)

FIGURE 15.29

Installing Corewall® insulated precast concrete curtain wall panels on an industrial building. See Figure 12.14 for photographs of the production of these panels. (Reproduced by permission of Corewall Limited)

Metal and Glass Curtain Walls

Two off-the-shelf stick systems for aluminum and glass curtain wall construction are illustrated here (Figures 15.30–15.41), along with some generic expansion joint details for aluminum curtain walls (Figures 15.42, 15.43) and photographs of several buildings clad with custom-designed aluminum and glass curtain walls (Figure 15.45). There are literally dozens of stock curtain wall systems available to the designer, and a number of manufacturers who specialize in custom-designed curtain walls; the designs shown here are offered as examples of some of the possibilities.

FIGURE   15.30
An aluminum and glass curtain wall, constructed using the stick system detailed in Figures 15.31 through 15.37. (Architect: Gruzen and Partners. Photo ©1980, E. Alan McGee, Atlanta, courtesy of Amarlite Architectural Products)

Facades are simply light membranes of isolating walls or windows. The facade is free...

Le Corbusier

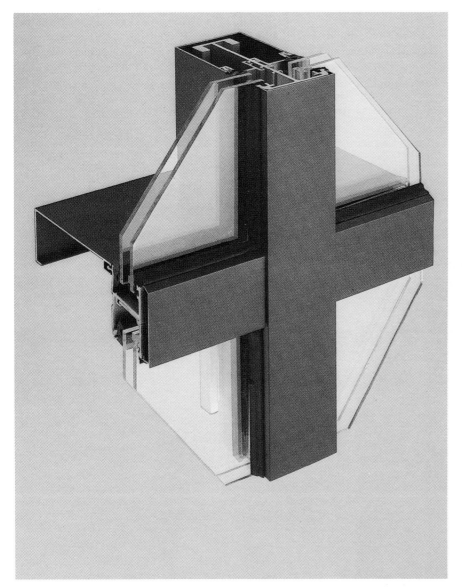

FIGURE 15.31
A sample segment of the curtain wall shown in Figure 15.30, showing the intersection of a vertical mullion and a horizontal mullion. The windows are double glazed for thermal efficiency. (Courtesy of Amarlite Architectural Products)

FIGURE 15.32
The first page of the manufacturer's catalog for the curtain wall system illustrated in Figure 15.30 shows the major details at one-quarter full size. Each section is keyed to the elevation view at the upper right. (Courtesy of Amarlite Architectural Products)

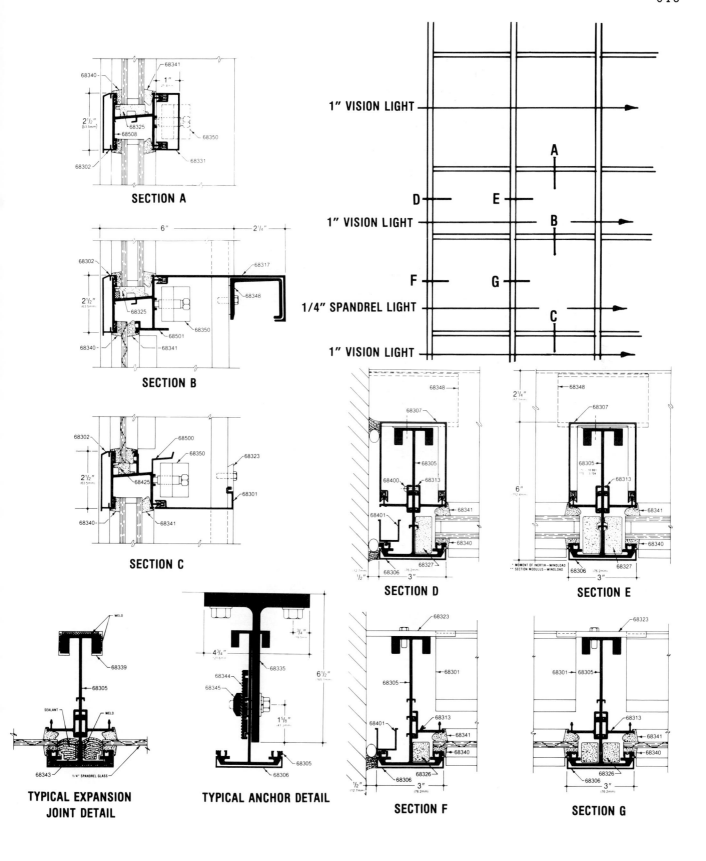

SECTION A

SECTION B

SECTION C

1" VISION LIGHT

1" VISION LIGHT

1/4" SPANDREL LIGHT

1" VISION LIGHT

SECTION D

SECTION E

TYPICAL EXPANSION JOINT DETAIL

TYPICAL ANCHOR DETAIL

SECTION F

SECTION G

SECTION SCALE = FULL SIZE

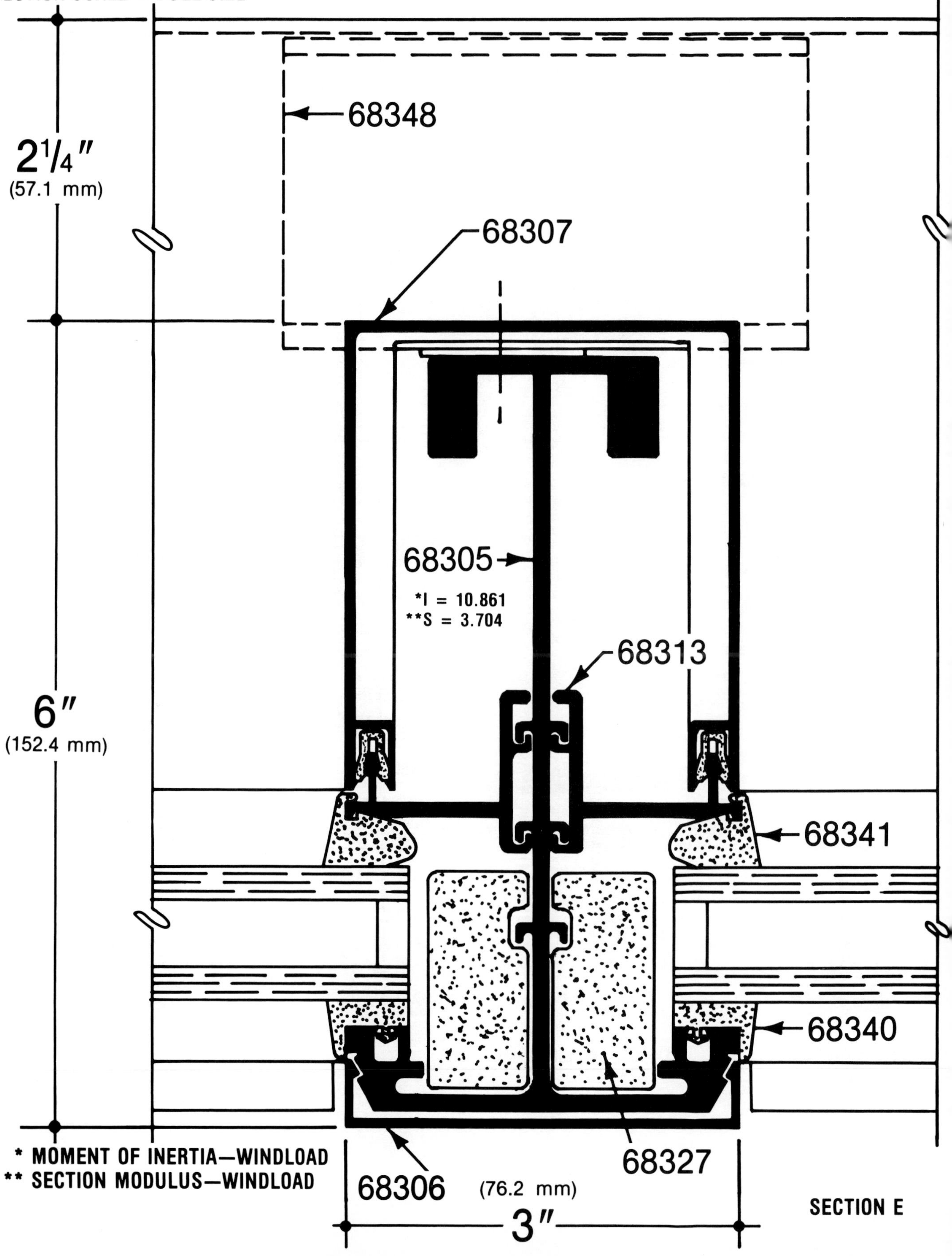

2¼"
(57.1 mm)

68348

68307

6"
(152.4 mm)

68305

*I = 10.861
**S = 3.704

68313

68341

68340

68306

68327

* **MOMENT OF INERTIA—WINDLOAD**
** **SECTION MODULUS—WINDLOAD**

(76.2 mm)

3"

SECTION E

SECTION SCALE = FULL SIZE

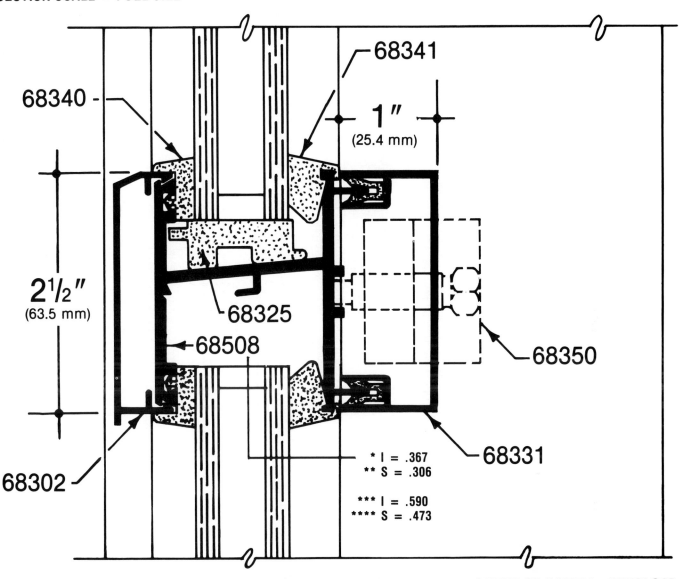

68340 –

68341

1"
(25.4 mm)

2¹/₂"
(63.5 mm)

68325

68508

68350

68302

68331

* I = .367
** S = .306

*** I = .590
**** S = .473

*** MOMENT OF INERTIA—DEADLOAD
** SECTION MODULUS—DEADLOAD**

***** MOMENT OF INERTIA—WINDLOAD
**** SECTION MODULUS—WINDLOAD**

FIGURE   15.34

A full-scale detail of a horizontal mullion, consisting of three aluminum extrusions, two extruded plastic thermal breaks, four synthetic rubber gaskets, and synthetic rubber setting blocks for the double glazing units (68325). Outdoors is to the left. Omitted from this drawing are the weep holes that are drilled through the mullion and the exterior cover to drain away any leakage or condensate that may accumulate in the frame. 68350 is an attachment between the horizontal and vertical mullions, as shown in Figure 15.35. (Courtesy of Amarlite Architectural Products)

FIGURE   15.33

A full-scale detail of a vertical mullion as shown in the manufacturer's catalog. Outdoors is toward the bottom of the page. The mullion is made up of five aluminum extrusions, two plastic extrusions, and synthetic rubber gaskets and glazing spacers. The actual structural mullion is part number 68305. 68340 and 68341 are continuous gaskets that hold the 1-inch-thick (25.4 mm) double glazing. 68313 is an aluminum extrusion that interlocks with the structural mullion and holds the interior glazing gasket. 68306 is an exterior snap-on decorative cover. 68307 is a snap-on interior cover, separated from the rest of the mullion parts by plastic extrusions (not numbered) that act as thermal breaks to retard the passage of heat through the aluminum. The parts numbered 68327 are spacers to maintain the position of the glass. (Courtesy of Amarlite Architectural Products)

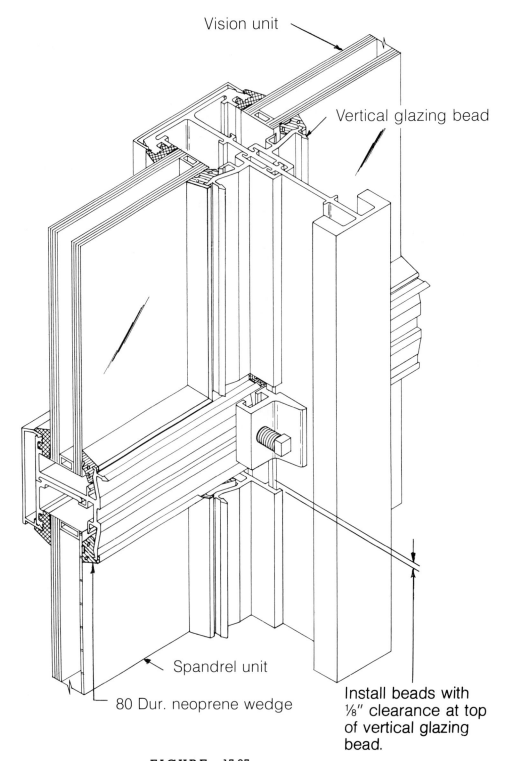

Vision unit

Vertical glazing bead

Spandrel unit

80 Dur. neoprene wedge

Install beads with
⅛" clearance at top
of vertical glazing
bead.

FIGURE 15.35
*A three-dimensional view of the
assembly of the horizontal and vertical
mullions. (Courtesy of Amarlite
Architectural Products)*

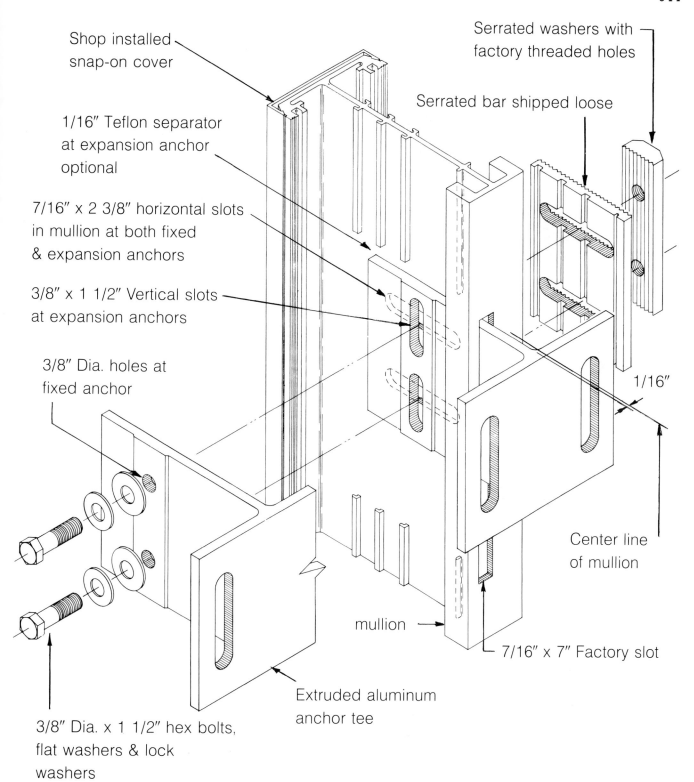

Shop installed
snap-on cover

1/16″ Teflon separator
at expansion anchor
optional

7/16″ x 2 3/8″ horizontal slots
in mullion at both fixed
& expansion anchors

3/8″ x 1 1/2″ Vertical slots
at expansion anchors

3/8″ Dia. holes at
fixed anchor

3/8″ Dia. x 1 1/2″ hex bolts,
flat washers & lock
washers

Serrated washers with
factory threaded holes

Serrated bar shipped loose

1/16″

Center line
of mullion

mullion

7/16″ x 7″ Factory slot

Extruded aluminum
anchor tee

FIGURE 15.36

*Vertical mullions are attached to the
edges of the floors of the building using
this detail. Slotted holes allow for
adjustment as necessary to assure
perfect alignment of the curtain wall.*
(Courtesy of Amarlite Architectural
Products)

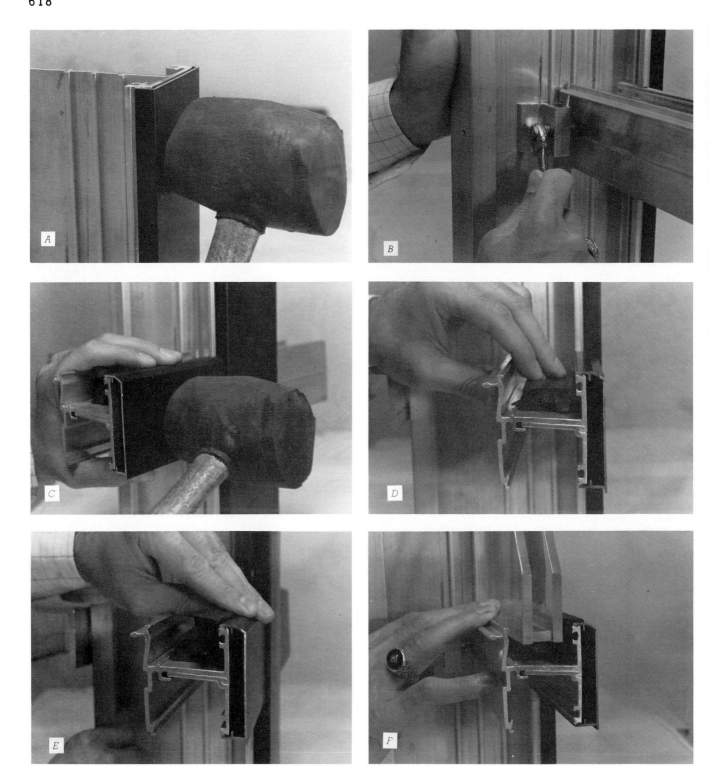

FIGURE 15.37

The sequence for assembling the curtain wall detailed in Figures 15.32 through 15.36: (a) The exterior cover is snapped on by tapping it firmly with a mallet. (b) The horizontal mullion is screwed to the vertical mullion. (c) The exterior cover is snapped on the horizontal mullion. (d) Setting blocks for the double glazing are inserted in the horizontal mullion. (e) The exterior glazing gasket is snapped into its slot in the aluminum mullion. (f) The double glazing unit is set into the frame working from the inside of the building, eliminating the need for exterior scaffolding. (g) The interior gasket is rolled into position, completing the glazing operation. (h) The support for the aluminum window stool is screwed to the vertical mullion. (i) The window stool is tapped into position. The stool shown here is an alternative to the inside cover shown in Figure 15.34. (j) The inside vertical mullion cover with its thermal breaks is snapped on. The curtain wall installation is now complete. (Courtesy of Amarlite Architectural Products)

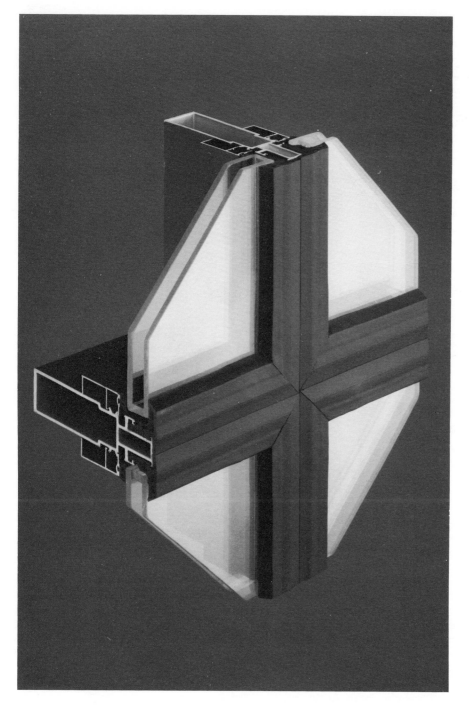

SECTION SCALE = FULL SIZE

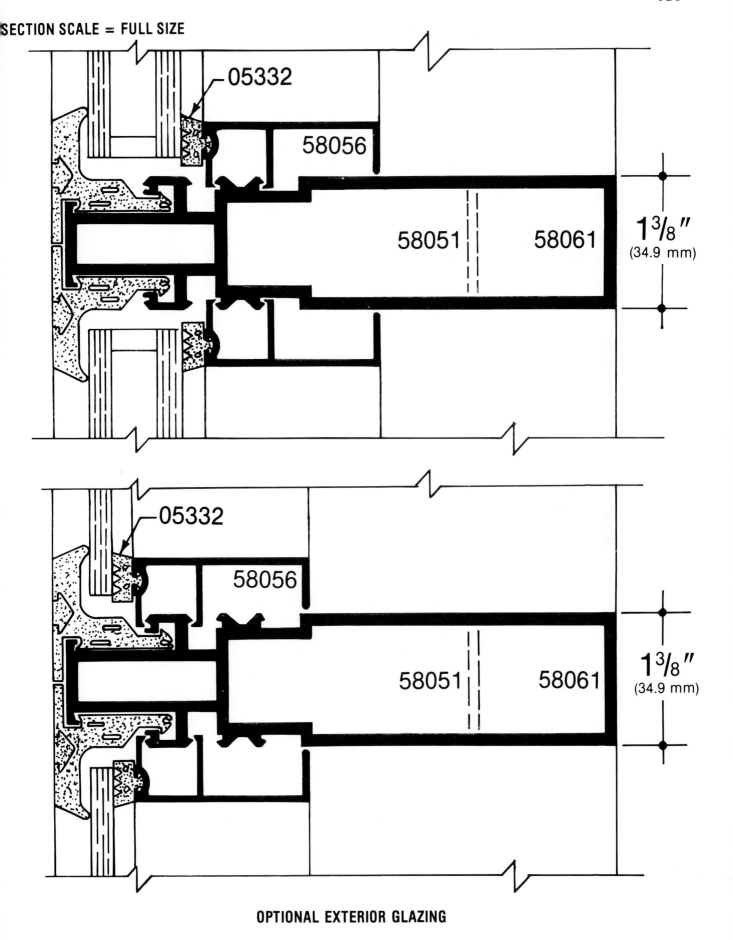

05332

58056

58051

58061

$1^3/_8''$
(34.9 mm)

05332

58056

58051

58061

$1^3/_8''$
(34.9 mm)

OPTIONAL EXTERIOR GLAZING

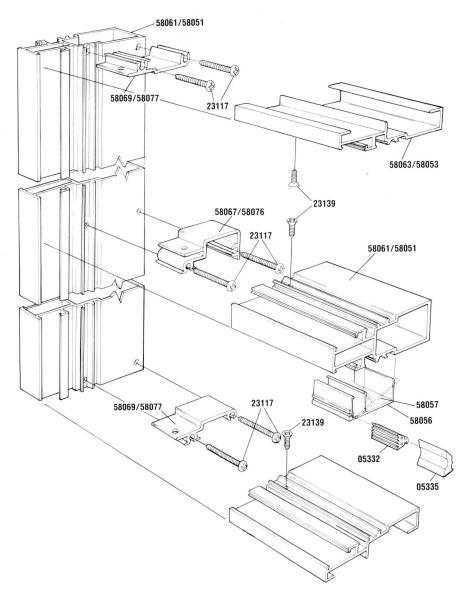

58061/58051

58069/58077

23117

58063/58053

23139

58067/58076

23117

58061/58051

23117

58069/58077

23139

58057

58056

05332

05335

FIGURE 15.40

An exploded assembly diagram of the aluminum components for the lockstrip gasket curtain wall. Short clips with screw ports are screwed to the vertical mullion to allow attachment of the horizontal mullions with flat-head screws (23139). (Courtesy of Amarlite Architectural Products)

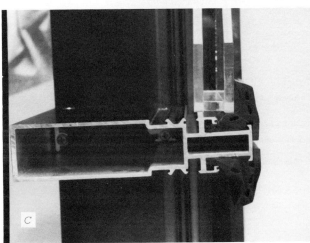

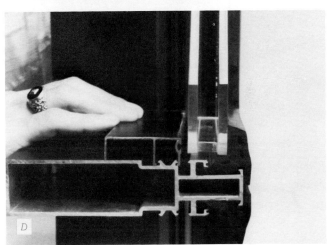

FIGURE 15.41

Steps in assembling the lockstrip gasket curtain wall: (a) The clip is screwed to holes drilled in the vertical mullion. (b) The horizontal mullion is screwed to holes drilled in the clip. The holes are countersunk so the screws lie flush with the top surface of the horizontal mullion. (c) The exterior gaskets are installed, and the glass is set in place, working from inside the building. Notice that the gaskets shown in this photograph have no lockstrips; lockstrips need be used in this system only if the designer wishes to retain the option of reglazing broken lights from outside the building. (d) The interior stop with its glazing gasket is snapped on, completing the installation. (Courtesy of Amarlite Architectural Products)

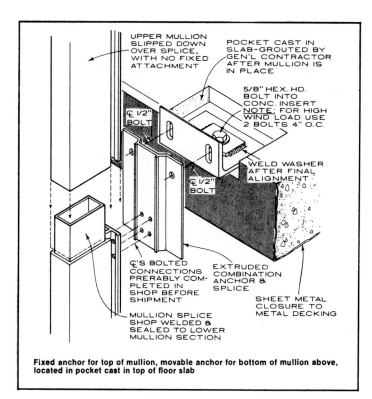

UPPER MULLION
SLIPPED DOWN
OVER SPLICE,
WITH NO FIXED
ATTACHMENT

POCKET CAST IN
SLAB-GROUTED BY
GEN'L CONTRACTOR
AFTER MULLION IS
IN PLACE

5/8" HEX. HD.
BOLT INTO
CONC. INSERT
NOTE: FOR HIGH
WIND LOAD USE
2 BOLTS 4" O.C.

₵ 1/2"
BOLT

₵ 1/2"
BOLT

WELD WASHER
AFTER FINAL
ALIGNMENT

₵'S BOLTED
CONNECTIONS
PRERABLY COM-
PLETED IN
SHOP BEFORE
SHIPMENT

EXTRUDED
COMBINATION
ANCHOR &
SPLICE

SHEET METAL
CLOSURE TO
METAL DECKING

MULLION SPLICE
SHOP WELDED &
SEALED TO LOWER
MULLION SECTION

Fixed anchor for top of mullion, movable anchor for bottom of mullion above, located in pocket cast in top of floor slab

ANGLES WELDED BACK-
TO-BACK AFTER ALIGNMENT

₵'S 1/2" BOLTS

WASHER PL. WELDED
TO ANGLE AFTER
ALIGNMENT

SLOTTED HOLES
IN ANGLE LEG

FACE OF
FLR. SLAB

UNISTRUT
ANCHOR IN
SLAB TOP

₵ 1/2"
BOLT

LOOSE
SPACERS

₵ 1/2" BOLT

ALUMINUM
MULLION

**Movable anchor (as shown), located on top of floor slab
(Fixed anchor if round holes in mullion stem)**

FIGURE 15.42
Expansion joints must be provided periodically in a metal and glass curtain wall. Illustrated here are two different ways of allowing for expansion and contraction in vertical mullions. (Reprinted with permission from AAMA Aluminum Curtain Wall Design Guide Manual)

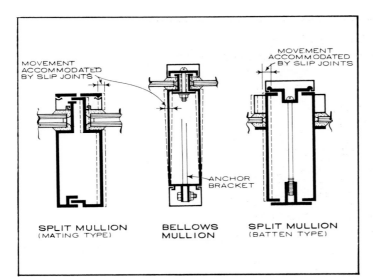

FIGURE 15.43
Aluminum mullion designs allowing for horizontal expansion and contraction. Such mullions are installed in place of regular mullions at intervals determined by the expected thermal and structural movement in the wall. (Reprinted with permission from AAMA Aluminum Curtain Wall Design Guide Manual)

FIGURE 15.44
An aluminum and glass entry made of stock parts. (Courtesy of Amarlite Architectural Products)

FIGURE 15.45

Three custom-designed metal and glass curtain walls: (a) Facets of aluminum and reflective glass on a San Francisco office building (Architects: Philip Johnson and John Burgee, with Kendall and Heaton) *(b) Fluoropolymer finished aluminum spandrel panels and exterior flush glazing of dark glass* (Architects: I.M. Pei & Partners). *(c) A polished stainless steel facing on an aluminum and glass curtain wall.* (Architects: Lloyd, Jones, Brewer. All photographs courtesy of Cupples Products, Division of H. H. Robertson Company)

Exterior Insulation and Modified Stucco

An entirely different approach to building cladding is offered by systems based on exterior plastic foam insulation and a thin, glass-fiber-reinforced weathering layer of polymer-modified stucco (Figures 15.46 through 15.48). These offer any required degree of thermal insulation without thermal bridging. The finish layer may be of several surface textures and any of a number of integral colors. These systems are unusually versatile, being appropriate to buildings from single-family residences of wood or masonry to the largest buildings of fire-resistant construction and are used both for new construction and for refacing and insulating existing buildings. The exterior surface appearance is indistinguishable from conventional stucco. The thin stucco layer is, however, easily damaged and should be specially reinforced as directed by the manufacturer in areas where it is exposed to abuse from vehicles or passersby. When damage does occur, it is easily and unobtrusively patched.

FIGURE 15.46

Four steps in installing an insulation and modified stucco cladding over a building with walls of masonry or solid sheathing: (a) A panel of foam is daubed with polymer-modified Portland cement mortar. The foam may be of any required thickness to achieve the desired thermal performance. (b) The foam panel is pressed into place, and held permanently by the daubs of mortar. (c) A thin base coat of polymer-modified stucco is applied to the surface of the foam panels, with an embedded mesh of glass fiber to act as reinforcing. (d) After the base coat has hardened, a finish coat in any desired color is troweled on. (Courtesy of Dryvit® System, Inc.)

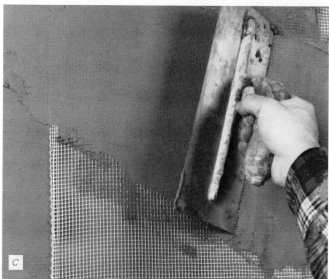

FIGURE 15.47
A new bank building clad in plastic foam and polymer-modified stucco. (Architect: Paul Thoryk. Photo by John Bare, courtesy of Dryvit® System, Inc.)

FIGURE 15.48
Foam-stucco cladding can also be shop fabricated and erected in panel form: (a) Steel studs are welded into panel frames. (b) Rigid sheathing is screwed to the panel frames, and finished with foam and stucco as shown in Figure 15.46. (c) The finished panels are installed in the same manner as other curtain wall panels. (Courtesy of Dryvit® System, Inc.)

CURTAIN WALL DESIGN: THE PROCESS

The architect and building owner may elect either of two methods for designing a prefabricated curtain wall. One curtain wall manufacturer may be selected and made a part of the building design team from the early stages of the project. Or the architect, often with the help of an independent cladding consultant, may prepare a rough design and performance specifications and submit these to one or several manufacturers for proposals. Each manufacturer will then submit a more detailed design and a financial proposal, and one will be selected on the basis of these proposals to proceed with the project.

The three key participants in the process of designing a curtain wall system are the architect, the structural engineer (who is involved in working out the attachments and structural support for the curtain wall), and the manufacturer. All should be closely involved in the design from the earliest stages. From conceptual drawings prepared by the architect and the engineer, and for large buildings, from wind tunnel tests on a building model that indicate expected wind pressures on the cladding, the manufacturer prepares a more detailed set of design drawings as a basis for reaching preliminary agreement on the design and then prepares a very detailed set of shop drawings and installation drawings, which should be checked carefully by the architect and engineer to assure compliance with design intentions and the structural capabilities of the building frame. The curtain wall manufacturer may wish to visit the construction site during the erection of the building frame to become familiar with the level of dimensional accuracy of the structural surfaces to which the curtain wall will be fastened.

For a new curtain wall design, it is advisable to build and test a full-scale section of wall to determine its resistance to air infiltration and water infiltration, and its structural performance under heavy wind loadings, using standardized ASTM testing procedures in commercial laboratories established for the purpose. The thermal and noise transmission coefficients of the wall can also be tested at this stage if desired, and problems of appearance and assembly techniques can be worked out. When final design adjustments have been made, production of the wall components begins, and deliveries to the site can commence as soon as the frame is ready to receive the cladding.

CLADDING AND THE BUILDING CODES

The major impact of building codes on the design of building cladding is in the areas of structural strength and fire resistance. Strength requirements relate to the strength and stiffness of the cladding itself and to the adequacy of its attachments to the building, with special reference to wind and earthquake loadings. Fire requirements are concerned with the combustibility of the cladding materials, the fireresistance ratings and vertical dimensions of parapets and spandrels, the fireresistance ratings of exterior walls facing other buildings that are near enough to raise questions of fire spread from one building to the other, and the closing off (*firestopping*) of any vertical openings in the cladding that are more than one story in height. The space inside column covers must be firestopped at each floor. The space between curtain wall panels and the edges of floors must also be firestopped, using either a steel plate and grout, metal lath and plaster, or mineral wool *safing* (Figure 15.23). Such firestopping must be permanently attached to the edge of the deck.

C.S.I./C.S.C Masterformat Section Numbers for Cladding	
03400	PRECAST CONCRETE
03450	Architectural Precast Concrete (Plant Cast)
	Faced Architectural Precast Concrete
	Glass-Fiber-Reinforced Precast Concrete
04200	UNIT MASONRY
04235	Preassembled Masonry Panels
04250	Ceramic Veneer
04255	Masonry Veneer
04400	STONE
04450	Stone Veneer
07400	PREFORMED ROOFING AND CLADDING/SIDING
07410	Preformed Roof and Wall Panels
07420	Composite Building Panels
07460	Cladding/Siding
08900	GLAZED CURTAIN WALLS
08910	Glazed Steel Curtain Walls
08920	Glazed Aluminum Curtain Walls
08930	Glazed Stainless Steel Curtain Walls
08940	Glazed Bronze Curtain Walls

ALUMINUM EXTRUSIONS AND ALUMINUM FINISHES

Aluminum Extrusions

Aluminum is the metal used in most stick-type curtain wall systems because it is the only common metal from which intricate sections can be produced easily in any desired length through the process of *extrusion*. The principle of extrusion is easily visualized: Imagine squeezing toothpaste from a tube. The column of toothpaste that is extruded is cylindrical because the orifice in the tube is round. If the shape of the orifice could be varied, one could produce many other sections of toothpaste as well—square, triangular, flat, and so on. In aluminum extrusion, a large billet of metal is heated to a temperature that allows the material to flow easily under pressure but still retain its shape

when not under load. The billet is then placed under enormous pressure in a large press that squeezes it through a shaped metal orifice called a *die* to produce any desired length of section.

Very intricate sections can be extruded for a variety of purposes, including not only curtain wall components but doors, windows, handrails, grillwork, and structural shapes. The precision of the process permits it to be used for close-tolerance details such as snap-in glazing beads, snap-on mullion covers, screw slots, and screw ports. The accompanying illustrations show some of the ways in which extruded aluminum details are utilized in building cladding. Extrusion dies are easily produced for custom-designed sections if there will be a long enough production run to amortize their expense.

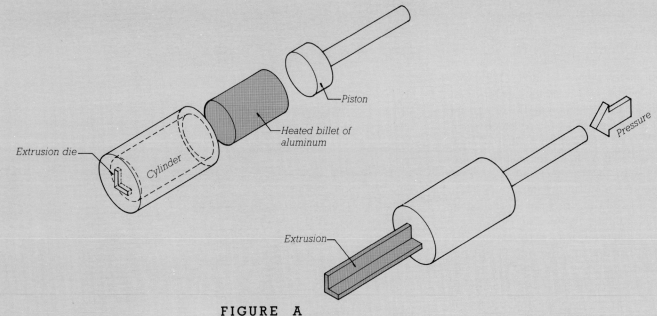

FIGURE A
The concept of extrusion: A piston forces a heated billet of aluminum through a shaped die.

Finishes for Aluminum

Ordinary aluminum, though chemically a very active metal, does not corrode away in service because it quickly protects itself with a thin, tenacious oxide coating that discourages further oxidation. While this oxide coating does an adequate job of protecting the aluminum, on buildings it takes on a chalky or spotty appearance that looks rather shabby.

Anodizing is a process that produces an integral oxide coating on aluminum that is thousands of times thicker and more durable than the natural oxide film that would otherwise form. The part to be anodized is immersed in an acid bath and becomes the anode in an electrolytic process that takes oxygen from the acid and combines it with the aluminum. Color can be added to the coating by means of dyes, pigments, special electrolytes, or special aluminum alloys. The colors most frequently used in buildings are the natural aluminum color, golds, bronzes, grays, and black, but other colors are possible. The advantages of anodic coatings are their extreme durability and, in most colors, extreme resistance to weather and fading.

Other finishes are also used on aluminum building cladding, including organic coatings and porcelain enamels. Perhaps the most widely used of the organic coatings are the *fluoropolymers*, which are based on highly inert synthetic resins that are exceptionally resistant to all forms of weathering, including ultraviolet deterioration. Fluoropolymers are applied in either two or three coats: a primer, a color coat, and, optionally, a clear finish coat. Aluminum sheet stock coated with fluoropolymers may be bent to shape after coating without damage to the finish. Fluoropolymer coatings are available in a broad spectrum of colors, including bright metallic finishes.

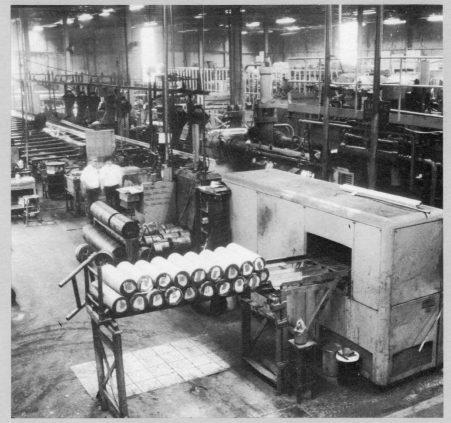

FIGURE B
Aluminum billets await the billet heater, in the foreground, and the extrusion press beyond. (Courtesy of Amarlite Architectural Products)

FIGURE C
*A finished aluminum shape emerges
from the die.* (Courtesy Amarlite
Architectural Products)

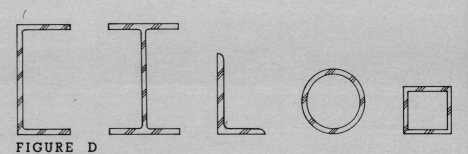

FIGURE D
*Standard structural shapes are readily
extruded in depths and diameters up to
12 inches (300 mm).*

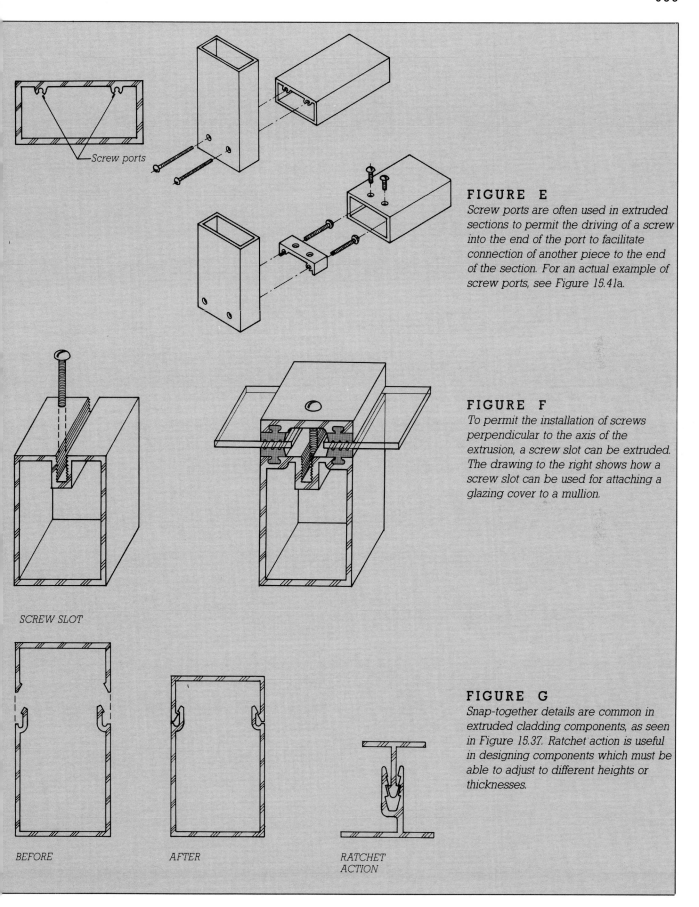

FIGURE E

Screw ports are often used in extruded sections to permit the driving of a screw into the end of the port to facilitate connection of another piece to the end of the section. For an actual example of screw ports, see Figure 15.41a.

FIGURE F

To permit the installation of screws perpendicular to the axis of the extrusion, a screw slot can be extruded. The drawing to the right shows how a screw slot can be used for attaching a glazing cover to a mullion.

SCREW SLOT

FIGURE G

Snap-together details are common in extruded cladding components, as seen in Figure 15.37. Ratchet action is useful in designing components which must be able to adjust to different heights or thicknesses.

—Screw ports

BEFORE

AFTER

RATCHET ACTION

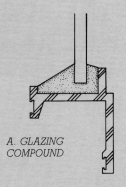

A. GLAZING
COMPOUND

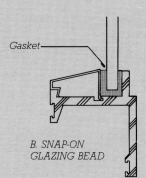

Gasket

B. SNAP-ON
GLAZING BEAD

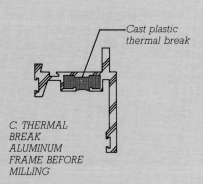

Cast plastic
thermal break

C. THERMAL
BREAK
ALUMINUM
FRAME BEFORE
MILLING

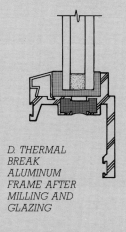

D. THERMAL
BREAK
ALUMINUM
FRAME AFTER
MILLING AND
GLAZING

FIGURE H

Aluminum extrusions are easily adapted
to different types of glazing details:
A. A simple aluminum sash section uses
glazing compound to hold the glass in
place. The two notches in the legs of
the section are for insertion of strips of
pile weatherstripping to seal against the
outer frame of the window. **B.** Far more
common than glazing compound in
aluminum sash is the snap-on glazing
bead, a simple form of which is shown
here. The bead is sprung into place to
hold the glass and is easily removed for
glass replacement when necessary. **C,
D.** To reduce the conduction of heat
through aluminum window frames,
thermal breaks of various types are
often employed. The type shown here is
produced by casting rigid plastic into a
channel in the aluminum extrusion, then
cutting away the aluminum connection
between the inside and outside portions
of the frame with a milling machine,
leaving only the plastic thermal break to
connect the two halves.

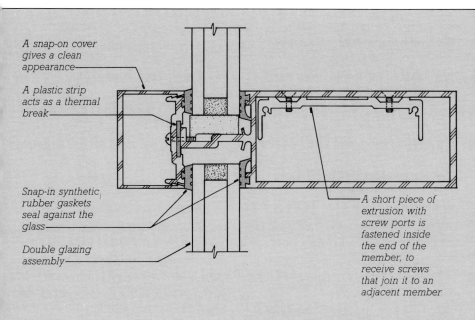

A snap-on cover gives a clean appearance

A plastic strip acts as a thermal break

Snap-in synthetic rubber gaskets seal against the glass

Double glazing assembly

A short piece of extrusion with screw ports is fastened inside the end of the member, to receive screws that join it to an adjacent member

FIGURE I

An aluminum curtain wall mullion summarizes some of the extrusion features presented here, including a snap-on exterior cover with ratcheting action, extruded channels to receive synthetic rubber glazing gaskets, and a connecting clip that employs screw ports. (Courtesy of Amarlite Architectural Products)

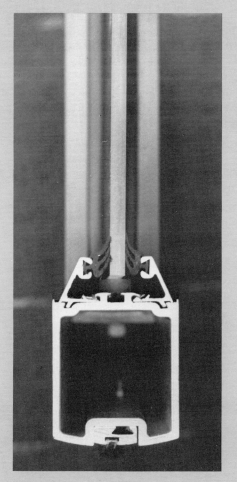

FIGURE J

Four aluminum extrusions comprise the frame of this commercial entry door: a tubular frame, two snap-in glazing beads with synthetic rubber glazing gaskets, and an edge insert incorporating a slot for pile weatherstripping. (Courtesy of Amarlite Architectural Products)

SELECTED REFERENCES

1. Architectural Aluminum Manufacturers Association. *Aluminum Curtain Wall Design Guide Manual*. Chicago, 1979.

This 182-page booklet covers its topic in exemplary fashion with clear text and beautifully prepared illustrations. (Address for ordering: 2700 River Road, Suite 118, Des Plaines, IL 60018.)

2. National Research Council Canada. *Construction Details for Airtightness*. Ottawa, 1980 (NRCC #18291).

3. National Research Council Canada. *Cracks, Movements, and Joints in Buildings*. Ottawa, 1976 (NRCC #15477).

4. Latta, J.K.: National Research Council Canada. *Walls, Windows, and Roofs for the Canadian Climate*. Ottawa, 1973 (NRCC #13487).

The Division of Building Research of the National Research Council Canada has done pioneering work in theorizing about cladding design and performing tests and field observations to back up the theory. These three books treat their subjects completely and understandably, without jargon or higher mathematics. (Address for ordering: NRCC, Ottawa, Ontario K1A OR6, Canada.)

5. Prestressed Concrete Institute. *Architectural Precast Concrete*. Chicago, 1973.

This is a well-illustrated, hardbound book that covers all aspects of the design, manufacture, and installation of precast concrete curtain walls. (Address for ordering: 201 North Wells Street, Chicago, IL 60606.)

6. Brick Institute of America. *Technical Notes on Brick Construction, Nos. 18, 18A, 18B*. McLean, Virginia, 1963–1980.

These three pamphlets cover differential movement in brick construction, including flexible anchorage systems for tying brick curtain walls to metal and concrete frames. (Address for ordering: 1750 Old Meadow Road, McLean, VA 22102.)

Also recommended as additional reading are the *Indiana Limestone Handbook*, which is listed as one of the references to Chapter 8, and trade literature from various producers of building stone and other cladding materials.

KEY TERMS AND CONCEPTS

cladding
thermal bridge
curtain wall
stick system
unit system
unit-and-mullion system
panel system
column-cover-and-spandrel system
internal drainage, secondary defense
momentum
adhesion
capillary
gravity
wind currents

wind pressure
labyrinth joint
capillary break
drip
rainscreen principle
pressure-equalized wall design
rainscreen
air barrier
pressure equalization chamber (PEC)
sealant
caulk
gunnable sealant
polysulfide
polymercaptan

silicone
polyurethane
gasket
lockstrip
preformed cellular tape sealant
preformed solid tape sealant
elongation
bond breaker
backup rod
shelf angle
safing
extrusion
anodizing
fluoropolymer

REVIEW QUESTIONS

1. Why is cladding so difficult to make watertight?

2. List the functions that cladding performs and list one or two ways in which each of these functions is typically satisfied in a cladding design.

3. Explain with a series of simple sketches the principles of sealant joint design. List several sealant materials suitable for use in the joints you have shown.

4. What are the relative advantages and disadvantages of a stick system and a panel system of curtain walling?

5. Under what conditions will water leak through cladding?

6. What are the forces that can move water through a joint in a curtain wall? How can each be neutralized?

EXERCISES

1. Make a photocopy of a metal curtain wall detail from a manufacturer's catalog. Paste the detail on a larger sheet of paper and add notes and arrows to explain every aspect of the detail—the features of the extrusions, the gaskets and sealants, the glazing materials and methods, drainage, insulation, thermal breaks, rainscreen features, and so on. Is the system glazed from inside the building or outside? How can you tell?

2. Examine the cladding of a building with which you are familiar. Look especially for features that have to do with insulation, condensation, drainage, and movement. Sketch a detail of how this cladding is installed and how it works. You will probably have to guess at some of the hidden features, but try to produce a complete, plausible detail. Add explanatory notes to make everything clear.

SELECTING INTERIOR FINISHES

*W*orkers complete an elaborate ceiling of gypsum board. The corners of the steplike construction have been reinforced with metal corner bead, and the nail heads and joints have been filled and sanded, ready for painting. *(Courtesy of United States Gypsum Company)*

When a building has been roofed and at least a part of its exterior cladding has been installed, its interior is sufficiently protected from the weather that work can begin on the mechanical and electrical systems. The waste lines and water supply lines of the plumbing system are installed along with the pipes for an automatic sprinkler system, if one is intended for the building. The major part of the work for the heating, ventilating, and air conditioning system is carried out, including the installation of boilers, chillers, cooling towers, pumps, fans, piping, and ductwork. Electrical, communications, and control wiring are routed through the building. Elevators and escalators are installed in the structural openings provided for them.

INSTALLATION OF MECHANICAL AND ELECTRICAL SERVICES

The vertical runs of pipes, ducts, wires, and elevators through a multistory building are made through vertical *shafts* whose sizes and locations were determined as the building was planned. By the time the building is finished, each shaft will be enclosed with fire-resistive walls, as described in Chapter 17, to prevent the vertical spread of fire (Figure 16.1). Horizontal runs of mechanical and electrical lines are usually located just below the structure of each floor to keep them up out of the way. These may be left exposed in the finished building, or hidden above *suspended ceilings*. Sometimes these services, especially wiring, are concealed within a hollow floor structure, such as cellular metal decking or hollow-core precast concrete planks. Sometimes services are

FIGURE 16.1
A worker constructs a fire-resistant wall around an elevator shaft, using gypsum panels and steel C-H studs. (Courtesy of United States Gypsum Company)

FIGURE 16.2
These diagrammatic plans for an actual three-story suburban office building show the principle arrangements for water, plumbing waste risers, communications, electricity, heating, and cooling. Heating and cooling are by means of air ducted downward through two shafts from equipment mounted on the roof. The conditioned air from the vertical ducts is distributed around each floor by a system of ducts running above a suspended ceiling, as shown on the plan of the intermediate floor. A double row of columns divides the building into two independent structures at the expansion joint, to allow for differential foundation settlement and thermal expansion and contraction. (Courtesy of ADD Incorporated, Architects)

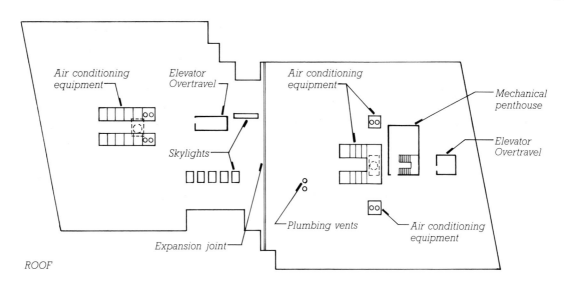

ROOF

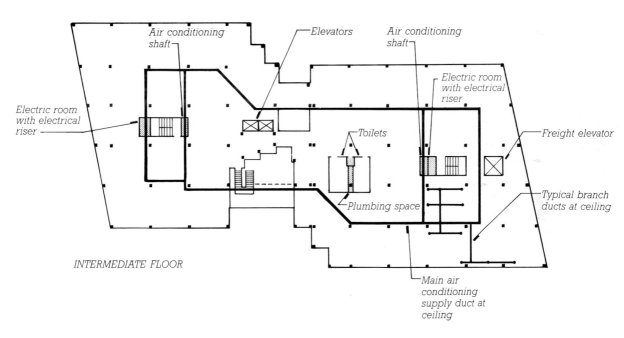

INTERMEDIATE FLOOR

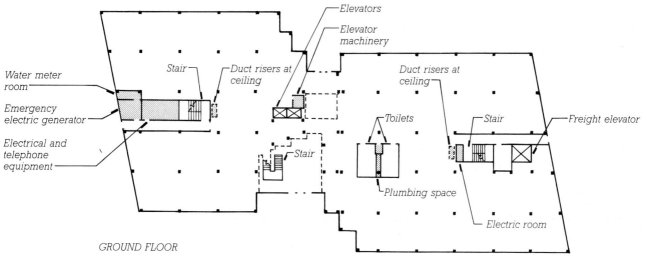

GROUND FLOOR

run between the structural floor deck and a raised *access flooring* system above. (For a more complete explanation of suspended ceilings, cellular floors, and access flooring, see Chapter 18.) Where several plumbing fixtures are lined up along a wall, the wall is usually constructed double, with space between to permit access to the pipes by maintenance personnel.

Specific floor areas are set aside for mechanical and electrical functions in larger buildings (Figure 16.2). Distribution boxes for electrical and communications wiring are housed in special rooms or closets. Rooms are often provided on each floor for air handling machinery. At the bottom of a multistory building, space is customarily set aside, usually at a basement or subbasement level, for pumps, boilers, chillers, electrical transformers, and other heavy equipment. At the roof are penthouses for elevator machinery and such components of the mechanical systems as air conditioning units, cooling towers, water tanks, and ventilating fans. In very tall buildings, one or two entire intermediate floors are set aside for mechanical equipment, and the building is zoned vertically into groups of floors that can be reached by ducts and pipes from each of the mechanical floors.

THE SEQUENCE OF INTERIOR FINISHING OPERATIONS

Simultaneously with the mechanical and electrical work, finishing operations are begun in a carefully ordered sequence that varies somewhat from one building to another depending on the specific requirements of each project. The first finish items to be installed are usually hanger wires for suspended ceilings, and full-height partitions and enclosures, especially those around mechanical and electrical shafts, elevator shafts, mechanical

equipment rooms, and stairways. When the major horizontal electrical conduits and air ducts have been installed at each ceiling, the grid for the suspended ceiling is installed so the lights and ventilating louvers can be mounted in it. Then, typically, the ceilings are finished, and the partitions that do not penetrate the finish ceiling are installed along with their electrical wiring and doors. The last major finishing operation is usually the installation of the finish flooring materials. This is delayed as long as possible, to let the other trades complete their work and get out of the building; otherwise the floor materials could be damaged by dropped tools, spilled paint, heavy construction equipment, weld spatter, coffee stains, and construction debris ground underfoot.

SELECTING INTERIOR FINISH SYSTEMS

Appearance

A major function of interior finish components is to make the interior of the building look neat and clean by covering at least the rougher and less organized portions of the framing, insulation, vapor retarder, electrical wiring, ductwork, and piping. Beyond this, the architect designs the finishes to carry out a particular concept of interior space, light, color, pattern, and texture. The form and height of the ceiling, changes in floor level, interpenetrations of space from one floor to another, and the configurations of the partitions are primary factors in determining the feeling of the interior space. Light originates from windows and electric lighting fixtures, and is propagated by successive reflections off the interior surfaces of the building. Lighter-colored materials raise interior levels of illumination; darker colors and heavier textures result in a darker interior. Patterns and textures

of interior finish materials are important in bringing the building down to a scale of interest that can be readily appreciated by the human eye and hand. No two buildings have the same requirements: Deep carpets and rich, polished marbles in muted tones may be chosen to give an air of affluence to a corporate lobby, brightly colored surfaces to create a happy atmosphere in a day-care center, or slick plastic and highly reflective surfaces to make a trendy ambience for the sale of designer clothing.

Durability and Maintenance

Expected levels of wear and tear must be considered carefully in selecting finishes for a building. Highly durable finishes generally cost more and are not always required. In a public corridor, a transportation terminal, a recreation building, or a retail store, traffic is intense, and long-wearing materials are essential, but in a private office or an apartment, more economical finishes are usually adequate. Water resistance is an important attribute of finish materials in kitchens, locker and shower rooms, some industrial buildings, and entrance lobbies. In hospitals, medical offices, kitchens, and laboratories, finish surfaces must not trap dirt and must be capable of being cleaned and disinfected. Maintenance procedures and costs should be considered in selecting finishes for any building: How often will each surface be cleaned, with what type of equipment, and how much will this procedure add to the cost of owning the building? How long will each surface last, and what will it cost to replace it?

Acoustic Criteria

Interior finish materials strongly affect the quality of listening conditions and the levels of acoustical privacy inside a building. In noisy environments, interior surfaces that are highly absorp-

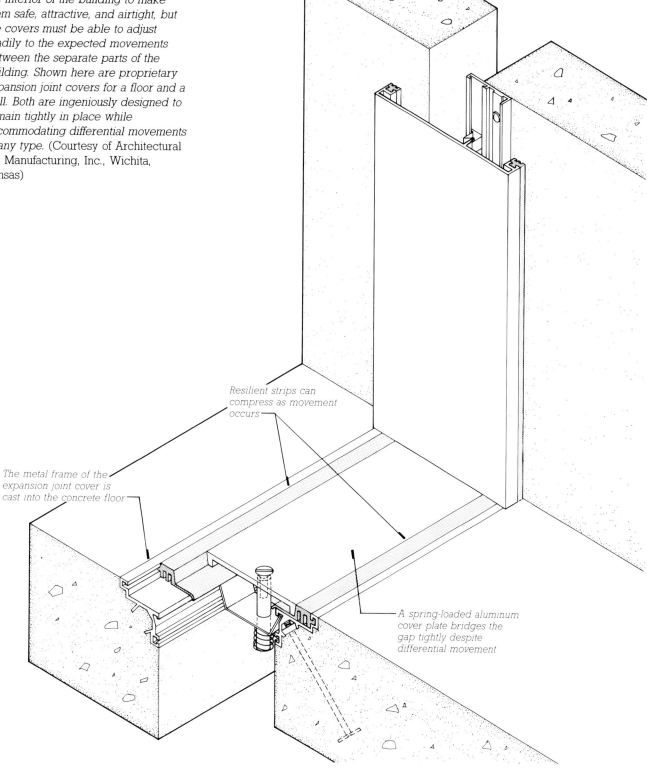

FIGURE 16.3
Expansion joints need to be covered on the interior of the building to make them safe, attractive, and airtight, but the covers must be able to adjust readily to the expected movements between the separate parts of the building. Shown here are proprietary expansion joint covers for a floor and a wall. Both are ingeniously designed to remain tightly in place while accommodating differential movements of any type. (Courtesy of Architectural Art Manufacturing, Inc., Wichita, Kansas)

Resilient strips can compress as movement occurs

The metal frame of the expansion joint cover is cast into the concrete floor

A spring-loaded aluminum cover plate bridges the gap tightly despite differential movement

tive of sound can decrease the noise level to a tolerable value. In lecture rooms, classrooms, meeting rooms, theaters, and concert halls, acoustically reflective and absorptive surfaces must be proportioned and placed so as to create optimum hearing conditions.

Between rooms, acoustic privacy is created most simply by partitions that are both heavy and airtight. The acoustic isolation properties of lighter-weight partitions can be enhanced by partition details that are airtight and damp the transmission of sound vibrations, using resilient mountings on one of the partition surfaces, and sound-absorbing batts of mineral wool in the interior cavity of the partition. Full-scale sample partitions of every type of material are tested for their ability to reduce the passage of sound between rooms in a procedure outlined in ASTM standard E90. The results of this test are converted to *Sound Transmission Class (STC)* numbers that can be related to accepted standards of acoustic privacy. But if the cracks around the edges of a partition are not completely sealed, or if a loosely fitted door or even an unsealed electric outlet is inserted into the partition, its airtightness is compromised and the STC value is meaningless. Similarly, a partition with a high STC is worthless if adjacent rooms are served by a common air duct that also serves incidentally as a conduit for sound, or if the partition reaches only to a lightweight, porous suspended ceiling that allows sound to pass over the top of the partition.

INTERIOR FINISH REQUIREMENTS[h]

Use groups	Required vertical exits and passageways[d]	Corridors providing exit access	Rooms or enclosed spaces[a]
A-1 Assembly, theaters	I	I[f]	II[b]
A-2 Assembly, night clubs	I	I[f]	II[b]
A-3 Assembly halls, terminals, restaurants	I	I[f]	II[b]
A-4 Assembly, churches	I	II	III
B Business	I	II	III
E Educational	I	II	III
F Factory and industrial	I	II	III
H High hazard	I	II	III[g]
I-1 Institutional, residential care	I	II	III
I-2 Institutional, incapacitated	I	I	I[c]
I-3 Institutional, restrained	I	I	I[c]
M Mercantile walls,	I	II	III
ceilings	I	II	II[e]
R-1 Residential, hotels	I	II	III
R-2 Residential, multi-family dwellings	I	II	III
R-3 Residential, 1 and 2 family dwellings	III	III	III
S-1 Storage, moderate hazard	II	II	III
S-2 Storage, low hazard	II	II	III

Note a. Requirements for rooms or enclosed spaces are based upon spaces enclosed in partitions of the building or structure, and where fireresistance rating is required for the structural elements, the enclosing partitions shall extend from the floor to the ceiling: Partitions which do not comply with this shall be considered as enclosing spaces and the rooms or spaces on both sides thereof shall be counted as one. In determining the applicable requirements for rooms or enclosed spaces, the specific use or occupancy thereof shall be the governing factor, regardless of the use group classification of the building or structure. When an approved automatic fire suppression system is provided, the interior finish of Class II or III materials may be used in place of Class I or II materials respectively, where required in the table.
Note b. Class III interior finish materials may be used in places of assembly with a capacity of 300 persons or less.
Note c. Class III interior finish material may be used in administrative areas. Class II interior finish materials may be used in individual rooms of not over 4 persons capacity. Provisions in Note a allowing a change in interior finish classes when fire suppression protection is provided shall not apply.
Note d. Class III interior finish materials may be used for wainscoting or paneling for not more than 1,000 square feet of applied surface area in the grade lobby when applied directly to a noncombustible base or over furring strips applied to a noncombustible base and firestopped as required by Section 1422.0.
Note e. Class III interior finish materials may be used in mercantile occupancies of 3,000 square feet or less gross area used for sales purposes on the street floor only (balcony permitted).
Note f. Lobby areas may be Class II.
Note g. Where building height is over two stories, Class II shall be required.
Note h. The classifications of interior finishes referred to herein correspond to flame spread ratings determined by ASTM E84 listed in Appendix A as follows. Class I flame spread, 0 - 25; Class II flame spread, 26 - 75; Class III flame spread, 76 - 200 (see Section 1421.5.3).
Note I. 1 square foot = 0.093 m².

FIGURE   16.4

A table of fireresistance requirements for interior finish materials, taken from the BOCA Basic/National Building Code. Note h below the table defines the Roman numeral rating system in terms of flame spread ratings. (Figures 16.4 through 16.7 are from the BOCA Basic/National Building Code/1984, Copyright 1984, Building Officials and Code Administrators International, Inc. Published by arrangements with the author. All rights reserved. No parts of the BOCA Code may be reproduced or transmitted in any form or by any means, electronic or mechanical, including photocopying, recording or by an information, storage and retrieval system without advance permission in writing from Building Officials and Code Administrators International, Inc. For information, address: BOCA, Inc., 4051 West Flossmoor Road, Country Club Hills, IL 60477)

Transmission of impact noise from footsteps and machinery through floor-ceiling assemblies can be a major problem. Impact noise transmission is measured by ASTM procedure E492, in which a standard machine taps on a floor above while instruments in a chamber below record sound levels. Impact noise transmission can be reduced by floor details that include soft materials that do not transmit vibration readily, such as carpeting and/or soft underlayment boards.

Fire Criteria

A typical building code devotes many pages to provisions that control the materials and details for interior finishes in buildings. Code requirements are aimed at several important characteristics of interior finishes with respect to fire.

Combustibility

The surface burning characteristics of interior finish materials are tested in accordance with ASTM procedure E84, also called the Steiner Tunnel Test, in which a sample of material 20 inches wide by 25 feet long (500 mm × 7620 mm) forms the roof of a rectangular furnace into which a controlled flame is introduced 1 foot (305 mm) from one end. The time for the flame to spread across the face of the material from one end of the furnace to the other is recorded, along with the amount of fuel contributed by the material and the density of smoke developed. The results of this test are given in trade literature in three forms: the *flame spread rating*, which indicates the rapidity with which fire can spread across a surface of a given material; the *fuel contributed rating*, which indicates the amount of combustible substances in the material; and the *smoke developed rating*, which classifies a material according to the amount of smoke it gives off when it burns. Materials with smoke developed ratings greater than 450 are not permitted to be used inside build-

ings. Figure 16.4 is a typical building code table that defines allowable flame spread ratings for interior finish materials for various use groups of buildings. Class I materials as defined in Note h of this table have flame spread ratings between 0 and 25, Class II between 26 and 75, and Class III between 76 and 200. (The scale of flame spread numbers is established arbitrarily by assigning a value of 0 to cement-asbestos board, and 100 to a red oak board.) Another table (Figure 16.5) gives specifications for the combusti-

bility of flooring materials. Interior trim materials, if their total surface area in a room does not exceed 10 percent of the total wall and ceiling area of the room, may be of Class I, II, or III materials in any type of building.

Fireresistance Ratings

A building code typically regulates the degree of fire resistance of partitions, ceilings, and floors that are used to protect the structure of the building, or to separate various parts of a build-

INTERIOR FLOOR FINISH REQUIREMENTS[c]

Use groups	Required vertical exits and passageways	Corridors providing exit access	Rooms or enclosed spaces[a]
A-1 Assembly, theaters	II	II	DOC FF-1[b]
A-2 Assembly, night clubs	II	II	DOC FF-1[b]
A-3 Assembly halls, terminals restaurants	II	II	DOC FF-1[b]
A-4 Assembly, churches	II	II	DOC FF-1[b]
B Business	II	II	DOC FF-1[b]
E Educational	II	II	DOC FF-1[b]
F Factory and Industrial	DOC FF-1[b]	DOC FF-1[b]	DOC FF-1[b]
H High hazard	DOC FF-1[b]	DOC FF-1[b]	DOC FF-1[b]
I-1 Institutional, residential care	II	II	DOC FF-1[b]
I-2 Institutional, incapacitated	I	I	DOC FF-1[b]
I-3 Institutional, restrained	II	II	DOC FF-1[b]
M Mercantile	II	II	DOC FF-1[b]
R-1 Residential, hotels	II	II	DOC FF-1[b]
R-2 Residential, multi-family dwellings	II	II	DOC FF-1[b]
R-3 Residential, 1 and 2 family dwellings	DOC FF-1[b]	DOC FF-1[b]	DOC FF-1[b]
S-1 Storage, moderate hazard	DOC FF-1[b]	DOC FF-1[b]	DOC FF-1[b]
S-2 Storage, low hazard	DOC FF-1[b]	DOC FF-1[b]	DOC FF-1[b]

Note a. Requirements for rooms or enclosed spaces are based upon the spaces being enclosed with partitions extending from the floor to the ceiling. Where partitions do not satisfy this criteria, the room or space is considered part of the corridor.

Note b. All carpet manufactured for sale in the United States is required by Federal regulations to pass the DOC FF-1 "pill test" (16 CFR, Part 1630) listed in Appendix A. If a material other than carpet is used, the material should be shown to be at least as resistant to flame propagation as a material which passes DOC FF-1 (minimum critical radiant flux of 0.04 Watts/cm²).

Note c. The classifications correspond to that determined by ASTM E648 listed in Appendix A as follows: Class I, 0.45 Watts/cm²; Class II, 0.22 Watts/cm² (see Section 1404.3).

FIGURE 16.5

Interior floor finish requirements, from the BOCA Basic/National Building Code. Note c defines the Roman numeral rating system. DOC FF-1 refers to United States Department of Commerce Standard FF1-70, a standard relating to the surface flammability of carpet and rugs.

ing from one another. Figure 16.6 is a table from the BOCA Basic/National Building Code that specifies the required *fireresistance rating* in hours of separation between different uses housed in the same building. Figure 1.2 gives required fireresistance ratings for various sorts of interior partitions, including shaft walls, exit hallways, exit stairs, dwelling unit separations, and other nonbearing partitions. These can be related to fireresistance ratings in manufacturers' literature similar to the examples shown in Figures 1.3 through 1.5.

Fireresistance ratings for building assemblies are determined by full-scale fire endurance tests conducted in accordance with ASTM E119, which applies not only to partitions and walls, but also to beams, girders, columns, and floor-ceiling assemblies. In this test, the assembly is constructed in a large laboratory furnace and subjected to the

structural load (if any) for which it is designed. The furnace is then heated according to a standard time-temperature curve, reaching more than 1700 degrees Fahrenheit (925° C) at one hour and 2000 degrees Fahrenheit (1093° C) after four hours. To achieve a given fireresistance rating in hours, an assembly must safely carry its design structural load for the designated period, must not develop any openings which permit the passage of flame or hot gases, and must insulate sufficiently against the heat of the fire to maintain surface temperatures on the side away from the fire within specified maximum levels. Wall and partition assemblies must also pass a hose test, in which a duplicate sample is subjected to half the rated fire exposure of the assembly, then sprayed with water from a calibrated fire nozzle for a specified period at a specified pressure in order to simulate the action of

a fire hose during an actual fire. To pass this test, the assembly must not allow passage of a stream of water.

Openings in ceilings and partitions with required fireresistance ratings are restricted in size by most codes and must be protected against the passage of fire in various ways. Doors must be rated for fire resistance in accordance with a table such as that shown in Figure 16.7. Ducts that pass through rated assemblies must be equipped with sheet metal dampers that close automatically if hot gases from a fire enter the duct. Penetrations for pipes and conduits must be closed tightly with fire-resistive material.

As an example of the use of these tables, consider a multistory vocational school building of Type 2A construction that includes both a number of classrooms and a woodworking shop. According to the table in Figure 16.6, the classrooms fall into use group E

FIRE GRADING OF USE GROUPS

Use group		Fire grading in hours
A-1	Assembly, theaters	3
A-2	Assembly, night clubs	3
A-3	Assembly, recreation centers, lecture halls, terminals, restaurants	2
A-4	Assembly, churches	1½
B	Business	2
E	Educational	1½
F	Factory and industrial	3
H	High hazard	4
I-1	Institutional, residential care	1
I-2	Institutional, incapacitated	2
I-3	Institutional, restrained	3
M	Mercantile	3
R-1	Residential, hotels	2
R-2	Residential, multi-family dwellings	1½
R-3	Residential, 1 and 2 family dwellings	1
S-1	Storage, moderate hazard	3
S-2	Storage, low hazard	2

FIGURE 16.6
Fire grading of use groups, from the BOCA Basic/National Building Code.

FIRE DOOR FIRERESISTANCE RATINGS

Location	Fireresistance rating in hours
Fire walls and fire separation walls of 3 or more hour construction	3
Fire walls, fire separation walls and exit enclosures of 1½ or 2 hour construction	1½
Shaft enclosures and elevator hoistways of 2 hour construction	1½
Shaft and exit enclosures of 1 hour construction	1
Exit access corridor enclosures of 1 hour construction	⅓
Other fire separation walls of 1 hour construction	¾

FIGURE 16.7
Required fireresistance ratings for fire doors, from the BOCA Basic/National Building Code.

and the shop into group F. The two uses will have to be separated by a wall of 3-hour construction, and the doors through this wall will also have to be rated at 3 hours (Figure 16.7). Figure 16.4 indicates that in required vertical exits (enclosed exit stairways), only Class I materials are allowed, while Class II materials are acceptable in exit corridors, and Class III materials in the individual classrooms.

Relationship to Mechanical and Electrical Services

Interior finish materials join the mechanical and electrical services of a building at the points of delivery of the services—the electrical outlets, the lights, the ventilating grills, the convectors, the lavatories and water closets. Beyond these points, the services may or may not be concealed by the finish materials, but spaces must be provided for the services, and the provision of these spaces is an important factor in the selection of finishes. If the service lines are to be concealed, the finish systems must provide space for them, and for maintenance access points as required by the services, in the form of access doors, panels, hatches, cover plates, or ceiling or floor components which can be lifted out to expose the lines. If service lines are to be left exposed, the architect will usually want to organize them to some extent and to specify a sufficiently high standard of workmanship in the installation of the services that their appearance will be satisfactory.

Changeability

How often are the use patterns of a building likely to change? In a concert hall, a chapel, or a hotel, major changes will be infrequent, and fixed, unchangeable interior finishes are appropriate. Fixed finishes include many of the heavier, more expensive, more luxurious types of materials such as tile, marble, masonry, and plaster, which are considered highly desirable by many building owners. In a rental office building, an art museum, or a retail store, changes will be frequent; lighting and partitions should be easily and economically adjustable to new use patterns without serious delay or disruption. The possibility of more frequent change can lead either to relatively inexpensive, easily demolished construction such as gypsum wallboard partitions, or to relatively expensive but durable and reusable construction such as proprietary systems of modular, relocatable partitions. The functional and financial choices must be weighed for each building.

Cost

The cost of interior finish systems is measured in two different ways. *First cost* is the installed cost, and is often of paramount importance when the construction budget is tight. *Life-cycle cost* is a cost figured by any of several formulas that take into account not only first cost, but the expected lifetime of the finish system, maintenance costs, fuel costs (if any), replacement cost, an assumed rate of economic inflation, and the time value of money. Life-cycle cost has become increasingly important to building owners who expect to retain ownership over an extended period of time. A material that is inexpensive to buy and install may be more costly over the lifetime of a building because of higher maintenance and replacement costs than a material which is initially more expensive.

TRENDS IN INTERIOR FINISH SYSTEMS

Interior finish systems have undergone a transformation over the last few decades. Formerly the finishes for a commercial office began with the laying up of partitions of heavy clay tiles or gypsum blocks. These were covered with two or three coats of plaster, and joined to a three-coat plaster ceiling. The floor was made of hardwood strips with a wood baseboard, or of poured terrazzo with an integral terrazzo base. Today the same office might be framed in light metal studs and walled with gypsum board. The ceiling might be a separate assembly of lightweight, acoustically absorbent tiles, and the floor a thin layer of vinyl-asbestos tile glued to a smooth concrete slab.

Several trends can be discerned in these changes. One trend is away from an integral, single-piece system of finishes toward a system made up of discrete components. In the old office, the walls, ceiling, and floor were all joined, and none could be changed without disrupting the others. In the new office, the ceiling and floor often run continuously from one side of the building to the other, and the partitions can be changed at will without affecting either the ceiling or floor. The trend toward discrete components is epitomized by partitions made of modular, demountable, relocatable panels.

Another discernible trend leads away from heavy finish materials toward lighter ones. A partition of metal studs and gypsum board weighs a fraction as much as one of clay tiles and plaster, and a traditional terrazzo installation is many times as heavy as an equal area of vinyl-asbestos tile. Lighter finishes dramatically reduce the dead load the structure of the building must carry, to enable the structure itself to be lighter and less expensive. Lighter finishes reduce shipping, handling, and installation costs, and are easier to move or remove when changes are required.

"Wet" systems of interior finish, made of materials mixed with water on the building site, are steadily being replaced by "dry" ones, which are installed in rigid form. Plaster is being replaced by gypsum board and ceiling tiles in most new buildings, and tile and terrazzo floors by plastic tiles and carpet. "Dry" systems are quicker to install and less dependent on weather conditions. They require less skill on the part of the installer than "wet" systems, by transferring the skilled work from the job site to the factory, where it is done by machines. All these differences tend to result in a lower installed cost.

The traditional finishes, nevertheless, are far from obsolete. Gypsum board cannot rival real plaster for surface quality or design flexibility. Tile and terrazzo flooring are unsurpassed for wearing quality and appearance. In many situations, the life-cycle costs of traditional finishes compare favorably with those of lighter weight finishes whose first cost is considerably less. And the aesthetic qualities of, for example, marble floors, wood wainscoting, and sculpted plaster ceilings, where such qualities are called for, cannot be imitated by any other material.

SELECTED REFERENCE

1. Building Officials & Code Administrators International, Inc. *The BOCA Basic/National Building Code 1984.* Flossmoor, Illinois.

The reader is referred to Article 14 of this model code, which deals with fire-resistive construction requirements, or to the corresponding section of the code that is in force locally. (Address for ordering: 4051 West Flossmoor Road, Country Club Hills, IL 60477.)

KEY TERMS AND CONCEPTS

suspended ceiling
shaft
Steiner Tunnel Test
Sound Transmission Class (STC)

flame spread rating
fuel contributed
smoke developed
fireresistance rating

hose test
first cost
life-cycle cost

REVIEW QUESTIONS

1. Draw a flow diagram of the approximate sequence in which finishing operations are carried out on a large building of Type 2A construction.

2. List the major considerations in selecting interior finish materials and systems.

3. What are the two major types of fire tests conducted on interior finish systems? What quantities are derived from each?

4. What is the difference between first cost and life-cycle cost?

EXERCISE

1. You are designing a thirty-one-story apartment building that faces Central Park in Manhattan. What types of construction are you permitted to use under the BOCA Code? What fire-resistance rating will be required for the separation between the apartment floors and the retail stores on the ground floor? What classes of finish materials can you use in the exit stairways? In the corridors to those stairways? Within the individual apartments? If a red oak board has a flame spread rating of 100, can you panel an apartment in red oak? What fireresistance ratings are required for partitions between apartments? Between an exit corridor and an apartment? What type of fire door is required between the apartment and the exit corridor? What is the required fireresistance rating for the walls around the elevator shafts?

## CHAPTER SEVENTEEN

INTERIOR WALLS AND PARTITIONS

Types of Interior Walls

Fire Walls

Fire Separation Walls

Shaft Walls

Other Nonbearing Partitions

Framed Partition Systems

Partition Framing

Plaster

Plaster Partition Systems

Gypsum Board

Masonry Partition Systems

Wall and Partition Facings

Interior Doors

A plasterer applies the scratch coat of gypsum plaster to expanded metal lath. The partition is framed with open-truss wire studs. Soft-annealed galvanized steel wire is used to make all the connections in this partition. *(Courtesy of United States Gypsum Company)*

There is more to interior walls and partitions than meets the eye. Behind their simple surfaces lie assemblies of materials carefully chosen and combined to meet specific performance requirements relating to structural strength, fire resistance, durability, and acoustical isolation. A partition may be framed with steel or wood studs and faced with plaster or gypsum board, or, alternatively, masons may construct it of concrete blocks or structural clay tiles. For improved appearance or durability, a partition may be faced with ceramic tiles or a masonry veneer.

TYPES OF INTERIOR WALLS

Fire Walls

A *fire wall* is a wall that forms a required separation to restrict the spread of fire through a building, and which extends continuously from the foundation to the roof. A fire wall is required either to meet a noncombustible roof structure at the top, or to extend through and above the roof by a specified minimum distance, usually 32 inches (813 mm). Openings in fire walls are restricted in size and must be closed with fire doors or wired glass. Fire walls are used to divide a building that is otherwise too large for its construction type into smaller units in order to become legal under certain code provisions. The required fireresistance ratings for fire walls are defined by line 2 of Figure 1.2, and by Figure 16.7.

Fire Separation Walls

A *fire separation wall*, like a fire wall, forms a required separation to restrict the spread of fire through a building but, unlike a fire wall, does not extend from foundation to roof. Openings in fire separation walls are restricted in size and must be closed with fire doors or wired glass. Fire separation walls are used to divide a building between mixed occupancies (as in the case of the woodworking shop in a classroom building that was discussed in the preceding chapter) and for enclosure of required stairways and exitway corridors. The required fireresistance ratings for fire separation walls are defined by lines 3, 4, 6, 7, and 8 of Figure 1.2 and by Figure 16.7.

Shaft Walls

A *shaft wall* is used to enclose a multistory open space in a building, such as an elevator shaft, or a shaft for ductwork, conduits, or pipes. The required fireresistance ratings for shaft walls are defined by line 5 of Figure 1.2. Walls for elevator shafts must be able to withstand the air pressure and suction loads placed on them by the movements of the elevator cars within the shaft, and should be designed to prevent the noise of the elevator machinery from reaching other areas of the building.

Other Nonbearing Partitions

Many of the partitions in a building neither bear a structural load nor are required as fire separation walls. These may be made of any material that meets the combustibility provisions of the code for the given type of construction.

FRAMED PARTITION SYSTEMS

Partition Framing

Partitions to be finished in plaster or gypsum board are usually framed in wood or metal (Figures 17.1 through 17.4). Wood may be used only as allowed by combustibility requirements of the building code. Fire-retardant treated wood is allowed for partition framing in all types of construction, provided it is used as part of a partition assembly that has been tested and certified to have the required overall fire resistance.

Metal framing is directly analogous to wood framing and may be of *light-gauge steel studs* and *runner channels*, or, for plaster partitions only, *open-truss wire studs*. Light-gauge steel studs are assembled with *self-drilling, self-tapping screws*. These are hardened steel screws of special design that drill their own holes through the metal and make their own screw threads in the holes. They are inserted with an electrically driven screwdriver. Neither the light-gauge steel stud nor the wire stud is very strong structurally until restrained from buckling by the finish surfaces of plaster or sheet material.

If plaster or gypsum board surfaces are to be applied over a masonry wall, they may be spaced away from the wall with either wood or metal *furring strips* (Figures 17.5 through 17.7). Furring allows for the installation of a flat wall finish over an irregular masonry surface, and provides a space for installing plumbing, wiring, and thermal insulation.

Plaster

Plaster is a generic term that refers to any of a number of cementitious substances that are applied to a surface in paste form, after which they harden into a solid material. Plaster may be applied directly to a masonry surface or, more generally, to any of a group of plaster bases known collectively as

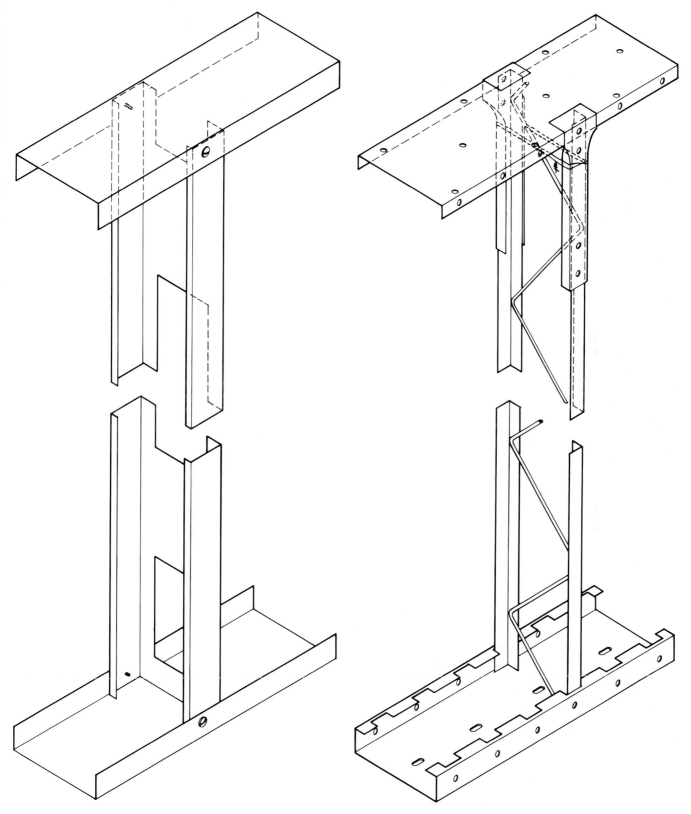

FIGURE 17.1

Partitions of light-gauge steel studs, at the left, and open-truss wire studs. The light-gauge steel studs are assembled with screws, and may be used with any type of lath or panel. The open-truss wire studs are made specifically for use with expanded metal or gypsum lath. The bottom track needs no fasteners to the studs, but metal shoes attached with wire are required at the tops of the studs. Two loops of wire are used to make a stronger connection to studs that support a door frame.

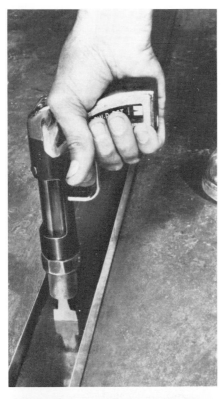

FIGURE 17.2
Attaching a runner to a concrete floor using powder-driven fasteners. The gun explodes a small charge of gunpowder to drive a steel pin through the metal and into the concrete to make a secure connection.

FIGURE 17.3
Inserting studs into the runners to frame a partition of light-gauge steel studs. The rectangular cutouts in the webs of the studs provide a passage for electrical conduits. To the right of the photograph, a stack of gypsum board awaits installation. (Courtesy of United States Gypsum Company)

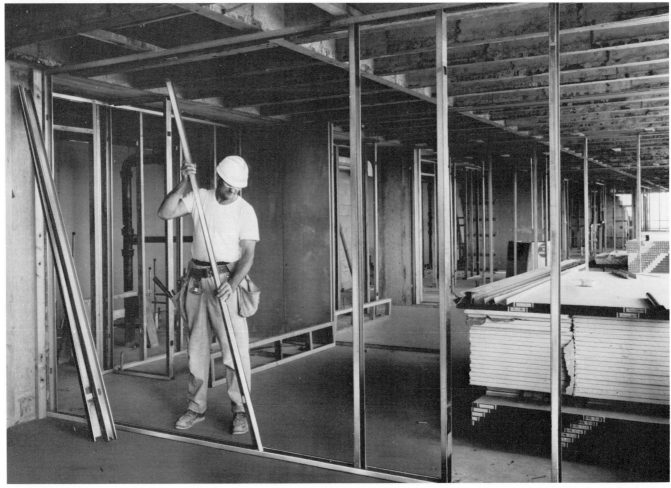

FIGURE 17.4
Inserting an open-truss wire stud into runner track. The stud is 3¼ inches (83 mm) deep; the tip of a shoe to the right of the photograph helps to give an idea of its size. (Courtesy of United States Gypsum Company)

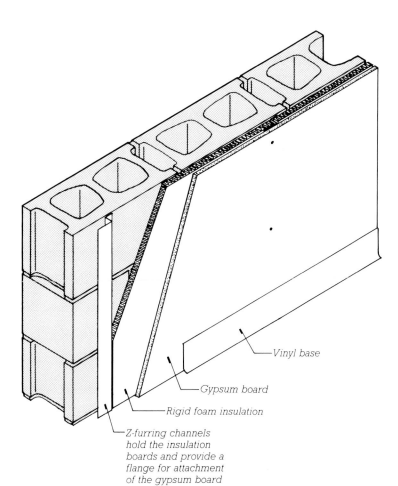

FIGURE 17.5
A furred gypsum board finish over a concrete block wall. The Z-furring channels are attached to the masonry with powder-driven fasteners. The plastic foam insulation is tucked in behind the flange of the channel, and the gypsum board is screwed to the face of the flange. Long slots punched from the web of the channel (not visible in this drawing) help to reduce the thermal bridging effect of the Z-furring channel.

Vinyl base

Gypsum board

Rigid foam insulation

Z-furring channels
hold the insulation
boards and provide a
flange for attachment
of the gypsum board

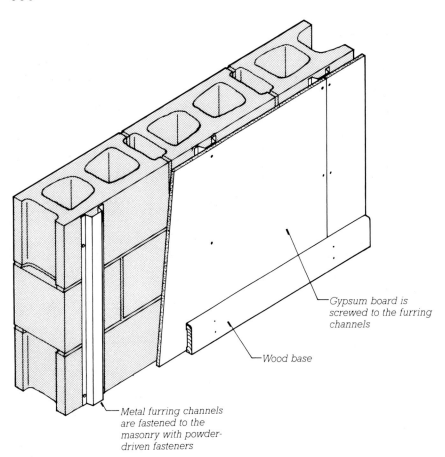

FIGURE 17.6
A furred gypsum board finish using a standard hat-shaped metal furring channel.

Gypsum board is screwed to the furring channels

Wood base

Metal furring channels are fastened to the masonry with powder-driven fasteners

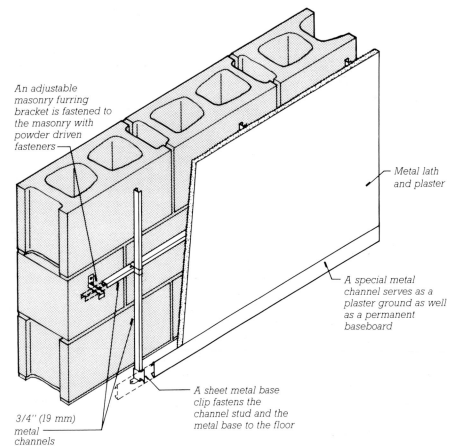

FIGURE 17.7
A furred plaster finish using adjustable furring brackets. Each bracket has a series of teeth along its upper edge, so a metal channel can be wired securely to it in any of a number of positions, allowing the lather to produce a flat wall regardless of the surface quality of the masonry.

An adjustable masonry furring bracket is fastened to the masonry with powder driven fasteners

Metal lath and plaster

A special metal channel serves as a plaster ground as well as a permanent baseboard

A sheet metal base clip fastens the channel stud and the metal base to the floor

3/4" (19 mm) metal channels

lath (rhymes with "math"). Plastering began in prehistoric times with the smearing of mud over masonry walls, or over a mesh of woven sticks and vines to produce a construction known as *wattle and daub*, the wattle being the mesh and the daub the mud. The early Egyptians and Mesopotamians invented finer, more durable plasters based on gypsum and lime. Portland cement plasters were developed in the nineteenth century. It is from these latter three materials—gypsum, lime, and portland cement—that the plasters used in buildings today are prepared.

Gypsum Plaster

Gypsum is an abundant mineral in nature, a crystalline hydrous calcium sulfate. It is quarried, crushed, dried, ground to a fine powder, and heated to 350 degrees Fahrenheit (175° C) in a process known as *calcining* to drive off about three-quarters of its water of hydration. The calcined gypsum, ground to a fine white powder, is known as *plaster of Paris*. When plaster of Paris is mixed with water it rehydrates and recrystallizes rapidly to return to its original, solid state. As it hardens, it gives off heat and expands slightly.

Gypsum is a major component of interior finish materials in most buildings. It has but one major disadvantage—its susceptibility to water damage. Among its advantages are that it is durable and light in weight as compared to many other materials. It resists the passage of sound better than most materials and can be fashioned into surfaces that range from smooth to heavily textured. But above all it is inexpensive, and it is highly resistant to the passage of fire.

When a gypsum building assembly is subjected to the intense heat of a fire, a thin surface layer is calcined and gradually disintegrates. In the process, it absorbs considerable heat and gives off steam, which further cools the fire (Figure 17.8). Layer by layer, the fire works its way through the gypsum, but the process is a slow one.

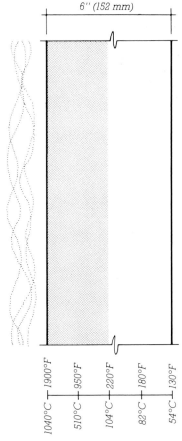

6" (152 mm)

1040°C — 1900°F
510°C — 950°F
104°C — 220°F
82°C — 180°F
54°C — 130°F

FIGURE 17.8

The effect of fire on gypsum, based on data from Underwriters Laboratories, Inc. After a two-hour exposure to heat following the ASTM E119 time-temperature curve, less than half the gypsum on the side toward the fire, shown here by shading, has calcined. The portions of the gypsum to the right of the line of calcination remain at temperatures below the boiling point of water.

The uncalcined gypsum never reaches a temperature more than a few degrees above the boiling point of water, so areas behind the gypsum assembly are well protected from the fire's heat. Any required degree of fire resistance can be created by increasing the thickness of the gypsum as necessary. The fire resistance of gypsum can also be increased by adding lightweight aggregates to reduce its thermal conductivity, and by adding reinforcing fibers to retain the calcined gypsum in place as a fire barrier.

For use in construction, calcined gypsum is carefully formulated with various substances to control its setting time and other properties. Gypsum plaster is made by mixing the appropriate dry plaster formulation with water and an aggregate, either fine sand or perlite. Because of its expansion during setting, it is very resistant to cracking.

For certain types of fine finish coats, the plaster mix is based on hydrated lime, which has superb qualities of workability. Lime plasters, however, are always mixed with quantities of gypsum to accelerate setting and prevent cracking.

Gypsum plasters are manufactured in accordance with ASTM C28. The more common types of gypsum plasters are:

Ordinary *gypsum plaster*, in various formulations for use with sand aggregate or lightweight aggregate, and for machine application

Gypsum plaster with wood fiber, for lighter weight and greater fire resistance

Gypsum plaster with perlite aggregate, for lighter weight and greater fire resistance

High-strength basecoat plaster, for use under high-strength finish coats such as Keenes cement

Gauging plaster, for mixing with lime putty to accelerate its set and eliminate cracking

Keenes cement, a high-strength gypsum plaster that produces an exceptionally strong, crack-resistant finish

Molding plaster, a fast-setting, fine-textured material for molded plaster ornament and drawn cornices (see Figures A and B in the sidebar).

Lime and Portland Cement Plasters

In addition to these gypsum materials, two other types of plasters are commonly used:

Finish lime is mixed with gauging plaster to make high-quality finish coat plasters

Portland cement-lime plaster, also known as *stucco*, is similar to masonry mortar. It is used where the plaster is likely to be subjected to moisture, as on exterior wall surfaces, or in commercial kitchens, industrial plants, and shower rooms. Freshly mixed stucco is not as buttery and smooth as gypsum and lime plasters, so it is not as easy to apply and finish. It shrinks slightly during curing, so should be installed with frequent control joints to regulate cracking.

Plastering

Plaster can be applied either by machine or by hand. Machine application is essentially a spraying process (Figure 17.9). Hand application is done with two very simple tools, a *hawk* in one hand to hold a small quantity of plaster ready for use, and a *trowel* in the other hand to apply the plaster to the surface and smooth it into place (Figures 17.10, 17.22). Plaster is transferred from the hawk to the trowel with a quick, practiced motion of both hands, and the trowel is moved up the wall or across the ceiling to spread the plas-

FIGURE 17.9

Spray-applying a scratch coat of plaster to gypsum lath. (Courtesy of United States Gypsum Company)

ter, much as one uses a table knife to spread soft butter. After a surface is covered, it is leveled by drawing a straightedge called a *darby* across it, after which the trowel is used again to smooth the surface.

Lathing

Until a few decades ago the most common form of lath was thin strips of wood nailed to wood framing with small spaces between the strips to allow keying of the plaster. Most lath today is made either of expanded metal or preformed gypsum boards. The skilled tradesperson who applies lath and trim accessories is known as a *lather*.

Expanded metal lath is made from thin sheets of steel alloy that are slit and stretched in such a way as to produce a mesh of diamond-shaped openings (Figure 17.11). It is applied to open-truss wire studs with intermittent ties of soft steel wire, to steel studs with self-drilling, self-tapping screws, or to wood studs with large-headed lathing nails.

Gypsum lath is made in sheets 16 inches by 48 inches (406 mm × 1220 mm) and in thicknesses of $\frac{3}{8}$ inch (9.5 mm) and $\frac{1}{2}$ inch (13 mm). It consists of sheets of hardened gypsum

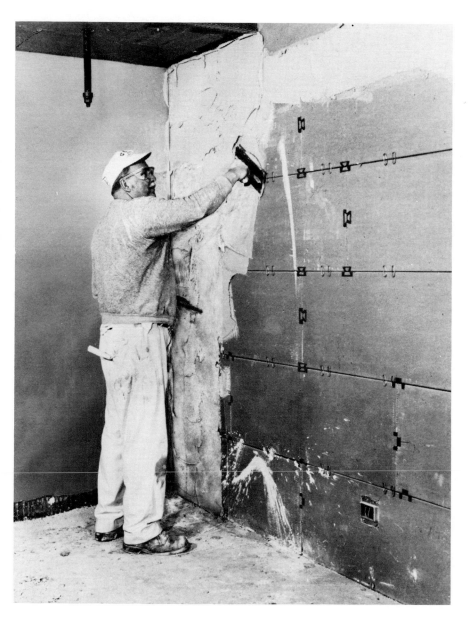

FIGURE 17.10

Applying a scratch coat over gypsum lath with a hawk (a corner of which is visible behind the plasterer's stomach) and trowel. Notice the wire clips that hold the lath to the open-truss studs, and the sheet metal clips that strengthen the end joints between panels. The end joints do not fall over studs. (Courtesy of United States Gypsum Company)

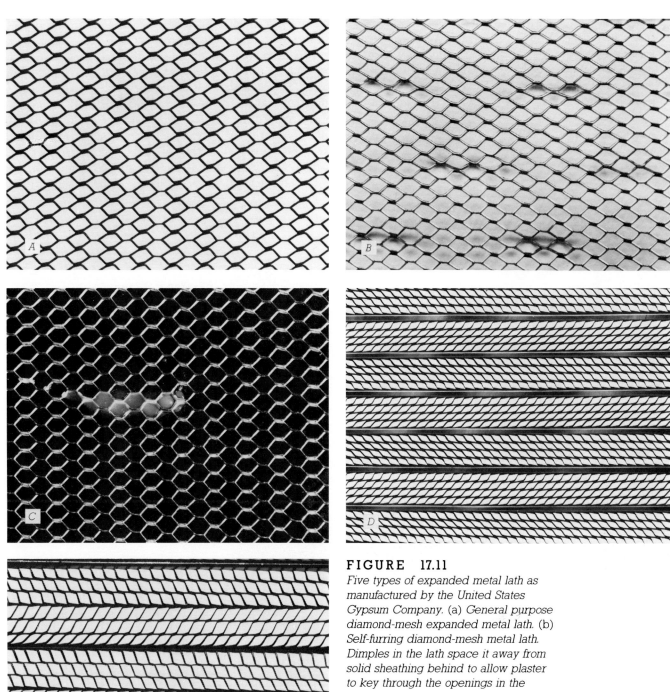

FIGURE 17.11

Five types of expanded metal lath as manufactured by the United States Gypsum Company. (a) General purpose diamond-mesh expanded metal lath. (b) Self-furring diamond-mesh metal lath. Dimples in the lath space it away from solid sheathing behind to allow plaster to key through the openings in the mesh. (c) Paper-back lath, used for backup walls beneath ceramic tile and for exterior stucco. (d) 4-mesh Z-Riblath is stiffer than ordinary diamond-mesh lath, making it suitable for ceilings. (e) Three-eighths-inch (10 mm) riblath has V-shaped ribs for exceptional rigidity; it is used for ceilings or concrete formwork where supports are widely spaced. (Courtesy of United States Gypsum Company)

plaster faced with outer layers of a special absorbent paper to which fresh plaster readily adheres, and inner layers of water resistant paper to protect the gypsum core. Gypsum lath is attached to truss studs with special metal clips, and to steel or wood studs with screws (Figure 17.12). Gypsum lath cannot be used as a base for lime plaster or portland cement stucco.

Veneer plaster base is a paper-faced gypsum board that comes in sheets 4 feet wide (1220 mm) and 8 to 14 feet long (2440 to 4270 mm). It is screw-applied to wood or steel studs, or nailed to wood studs, as a base for the application of *veneer plaster*.

Various lathing *trim accessories* are used at the edges of a plaster surface to make a neat, durable edge or corner (Figure 17.13). These are installed by the lather at the same time as the lath. In very long or tall plaster surfaces, metal control joint accessories are mounted over seams in the lath to control cracking. Trim accessories are also designed to act as lines which gauge the proper thickness and plane of a plaster surface. A straightedge may be run across them to level the wet plaster. In this role the trim accessories are known collectively as *grounds*. Trim accessories are made in several different thicknesses to match the required plaster thicknesses over the different types of lath.

FIGURE   17.12
Installing gypsum lath over light-gauge steel studs with self-drilling, self-tapping screws. The electric screw gun disengages automatically from the screw head when the screw has reached the proper depth. (Courtesy of United States Gypsum Company)

USG Corner Beads, Trim, Control Joints

description

USG Corner Beads and Trim, made from top-quality galvanized steel, enjoy the industry's top acceptance because of their dependability and continual improvement in design. Corner beads are available in 8 and 10-ft. lengths, metal trim in 7 and 10-ft. lengths, casing beads in 7, 8 and 10-ft. lengths.

1-A Expanded Corner Bead has 2⅞″ wide expanded flanges that are easily flexed. Preferred for irregular corners. Provides increased reinforcement close to nose of bead. Made from galvanized steel or zinc alloy for exterior applications.

X-2 Corner Bead has full 3¼″ flanges easily adjusted for plaster depth on columns. Ideal for finishing corners of structural tile and rough masonry. Has perforated stiffening ribs along expanded flange.

4-A Flexible Corner Bead is an economical general purpose bead. By snipping flanges, this bead may be bent to any curved design (for archways, telephone niches, etc.).

800 Corner Bead gives ¹/₁₆″ grounds needed for one-coat veneer finishes. Approx. 90 keys per lin. in. provide superior bonding and strong, secure corners. The 1¼″ fine-mesh flange eliminates shadowing, is easily nailed or stapled.

900 Corner Bead is used with two-coat veneer systems, gives ³/₃₂″ grounds. Its 1¼″ fine-mesh flange can be either stapled or nailed. Provides superior plaster key and eliminates shadowing.

Cornerite and Striplath are strips of painted Diamond Mesh Lath used as reinforcement. **Cornerite** is bent lengthwise in the center to form a 100° angle. It should be used in all interior angles where metal lath is not lapped or carried around, over non-ferrous lath anchored to the lath, and over internal angles of masonry constructions to reduce plaster cracking. **Sizes:** 2″ x 2″ x 96″ and 3″ x 3″ x 96″. **Striplath** is a similar flat strip, used as a plaster reinforcement over joints of non-metallic lathing bases and where dissimilar bases join; also to span pipe chases. **Size:** 4″ x 96″.

USG Metal Trim comes in two styles and two grounds to provide neat edge protection for veneer finishing at cased openings and ceiling or wall intersections. All have fine-mesh expanded flanges to strengthen plaster bond and eliminate shadowing. **No. 701-A,** channel-type, and **No. 701-B,** angle edge trim, provide ³/₃₂″ grounds for two-coat systems. **No. 801-A,** channel-type, and **No. 801-B,** angle edge trim, provide ¹/₁₆″ grounds for one-coat systems. **Sizes:** for ½″ and ⅝″ IMPERIAL Gypsum Base.

USG P-1 Vinyl Trim is a channel-shaped rigid trim with flexible vinyl fins which compress on installation to provide a positive acoustical seal comparable in performance to one bead of acoustical sealant. For veneer finish partition perimeters. **Lengths:** 8, 9 and 10 ft. **Sizes:** for ½″ and ⅝″ gypsum base.

USG P-2 Vinyl Trim is a channel-shaped vinyl trim with a pressure-sensitive adhesive backing for attachment to the wall at wall-ceiling intersections. Provides positive perimeter relief in radiant heat and veneer finish systems. Allow ⅛″ to ¼″ clear space for insertion. **Length:** 10 ft.

USG Control Joint relieves stresses of expansion and contraction in large plastered areas. Made from roll-formed zinc, it is resistant to corrosion in both interior and exterior uses with gypsum or portland cement plaster. An open slot, ¼″ wide and ½″ deep, is protected with plastic tape which is removed after plastering is completed. The perforated short flanges are wire-tied to metal lath, screwed or stapled to gypsum lath. Thus the plaster is key-locked to the control joint, which not only provides plastering grounds but can also be used to create decorative panel designs. **Limitations:** Where sound and/or fire ratings are prime considerations, adequate protection must be

provided behind the control joint. USG Control Joints should not be used with magnesium oxychloride cement stuccos or stuccos containing calcium chloride additives. **Sizes and grounds: No. 50,** ½″; **No. 75,** ¾″; **No. 100,** 1″ (for exterior stucco curtain walls)—10-ft. lengths.

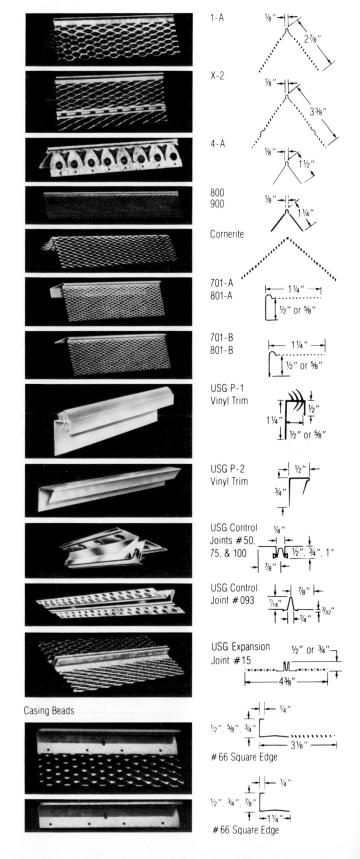

Casing Beads

Plaster Systems

Plaster Over Expanded Metal Lath

Plaster is applied over expanded metal lath in three coats (Figure 17.14). The first, called the *scratch coat*, is troweled on rather roughly and cannot be made completely flat be-cause the uncoated lath is not very rigid under the pressure of the trowel. This first coat is scratched while still wet, using a notched darby, a broom, or a special rake, to create a rough surface to which the second coat can bond mechanically (Figure 17.15).

When the scratch coat has hard-

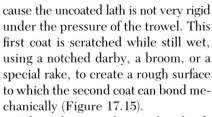

FIGURE 17.13
Trim accessories for lath and plaster construction, as manufactured by the United States Gypsum Company. (Courtesy of United States Gypsum Company)

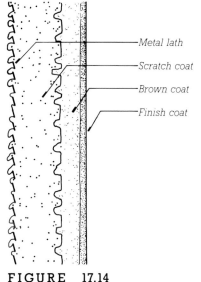

Metal lath
Scratch coat
Brown coat
Finish coat

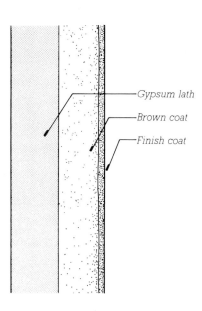

Gypsum lath
Brown coat
Finish coat

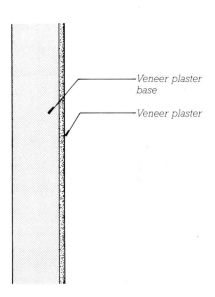

Veneer plaster base
Veneer plaster

FIGURE 17.14
Sections through the three common lath-and-plaster systems, reproduced at full scale. Metal lath (left) requires three coats of plaster, with the first coat customarily scratched for a better bond. Gypsum lath (middle) may be finished with three coats, or with two, as shown. Veneer plaster (right) uses only a thin coating of plaster.

FIGURE 17.15
Scratching the scratch coat while still wet to create a better bond to the brown coat. (Courtesy of United States Gypsum Company)

ened, it works together with the lath to create a rigid base for the second application of plaster, which is called the *brown coat*. The purpose of the brown coat is to build strength and thickness, and to present a level surface for the application of the third or *finish coat*. The level surface is produced by drawing a long straight-edge across the surfaces of the lathing accessories (the grounds, edge beads, corner beads, and control joints) to strike off the wet plaster. On large, uninterrupted plaster surfaces, *plaster screeds*, intermittent spots or strips of plaster, are leveled up to the grounds in advance of brown coat plastering to serve as intermediate reference points for setting the thickness of the plaster during the striking-off operation.

The finish coat is very thin, about $\frac{1}{16}$ inch (1.5 mm) in thickness. It may be troweled smooth, or worked into any desired texture (Figures 17.16, 17.17). The total thickness of plaster that results from this three-coat process, as measured from the face of the lath, is about $\frac{5}{8}$ inch (16 mm). Three-coat work over metal lath is the premium quality plaster system, extremely strong and resistant to fire. The only disadvantage of three-coat plaster work is its cost, which can be attributed largely to the labor involved in applying the lath and the three separate coats of plaster.

Plaster Over Gypsum Lath   The best plaster work over gypsum lath is applied in three coats, but gypsum lath is sufficiently rigid that if it is firmly mounted to the studs, only a brown coat and a finish coat need be applied. The elimination of the scratch coat has obvious economic advantages. Even with three coats of plaster, gypsum lath is often economically advantageous because the gypsum in the lath replaces much of the plaster that would otherwise have to be mixed and applied by hand in the scratch coat. The total thickness of plaster applied over gypsum lath is $\frac{1}{2}$ inch (13 mm).

Plaster Applied to Masonry
Where plaster is applied directly

FIGURE   17.16
A sponge-faced float can be used to create various surface textures on plaster. (Courtesy of Portland Cement Association, Skokie, Illinois)

In lathing I was pleased to be able to send home each nail with a single blow of the hammer, and it was my ambition to transfer the plaster from the board to the wall neatly and rapidly...I admired anew the economy and convenience of plastering, which so effectually shuts out the cold and takes a handsome finish...I had the previous winter made a small quantity of lime by burning the shells of the *Unio fluviatilis,* which our river affords...

Henry David Thoreau, *Walden*

over brick, concrete block, or poured concrete walls, the walls should be dampened thoroughly in advance of plastering to prevent premature dehydration of the plaster. A bonding agent may have to be applied to some types of smooth masonry surfaces to ensure good adhesion of the plaster. The number of coats of plaster required to cover a wall is determined by the degree of unevenness of the masonry surface. For the best work three coats totaling $\frac{5}{8}$ inch (16 mm) should be applied, but for many walls two coats will suffice (Figure 17.18).

Veneer Plaster   Veneer plaster is the least expensive of the plaster systems and is competitive in price with gypsum board finishes in many regions. The veneer base and accessories create a very flat surface that can be finished with a layer of a specially formulated dense plaster that is only about $\frac{1}{8}$ inch (3 mm) thick (Figures 17.19 through 17.22). The plaster is applied in a "double-back" process in which a thin first coat is followed immediately by a second "skim" coat that is finish troweled to the desired texture. The plaster veneer hardens and dries so rapidly that it may be painted the following day.

Stucco   Stucco is applied over galvanized lath, using galvanized or plastic accessories, to prevent rusting in damp locations. While gypsum plaster expands during hardening, and is therefore highly resistant to cracking, portland cement stucco shrinks and is prone to cracking. Stucco walls should be provided with control joints at frequent intervals to channel the shrinkage into predetermined lines rather than allowing it to cause random cracks. The curing reaction in stucco is the same as that of concrete and is relatively slow as compared to that of gypsum plaster. Stucco must be kept moist for a period of at least a week before it is allowed to dry, to attain maximum hardness and strength through full hydration of its portland cement binder.

In exterior applications over metal or wood studs, stucco may be applied

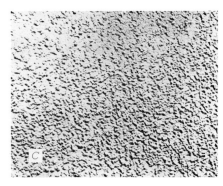

FIGURE 17.17
Three different plaster surface textures from among many: (a) Float finish. (b) Spray finish. (c) Texture finish.

(Courtesy of United States Gypsum Company)

FIGURE 17.18
Applying a finish coat of portland cement plaster over a concrete block partition. The block joints are visible in the base coat of plaster because of a difference in the rate of water absorption between the blocks and the mortar joints. (Courtesy of Portland Cement Association, Skokie, Illinois)

PLASTER ORNAMENT

Gypsum plaster, with its fine grain and even texture, has more sculptural potential than any other material used in architecture. While wet, it is easily formed with trowels and spatulas, with molds, or with templates. When dry, it is readily worked by sawing, sanding, machining, and carving. Plaster building ornament has been made for many centuries by two economical but powerful techniques, *casting* and *drawing,* and is seen in buildings of every historic style, from small houses to large public buildings.

Cast ornament is made in a shop by pouring soupy plaster into molds. The plaster hardens in a few minutes, allowing the mold to be stripped and reused. Both rigid molds and soft rubber molds are used. The rubber molds are very flexible, so even undercut shapes can be cast without encountering difficulties in mold removal. Traditional rubber molds are created by first carving a plaster original, then brushing layers of latex over the original to build up the required wall thickness. More recently, two-component synthetic rubber compounds have replaced latex in most applications; their advantage is that they can be spread over the original in a single application rather than in layers.

Traditionally, cast ornament is adhered to the brown coat of plaster in walls and ceilings with gobs of wet plaster. Once the ornament is securely in place, the finish coat of plaster is applied, and the plaster surfaces and adjacent pieces of ornament are merged by skillful trowel work and sanding to create a single-piece finish.

Drawing is used to make linear ornament such as classic cornice moldings. A sheet metal or rigid sheet plastic template bearing the profile of the molding is run back and forth along a guide strip mounted temporarily in gobs of plaster on the wall or ceiling, and a mixture of lime putty and gauging plaster is inserted in front of the template, to be struck off to the desired profile. Repeated passes of the template are required to finish the molding smoothly and perfectly. These must be completed swiftly before the plaster begins to harden, or the setting expansion of the gypsum will cause the template to chatter and spoil the plaster surface.

Cast plaster ornament can be reinforced with alkali-resistant glass fibers. These greatly increase its strength and toughness and allow it to be produced in much thinner sections and much larger pieces than unreinforced plaster. This recent development has dramatically changed the economics and methods of ornamentation in plaster. Much of the on-site assembly work for elaborate ornaments can be eliminated by combining what were formerly a number of small, thick, brittle castings into a single, larger, thinner casting that is light in weight and highly resistant to breakage. Drawn ornaments, which formerly were done entirely in place, can instead be cast in the shop in long, hollow sections. On the job site, the lightweight castings are glued in place over gypsum board or veneer plaster base, using an ordinary mastic adhesive. The edges of the ornaments are feathered into the wall or ceiling surfaces with joint compound or veneer plaster, and the joints between pieces of ornament are smoothed over and sanded.

Casting and drawing can be used to reproduce plaster ornament during restoration of historic buildings. Rubber molds for casting can be made directly from existing ornaments, and templates for drawing duplicates of existing profiles are easily and cheaply produced.

New designs for plaster ornament are readily translated from the architect's paper sketches into carved plaster originals, from which rubber molds are made and duplicates cast. The possibilities are almost limitless, yet few contemporary architects have chosen to explore them. This is surprising because, as compared to ornament carved from wood or stone, plaster ornament is inexpensive, and there are few technical constraints on what can be accomplished.

FIGURE A
Removing the mold from a cast plaster ornament. (Courtesy of Dovetail, Inc.)

FIGURE B
Drawing a plaster cornice during restoration of a historic building. Notice the large cast plaster ornament behind the worker. (Courtesy of Dovetail, Inc.)

FIGURE 17.19
Installing veneer plaster base with a screw gun. (Courtesy of United States Gypsum Company)

FIGURE 17.20
Stapling a corner bead to veneer plaster base to create a straight, durable corner. (Courtesy of United States Gypsum Company)

FIGURE 17.21
Reinforcing the panel joints in veneer plaster base with a self-adhesive glass fiber mesh tape. A panel for access to mechanical equipment is being provided behind the installer. (Courtesy of United States Gypsum Company)

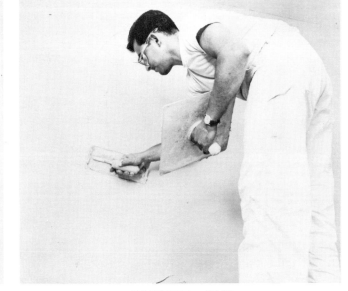

FIGURE 17.22
Applying veneer plaster with hawk and trowel. (Courtesy of United States Gypsum Company)

either over sheathing or without sheathing. Over sheathing, asphalt-saturated felt paper is first applied as an air and moisture barrier. Then *self-furring* metal lath, which is formed with "dimples" that hold the lath away from the surface of the wall a fraction of an inch to allow the stucco to key to the lath, is attached with nails or screws (Figure 17.11*b*). If no sheathing is used, the wall is laced tightly with strands of *line wire* a few inches apart, and paper-backed metal lath is attached to the line wire. The stucco bonds to the lath and line wire to encase the building in a thin layer of reinforced concrete.

Stucco is usually applied in three coats, either with a hawk and trowel, or by spraying (Figure 6.38). In exterior work pigments or dyes are often added to stucco to give an integral color, and rough textures are frequently used.

Plaster Partition Systems

Several types of plaster partitions are detailed in Figures 17.23 and 17.24. These diagrams show some of the ways in which the various trim accessories are used, and the precautions taken to isolate the partitions from structural or thermal movements in the loadbearing frame of the building.

Gypsum Board

Gypsum board is a prefabricated plaster sheet material that is manufactured in widths of 4 feet (1220 mm) and lengths of 8 to 14 feet (2440 to 4270 mm). It is also known as *gypsum wallboard, plasterboard,* and *drywall*. (The term *sheetrock* is a registered trademark of one manufacturer of gypsum board, and should not be used in a generic sense.)

Gypsum board is the least expensive of all interior wall finishing materials. For this reason alone it has found wide acceptance throughout North America as a substitute for plaster, not only in residential construction, but in buildings of every type. It retains the fire-resistive characteristics of gypsum plaster, but it is installed with less labor by less skilled workers than lathers and plasterers. And because it brings very little water into a building, it eliminates some of the waiting that is associated with the curing and drying of plaster.

The core of gypsum board is formulated as a slurry of calcined gypsum, starch, water, pregenerated foam to reduce the density of the mixture, and various additives. This is sandwiched between special paper faces and passed between sets of rollers to reduce it to the desired thickness. Within two to three minutes the core material has hardened and bonded to the paper faces. The board is cut to length and heated to drive off residual moisture, then bundled for shipping (Figure 17.25).

Types of Gypsum Board

Gypsum board is manufactured in a number of different types in accordance with ASTM C36.

- *Regular gypsum board* is used for the majority of applications.

- In locations exposed to moderate amounts of moisture, a special *water-resistant gypsum board* is used, with a water repellant paper facing and a moisture-resistant core formulation.

- For many types of fire-rated assemblies, *Type X gypsum board* is required. The core material of Type X board is reinforced with short glass fibers. In a severe fire, the fibers hold the calcined gypsum in place to continue to act as a barrier to fire, rather than permitting it to erode or fall out.

- *Foil-backed gypsum board* can be used to eliminate the need for a separate vapor retarder in outside wall assemblies. If the back of the board faces a dead airspace at least $\frac{3}{4}$ inch (19 mm) wide, the bright foil also acts as a thermal insulator.

- *Predecorated gypsum board* is board that has been covered with paint, printed paper, or decorative plastic film. If handled carefully and installed with small nails, it needs no further finishing.

- *Gypsum backing board* is a less expensive board that is used for a backup layer in multilayer finishing systems.

- *Coreboard* is a 1-inch-thick panel (25.4 mm) used for shaftwalls and solid gypsum board partitions. To facilitate handling it is fabricated in sheets 24 inches (610 mm) wide.

Gypsum board is manufactured with a variety of edge profiles, but the most common is the *tapered edge,* which permits sheets to be joined with a flush, invisible seam by means of subsequent joint finishing operations. Rounded and beveled edges are useful in predecorated panels, and tongue-and-groove edges serve to join coreboard panels in concealed locations.

A number of different thicknesses of board are produced:

- $\frac{1}{4}$ inch board (6.4 mm) is used as a backing board in certain sound control applications.

- $\frac{5}{16}$ inch board (8 mm) is made for manufactured housing, where weight reduction to facilitate shipping is an important consideration.

- $\frac{3}{8}$ inch board (9.5 mm) is largely used in double-layer wall finishes, but it is permitted to be used over wall studs 16″ (400 mm) apart by some building codes.

- $\frac{1}{2}$ inch board (12.7 mm) is the most common thickness. It is used for stud spacings up to 24″ (610 mm).

- $\frac{5}{8}$ inch board (16 mm) is also limited to stud spacings of not more than 24″, but it is often used where additional fire resistance or structural stiffness is required.

- 1-inch-thick board (25.4 mm) is made only as coreboard.

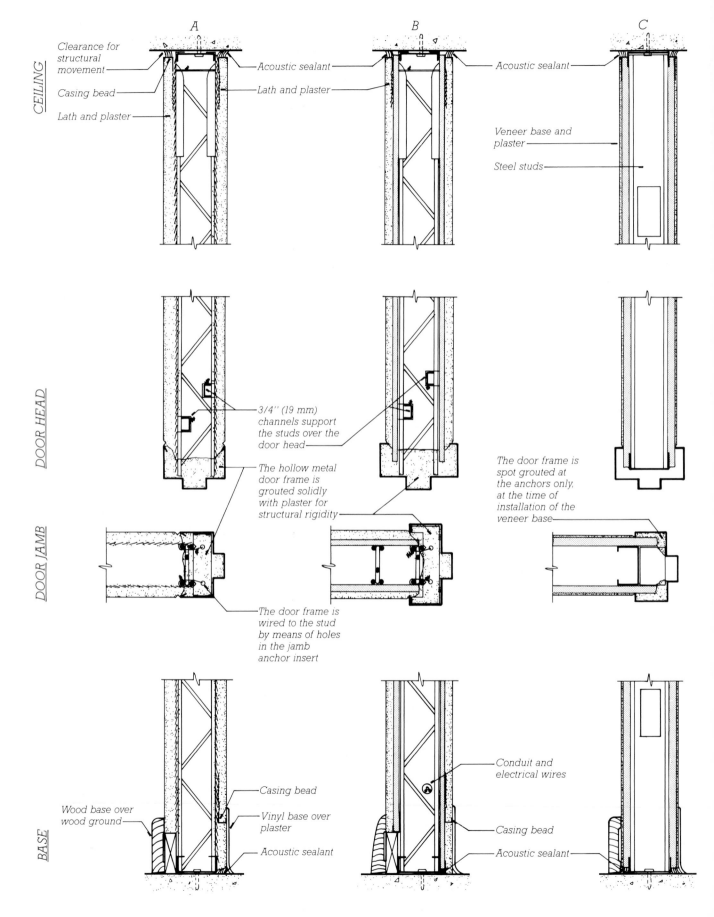

A   B   C

CEILING

Clearance for structural movement

Casing bead

Lath and plaster

Acoustic sealant

Lath and plaster

Acoustic sealant

Veneer base and plaster

Steel studs

DOOR HEAD

3/4″ (19 mm) channels support the studs over the door head

DOOR JAMB

The hollow metal door frame is grouted solidly with plaster for structural rigidity

The door frame is spot grouted at the anchors only, at the time of installation of the veneer base

The door frame is wired to the stud by means of holes in the jamb anchor insert

BASE

Wood base over wood ground

Casing bead

Vinyl base over plaster

Acoustic sealant

Conduit and electrical wires

Casing bead

Acoustic sealant

FIGURE 17.23

*Three plaster partition systems. Left **A**, three coats of plaster on metal lath and open-truss wire studs, rated at 1 hour of fire resistance and STC 39. Center **B**, two coats of plaster over gypsum lath and open-truss wire studs, also rated at 1 hour and STC 41. Right **C**, veneer plaster on light-gauge steel studs, STC 40, 1 hour. Notice especially the provisions for airtightness and structural movement at the top and bottom of each partition, and the methods of attachment used for hollow metal door frames. Doors weighing more than 50 pounds (23 kg) require special reinforcing details around the frames.*

FIGURE 17.24

Two plaster partition systems. The partition shown at left increases its STC to 51 by mounting the gypsum lath on one side with resilient metal clips, and filling the hollow space in the partition with a sound attenuation blanket. The solid plaster partition, right, has an STC of only 38 but is used in situations where floor space must be conserved. Both these partitions are rated at 1 hour.

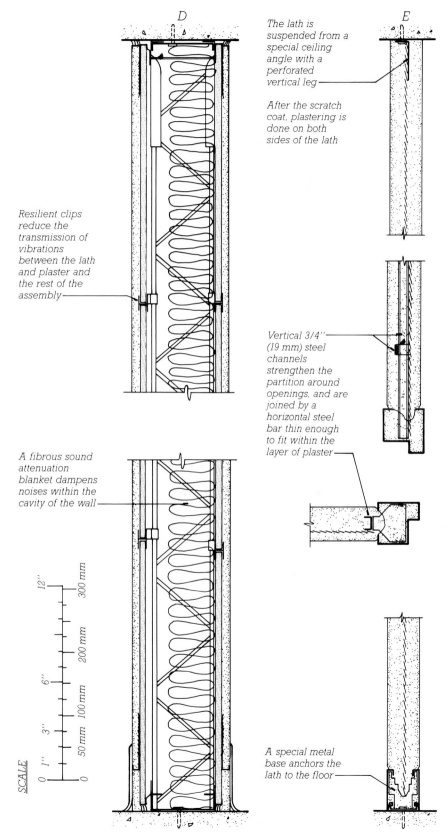

Resilient clips reduce the transmission of vibrations between the lath and plaster and the rest of the assembly

A fibrous sound attenuation blanket dampens noises within the cavity of the wall

The lath is suspended from a special ceiling angle with a perforated vertical leg

After the scratch coat, plastering is done on both sides of the lath

Vertical 3/4'' (19 mm) steel channels strengthen the partition around openings, and are joined by a horizontal steel bar thin enough to fit within the layer of plaster

A special metal base anchors the lath to the floor

SCALE

12''   300 mm

6''   200 mm

3''   100 mm

1''   50 mm

0   0

FIGURE   17.25
Sheets of gypsum board roll off the manufacturing line, trimmed and ready   *for packaging.* (Courtesy of United States Gypsum Company)

Installing Gypsum Board

Hanging the Board   Gypsum board may be installed over either steel or wood studs, using self-drilling, self-tapping screws to fasten to steel, and either screws or nails to fasten to wood. Wood studs can be troublesome with gypsum board because they usually shrink somewhat after the board is installed, which can cause nails to loosen slightly and "pop" through the finished surface of the board (Figure 17.26). *Nail popping* can be minimized by using only fully dried lum-ber, by using the shortest nail that will do the job, and by using ring-shank nails that have extra gripping power in the wood. Screws have less of a tendency to pop than nails. When screws or nails are driven into gypsum board, their heads are driven to a level slightly below the surface of the board, but not enough to tear the paper surface.

To minimize the length of joints that must be finished, and to create the stiffest wall possible, gypsum board is usually installed with the long dimension of the boards horizontal. The longest possible boards are used, to eliminate or at least minimize end joints between boards, which are difficult to finish because ends of boards are not tapered. Gypsum board is cut rapidly and easily by scoring one paper face with a sharp knife, snapping the brittle core along the score line with a blow from the heel of the hand, and cutting the other paper face along the fold created by the snapped core (Figure 17.27). Notches, irregular cuts, and holes for electric boxes are made with a small saw.

When two or more layers of gypsum board are installed on a surface, the

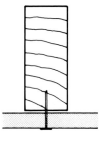
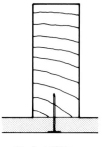

Stud at 19%
moisture content

Stud at 10%
moisture content

FIGURE 17.26
*When wood studs dry and shrink during
a building's first heating season, nail
heads may pop through the surface of
gypsum board walls.*

FIGURE 17.27
Cutting gypsum board: (a) *A sharp knife
and metal T-square are used to score a
straight line through one paper face of
the panel.* (b) *The scored board is
easily "snapped," and the knife used a
second time to slit the second paper
face.* (Courtesy of United States Gypsum
Company)

FIGURE 17.28
Attaching gypsum board to studs with a screw gun. (Courtesy of United States Gypsum Company)

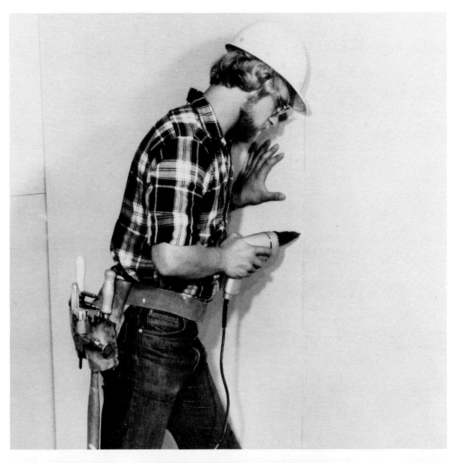

FIGURE 17.29
Gypsum board can be curved to a large radius by simply bending it around a curving line of studs. (Courtesy of United States Gypsum Company)

joints between layers are staggered to create a stiffer wall, and a mastic adhesive is often used to join the layers to one another. Adhesive is sometimes used between the studs and the gypsum board in single-layer installations to make a stronger joint.

Gypsum board can be curved when a design requires it. For gentle curves the board can be bent into place dry (Figure 17.29). For sharper curves, the paper faces are moistened to decrease the stiffness of the board before it is installed. When the paper dries, the board is as stiff as before.

Metal trim accessories are required at exposed edges and external corners to protect the brittle board and present a neat edge (Figure 17.30).

Finishing the Joints and Fastener Holes Joints and holes in gypsum board are finished to create the appearance of a monolithic surface, indistinguishable from plaster. The finishing process is based upon use of a joint compound that resembles a smooth, sticky plaster. For most purposes a drying-type all-purpose joint compound is used; this is a mixture of marble dust, binder, and additives, furnished either as a dry powder to be mixed with water, or as a premixed paste. In some high-production commercial work, joint compounds that cure rapidly by chemical reaction are

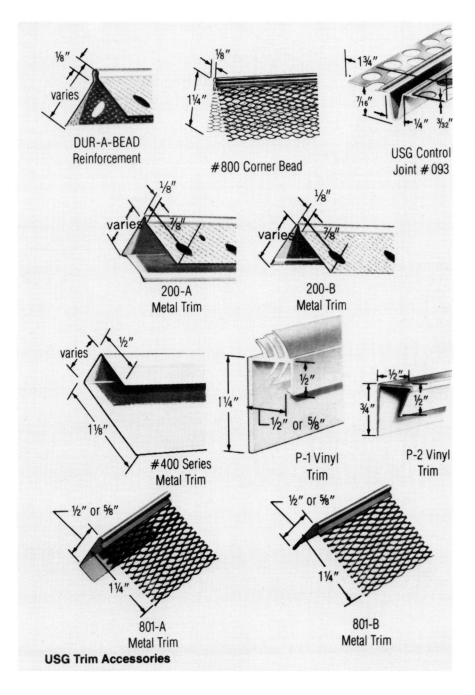

FIGURE   **17.30**
Accessories for gypsum board construction, as manufactured by the United States Gypsum Company. (Courtesy of United States Gypsum Company)

used to minimize waiting time between applications.

The finishing of a joint between panels of gypsum board begins with the troweling of a layer of joint compound into the tapered edge joint, and the bedding of a paper or glass fiber reinforcing tape in the compound (Figures 17.31 through 17.33). After overnight drying, a second layer of compound is applied to the joint, to bring it level with the face of the board and to fill the space left by the slight drying shrinkage of the joint compound. Compound is also troweled over the nail or screw holes. When this second coat is dry, the joints are lightly sanded before a very thin final coat is applied to fill any remaining voids and feather out to an invisible edge. Before painting, the wall is again sanded lightly to remove any roughness or ridges. If the finishing is done properly, the painted or papered wall will show no signs at all that it is made of discrete panels of material.

Gypsum board has a smooth surface finish, but a number of spray-on textures and textured paints can be applied to give a rougher surface. Most gypsum board contractors prefer to texture ceilings; the texture conceals the minor irregularities in workmanship that are likely to occur because of the difficulty of working in the overhead position.

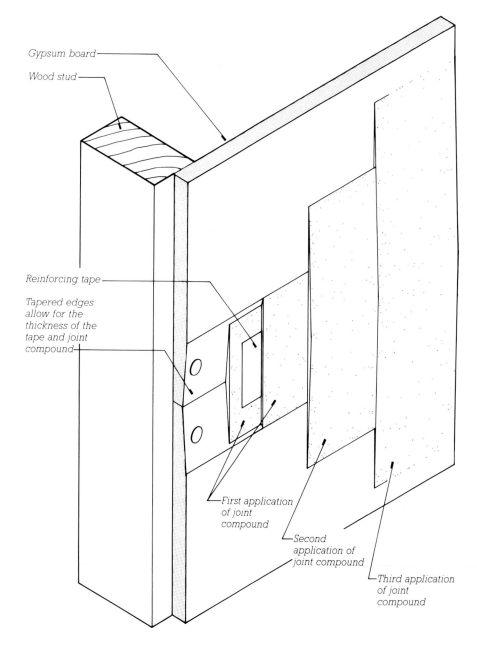

Gypsum board

Wood stud

Reinforcing tape

Tapered edges allow for the thickness of the tape and joint compound

First application of joint compound

Second application of joint compound

Third application of joint compound

FIGURE 17.31

Finishing a joint between panels of gypsum board.

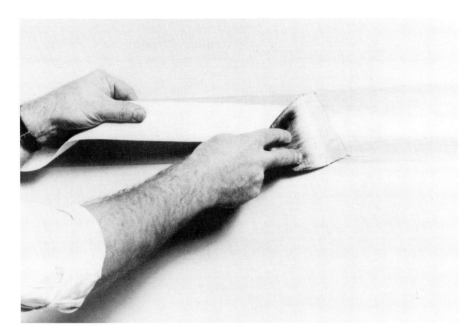

FIGURE 17.32
Applying paper joint reinforcing tape to gypsum board. (Courtesy of United States Gypsum Company)

FIGURE 17.33
An automatic taper applies tape and joint compound to gypsum board joints. (Courtesy of United States Gypsum Company)

CEILING

DOOR HEAD

DOOR JAMB

BASE

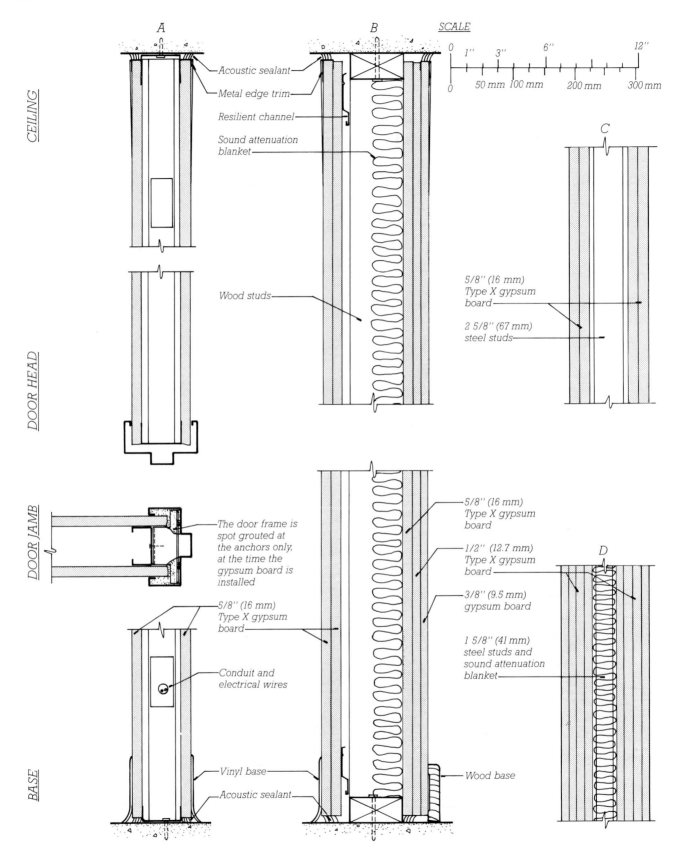

A

B

SCALE

C

D

Acoustic sealant

Metal edge trim

Resilient channel

Sound attenuation
blanket

Wood studs

5/8'' (16 mm)
Type X gypsum
board

2 5/8'' (67 mm)
steel studs

The door frame is
spot grouted at
the anchors only,
at the time the
gypsum board is
installed

5/8'' (16 mm)
Type X gypsum
board

Conduit and
electrical wires

Vinyl base

Acoustic sealant

5/8'' (16 mm)
Type X gypsum
board

1/2'' (12.7 mm)
Type X gypsum
board

3/8'' (9.5 mm)
gypsum board

1 5/8'' (41 mm)
steel studs and
sound attenuation
blanket

Wood base

0   1''   3''   6''   12''

0   50 mm 100 mm   200 mm   300 mm

FIGURE 17.34
*Four gypsum board partition systems: **A**.
A 1-hour partition, STC 40, using Type X
gypsum board over light-gauge steel
studs. **B**. This 1-hour partition on wood
studs achieves an STC of 60 to 64
through heavy laminations of gypsum
board, a sound attenuation blanket, and
resilient channel mounting for one face
of the partition. **C**. A 2-hour partition
with an STC of 48. **D**. A 4-hour partition,
STC 58. These are representative of a
large number of gypsum board partition
systems in a range of fire ratings and
sound transmission classes.*

Gypsum Board Partition Systems

Gypsum board partition assemblies have been designed and tested with fireresistance ratings up to 4 hours and acoustic performance to STC 59. A selection of these partitions is shown in Figure 17.34. Notice in these details how provisions are made to prevent the gypsum panels from being subjected to structural loadings caused by movements in the loadbearing frame of the building—structural deflections, concrete creep, moisture expansion, and temperature expansion and contraction. Notice also the use of sealant to eliminate sound transmission around the edges of the partitions.

Demountable partition systems of gypsum board have also been developed, using concealed mechanical fasteners that can be disassembled and reassembled easily without damage to the panels. These systems are of use in buildings whose partitions must be rearranged at frequent intervals.

Gypsum Shaft Wall Systems

Walls around elevator shafts, stairways, and mechanical chases can be made of any masonry, lath and plaster, or gypsum panel assembly that meets fireresistance and structural requirements. Gypsum shaft wall systems offer several advantages over the alternatives: They are lighter in weight, they are installed dry, and they are erected entirely from the floor outside the shaft,

FIGURE 17.35
Installing a sound attenuation blanket.
(Courtesy of United States Gypsum Company)

with no need to erect scaffolding inside. Depending on the requirements for fireresistance rating, air pressure resistance, STC, and floor-to-ceiling height, any of a number of designs is used. Figure 17.36 shows representative shaft wall details, and Figure 16.1 shows a shaft wall being installed.

MASONRY PARTITION SYSTEMS

A century ago, interior partitions were often made of common brick plastered on both sides. These had excellent STC and fireresistance ratings, but were labor intensive and heavy. Partition systems of hollow clay tile and hollow gypsum tile (Figure 17.37) were developed to meet these objections and continued in use until recent times, but both have now become obsolete, replaced by plaster, gypsum board, and concrete block.

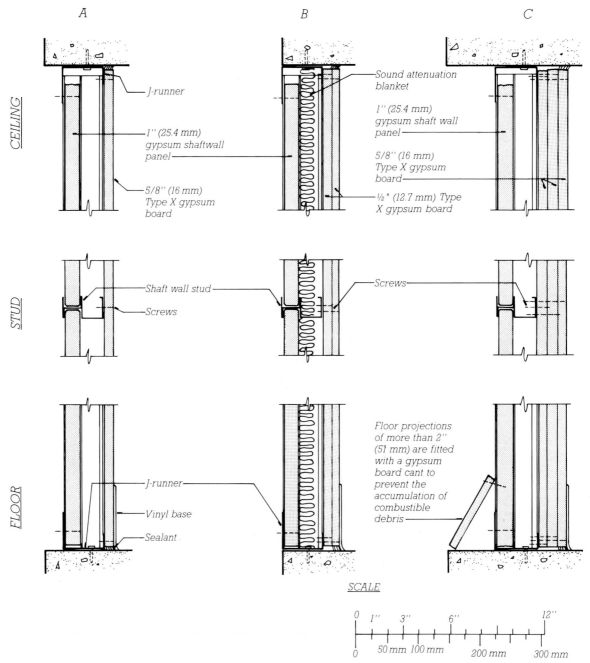

FIGURE 17.36
Gypsum shaft wall systems, all three framed with steel C-H studs. The H portion of the stud holds the 1-inch (25-mm) shaftwall panel, while the C portion *accepts the screws used to attach the finish layers of gypsum board.* **A.** *A 1-hour system.* **B.** *A 2-hour system, STC 47.* **C.** *A 3-hour system.*

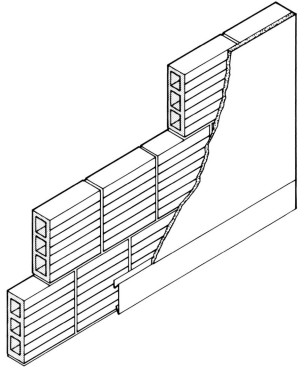
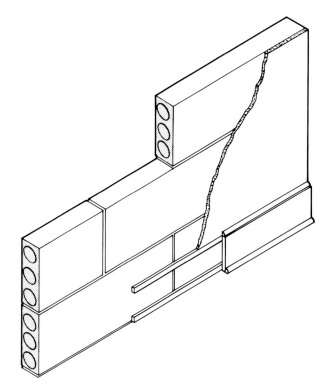

FIGURE 17.37
Obsolete partition systems often found in existing buildings: hollow clay tiles and plaster (left), and gypsum tiles and plaster (right).

FIGURE 17.38
A glazed structural clay tile partition installation. The floor is finished with glazed ceramic tiles. (Courtesy of Stark Ceramics, Inc.)

Concrete block partitions may be plastered or faced with gypsum board but are more often left exposed, with or without painting. Several types of lightweight aggregate are used to reduce the dead weight of partition blocks. Electrical wiring is relatively difficult to conceal in concrete block partitions; the electrician and the mason must coordinate their work closely, or the wiring must be mounted on the surface of the wall after the mason has finished.

Glazed structural clay tiles make excellent partitions, especially in areas with heavy wear, moisture problems, or strict sanitation requirements (Figure 17.38). The ceramic glazes are nonfading and virtually indestructible.

WALL AND PARTITION FACINGS

Ceramic tile facings are often added to walls for reasons of appearance, durability, sanitation, or moisture resistance. The best quality tile work is done over a base of metal lath and portland cement stucco (Figure 17.39). Lower-cost tile facing systems eliminate the stucco base in favor of special tile base panels of fiber-reinforced lightweight concrete or, less optimally, water-resistant gypsum board. The tiles are mounted to the stucco or concrete panel base with mortar, and to the gypsum with organic adhesive. After the tiles are fully adhered, a cementitious grout of any desired color is wiped into the tile joints with a rubber trowel. For maximum speed of installation, sheets of tile can be obtained with the joints pregrouted with an elastomeric sealant material. The sheets are fixed in place with adhesive, and the joints between sheets are grouted with matching sealant.

Facings of granite, limestone, marble, or slate are sometimes used in public areas of major buildings. These are mounted in much the same way as exterior stone facing panels.

Wood wainscoting and paneling may be used in limited quantities in fire-resistant buildings. They are mounted over a backing of plaster or gypsum board to retain the fire-resistive qualities of the partition.

INTERIOR DOORS

Depending on their respective fire-resistive requirements, interior doors may be of wood or steel, mounted in either wood or hollow steel frames. Both wood and steel doors are available with mineral cores in a range of fire ratings.

Hollow steel door frames are mounted solidly to the floor to absorb as much of the shock of door closings as possible. The jambs are attached to the studs or masonry of the partition in various ways (Figure 17.40). For the best mechanical attachment and fire resistance, hollow steel frames are grouted full with mortar or gypsum plaster.

Large doors, heavy doors, and doors that are heavily used should be equipped with hydraulic *closers* to re-

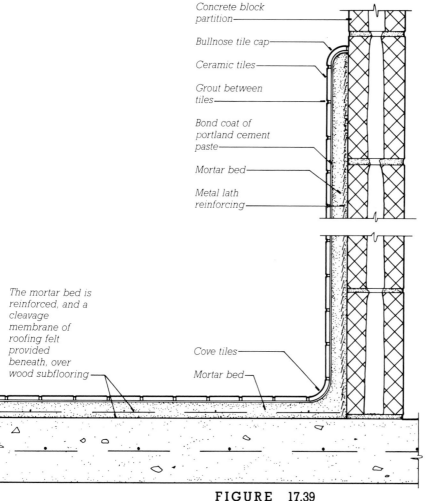

Concrete block partition

Bullnose tile cap

Ceramic tiles

Grout between tiles

Bond coat of portland cement paste

Mortar bed

Metal lath reinforcing

The mortar bed is reinforced, and a cleavage membrane of roofing felt provided beneath, over wood subflooring

Cove tiles

Mortar bed

FIGURE 17.39
Typical details of ceramic tile installation.

Anchor clips

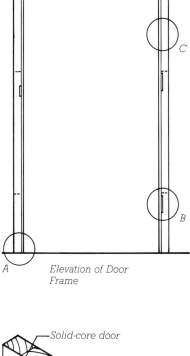

A   Elevation of Door Frame

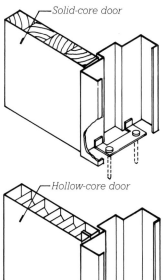

Solid-core door

A. Jambs can be attached to the floor with powder-driven fasteners

Hollow-core door

A. Jambs can be attached to the floor by pouring floor topping concrete around the door frame

B. Reinforcement of jamb at hinge attachments

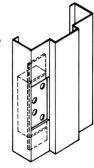

C. For jamb anchorage to masonry walls, loose sheet metal tees are inserted into the frame and built into mortar joints

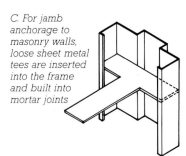

C. For jamb anchorage to steel studs, sheet metal zees are factory-welded to the jambs to receive screws driven through the studs

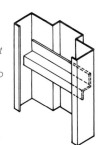

C. For jamb anchorage to wire truss studs, notches key to the vertical members of the studs, and holes provide for tie wires

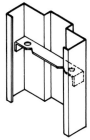

C. Jambs are anchored to wood studs by nailing through holes in the jamb inserts

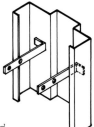

FIGURE 17.40
Details of hollow metal door frames. The lettered circles on the elevation at the upper left correspond to the details on the rest of the page.

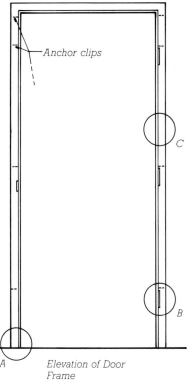

duce the shocks transmitted to the door, the hardware, and the surrounding partition. Required fire doors must always have self-closing devices. Fire doors must, of course, swing in the direction of exit flow, and under many code-defined circumstances must be equipped with *panic hardware* which opens the door automatically if people push against it. Wired glass windows are permitted in all but the highest-rated fire doors; the designer is referred to the local building code for the applicable limitations on dimension and area.

C.S.I./C.S.C. Masterformat Section Numbers for Interior Walls and Partitions	
08100	METAL DOORS AND FRAMES
08110	Steel Doors and Frames
08200	WOOD AND PLASTIC DOORS
08210	Wood Doors
09100	METAL SUPPORT SYSTEMS
09110	Nonloadbearing Wall Framing Systems
09200	LATH AND PLASTER
09205	Furring and Lathing
09210	Gypsum Plaster
09215	Veneer Plaster
09220	Portland Cement Plaster
09250	GYPSUM BOARD
09300	TILE
09310	Ceramic Tile
10600	PARTITIONS
10615	Demountable Partitions
10630	Portable Partitions, Screens, and Panels

SELECTED REFERENCES

1. Gypsum Association. *Fire Resistance Design Manual*, 10th Edition. Evanston, Illinois, 1983.

Fireresistance ratings and STCs are given in this booklet for a large number of wall and ceiling assemblies that use either gypsum plaster or gypsum board. (Address for ordering: 1603 Orrington Avenue, Evanston, IL 60201.)

2. United States Gypsum Company. *Gypsum Construction Handbook*, 2nd Edition. Chicago, 1982.

This manual represents manufacturers' literature at its best—more than 400 well-illustrated pages crammed with every important fact about gypsum wallboard, gypsum plaster, and associated products. (Address for ordering: 101 South Wacker Drive, Chicago, IL 60606.)

3. Portland Cement Association. *Portland Cement Plaster (Stucco) Manual*. Skokie, Illinois, 1980.

A complete, illustrated guide to stucco.

(Address for ordering: 5420 Old Orchard Road, Skokie, IL 60077.)

4. Gypsum Association. *Using Gypsum Board for Walls and Ceilings*. Evanston, Illinois, 1980.

This is a clear, comprehensive how-to guide for both the designer and the installer of gypsum board. (Address for ordering: see reference 1.)

KEY TERMS AND CONCEPTS

fire door
fire wall
fire separation wall
shaft wall
nonbearing partition
light-gauge steel studs
runner channels

open-truss wire studs
self-drilling, self-tapping screw
furring strip
plaster
lath
wattle and daub
gypsum

calcining
plaster of Paris
gypsum plaster
gypsum plaster with wood fiber
gypsum plaster with perlite aggregate
high-strength basecoat plaster
gauging plaster

Keenes cement
molding plaster
finish lime
portland cement-lime plaster or stucco
hawk
trowel
darby
lather
expanded metal lath
gypsum lath
veneer plaster base
trim accessories

veneer plaster
grounds
scratch, brown, finish coats
plaster screed
self-furring metal lath
line wire
cast plaster ornament
drawn plaster ornament
gypsum board, gypsum wallboard,
    plasterboard, drywall
regular gypsum board
type X gypsum board

foil-backed gypsum board
predecorated gypsum board
gypsum backing board
coreboard
tapered edge
nail popping
glazed structural clay tile
ceramic tile
hollow steel door frame
closer
panic hardware

REVIEW QUESTIONS

1.   What are the major types of interior walls and partitions in a larger building, such as a hospital, classroom building, apartment, or office building?

2.   Why is gypsum used so much in interior finishes?

3.   Name the coats of plaster used over expanded metal lath, and explain the role of each.

4.   Under what circumstances would you specify the use of portland cement plaster? Keenes cement plaster?

5.   Describe step by step how the joints between sheets of gypsum board are made invisible.

EXERCISES

1.   Determine the construction of a number of partitions in the places where you live and work. What materials are used? What accessories? Why were these chosen for their particular situations? Sketch a detail of each partition.

2.   Sketch typical details showing how the various metal lath trim accessories are used in a plaster wall.

3.   Repeat Exercise 2 for gypsum board trim accessories.

4.   What type of gypsum wall finish

system would you specify for a major art museum? For a low-cost rental office building? Outline a complete specification of wall and partition construction for a building you are presently working on.

FINISH CEILINGS AND FLOORS

A suspended ceiling of aluminum strips, prefinished in five shades ranging from white to dark blue, covers a corridor in a hospital designed by William Kessler and Associates, Architects. *(Courtesy of Alcan Building Products, Division of Alcan Aluminum Corporation)*

As the ceilings and finish floors are installed, the construction of a building is drawing to a close. The remaining exposed components of the mechanical and electrical systems are finished or concealed, intersections of interior surfaces are neatly trimmed, and painters work their magic to reveal for the first time the interior character of the building. The architect and engineers make their last inspections and, following last-minute corrections of minor defects, the contractor turns the building over to its owner.

Finish Ceilings

Functions of Finish Ceilings

The ceiling surface is an important functional component of a room. It helps control the diffusion of light and sound about the room. It may play a role in preventing the passage of sound vertically between the rooms above and below, and often it is expected to resist the passage of fire. Frequently it is called upon to assist in the distribution of conditioned air, artificial light, and electrical energy. In many buildings it must accommodate sprinkler heads for fire suppression. And its color, texture, pattern, and shape figure prominently in the overall visual impression of the room. A ceiling can be a simple, level plane, one or more sloping planes that give a sense of the roof above, a luminous surface, a richly coffered ornamental ceiling, even a frescoed plaster vault such as Michelangelo's famous ceiling in the Sistine Chapel in Rome; the possibilities are endless.

Types of Ceilings

Exposed Structural and Mechanical Components

In many buildings, it makes sense to omit a finished ceiling surface altogether and simply expose the structural and mechanical components of the floor or roof above (Figure 18.1). In industrial and agricultural buildings, where appearance is not of prime importance, this approach offers the advantages of economy and ease of access for maintenance. Many types of floor and roof structures are inherently attractive if left exposed, such as heavy timber beams and decking, concrete waffle slabs, steel trusses and space trusses. Some types of structures, such as concrete flat plates and precast concrete planks, have little visual interest. These can be painted and left exposed as finished ceilings in apartment buildings and hotels, which have little need for mechanical services at the ceiling; this saves money and reduces the overall height of the building. And in some buildings, the structural and mechanical elements at the ceiling, if carefully designed, installed, and painted, can create a powerful aesthetic of their own.

Exposing structural and mechanical components rather than covering them with a finished ceiling does not necessarily save money. Mechanical and structural work are not normally done in a precise, attractive fashion because they are not usually expected to be visible, and it is less expensive for workers to take only as much care in installation as is required for satisfactory functional performance. To achieve perfectly straight ductwork free of dents, steel decks without rust and weld spatter, and neat runs of electrical conduit and plumbing, the drawings and specifications for these trades must tell exactly the results that are expected, and a higher labor cost must be anticipated.

Tightly Attached Ceilings

Ceilings of any material may be attached tightly to wood joists, wood rafters, steel joists, or concrete slabs (Figure 18.2). Special finishing arrangements must be worked out for any beams and girders that protrude through the plane of the ceiling, and for ducts, conduits, pipes, and sprinkler heads that fall below the ceiling.

Suspended Ceilings

A ceiling suspended on wires some distance below the floor or roof structure can hang level and flat despite varying sizes of girders, beams, joists, and slabs above, even under a roof structure that is pitched toward roof drains. Ducts, pipes, and conduits can be contained entirely in the space between the ceiling and the structure. Lighting fixtures, sprinkler heads, loudspeakers, and fire-detection devices may be recessed into the ceiling. Many such ceilings can also serve as *membrane fire protection* for the floor or roof structure above, eliminating the need for fussy individual fireproofing of steel joists, or imparting a higher fireresistance rating to wood structures. For these reasons, *suspended ceilings* have become a popular and economical feature in many types of buildings, especially office and retail structures.

Suspended ceilings can be made of almost any material; the most widely used are gypsum board, plaster, and various proprietary boards composed of incombustible fibers. Each of these materials is supported on its own special system of small steel framing members, and the framing members are hung from the structure on heavy steel wires. Gypsum board suspended ceilings are screwed to ordinary light-

FIGURE 18.1

Sprinkler pipes, air conditioning ductwork, electrical conduits, and lighting fixtures are exposed in a roof structure of painted open-web steel joists and corrugated steel decking. The office space behind has a suspended acoustical ceiling with recessed lighting fixtures. The floor is of brick with a trowel finish concrete border. (Architects: Woo and Williams. Photo by Richard Bonarrigo, courtesy of the architects)

gauge steel cee channels suspended on wires. Suspended plaster ceilings have been in use for many decades; some typical details are shown in Figure 18.3. While most suspended plaster ceilings are flat, the lather is capable of constructing ceilings that are richly sculpted, from configurations resembling highly ornamented Greek or Roman coffered ceilings to nearly any form that the contemporary designer can draw. This capability is especially useful in auditoriums, theaters, lobbies of public buildings, and other uniquely shaped rooms.

Fibrous ceiling materials are delivered to the job site as lightweight tiles or panels. Ceilings made from fibrous materials are customarily referred to as *acoustical ceilings* because most of them are highly absorptive of sound energy, unlike plaster and gypsum board, which are highly reflective of sound. Figures on the sound absorption performance of acoustical ceiling systems are published in manufacturers' literature. Acoustical ceilings are often less costly than plaster or even gypsum board ceilings, and are available in hundreds of different designs, many of which are fireresistance rated. Figures 18.4 and 18.5 show some basic details for acoustical ceilings, and 18.6 through 18.11 some typical examples.

Figures 18.12 through 18.15 illustrate suspended ceilings of several other materials.

Where a suspended ceiling is used as membrane fireproofing for the structure above, or where it is part of a fire-resistive assembly, penetrations of the ceiling must be detailed so as to maintain the required degree of fire resistance. Lighting fixtures must be backed up with fire-resistive material, air conditioning grills must be isolated from the ducts that feed them by means of automatic fire dampers, and any access panels provided for maintenance of above-ceiling services must meet code requirements for fire resistance.

FIGURE 18.2
Spraying a textured finish onto an exposed ceiling in a residential structure, where there are no pipes, ducts, or wires to be concealed below the plane of the floor structure. (Courtesy of United States Gypsum Company)

FIGURE 18.3
*A suspended plaster ceiling on metal lath. At the top of the page is a cutaway isometric drawing, as viewed from below, of the essential components of the ceiling. Across the center of the page are details of six ways of supporting the hanger wires: **A.** A powder-driven pin into a concrete structure; **B.** A corrugated sheet metal tab with a hole punched in it is nailed to the formwork before the concrete is poured. When the formwork is stripped, the tab bends down and the hanger wire is threaded through the hole; **C.** A sharp, dagger-like tab of sheet metal is driven through sheet metal decking before the concrete topping is poured; **D.** A sheet-metal hook is hung onto the lap joints in the metal decking; **E.** The hanger wire is wrapped around the lower chord of an open web steel joist; **F.** The hanger wire is passed through a hole drilled near the bottom of a wood joist. At the bottom of the page is a section through a furred, insulated plaster wall and a suspended plaster ceiling.*

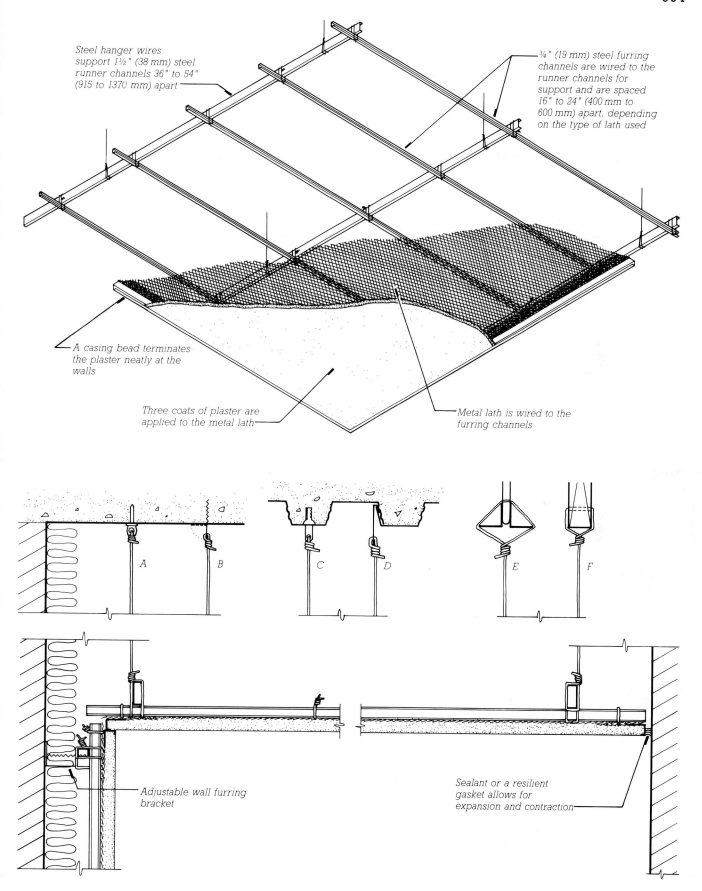

Steel hanger wires support 1½" (38 mm) steel runner channels 36" to 54" (915 to 1370 mm) apart

¾" (19 mm) steel furring channels are wired to the runner channels for support and are spaced 16" to 24" (400 mm to 600 mm) apart, depending on the type of lath used

A casing bead terminates the plaster neatly at the walls

Three coats of plaster are applied to the metal lath

Metal lath is wired to the furring channels

A   *B*   *C*   *D*   *E*   *F*

Adjustable wall furring bracket

Sealant or a resilient gasket allows for expansion and contraction

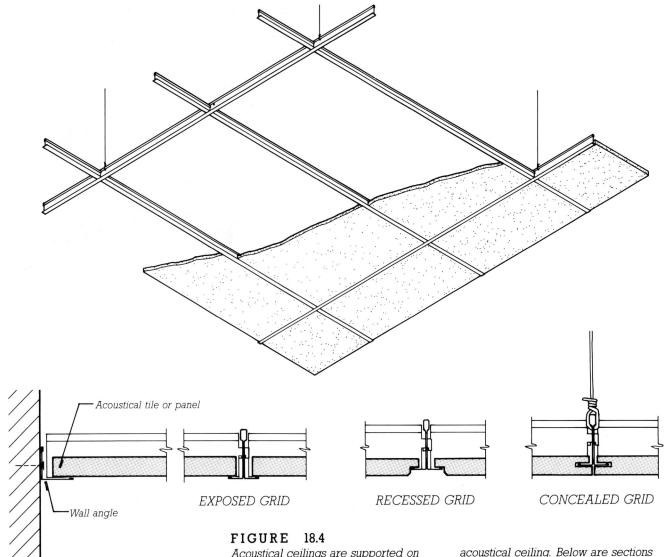

Acoustical tile or panel

Wall angle

EXPOSED GRID   RECESSED GRID   CONCEALED GRID

FIGURE 18.4
Acoustical ceilings are supported on suspended grids of tees formed from sheet metal. At the top of the page is a cutaway view looking up at an acoustical ceiling. Below are sections illustrating how the grid may be exposed, recessed, or concealed, for different visual appearances.

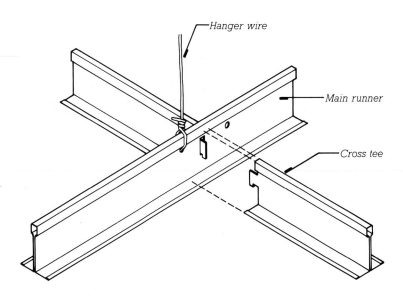

Hanger wire

Main runner

Cross tee

FIGURE 18.5
The grid for an acoustical ceiling is assembled with a simple interlocking joint.

FIGURE 18.6
Some acoustical ceilings are manufactured as systems that integrate the lighting fixtures and air conditioning outlets into the module of the grid. In this suspended ceiling, viewed from above, the hanger wires, grid, acoustical panels, fluorescent lighting fixture, and distribution boot for conditioned air have been installed. The boot will be connected to the main ductwork with a flexible oval duct. (Courtesy of Armstrong World Industries)

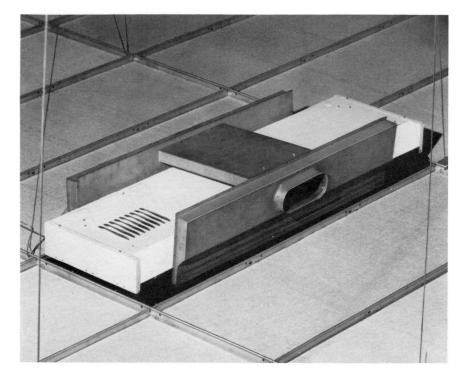

FIGURE 18.7
As viewed from below, the ceiling shown in Figure 18.6 has a slot around the lighting fixture for air distribution from the boot above. The acoustical panels used in this example are roughly textured, and patterned with two cross grooves that work with the recessed grid to create the look of a ceiling composed of smaller, square panels. (Courtesy of Armstrong World Industries)

FIGURE 18.8

Air distribution in this exposed grid acoustical ceiling is through slots that occur at three-panel intervals. Sprinkler heads are recessed into the panels. (Architect: Brubaker/Brandt, Inc. Courtesy of Armstrong World Industries)

FIGURE 18.9

Coffered acoustical ceilings that act as light diffusers are designed and marketed as single integrated systems. (Courtesy of Armstrong World Industries)

FIGURES 18.10, 18.11
A variety of patterns and textures are available in acoustical ceilings.
(Courtesy of Armstrong World Industries)

FIGURE 18.12
A mirror finish aluminum strip suspended ceiling, with an exposed steel space truss structure beyond.
(Architect: DeWinter and Associates. Courtesy of Alcan Building Products, Division of Alcan Aluminum Corporation)

FIGURE 18.13
A fully luminous suspended ceiling.
(Courtesy of Integrated Ceilings, Inc.)

FIGURE 18.14
A mirror finish suspended ceiling with small incandescent bulbs creates sparkle in a restaurant. (Courtesy of Integrated Ceilings, Inc.)

FIGURE 18.15
A mirrored ceiling in a shopping mall reflects the glazed quarry tile floor. (Courtesy of Integrated Ceilings, Inc.)

Economy has worked so great a change in our dwellings, that their ceilings are, of late years, little more than miserable naked surfaces of plaster. [A discussion of ceiling design will] possess little interest in the eye of speculating builders of the wretched houses erected about the suburbs of the metropolis, and let to unsuspecting tenants at rents usually about three times their actual value. To the student it is more important, inasmuch as a well-designed ceiling is one of the most pleasing features of a room.

Joseph Gwilt, *The Encyclopedia of Architecture*, London, 1842

Interstitial Ceilings   Many hospital and laboratory buildings have extremely elaborate mechanical and electrical systems, including not just the usual air conditioning ducts, water and waste piping, and electrical and communications wiring, but also such services as fume hood ducting, fuel gas lines, compressed air lines, oxygen piping, chilled water piping, vacuum piping, and chemical waste piping. These ducts and tubes occupy a considerable amount of space in the building, often in an amount that virtually equals the inhabited space. Furthermore, all these systems require continual maintenance and are subject to frequent change. As a consequence, many such buildings are designed with *interstitial ceilings*. An interstitial ceiling is suspended at a level that allows workers to travel freely in the ceiling space, usually while walking erect, and is structured strongly enough to support safely the weight of the workers and their tools. In effect, it is another floor of the building, slipped in between the other floors, and the overall height of the building must be increased accordingly. Its advantage is that maintenance and updating work can be carried on without interruption to the activities below. Interstitial ceilings are made of gypsum or lightweight concrete and combine the construction details of poured gypsum roof decks and suspended plaster ceilings. Figure 18.16 shows the installation of an interstitial ceiling.

FINISH FLOORING

Functions of Finish Flooring

Floors have a lot to do with one's visual and tactile appreciation of a building—people sense their colors, patterns, and textures, their "feel" underfoot, and

FIGURE 18.16

The final steps in constructing an interstitial ceiling: The ceiling plane consists of gypsum reinforced with hexagonal steel mesh. It is framed with steel truss tee subpurlins, which are visible at the right of the picture, supported by steel wide flange beams suspended on rods from the sitecast concrete framing above. The final layer of gypsum is being pumped onto the ceiling from the hose in the center of the picture. The wet gypsum is struck off level with the straightedge seen here hanging on the beams, and troweled to a smooth walking surface. When the gypsum has hardened, installation of the ductwork, piping, and wiring in the interstitial space can begin, with workers using the gypsum ceiling as a walking surface. (Courtesy of Keystone Steel and Wire Company)

the noises they make in response to footsteps. Floors affect the acoustics of a room, contributing to a noisy quality or a hushed quality, depending on whether a hard flooring or carpeting is used. Floors also react in various ways with light: Some floor materials give mirrorlike reflections, some give diffuse reflections or none at all; dark flooring materials absorb most of the light incident upon them and contribute to the creation of a darker room, while light materials reflect most incident light and help create a brighter room.

Floors are also a major functional component of a building. They are its primary wearing surfaces, subject to water, grit, dust, and the abrasive and penetrating actions of feet and furniture. They require more cleaning and maintenance effort than any other component of a building. They must be designed to deal with problems of skid resistance, sanitation, noise reduction between floors of a building, even electrical conductivity in occupancies such as computer rooms and hospital operating rooms where static electricity would pose a threat. And like other interior finish components, floors must be selected with an eye to combustibility, fireresistance ratings, and the structural loads they will place on the frame of the building.

Floor finish materials are classified for flame resistance by ASTM test procedure E648, and are grouped into Class I and Class II finishes. A typical building code table defining allowable types of floor finishes for various occupancies is shown in Figure 16.5.

Underfloor Services

Floors are frequently used for the distribution of electrical and communications wiring, especially in offices and computer facilities. Cellular steel decking, in combination with special fittings, is often used in this way (Figures 18.17 through 18.19). Conduits can also be cast into concrete toppings or structural concrete slabs, with junction boxes installed flush in the floor.

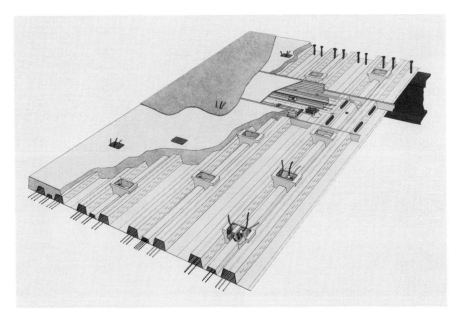

FIGURE 18.17
Cellular steel decking used for underfloor electrical and communications wiring. A transverse feeder trench, near the top of the picture, brings the wiring across the floor from the electrical risers to the cells in the deck. Boxes cast into the topping give access to the cells for the installation of electrical outlets. Notice the shear studs on the steel beam at the top of the picture. (Courtesy of H. H. Robertson Company)

FIGURE 18.18
A sectional view of cellular steel decking, showing telephone and other low-voltage communications wiring in the cell to the left, electrical power wiring in the cell to the right, and access to these services through the box in the middle. (Courtesy of H. H. Robertson Company)

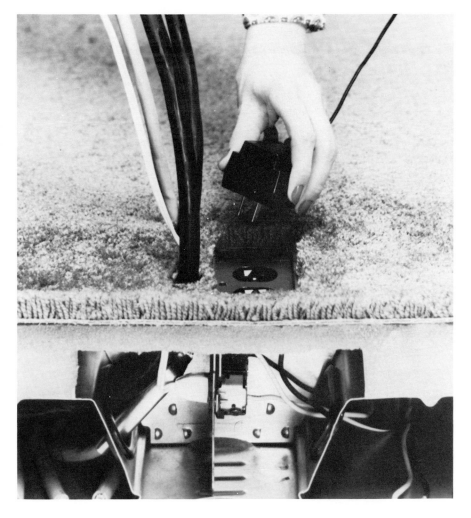

FIGURE 18.19
Plugging in to a floor-mounted electric receptacle served by wires from the cells of the floor decking. (Courtesy of H. H. Robertson Company)

Where wiring changes are frequent and unpredictable, as in computer rooms and offices with a large number of electronic machines, *raised access flooring* is often used (Figures 18.20, 18.21). If the access floor is raised high enough, ductwork for air distribution can be run over the structural floor, possibly eliminating the need for a suspended ceiling below. Raised access flooring has a virtually unlimited capacity to meet future wiring needs, and changes in wiring are extremely easy and inexpensive. *Undercarpet wiring systems* using flat conductors are appropriate in many buildings, for both electrical power and communications wiring (Figures 18.22, 18.23).

FIGURE 18.20
Raised access flooring provides unlimited capacity for wiring, piping, and ductwork. The space below the flooring can serve as a plenum for air distribution. Changes in any of the underfloor systems are easily made, and wiring outlets can be installed at any point in the floor. (Courtesy of Tate Architectural Products, Inc.)

FIGURE 18.21

Conditioned air is supplied to this computer room through the space below the raised access flooring and is fed upward through perforated floor panels. Air is returned through slots in the suspended ceiling. (Courtesy of Armstrong World Industries)

FIGURE 18.22

The flat conductors for an undercarpet wiring system lie unseen beneath the carpet, and are accessed through projecting boxes. (Courtesy of Burndy Corporation)

FIGURE 18.23

The flat conductors are ribbons of copper laminated between insulating layers of plastic sheet. These conductors are connected as necessary with the splicing tool shown in this photograph, and covered with a grounded metallic shield before being taped to the floor. (Courtesy of Burndy Corporation)

Reducing Noise Transmission Through Floors

In multistory buildings it is sometimes necessary to take precautions to reduce the amount of *impact noise* transmitted through a floor to the room below. This is particularly true of hotels and apartment buildings where people are sleeping below others who may be awake and moving about. Impact noise is generated by footsteps or machinery and is transmitted as structureborne vibration through the material of the floor to become airborne noise in the room below.

Soft flooring materials such as carpet or cushioned resilient flooring do an excellent job of reducing impact noise transmission. Further reductions can be made by mounting the plaster ceiling below on resilient clips, or on hanger wires with springs. With hard finish flooring materials, a resiliently mounted, airtight ceiling is often the only effective way of dealing with impact noise.

Many floor-ceiling assemblies have been tested for sound transmission and rated for both *Sound Transmission Class (STC)*, which is concerned with transmission of airborne sound, and *Impact Insulation Class (IIC)*. These ratings will be found in trade literature concerned with various types of floor construction, and offer a ready comparison of acoustical performance.

Hard Flooring Materials

Hard finish flooring materials (concrete, stone, brick, tile, and terrazzo) are often chosen for their resistance to

FIGURE 18.24
A slate floor in an automobile showroom. (Photo by Bill Engdahl, Hedrich-Blessing. Courtesy of Buckingham-Virginia Slate Company)

wear and moisture. Being rigid and unyielding, they are not comfortable to stand upon for extended periods of time, and they contribute to a live, noisy acoustic environment. But many of these materials are so beautiful in their colors and patterns, and so durable, that they are considered among the most desirable types of flooring by designers and building owners alike.

Concrete

With a broomed or wood float finish, concrete makes an excellent finish floor for parking garages and many types of agricultural and industrial buildings. With a steel trowel finish, concrete finds its way into a vast assortment of commercial and institutional buildings, and even into homes and offices. Color can be added with a colorant admixture or a couple of coats of floor paint. Concrete's chief advantages as a finish flooring material are its low initial cost and its durability. On the minus side, extremely good workmanship is required to make an acceptable floor finish, and even the best of concrete surfaces is likely to sustain some damage and staining during construction.

Stone

All the common building stones are used as flooring materials, in surface textures ranging from mirror-polished marble and granite to split face slate, sandstone, and limestone (Figures 18.24, 18.25). Installation is a relatively simple but highly skilled procedure of bedding the stone in mortar and filling the joints with grout (Figure 18.26). Most stone floorings are coated

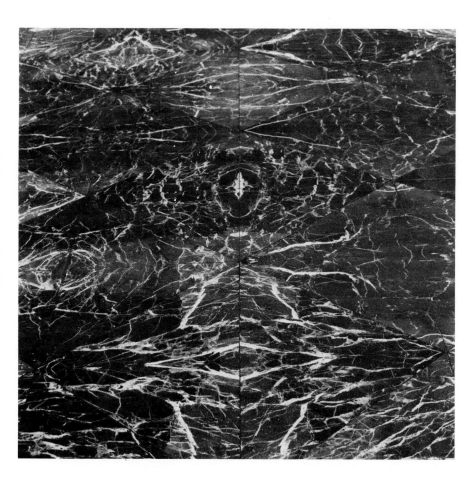

FIGURE 18.25

Flooring of matched, polished triangles of white-veined red marble gives a kaleidoscopic effect. (Architect: The Architects Collaborative. Photograph by the author)

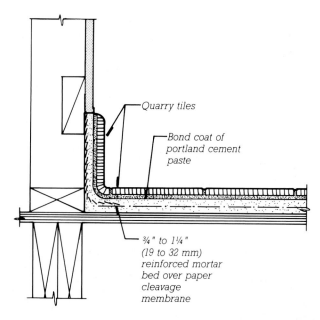

Quarry tiles

Bond coat of
portland cement
paste

¾" to 1¼"
(19 to 32 mm)
reinforced mortar
bed over paper
cleavage
membrane

*QUARRY TILE ON WOOD
SUBFLOOR*

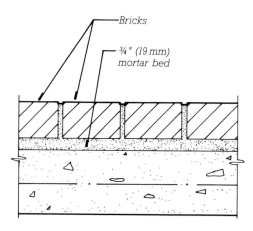

Bricks

¾" (19 mm)
mortar bed

BRICKS ON CONCRETE SLAB

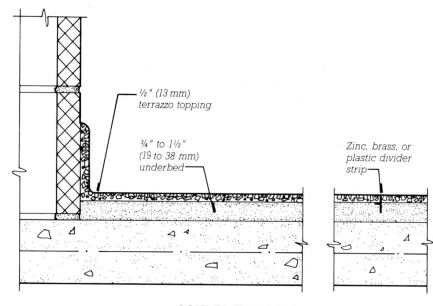

½" (13 mm)
terrazzo topping

¾" to 1½"
(19 to 38 mm)
underbed

Zinc, brass, or
plastic divider
strip

BONDED TERRAZZO

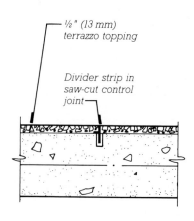

½" (13 mm)
terrazzo topping

Divider strip in
saw-cut control
joint

MONOLITHIC TERRAZZO

FIGURE 18.26

*Typical details of tile, brick, stone, and
terrazzo flooring. Traditional sand
cushion terrazzo flooring is not
illustrated. The terrazzo systems shown
here are thinner and lighter than sand
cushion terrazzo, but perform as well on
a properly engineered floor structure.
Thin set terrazzo, the lightest of the
systems, uses very small stone chips in
a mortar that is strengthened with
epoxy, polyester, or polyacrylate.*

with multiple applications of a clear sealer compound, and are waxed periodically throughout the life of the building to bring out the color and figure of the stone.

Brick and Brick Pavers

Both bricks and half-thickness bricks called *pavers* are used for finish flooring, with pavers often preferred because they add less thickness and dead weight to the floor (Figure 18.27). Bricks are usually laid with their largest surface horizontal, but may also be laid on edge. As with stone and tile flooring, decorative joint patterns can be designed specially for each installation.

Quarry Tiles

Quarry tiles are not, as might be guessed from their name, made of

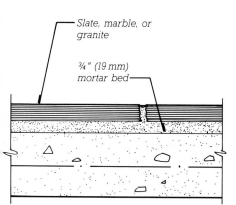

Slate, marble, or granite

¾" (19 mm) mortar bed

STONE ON CONCRETE SLAB

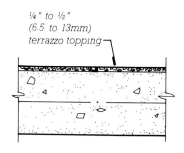

¼" to ½" (6.5 to 13mm) terrazzo topping

THIN SET TERRAZZO

FIGURE 18.27
A floor of glazed brick pavers meets a planting bed constructed of brick.
(Courtesy of Stark Ceramics, Inc.)

FIGURE 18.28
*Unglazed quarry tiles used as flooring
and as a facing for columns and railings.*
(Architect: Skidmore, Owings & Merrill.
Interior designer: Duffy, Inc. Courtesy of
American Olean Tile Company)

The brick floors, because the bricks may be made of diverse forms and of diverse colors by reason of the diversity of the chalks, will be very agreeable and beautiful to the eye...The ceilings are also diversely made, because many take delight to have them of beautiful and well-wrought beams...these beams ought to be distant one from another one thickness and a half of the beam, because the ceilings appear thus very beautiful to the eye.

Andrea Palladio,
*The Four Books of
Architecture,* 1570

stone. They are simply large, fired clay tiles, usually unglazed, usually square but sometimes rectangular, hexagonal, octagonal, or other shapes (Figures 18.28, 18.29). Sizes range from about 4 inches (100 mm) to 12 inches (300 mm) square, with thicknesses ranging from $\frac{3}{8}$ inch (9 mm) to a full inch (25 mm) for some handmade tiles. Quarry tiles are available in a myriad of earth colors, as well as certain kiln-applied colorations. They are usually set in a reinforced mortar bed, although in residential work they are

sometimes applied to a strong, stiff wood or plywood subfloor with epoxy mortar or organic adhesives.

Ceramic Tiles

Fired clay tiles smaller than quarry tiles are referred to collectively as *ceramic tiles*. Ceramic tiles are usually glazed. The most common shape is square, but rectangles, hexagons, circles, and more elaborate shapes are also available (Figure 18.30). Sizes range from $\frac{1}{2}$ inch (13 mm) to 4 inches (100 mm) and more.

The smaller sizes of tiles are shipped from the factory adhered to large backing sheets of plastic mesh or perforated paper; the tilesetter is thus able to lay a hundred or more tiles together in a single operation, rather than as individual units. Grout color has a strong influence on the appearance of tile installations, as it does for brick and stone. Many different premixed colors are available, or the tilesetter may color a grout with pigments. The use of ceramic tiles on interior wall surfaces is mentioned in the previous

FIGURE 18.29
Square and rectangular quarry tiles in contrasting colors create a pattern on the floor of a retail mall. (Architect: Edward J. DeBartolo Corp. Courtesy of American Olean Tile Company)

chapter, and typical details of ceramic tile installation on floors and walls are shown in Figure 17.39.

Terrazzo

Terrazzo is an exceptionally durable flooring. It is made by grinding and polishing a concrete that consists of marble or granite chips selected for size and color, set in a matrix of colored portland cement or other binding agent. The polishing brings out the pattern and color of the stone chips, and a sealer is usually applied to further enhance the appearance of the floor (Figure 18.31). Terrazzo may be formed in place, or may be installed as factory-made tiles. For stair treads, window sills, and other large components, terrazzo is often precast.

The endless variety of colors and textures of terrazzo leads to its use in decorative flooring patterns, where the colors are separated from one another by *divider strips* of metal, plastic, or marble. The divider strips are installed in the underbed prior to placing the terrazzo, and are ground and polished flush in the same operation as the terrazzo itself.

Traditionally, terrazzo is installed over a thin bed of sand that isolates it from the structural floor slab, thus protecting it to some extent from movements in the building frame. This *sand cushion terrazzo* is thick ($2\frac{1}{2}$ inches or 64 mm) and heavy, however. For greater economy and reduced thickness, the sand bed may be eliminated to produce *bonded terrazzo*, or both the sand bed and underbed may be eliminated with *monolithic terrazzo* (Figure 18.26). In any of these systems, a terrazzo base can be formed and finished as an integral part of the floor, thus eliminating a dirt-catching seam where the floor meets the wall.

FIGURE 18.30

Ceramic tile wainscoting and flooring in a bar. (Architect: Daughn/Salisbury, Inc. Designer: Morris Nathanson Design. Courtesy of American Olean Tile Company)

Wood Flooring

Wood is used in several different forms as a finish flooring material, the most common of which is wood *strip flooring* of white oak, red oak, pecan, or maple (Figures 18.32, 18.33). The strips are driven tightly together and *blind nailed* as shown, after which the entire floor is sanded smooth, stained if desired, and finished with a varnish or other clear coating. When its surface becomes worn, the flooring can be restored to a new appearance by sanding and refinishing.

For greater economy, wood flooring is available in square-edged strips about $\frac{5}{16}$ inch (8 mm) thick which are face-nailed to the subfloor. The nail holes are filled before the floor is sanded and finished. The initial appearance is es-

FIGURE 18.31
A terrazzo floor in a residential entry uses divider strips and contrasting colors to create a custom floor pattern. (Courtesy of National Terrazzo and Mosaic Association, Inc.)

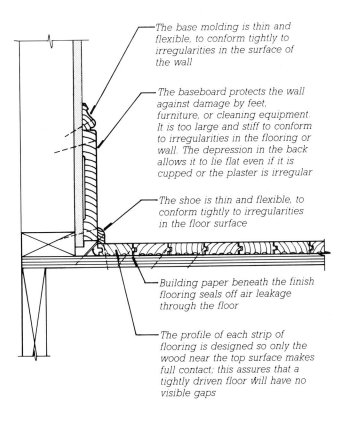

The base molding is thin and flexible, to conform tightly to irregularities in the surface of the wall

The baseboard protects the wall against damage by feet, furniture, or cleaning equipment. It is too large and stiff to conform to irregularities in the flooring or wall. The depression in the back allows it to lie flat even if it is cupped or the plaster is irregular

The shoe is thin and flexible, to conform tightly to irregularities in the floor surface

Building paper beneath the finish flooring seals off air leakage through the floor

The profile of each strip of flooring is designed so only the wood near the top surface makes full contact; this assures that a tightly driven floor will have no visible gaps

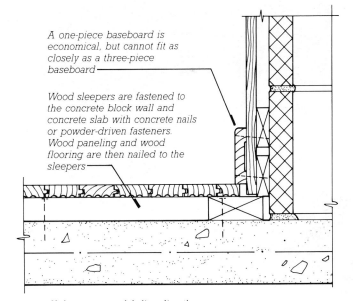

A one-piece baseboard is economical, but cannot fit as closely as a three-piece baseboard

Wood sleepers are fastened to the concrete block wall and concrete slab with concrete nails or powder-driven fasteners. Wood paneling and wood flooring are then nailed to the sleepers

If the concrete slab lies directly on grade, a sheet of polyethylene is laid beneath the sleepers to prevent moisture from entering the building

FIGURE 18.32
Details of hardwood strip flooring installation. At left, the flooring is applied to a wood joist floor, and at right, to wood sleepers over a concrete slab. The blind nailing of the flooring is shown only for the first several strips of *flooring at the left. The baseboard makes a neat junction between the floor and the wall, covering up the rough edges of the wall material and the flooring. The three-piece baseboard shown at the left does this job somewhat* *better than the one-piece baseboard, but it is more expensive and elaborate and has fallen out of favor with most designers.*

sentially the same as for the thicker flooring, but the thinner floor cannot be sanded and refinished as many times, so has a shorter lifetime. Another less costly form of wood flooring consists of factory-made, prefinished wood planks and parquet tiles. These are furnished in many different woods and patterns, and are fastened to the subfloor with a mastic adhesive. Most are too thin to be able to withstand subsequent sanding and refinishing.

Exceptionally long-wearing industrial wood floors are made of small blocks of wood set in adhesive with their grain oriented vertically. While this type of floor is relatively high in first cost, it is economical for heavily used floors and is sometimes chosen for use in public spaces because of the beauty of its pattern and grain.

FIGURE 18.33
Oak strip flooring in a hair salon.
(Architects: Michael Rubin and Henry Smith-Miller in association with Kenneth Cohen. Courtesy of Oak Flooring Institute)

Resilient Flooring

The original resilient flooring was *linoleum*, a sheet material made of ground cork in a linseed oil binder over a burlap backing. Asphalt tiles were later developed as an alternative to linoleum, but most of today's resilient sheet floorings and tiles are made of vinyl (polyvinyl chloride), often in combination with asbestos reinforcing fibers. Thicknesses are on the order of $\frac{1}{8}$ inch (3 mm), slightly thinner for lighter-duty floorings, and slightly thicker if a cushioned back is added to the product. Most resilient flooring materials are glued to the concrete or wood of the structural floor (Figures 18.34 through 18.36). The primary advantages of resilient floorings are the wide range of available colors and patterns, a moderately high degree of durability, and low initial cost. Vinyl-asbestos tile has the lowest installed cost of any flooring material except concrete and is used in vast quantities on the floors of residences, offices, classrooms, and retail space. Resilient tiles are usually 12 inches (305 mm) square. Sheet flooring is furnished in rolls 6 to 12 feet (1.83 to 3.66 m) wide. If skillfully made, the seams between strips of sheet flooring are virtually invisible.

Resilient floorings of rubber, cork, and other materials are also available. Each offers particular advantages of durability or appearance.

Most resilient flooring materials are so thin that they show even the slightest irregularities in the floor deck beneath. Concrete surfaces must be scraped clean of construction debris and spatters. Plywood decks are covered with a layer of smooth *underlayment* panels, usually of hardboard, particle board, or sanded plywood. Joints between underlayment panels are offset from joints in the subfloor to eliminate soft spots. The thickness of the underlayment is chosen to make the surface of the resilient flooring level with the surfaces of surrounding flooring materials such as hardwood and ceramic tile.

Carpet

Carpet is manufactured in fibers, styles, and patterns to meet almost any flooring requirement, indoors or out, except for rooms that need thorough sanitation, such as hospital rooms, food processing facilities, and toilet rooms. Some carpets are tough enough to wear for years in public corridors (Figure 18.37), yet others are soft enough for intimate residential interiors. Costs of carpeting are often competitive with those of other flooring materials of similar quality, whether measured on an installed-cost or life-cycle-cost basis.

Carpets are either glued directly to the floor deck, or stretched over a carpet pad and attached around the pe-

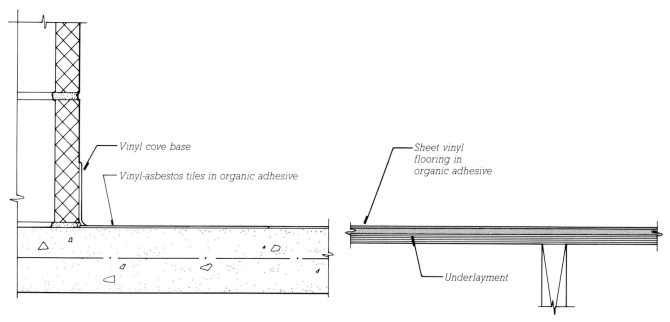

FIGURE 18.34
Typical installation details for resilient flooring. At left, vinyl-asbestos tiles applied directly to a steel-trowel-finish concrete slab, with a vinyl cove base adhered to a concrete block partition. To the right, sheet vinyl flooring on underlayment and a wood joist floor structure.

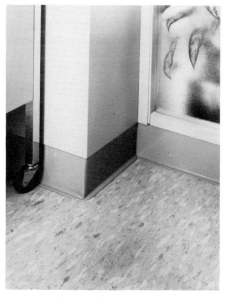

FIGURE 18.35
Vinyl-asbestos tile and a vinyl cove base. Notice how the base is simply folded around the outside corner of the partition. (Courtesy of Armstrong World Industries)

FIGURE 18.36
Sheet vinyl flooring can be flash-coved to create an integral base that is easily cleaned, for use in health care facilities, kitchens, and bathrooms. The seams are welded to eliminate dirt-catching cracks. (Courtesy of Armstrong World Industries)

FIGURE 18.37

*Wall-to-wall carpeting in a commercial
installation.* (Courtesy of Armstrong
World Industries)

rimeter of the room by means of a *tackless strip*, a continuous length of wood fastened to the floor, with protruding spikes along the top that catch the backing of the carpet and hold it taut.

If carpet is laid directly over a wood panel subfloor such as plywood, the panel joints perpendicular to the floor joists should be blocked beneath to prevent movement between sheets. Tongue-and-groove plywood subflooring accomplishes the same result without blocking. Alternatively, a layer of underlayment panels can be nailed over the subfloor with its joints offset from those in the subfloor.

FLOORING THICKNESS

Thicknesses of floor finishes vary from the $\frac{1}{8}$ inch (3 mm) of resilient flooring to 3 inches (76 mm) or more for brick flooring. Frequently, several different types of flooring are used in the same building. If the differences in the thicknesses of the flooring materials are not great, they can be resolved by using tapered edgings or thresholds at changes of material. Otherwise, the level of the top of the floor deck must be adjusted from one part of the building to the next, to bring the finish floor surfaces to the same elevation. The architect should work out the necessary level changes in advance and indicate them clearly on the construction drawings. In many cases special structural details must be drawn to indicate how the level changes should be made. In wood framing they can usually be accomplished either by notching the ends of the floor joists to lower the subfloor in parts of the building with thicker floor materials, or by adding sheets of underlayment material of the proper thickness to areas of thinner flooring. In steel and concrete buildings, slab or topping thicknesses can change, or whole areas of structure can be raised or lowered the necessary amount.

C.S.I./C.S.C. Masterformat Section Numbers for Ceiling and Floor Finishes	
09100	METAL SUPPORT SYSTEMS
09120	Ceiling Suspension Systems
09130	Acoustical Suspension Systems
09200	LATH AND PLASTER
09250	GYPSUM BOARD
09300	TILE
09310	Ceramic Tile
09330	Quarry Tile
09400	TERRAZZO
09500	ACOUSTICAL TREATMENT
09510	Acoustical Ceilings
09515	Metal Ceiling Systems
09550	WOOD FLOORING
09560	Wood Strip Flooring
09565	Wood Block Flooring
09570	Wood Parquet Flooring
09600	STONE FLOORING
09610	Flagstone Flooring
09615	Marble Flooring
09620	Granite Flooring
09625	Slate Flooring
09630	UNIT MASONRY FLOORING
09635	Brick Flooring
09650	RESILIENT FLOORING
09660	Resilient Tile Flooring
09665	Resilient Sheet Flooring
09680	CARPET

SELECTED REFERENCES

Because so many ceiling and flooring materials and systems are proprietary, much of the best information on ceiling and floor finishes is to be found in manufacturers' literature. Certain generic products are, however, well documented in trade association literature:

1. Tile Council of America, Inc. *Handbook for Ceramic Tile Installation*. Princeton, New Jersey.

Crammed into this booklet are details and specifications for ceramic tile floors and walls of every conceivable type, as applied to various kinds of underlying structures. (Address for ordering: P.O. Box 326, Princeton, NJ 08540.)

2. National Oak Flooring Manufacturers Association. *Hardwood Flooring Installation Manual* and *Hardwood Floors, Walls, and Ceilings*. Memphis, Tennessee.

Oak floor grades and installation procedures are covered by these two pamphlets. (Address for ordering: 804 Sterick Building, Memphis, TN 38103.)

3. The National Terrazzo & Mosaic Association, Inc. *Terrazzo Design/Technical Data Book*. Des Plaines, Illinois.

Terrazzo details, specifications, colors, and patterns are treated thoroughly in this looseleaf binder of information. (Address for ordering: 3166 Des Plaines Avenue, Suite 15, Des Plaines, IL 60018.)

KEY TERMS AND CONCEPTS

suspended ceiling
membrane fire protection
runner channel
furring channel
acoustical ceiling
membrane fireproofing
exposed grid
concealed grid
interstitial ceiling

raised access flooring
undercarpet wiring system
impact noise
Sound Transmission Class (STC)
Impact Insulation Class (IIC)
brick paver
quarry tile
ceramic tile
divider strips
terrazzo

sand cushion terrazzo
bonded terrazzo
monolithic terrazzo
wood strip flooring
blind nailing
resilient flooring
underlayment panels
carpet
tackless strip

REVIEW QUESTIONS

1. List the potential range of functions of a finish ceiling, and of a finish floor.

2. What are the advantages and disadvantages of a suspended ceiling?

3. When designing a building with its structure and mechanical equipment left exposed at the ceiling, what sorts of precautions should you take to assure a satisfactory appearance?

4. What does underlayment do?

5. List several different approaches to the problem of running electrical and communications wiring beneath a floor.

EXERCISES

1. Visit some rooms you happen to like in nearby buildings: classrooms, auditoriums, theaters, restaurants, bars, museums, shopping centers, both new buildings and old. For each room, list the ceiling and floor finishes used. Why was each material chosen? Sketch details of critical junctions between materials. What does the room sound like—noisy, hushed, "live"? How does this acoustical quality relate to the floor and ceiling materials? What is the quality of the illumination in the room, and what role do ceiling and floor materials play in creating this quality?

2. Look in current architectural magazines for interior photographs of buildings that appeal to you. What ceiling and floor materials are used in each? Why?

APPENDIX

—

Densities and Coefficients of Thermal Expansion of Common Building Materials

Material		Density		Coefficient of Thermal Expansion	
		lb/ft^3	kg/m^3	in/in/°F	mm/mm/°C
Wood (seasoned)					
Douglas fir	parallel to grain	32	510	0.0000021	0.0000038
	perpendicular to grain			0.000032	0.000058
Pine	parallel to grain	26	415	0.0000030	0.0000054
	perpendicular to grain			0.000019	0.000034
Oak, red or white	parallel to grain	41–46	655–735	0.0000027	0.0000049
	perpendicular to grain			0.000030	0.000054
Masonry					
Limestone		160	2560	0.0000044	0.0000079
Granite		165	2640	0.0000047	0.0000085
Marble		165	2640	0.0000073	0.0000131
Brick (average)		100–140	1600–2240	0.0000036	0.0000065
Concrete masonry units		100–140	1600–2240	0.0000052	0.0000094
Concrete					
Normal weight concrete		145	2320	0.0000055	0.0000099
Metals					
Steel		490	7850	0.0000065	0.0000117
Stainless steel, 18–8		490	7850	0.0000099	0.0000178
Aluminum		165	2640	0.0000128	0.0000231
Copper		556	8900	0.0000093	0.0000168
Finish Materials					
Gypsum board		43–50	690–800	0.000009	0.0000162
Gypsum plaster, sand		105	1680	0.000007	0.0000126
Glass		156	2500	0.0000050	0.0000090
Acrylic glazing sheet		72	1150	0.0000410	0.0000742
Polycarbonate glazing sheet		75	1200	0.0000440	0.0000796

English/metric conversions

English	Metric	English	Metric
1″	25.4 mm	0.0394″	1 mm
1′	304.8 mm	39.37″	1 m
1′	0.3048 m	3.2808′	1 m
1 psi (lb/in^2)	0.0068948 MN/m^2	145.14 psi (lb/in^2)	1 MN/m^2
1 ft^2	0.0929 m^2	10.76 ft^2	1 m^2
1 lb/ft^3	16.019 kg/m^3	0.0624 lb/ft^3	1 kg/m^3
1 lb	0.454 kg	2.205 lb	1 kg

Note: Units are converted in the text to a degree of precision consistent with the precision of the number being converted.

GLOSSARY

A

Access flooring A raised finish floor surface consisting entirely of small, individually removable panels beneath which wiring, ductwork, and other services may be installed.

Acoustical ceiling A ceiling of fibrous tiles that are highly absorbent of sound energy.

Acrylic A transparent plastic material widely used in sheet form for glazing windows and skylights.

Admixture A substance other than cement, water, and aggregates included in a concrete mixture for the purpose of altering one or more properties of the concrete.

Aggregate Inert particles such as sand, gravel, crushed stone, or expanded minerals, in a concrete or plaster mixture.

Air-entraining cement A portland cement with an admixture that causes a controlled quantity of stable, microscopic air bubbles to form in the concrete during mixing.

Air-to-air heat exchanger A device that exhausts air from a building while recovering much of the heat from the exhausted air and transferring it to the incoming air.

AISC The American Institute of Steel Construction.

Alloy A substance composed of two or more metals, or of a metal and a nonmetallic constituent.

Anchor bolt A bolt embedded in concrete for the purpose of fastening a building frame to a concrete or masonry foundation.

Angle A structural section of steel or aluminum whose profile resembles the letter *L*.

Annealed Cooled under controlled conditions to minimize internal stresses.

Anodizing An electrolytic process that forms a permanent, protective oxide coating on aluminum, with or without added color.

Anticlastic Saddle-shaped, or having curvature in two opposing directions.

Apron The finish piece that covers the joint between a window stool and the wall finish below.

Arch A structural device that supports a vertical load by translating it into axial, inclined forces.

Arc welding A process of joining two pieces of metal by melting them together at their interface with a continuous electric spark and adding a controlled additional amount of molten metal from a metallic electrode.

Area divider A curb used to partition a large roof membrane into smaller areas, to allow for expansion and contraction in the deck and membrane.

Ash dump A door in the underfire of a fireplace that permits ashes from the fire to be swept into a chamber beneath, from which they may be removed at a later time.

Ashlar Squared stonework.

Asphalt A tarry brown or black mixture of hydrocarbons.

Asphalt roll roofing A continuous sheet of the same roofing material used in asphalt shingles. *See* **asphalt shingle.**

Asphalt-saturated felt A moisture-resistant sheet material, available in several different thicknesses, usually consisting of a heavy paper that has been impregnated with asphalt.

Asphalt shingle A roofing unit composed of a heavy organic or inorganic felt saturated with asphalt and faced with mineral granules.

ASTM American Society for Testing and Materials, an organization that promulgates standard methods of testing the performance of building materials and components.

Auger A helical tool for creating cylindrical holes.

Awning window A window that pivots on an axis at or near the top edge of the sash and projects toward the outdoors.

Axial In a direction parallel to the long axis of a structural member.

B

Backfill Earth or earthen material used to fill the excavation around a foundation; the act of filling around a foundation.

Backup bar A small rectangular strip of steel applied beneath a joint to provide a solid base for beginning a weld between two steel structural members.

Backup rod A flexible, compressible strip of plastic foam inserted into a joint to limit the depth to which sealant can penetrate.

Ballast A heavy material installed over a roof membrane to prevent wind uplift and shield the membrane from sunlight.

Balloon frame A wooden building frame composed of closely spaced members nominally 2 inches (51 mm) in thickness, in which the wall members are single pieces running from the sill to the top plates at the eave.

Baluster A small, vertical member that serves to fill the opening between a handrail and a stair or floor.

Band joist A wooden joist running perpendicular to the primary direction of the joists in a floor and closing off the floor platform at the outside face of the building.

Bar A small rolled steel shape, usually round or rectangular in cross section; a rolled steel shape used for reinforcing concrete.

Barrel shell A scalloped roof structure of reinforced concrete that spans in one direction as a barrel vault and in the other as a folded plate.

Barrel vault A segment of a cylinder that spans as an arch.

Baseboard A strip of finish material placed at the junction of a floor and a wall to create a neat intersection and protect the wall against damage from feet, furniture, and floor-cleaning equipment.

Baseplate A steel plate inserted between a column and a foundation to spread the concentrated load of the column across a larger area of the foundation.

Batten A strip of wood or metal used to

cover the crack between two adjoining boards or panels.

Batten seam A seam in a sheet metal roof that encloses a wood batten.

Bay A rectangular area of a building defined by four adjacent columns; a portion of a building that projects from a facade.

Bead A narrow line of weld metal or sealant; a strip of metal or wood used to hold a sheet of glass in place; a narrow, convex molding profile; a metal edge or corner accessory for plaster.

Beam A straight structural member that acts primarily to resist nonaxial loads.

Bearing A point at which one building element rests upon another.

Bearing block A piece of wood fastened to a column to provide support for a beam or girder.

Bearing pad A block of plastic or synthetic rubber used to cushion the point at which one precast concrete element rests upon another.

Bearing wall A wall that supports floors or roofs.

Bed *See* **casting bed.**

Bed joint The horizontal layer of mortar beneath a masonry unit.

Bedrock A solid stratum of rock.

Bending moment The moment that causes a beam or other structural member to bend.

Bending stress A compressive or tensile stress resulting from the application of a nonaxial force to a structural member.

Bent A plane of framing consisting of beams and columns joined together, often with rigid joints.

Bentonite clay An absorptive, colloidal clay that swells to several times its dry volume when saturated with water.

Bevel An end or edge cut at an angle other than a right angle.

Bevel siding Wood cladding boards that taper in cross section.

Billet A large cylinder or rectangular solid of metal produced from an ingot as an intermediate step in converting it into rolled or extruded metal products.

Bite The depth to which the edge of a piece of glass is held by its frame.

Bitumen A tarry mixture of hydrocarbons, such as asphalt or coal tar.

Blind nailing Attaching boards to a frame or sheathing with toenails driven through the edge of each piece so as to be completely concealed by the adjoining piece.

Blocking Pieces of wood inserted tightly between joists, studs, or rafters in a building frame to stabilize the structure, inhibit the passage of fire, provide a nailing surface for finish materials, or retain insulation.

Bloom A rectangular solid of steel formed from an ingot as an intermediate step in creating rolled steel structural shapes.

Blooming mill A set of rollers used to transform an ingot into a bloom.

Bluestone A sandstone that is gray to blue-gray in color and splits readily into thin slabs.

Board foot A unit of lumber volume, a rectangular solid nominally twelve square inches in cross-sectional area and one foot long.

Board siding Wood cladding made up of boards, as differentiated from shingles or manufactured wood panels.

BOCA Building Officials and Code Administrators International, Inc., an organization that publishes a model building code.

Bolster A long chair used to support reinforcing bars in a concrete slab.

Bolt A fastener consisting of a cylindrical metal body with a head at one end and a helical thread at the other, intended to be inserted through holes in adjoining pieces of material and closed with a threaded nut.

Bond In masonry, the adhesive force between mortar and masonry units, or the pattern in which masonry units are laid to tie two or more wythes together into a structural unit. In reinforced concrete, the adhesion between the surface of a reinforcing bar and the surrounding concrete.

Bond breaker A strip of material to which sealant does not adhere.

Bonded construction Posttensioned construction in which the tendons are grouted solidly into the mass of the beam or slab.

Bonded terrazzo Terrazzo flooring whose underbed is poured directly upon the structural floor.

Bottom bars The reinforcing bars that lie close to the bottom of a beam or slab.

Box beam A bending member of metal or plywood whose cross section resembles a closed rectangular box.

Bracing Diagonal members, either temporary or permanent, installed to stabilize a structure against lateral loads.

Brad A small finish nail.

Bridging Bracing or blocking installed between steel or wood joists at midspan to stabilize them against buckling and, in some cases, to permit adjacent joists to share loads.

British thermal unit (Btu) The quantity of heat required to raise one pound of water one degree Fahrenheit.

Broom finish A skid-resistant texture imparted to an uncured concrete surface by dragging a stiff-bristled broom across it.

Brown coat The second coat of plaster in a three-coat application.

Brownstone A brownish or reddish sandstone.

Btu *See* **British thermal unit.**

Buckling Structural failure by gross lateral deflection of a slender element under compressive stress, such as the sideward buckling of a long, slender column or the upper edge of a long, thin floor joist.

Building code A set of legal restrictions intended to assure a minimum standard of health and safety in buildings.

Built-up roof (BUR) A roof membrane laminated from layers of asphalt-saturated felt or other fabric, bonded together with bitumen or pitch.

Buoyant uplift The force of water or liquefied soil that tends to raise a building foundation out of the ground.

BUR *See* **built-up roof.**

Butt The thicker end, as the lower edge of a wood shingle or the lower end of a tree trunk; a joint between square-edged pieces; a weld between square-edged pieces of metal that lie in the same plane; a type of door hinge that attaches to the edge of the door.

Butt-joint glazing A type of glass installation in which the vertical joints between lights of glass do not meet at a mullion, but are made weathertight with a sealant.

Button head A smooth, convex bolt head with no provision for engaging a wrench.

Buttress A structural device of masonry or concrete that resists the diagonal forces from an arch or vault.

Butyl rubber A synthetic rubber compound.

C

Caisson A cylindrical sitecast concrete foundation that penetrates through unsatisfactory soil to rest upon an underlying stratum of rock or satisfactory soil; an enclosure that permits excavation work to be carried out underwater.

Calcining The driving off of the water of hydration from gypsum by the application of heat.

Camber A slight initial curvature in a beam or slab.

Cambium The thin layer beneath the bark of a tree that manufactures cells of wood and bark.

Cantilever A beam, truss, or slab that extends beyond its last point of support.

Cant strip A strip of material with a sloping face used to ease the transition from a horizontal to a vertical surface at the edge of a membrane roof.

Capillary action The pulling of water through a small orifice or fibrous material by the adhesive force between the water and the material.

Capillary break A slot or groove intended to create an opening too large to be bridged by a drop of water, and thereby to eliminate the passage of water by capillary action.

Carbide-tipped tools Drill bits, saws, and other tools with cutting edges made of an extremely hard alloy.

Carbon steel Low-carbon or mild steel.

Carpenter One who makes things of wood.

Casement window A window that pivots on an axis at or near a vertical edge of the sash.

Casing The wood finish pieces surrounding the frame of a window or door; a cylindrical steel tube used to line a drilled or driven hole in foundation work.

Castellated beam A steel wide-flange section whose web has been cut along a zigzag path and reassembled by welding in such a way as to create a deeper section.

Casting Pouring a liquid material or slurry into a mold whose form it will take as it solidifies.

Casting bed A permanent, fixed form in which precast concrete elements are produced.

Cast-in-place Concrete that is poured in its final location; sitecast.

Cast iron Iron with too high a carbon content to be classified as steel.

Caulk A low-range sealant.

Cavity wall A masonry wall that includes a continuous airspace between its outermost wythe and the remainder of the wall.

Cellular decking Panels made of steel sheets corrugated and welded together in such a way that hollow longitudinal cells are created within the panels.

Cellulose The material of which the structural fibers in wood are composed; a complex polymeric carbohydrate.

Celsius A temperature scale on which the freezing point of water is established as zero and the boiling point as 100 degrees.

Cement A substance used to adhere materials together; in concrete work, the dry powder that, when it has combined chemically with the water in the mix, cements the particles of aggregate together to form concrete.

Cementitious Having cementing properties; usually used with reference to inorganic substances, such as portland cement and lime.

Centering Temporary formwork for an arch, dome, or vault.

Centering shims Small blocks of synthetic rubber or plastic used to hold a sheet of glass in the center of its frame.

Ceramic tile Small, flat, thin clay tiles intended for use as wall and floor facings.

Chair A device used to support reinforcing bars.

Chamfer A flattening of a longitudinal edge of a solid member on a plane that lies at an angle of 45 degrees to the adjoining planes.

Channel A steel or aluminum section shaped like a rectangular box with one side missing.

Chlorinated polyethylene A plastic material used in roof membranes.

Chlorosulfonated polyethylene A plastic material used in roof membranes.

Chord A member of a truss.

C-H stud A steel wall framing member whose profile resembles a combination of the letters C and H, used to support gypsum panels in shaft walls.

Chuck A device for holding a steel wire, rod, or cable securely in place by means of steel wedges in a tapering cylinder.

Churn drill A steel tool used with an up-and-down motion to cut through rock at the bottom of a steel pipe caisson.

Cladding A material used as the exterior wall enclosure of a building.

Clamp A tool for holding two pieces of material together temporarily; unfired bricks piled in such a way that they can be fired without using a kiln.

Class A, B, C roofing Roof covering materials classified according to their resistance to fire when tested in accordance with ASTM E108. Class A is the highest, and class C is the lowest.

Cleanout hole An opening at the base of a masonry wall through which mortar droppings and other debris can be removed prior to grouting the interior cavity of the wall.

Clear dimension, clear opening The dimension between opposing inside faces of an opening.

Climbing crane A heavy-duty lifting machine that is supported on its own steel tower and raises itself as the building rises.

Clinker A fused mass that is an intermediate product of cement manufacture; a brick that is overburned.

Closer The last masonry unit laid in a course; a partial masonry unit used at the corner of a header course to adjust the joint spacing; a mechanical device for regulating the closing action of a door.

CMU *See concrete masonry unit.*

Cohesive soil A soil such as clay whose particles are able to adhere to one another by means of cohesive and adhesive forces.

Cold-rolled steel Steel rolled to its final form at a temperature at which it is no longer plastic.

Cold-worked steel Steel formed at a temperature at which it is no longer plastic, as by rolling or forging.

Collar joint The vertical mortar joint between wythes of masonry.

Collar tie A piece of wood nailed across two opposing rafters near the ridge to resist wind uplift.

Column An upright structural member acting primarily in compression.

Column cage An assembly of vertical reinforcing bars and ties for a concrete column.

Column-cover-and-spandrel system A system of cladding in which panels of material cover the columns and spandrels, with horizontal strips of windows filling the remaining portion of the wall.

Column spiral A continuous coil of steel reinforcing used to tie a concrete column.

Column strip The zone of a two-way concrete floor or roof structure that is centered on a line of columns.

Column tie A single loop of steel bar, usually bent into a rectangular configuration, used to tie a concrete column.

Common bolt An ordinary carbon steel bolt.

Common bond Brickwork laid with each five courses of stretchers followed by one course of headers.

Composite metal decking Corrugated steel decking manufactured in such a way that it bonds securely to the concrete floor fill to form a reinforced concrete deck.

Composite wall A masonry wall that incorporates two or more different types of masonry units, such as clay bricks and concrete blocks.

Compression A squeezing force.

Compression gasket A synthetic rubber strip that seals around a sheet of glass or a wall panel by being squeezed tightly against it.

Compressive strength The ability of a structural material to withstand squeezing forces.

Concave joint A mortar joint tooled into a curved, indented profile.

Concealed grid A suspended ceiling framework that is completely hidden by the tiles or panels it supports.

Concrete A structural material produced by mixing predetermined amounts of portland cement, aggregates, and water, and allowing this mixture to cure under controlled conditions.

Concrete block A concrete masonry unit, usually hollow, that is larger than a brick.

Concrete brick A solid concrete masonry unit the same size and proportions as a modular clay brick.

Concrete masonry unit (CMU) A block of hardened concrete, with or without hollow cores, designed to be laid in the same manner as a brick or stone; a concrete block.

Condensate Water formed as a result of condensation.

Condensation The process of changing from a gaseous to a liquid state, especially as applied to water.

Conduit A steel or plastic tube through which electrical wiring is run.

Continuous ridge vent A screened, water-shielded ventilation opening that runs continuously along the ridge of a gable roof.

Contractor A person or organization that undertakes a legal obligation to do construction work.

Control joint An intentional, linear discontinuity in a structure or component, designed to form a plane of weakness where cracking can occur in response to various forces so as to minimize or eliminate cracking elsewhere in the structure.

Convector A heat exchange device that uses the heat in steam, hot water, or an electric resistance element to warm the air in a room; often called, inaccurately, a *radiator*.

Coping A protective cap on the top of a masonry wall.

Coping saw A handsaw with a thin, very narrow blade, used for cutting detailed shapes in the ends of wood moldings and trim.

Corbel A spanning device in which masonry units in successive courses are cantilevered slightly over one another; a projecting bracket of masonry or concrete.

Coreboard A thick gypsum panel used primarily in shaft walls.

Corner bead A metal or plastic strip used to form a neat, durable edge at an outside corner of two walls of plaster or gypsum board.

Cornice The exterior detail at the meeting of a wall and a roof overhang; a decorative molding at the intersection of a wall and a ceiling.

Corrosion Oxidation, such as rust.

Corrugated Pressed into a fluted or ribbed profile.

Counterflashing A flashing turned down from above to overlap another flashing turned up from below so as to shed water.

Course A horizontal layer of masonry units one unit high; a horizontal line of shingles.

Coursed Laid in courses with straight bed joints.

Cove base A flexible strip of plastic or synthetic rubber used to finish the junction between resilient flooring and a wall.

Crawlspace A space that is not tall enough to stand in, located beneath the bottom floor of a building.

Creep A permanent inelastic deformation in a material due to changes in the material caused by the prolonged application of structural stress.

Cripple stud A wood wall-framing member that is shorter than full-length studs because it is interrupted by a header or sill.

Cross-grain wood Wood incorporated into a structure in such a way that its direction of grain is perpendicular to the direction of the principal loads on the structure.

Crosslot bracing Horizontal compression members running from one side of an excavation to the other, used to support sheeting.

Crown glass Glass sheet formed by spinning an opened globe of heated glass.

Cruck A framing member cut from a bent tree so as to form one-half of a rigid frame.

Cup A curl in the cross section of a board or timber caused by unequal shrinkage or expansion between one side of the board and the other.

Curing The hardening of concrete, plaster, gunnable sealant, or other wet materials. Curing can occur through evaporation of water or a solvent, hydration, polymerization, or chemical reactions of various types, depending on the formulation of the material.

Curing compound A liquid that, when sprayed on the surface of newly placed concrete, forms a water-resistant layer to prevent premature dehydration of the concrete.

Curtain wall An exterior building wall that is supported entirely by the frame of the building, rather than being self-supporting or loadbearing.

Cylinder glass Glass sheet produced by blowing a large, elongated glass cylinder, cutting off its ends, slitting it lengthwise, and opening it into a flat rectangle.

D

Damper A flap to control or obstruct the flow of gases; specifically, a metal control flap in the throat of a fireplace, or in an air duct.

Dap A notch at the end of a piece of material.

Darby A stiff straightedge of wood or

metal used to level the surface of wet plaster.

Daylighting   Illuminating the interior of a building by natural means.

Dead load   The weight of the building itself.

Deadman   A large and/or heavy object buried in the ground as an anchor.

Decking   A material used to span across beams or joists to create a floor or roof surface.

Deformation   A change in the shape of a structure or structural element caused by a load or force acting on the structure.

Derrick   Any of a number of devices for hoisting building materials on the end of a rope or cable.

Dew point   The temperature at which water will begin to condense from a mass of air at a given temperature and moisture content.

Diagonal bracing   *See* **bracing.**

Diaphragm action   A bracing action that derives from the stiffness of a thin plane of material when it is loaded in a direction parallel to the plane. Diaphragms in buildings are typically floor, wall, or roof surfaces of plywood, reinforced masonry, steel decking, or reinforced concrete.

Die   An industrial tool for giving identical form to repeatedly produced units, such as a shaped orifice for giving form to a column of clay or an aluminum extrusion, a shaped punch for making cutouts of sheet metal or paper, or a mold for casting plastic or metal.

Die-cut   Manufactured by punching from a sheet material.

Dimension lumber   Lengths of wood, rectangular in cross section, sawed directly from the log.

Dimension stone   Building stone cut to a rectangular shape.

Distribution rib   A transverse thickening of a one-way concrete joist structure used to allow the joists to share concentrated loads.

Divider strip   A strip of metal or plastic embedded in terrazzo to form control joints and decorative patterns.

Dome   An arch rotated about its vertical axis to produce a structure shaped like an inverted bowl; a form used to make one of the cavities in a concrete waffle slab.

Dormer   A structure protruding through the plane of a sloping roof, usually with a window and its own smaller roof.

Double glazing   Two parallel sheets of glass with an airspace between.

Double-hung window   A window with two overlapping sashes that slide vertically in tracks.

Double-strength glass   Glass that is approximately $\frac{1}{8}$ inch (3 mm) in thickness.

Double tee   A precast concrete slab element that resembles the letters *TT* in cross section.

Dowel   A short cylindrical rod of wood or steel; a steel reinforcing bar that projects from a foundation to tie it to a column or wall.

Downspout   A vertical pipe for conducting water from a roof to a lower level.

Drainage   Removal of water.

Drawing   Shaping a material by pulling it through an orifice, as in the drawing of steel wire, or the drawing of a sheet of glass; shaping plaster moldings by pulling a template across the wet plaster.

Drawn glass   Glass sheet pulled directly from a container of molten glass.

Drift   Lateral deflection of a building caused by wind or earthquake loads.

Drift pin   A tapered steel shaft used to align bolt holes in steel connections during erection.

Drip   A discontinuity formed into the underside of a window sill or wall component to force adhering drops of water to fall free of the face of the building rather than move farther toward the interior.

Drop panel   A thickening of a two-way concrete structure at the head of a column.

Dry pack   A low-slump grout tamped into the space in a connection between precast concrete members.

Dry press process   A method of molding slightly damp clays and shales into bricks by forcing them into molds under high pressure.

"Dry" systems   Systems of construction that use little or no water during construction, as differentiated from systems such as masonry, plastering, and ceramic tile work.

Drywall   *See* **gypsum board.**

Dry well   An underground pit filled with broken stone or other porous material, from which rainwater from a roof drainage system can seep into the surrounding soil.

Duct   A hollow conduit, commonly of sheet metal, through which air can be circulated; a tube used to establish the position of a posttensioning tendon in a concrete structure.

DWV   Drain-waste-vent pipes, the part of the plumbing system of a building that removes liquid wastes and conducts them to the sewer or sewage disposal system.

E

Eave   The horizontal edge at the low side of a sloping roof.

Edge bead   A strip of metal or plastic used to make a neat, durable edge where plaster or gypsum board abuts another material.

Efflorescence   A powdery deposit on the face of a structure of masonry or concrete, caused by the leaching of chemical salts by water migrating from within the structure to the surface.

Elastic   Able to return to its original size and shape after removal of stress.

Elastomeric/plastomeric membrane   A rubberlike sheet material used as a roof covering.

Electrode   A consumable steel wire or rod used to maintain an arc and furnish additional weld metal in electric arc welding.

Elevation   A drawing that views a building from any of its sides; a vertical height above a reference point such as sea level.

Elongation   Stretching under load; growing longer because of temperature expansion.

End nail   A nail driven through one piece of lumber and into the end of another.

Engineered fill   Earth compacted into place in such a way that it has predictable physical properties, based on laboratory tests and specified, supervised installation procedures.

English bond   Brickwork laid with alternating courses each consisting entirely of headers or stretchers.

EPDM   Ethylene propylene diene monomer, a synthetic rubber material used in roofing membranes.

Erector   The subcontractor who raises, connects, and plumbs up a building frame from fabricated steel or precast concrete components.

Expanded metal lath   A thin sheet of metal

that has been slit and stretched to transform it into a metal mesh used as a base for the installation of plaster.

Expansion joint A discontinuity extending completely through the foundation, frame, and finishes of a building to allow for gross movement due to thermal stress, material shrinkage, or foundation settlement.

Exposed aggregate finish A concrete surface in which the coarse aggregate is revealed.

Exposed grid A framework for an acoustical ceiling that is visible from below after the ceiling is completed.

Extrados The convex surface of an arch.

Extrusion The process of squeezing a material through a shaped orifice to produce a linear element with the desired cross section; an element produced by this process.

F

Fabricator The company that prepares structural steel members for erection.

Facade An exterior face of a building.

Face brick A brick selected on the basis of appearance and durability for use in the exposed surface of a wall.

Face nail A nail driven through the side of one wood member into the side of another.

Fahrenheit A temperature scale on which the boiling point of water is fixed at 212 degrees and the freezing point at 32.

Fanlight A semicircular or semielliptical window above an entrance door, often with radiating muntins that resemble a fan.

Fascia The exposed vertical face of an eave.

Felt A thin, flexible sheet material made of soft fibers pressed and bonded together. In building practice, a thick paper, or a sheet of glass or asbestos fibers.

Figure The surface pattern of the grain of a piece of smoothly finished wood or stone.

Fillet A rounded inside intersection between two surfaces that meet at an angle.

Fillet weld A weld at the inside intersection of two metal surfaces that meet at right angles.

Finger joint A glued end connection between two pieces of wood, using an interlocking pattern of deeply cut "fin-

gers." A finger joint creates a large surface for the glue bond, to allow it to develop the full tensile strength of the wood it connects.

Finial A slender ornament at the top of a roof or spire.

Finish Exposed to view; material that is exposed to view.

Finish carpenter One who does finish carpentry.

Finish carpentry The wood components exposed to view on the interior of a building, such as window and door casings, baseboards, bookshelves, and the like.

Finish coat The final coat of plaster.

Finish floor The floor material exposed to view, as differentiated from the *subfloor*, which is the loadbearing floor surface beneath.

Finish lime A hydrated lime used in finish coats of plaster and in ornamental plaster work.

Firebrick A brick made from special clays to withstand very high temperatures, as in a fireplace, furnace, or industrial chimney.

Firecut A sloping end cut on a wood beam or joist where it enters a masonry wall. The purpose of the firecut is to allow the wood member to rotate out of the wall without prying the wall apart, if the floor or roof structure should burn through in a fire.

Fireproofing Material used around a steel structural element to insulate it against excessive temperatures in case of fire.

Fireresistance rating The time, in hours or fractions of an hour, that a material or assembly will resist fire exposure as determined by ASTM E119.

Fire resistant Incombustible; slow to be damaged by fire; forming a barrier to the passage of fire.

Fire separation wall A wall required under the building code to divide two parts of a building as a deterrent to the spread of fire.

Firestop A wood or masonry baffle used to close an opening between studs or joists in a balloon or platform frame in order to retard the spread of fire through the opening.

Fire wall A wall extending from foundation to roof, required under the building code to divide two parts of a building as a deterrent to the spread of fire.

Firing The process of converting dry clay into a ceramic material through the application of intense heat.

First cost The cost of construction.

Fixed window Glass that is immovably mounted in a wall.

Flame spread rating A measure of the rapidity with which fire will spread across the surface of a material as determined by ASTM E84.

Flange A projecting crosspiece of a wide-flange or channel profile.

Flash-cove A detail in which a sheet of resilient flooring is turned up at the edge and finished against the wall to create an integral baseboard.

Flashing A thin, continuous sheet of metal, plastic, rubber, or waterproof paper used to prevent the passage of water through a joint in a wall, roof, or chimney.

Flat seam A sheet metal roofing seam that is formed flat against the surface of the roof.

Flemish bond Brickwork laid with each course consisting of alternating headers and stretchers.

Flitch-sliced veneer A thin sheet of wood cut by passing a block of wood vertically against a long, sharp knife.

Float A small platform suspended on ropes from a steel building frame to permit ironworkers to work on a connection; a trowel with a slightly rough surface used in an intermediate stage of finishing a concrete slab; as a verb, to use a float for finishing concrete.

Float glass Glass sheet manufactured by cooling a layer of liquid glass on a bath of molten tin.

Flocculated A "fluffy" microstructure of clay particles in which the platelets are randomly oriented.

Flue A passage for smoke and combustion products from a furnace, stove, water heater, or fireplace.

Fluid-applied roof membrane A roof membrane applied in one or more coats of a liquid that cure to form an impervious sheet.

Fluoropolymer A highly stable organic compound used as a finish coating for building cladding.

Flush Smooth, lying in a single plane.

Flush door A door with smooth, planar faces.

Flux A material added to react chemically with impurities and remove them from molten metal. Fluxes are used both in steel making and in welding. Welding fluxes serve the additional purpose of shielding the molten weld metal from the air to reduce oxidation and other undesirable effects.

Flying formwork Large sections of slab formwork that are moved by crane.

Fly rafter A rafter in a rake overhang.

Foil-backed gypsum board Gypsum board with aluminum foil laminated to its back surface as a vapor retarder and thermal insulator.

Folded plate A roof structure whose strength and stiffness derive from a pleated or folded geometry.

Footing The widened part of a foundation that spreads a load from the building across a broader area of soil.

Form deck Thin, corrugated steel decking that serves as permanent formwork for a reinforced concrete deck.

Form tie A steel rod with fasteners on each end, used to hold together the formwork for a concrete wall.

Formwork Temporary structures of wood, steel, or plastic that serve to give shape to poured concrete, and to support it and keep it moist as it cures.

Foundation The portion of a building that has the sole purpose of transmitting structural loads from the building into the earth.

Framed connection A shear connection between steel members made by means of steel angles or plates connecting to the web of the beam or girder.

Framing plan A diagram showing the arrangement and sizes of the structural members in a floor or roof.

Freestone Fine-grained sedimentary rock that has no planes of cleavage or sedimentation along which it is likely to split.

French door A symmetrical pair of glazed doors hinged to the jambs of a single frame and meeting at the center of the opening.

Frictional soil A soil such as sand that has little or no attraction between its particles, and derives its strength from geometric interlocking of the particles; a noncohesive soil.

Friction connection Two or more structural steel members clamped together by high-strength bolts with sufficient force

that the loads on the members are transmitted between them by friction along their mating surfaces.

Frost line The depth in the earth to which the soil can be expected to freeze during a severe winter.

Fuel contributed rating A measure of the extent to which a building material will provide additional energy to a fire.

Furring channel A formed sheet metal furring strip.

Furring strip A length of wood or metal attached to a masonry or concrete wall to permit the attachment of finish materials to the wall using screws or nails.

G

Gable The triangular wall beneath the end of a gable roof.

Gable roof A roof consisting of two oppositely sloping planes that intersect at a level ridge.

Gable vent A screened, louvered opening in a gable, used for exhausting excess heat and humidity from an attic.

Galling Chafing of one material against another under extreme pressure.

Galvanizing The application of a zinc coating to steel as a means of preventing corrosion.

Gambrel A roof shape consisting of two roof planes at different pitches on each side of a ridge.

Gasket A dry, resilient material used to seal a joint between two rigid assemblies by being compressed between them.

Gauged brick A brick that has been rubbed on an abrasive stone to reduce it to a trapezoidal shape for use in an arch.

Gauging plaster A gypsum plaster formulated for use in combination with finish lime in finish coat plaster.

General contractor A contractor who has responsibility for the overall conduct of a construction project.

GFRC *See·* **glass-fiber-reinforced concrete.**

Girder A beam that supports other beams; a very large beam, especially one that is built up from smaller elements.

Girt A beam that supports wall cladding between columns.

Glass block A hollow masonry unit made of glass.

Glass fiber batt A thick, fluffy, non-

woven insulating blanket of filaments spun from glass.

Glass-fiber-reinforced concrete (GFRC) Concrete with a strengthening admixture of short alkali-resistant glass fibers.

Glass mullion system A method of constructing a large glazed area by stiffening the sheets of glass with perpendicular glass ribs.

Glaze A glassy finish on a brick or tile; a verb meaning to "install glass."

Glazed structural clay tile A hollow clay block with glazed faces, used for constructing interior partitions.

Glazier One who installs glass.

Glazier's points Small pieces of metal driven into a wood sash to hold glass in place.

Glazing The act of installing glass; installed glass; an adjective referring to materials used in installing glass.

Glazing compound Any of a number of types of mastic used to bed small lights of glass in a frame.

Glue laminated timber A timber made up of a large number of small strips of wood glued together.

Glulam A short expression for glue laminated timber.

Grade A classification of size or quality for an intended purpose; to classify as to size or quality.

Grade The surface of the ground; to move earth for the purpose of bringing the surface of the ground to an intended level or profile.

Grade beam A reinforced concrete beam that transmits the load from a bearing wall into spaced foundations such as pile caps or caissons.

Grain In wood, the direction of the longitudinal axes of the wood fibers, or the figure formed by the fibers.

Granite Igneous rock with visible crystals of quartz and feldspar.

Groove weld A weld made in a groove created by beveling or milling the edges of the mating pieces of metal.

Ground A strip attached to a wall or ceiling to establish the level to which plaster should be applied.

Grout A high-slump mixture of portland cement, aggregates, and water, which can be poured or pumped into cavities in concrete or masonry for the purpose of embedding reinforcing bars and/or increasing the amount of loadbearing material in a wall; a specially formulated, mortarlike material for filling under steel baseplates and around connections in precast concrete framing; a mortar used to fill joints between ceramic tiles or quarry tiles.

Gunnable sealant   A sealant material that is extruded in liquid or mastic form from a sealant gun.

Gusset plate   A flat steel plate to which the chords are connected at a joint in a truss; a stiffener plate.

Gutter   A channel to collect rainwater and snowmelt at the eave of a roof.

Guy derrick   A hoisting device consisting of a vertical mast braced with cables (guys), a rotating boom, and a hoisting cable.

Gypsum   Hydrous calcium sulfate.

Gypsum backing board   A lower-cost gypsum panel intended for use as an interior layer in multilayer constructions of gypsum board.

Gypsum board   An interior facing panel consisting of a gypsum core sandwiched between paper faces. Also called *drywall, plasterboard*.

Gypsum lath   Small sheets of gypsum board manufactured specifically for use as a plaster base.

Gypsum plaster   Plaster whose cementing substance is gypsum.

Gypsum wallboard   *See* **gypsum board.**

H

Hardboard   A very dense panel product, usually with at least one smooth face, made of highly compressed wood fibers.

Hawk   A metal square with a handle below, used by a plasterer to hold a small quantity of wet plaster and transfer it to a trowel for application to a wall or ceiling.

Header   Lintel; band joist; a joist that supports other joists; in steel construction, a beam that spans between girders; a brick or other masonry unit that is laid across two wythes with its end exposed in the face of the wall.

Head joint   The vertical layer of mortar between ends of masonry units.

Hearth   The incombustible floor area outside a fireplace opening.

Heartwood   The dead wood cells nearer the center of a tree trunk.

Heat-fuse   To join by softening or melting the edges with heat and pressing them together.

Heat of hydration   The thermal energy given off by concrete or gypsum as it cures.

Heat-strengthened glass   Glass that has been strengthened by heat treatment, though not to as great an extent as tempered glass.

High-lift grouting   A method of constructing a reinforced masonry wall in which the reinforcing bars are grouted into the wall in story-high increments.

High-range sealant   A sealant that is capable of a high degree of elongation without rupture.

High-strength bolt   A bolt designed to connect steel members by clamping them together with sufficient force that the load is transferred between them by friction.

Hip   The diagonal intersection of planes in a hip roof.

Hip roof   A roof consisting of four sloping planes that intersect to form a pyramidal or elongated pyramid shape.

Hollow concrete masonry   Concrete masonry units that are manufactured with open cores, such as ordinary concrete blocks.

Hollow-core door   A door consisting of two face veneers separated by an airspace, with solid wood spacers around the four edges. The face veneers are connected by a grid of thin spacers running through the airspace.

Hollow-core slab   A precast concrete slab element that has internal longitudinal cavities to reduce its self-weight.

Hook   A semicircular bend in the end of a reinforcing bar, made for the purpose of anchoring the end of the bar securely into the surrounding concrete.

Hopper window   A window whose sash pivots on an axis along the sill, and that opens by tilting toward the interior of the building.

Horizontal force   A force whose direction of action is horizontal or nearly horizontal; *see also* **lateral force.**

Hose test   A standard laboratory test to determine the relative ability of an interior building assembly to stand up to water from a fire hose during a fire.

Hot-rolled steel   Steel formed into its final shape by passing it between rollers while it is very hot.

Hydrated lime   Calcium hydroxide produced by burning calcium carbonate to form calcium oxide (quicklime), then allowing the calcium oxide to combine chemically with water.

Hydration   A process of combining chemically with water to form molecules or crystals that include hydroxide radicals or water of crystallization.

Hydronic heating system   A system that circulates warm water through convectors to heat a building.

Hydrostatic pressure   Pressure exerted by standing water.

Hygroscopic   Readily taking up and retaining moisture.

Hyperbolic paraboloid shell   A concrete roof structure with a saddle shape.

I

I-beam   (obsolete term) An American Standard section of hot-rolled steel.

Ice dam   An obstruction along the eave of a roof caused by the refreezing of water emanating from melting snow on the roof surface above.

Igneous rock   Rock formed by the solidification of magma.

IIC   *See* **Impact Insulation Class.**

Impact Insulation Class (IIC)   An index of the extent to which a floor assembly transmits impact noise from a room above to the room below.

Impact noise   Noise generated by footsteps or other impacts on a floor.

Impact wrench   A device for tightening bolts and nuts by means of rapidly repeated torque impulses produced by electrical or pneumatic energy.

Ingot   A large block of cast metal.

Insulating glass   Double or triple glazing.

Internal drainage   Providing a curtain wall with hidden channels and weep holes to remove any water that may penetrate the exterior layers of the wall.

Interstitial ceiling   A suspended ceiling with sufficient structural strength to support workers as they install and maintain mechanical and electrical installations above the ceiling.

Intrados   The concave surface of an arch.

Intumescent coating   A paint or mastic that

expands to form a stable, insulating char when exposed to fire.

Inverted roof A membrane roof assembly in which the thermal insulation lies above the membrane.

Iron In pure form, a metallic element. In common usage, ferrous alloys other than steels, including cast iron and wrought iron.

Iron dog A heavy U-shaped staple used to tie the ends of heavy timbers together.

Ironworker A skilled laborer who erects steel building frames, or places reinforcing bars in concrete construction.

Isocyanurate foam A thermosetting plastic foam with thermal insulating properties.

J

Jack A device for exerting a large force over a short distance, usually by means of screw action or hydraulic pressure.

Jack rafter A shortened rafter that joins a hip or valley rafter.

Jamb The vertical side of a door or window.

Joist One of a group of light, closely spaced beams used to support a floor deck or flat roof.

Joist-band A broad, shallow concrete beam that supports one-way concrete joists whose depths are identical to its own.

Joist girder A light steel truss used to support open-web steel joists.

K

Keenes cement A strong, crack-resistant gypsum plaster formulation.

Key A slot formed into a concrete surface for the purpose of interlocking with a subsequent pour of concrete; a slot at the edge of a precast member into which grout will be poured to lock it to an adjacent member.

Kiln A furnace for firing clay or glass products; a heated chamber for seasoning wood.

Knee wall A short wall under the slope of a roof.

Knife connection A steel beam/column connection in which the web of the beam is inserted from above between a pair of connecting angles attached to the column.

L

Labyrinth A cladding joint design in which a series of interlocking baffles prevents drops of water from penetrating the joint by momentum.

Lagging Planks placed between soldier beams to retain earth around an excavation.

Lag screw A large-diameter wood screw with a square or hexagonal head.

Laminate As a verb, to bond together in layers; as a noun, a material produced by bonding together layers of material.

Laminated glass A glazing material consisting of outer layers of glass laminated to an inner layer of transparent plastic.

Landing A platform in or at either end of a stair.

Lap joint A connection in which two pieces of material are overlapped before fastening.

Lateral force A force acting generally in a horizontal direction, such as wind, earthquake, or soil pressure against a foundation wall.

Lateral thrust The horizontal component of the force produced by an arch, dome, vault, or rigid frame.

Latex caulk A low-range sealant based on a synthetic latex.

Lath (Rhymes with "math.") A base material to which plaster is applied.

Lathe (Rhymes with "bathe.") A machine in which a piece of material is rotated against an edged tool to produce a shape with cross sections that are all circles; a machine in which a log is rotated against a long knife to peel a continuous sheet of veneer.

Lather (Rhymes with "rather.") One who applies lath.

Lead (Rhymes with "bead.") In masonry work, a corner or wall end accurately constructed with the aid of a spirit level to serve as a guide for placing the bricks in the remainder of the wall.

Leader A vertical pipe for conducting water from a roof to a lower level.

Lehr A chamber in which glass is annealed.

Let-in bracing Diagonal bracing nailed into notches cut in the face of the studs so it does not increase the thickness of the wall.

Level cut A saw cut that produces a level surface at the lower end of a sloping rafter.

Leveling plate A steel plate placed in grout on top of a concrete foundation to create a level bearing surface for the lower end of a steel column.

Lewis A device for lifting a block of stone by means of friction exerted against the sides of a hole drilled in the top of the block.

Life-cycle cost A cost that takes into account both the first cost and costs of maintenance, replacement, fuel consumed, monetary inflation, and interest over the life of the object being evaluated.

Lift-slab construction A method of building multistory sitecast concrete buildings by casting all the slabs in a stack on the ground, then lifting them up the columns with jacks and welding them in place.

Light A sheet of glass.

Light-gauge steel stud A length of thin sheet metal formed into a stiff shape and used as a wall framing member.

Lignin The natural cementing substance that binds together the cellulose in wood.

Limestone A sedimentary rock consisting of calcium carbonate, magnesium carbonate, or both.

Line wire Wire stretched across wall studs as a base for the application of metal mesh and stucco.

Linoleum A resilient floorcovering material composed primarily of ground cork and linseed oil on a burlap or canvas backing.

Lintel A beam that carries the load of a wall across a window or door opening.

Liquid sealant Gunnable sealant.

Lite *See* **light.**

Live load The weight of snow, people, furnishings, machines, vehicles, and goods in or on a building.

Load A weight or force acting on a structure.

Loadbearing Supporting a superimposed weight or force.

Load indicator washer A disk placed under the head or nut of a high-strength bolt to indicate sufficient tensioning of the bolt by means of the deformation of ridges on the surface of the disk.

Lockstrip gasket A synthetic rubber strip compressed around the edge of a piece of glass or a wall panel by inserting a spline (lockstrip) into a groove in the strip.

Lookout   A short rafter, running at an angle to the other rafters in the roof, which supports a rake overhang.

Louver   A construction of numerous sloping, closely spaced slats used to prevent the entry of rainwater into a ventilating opening.

Low-emissivity coating   A surface coating for glass that permits the passage of most shortwave electromagnetic radiation (light and heat), but reflects most longer-wave radiation (heat).

Low-iron glass   Glass formulated with a low iron content so as to have a maximum transparency to solar energy.

Low-lift grouting   A method of constructing a reinforced masonry wall in which the reinforcing bars are grouted into the wall in increments not higher than 4 feet (1200 mm).

Low-range sealant   A sealant that is capable of only a slight degree of elongation prior to rupture.

M

Mandrel   A stiff steel core placed inside the thin steel shell of a sitecast concrete pile to prevent it from collapsing during driving.

Mansard   A roof shape consisting of two superimposed levels of *hip roofs* with the lower level at a steeper pitch.

Marble   A metamorphic rock formed from limestone by heat and pressure.

Mason   One who builds with bricks, stones, or concrete masonry units; one who works with concrete.

Masonry   Brickwork, blockwork, and stonework.

Masonry cement   Portland cement with dry admixtures designed to increase the workability of mortar.

Masonry opening   The clear dimension required in a masonry wall for the installation of a specific window or door unit.

Masonry unit   A brick, stone, concrete block, glass block, or hollow clay tile intended to be laid in mortar.

Masonry veneer   A single wythe of masonry used as a facing over a frame of wood or metal.

Masterformat   The copyrighted title of a uniform indexing system for construction specifications, as created by the Construction Specifications Institute and Construction Specifications Canada.

Mastic   A viscous, doughlike, adhesive substance; can be any of a large number of formulations for different purposes such as sealants, adhesives, glazing compounds, or roofing membranes.

Mat foundation   A single concrete footing that is essentially equal in area to the area of ground covered by the building.

Medium-range sealant   A sealant material that is capable of a moderate degree of elongation before rupture.

Meeting rail   The wood or metal bar along which one sash of a double-hung or sliding window seals against the other.

Member   An element of a structure such as a beam, a girder, a column, a joist, a piece of decking, a stud, or a chord of a truss.

Membrane   A sheet material that is impervious to water or water vapor.

Membrane fire protection   A ceiling used to provide fire protection to the structural members above.

Metal lath   A steel mesh used primarily as a base for the application of plaster.

Metamorphic rock   A rock created by the action of heat or pressure on a sedimentary rock or soil.

Middle strip   The zone of a two-way concrete slab that lies midway between columns.

Mild steel   Ordinary structural steel, containing less than three-tenths of one percent carbon.

Mill construction   A building type with exterior masonry bearing walls and an interior framework of heavy timbers and solid timber decking.

Milling   Shaping or planing using rotating cutters.

Millwork   Wood interior finish components of a building, including moldings, windows, doors, cabinets, stairs, mantels, and the like.

Miter   A diagonal cut at the end of a piece; the joint produced by joining two diagonally cut pieces at right angles.

Mobile home   Euphemism for a portable house that is entirely factory built on a steel underframe supported by wheels.

Modular   Conforming to a multiple of a fixed dimension.

Modular home   Euphemism for a house assembled on the site from two or more boxlike factory-built sections.

Modulus of elasticity   An index of the stiffness of a material, derived by measuring the elastic deformation of the material as it is placed under stress, and then dividing the stress by the deformation.

Moisture barrier   A membrane used to prevent the migration of liquid water through a floor or wall.

Molding   A strip of wood, plastic, or plaster with an ornamental profile.

Molding plaster   A fast-setting gypsum plaster used for the manufacture of cast ornament.

Moment   A twisting action; a torque; a force acting at a distance from a point in a structure so as to cause a tendency of the structure to rotate about that point. *See also* **bending moment, moment connection.**

Moment connection   A connection between two structural members that is highly resistant to rotation between the members, as differentiated from a **shear connection,** which allows rotation.

Momentum   The energy possessed by a moving body.

Monolithic   Of a single massive piece.

Monolithic terrazzo   A thin terrazzo topping applied to a concrete slab without an underbed.

Mortar   A substance used to join masonry units, consisting of cementitious materials, fine aggregate, and water.

Mortise-and-tenon   A joint in which a tongue-like protrusion (tenon) on the end of one piece is tightly fitted into a rectangular slot (mortise) in the side of the other piece.

Mullion   A vertical or horizontal bar between adjacent window or door units.

Muntin   A small vertical or horizontal bar between small lights of glass in a sash.

Muriatic acid   Hydrochloric acid.

Mushroom capital   A flaring, conical head on a concrete column.

N

Nail-base sheathing   A sheathing material, such as wood boards or plywood, to which siding can be attached by nailing, as differentiated from one, such as cane fiber board or plastic foam board, that is too soft to hold nails.

Nail popping   The loosening of nails holding gypsum board to a wall, caused by drying shrinkage of the studs.

Nail set A hardened steel punch used to drive the head of a nail to a level below the surface of the wood.

Needle beam A steel or wood beam threaded through a hole in a bearing wall and used to support the wall and its superimposed loads during underpinning of its foundation.

Needling The use of needle beams.

Neoprene A synthetic rubber.

Nominal dimension An approximate dimension assigned to a piece of material as a convenience in referring to the piece.

Nonaxial In a direction not parallel to the long axis of a structural member.

Nonbearing Not carrying a load.

Nonworking joint A joint that is not subjected to significant deformations.

Nosing The projecting forward edge of a stair tread.

O

Ogee An S-shaped curve.

Oil-based caulk A low-range sealant made with linseed oil.

One-way action The structural action of a slab that spans between two parallel beams or bearing walls.

One-way concrete joist system A reinforced concrete framing system in which closely spaced concrete joists span between parallel beams or bearing walls.

One-way solid slab A reinforced concrete floor or roof slab that spans between parallel beams or bearing walls.

Open-truss wire stud A wall framing member in the form of a small steel truss.

Open-web steel joist A prefabricated, welded steel truss used at closely spaced intervals to support floor or roof decking.

Ordinary construction A building type with exterior masonry bearing walls and an interior structure of balloon framing.

Organic soil Soil containing decayed vegetable and/or animal matter; topsoil.

Oriented strand board (OSB) A building panel composed of long shreds of wood fiber oriented in specific directions and bonded together under pressure.

Oxidation Corrosion; rusting; rust.

P

Pan A form used to produce the cavity between joists in a one-way concrete joist system.

Panel A broad, thin piece of wood; a sheet of building material such as plywood or particle board; a prefabricated building component that is broad and thin, such as a curtain wall panel; an area within a truss consisting of an opening and the members that surround it.

Panel door A wood door in which two or more thin panels are held by stiles and rails.

Panic hardware A mechanical device that opens a door automatically if pressure is exerted against the device from the interior of the building.

Parapet The region of an exterior wall that projects above the level of the roof.

Parging Portland cement plaster applied over masonry to make it less permeable to water.

Particle board A building panel composed of small particles of wood bonded together under pressure.

Parting compound A substance applied to concrete formwork to prevent concrete from adhering.

Partition An interior nonloadbearing wall.

Patterned glass Glass into which a texture has been rolled during manufacture.

Paver A half-thickness brick used as finish flooring.

Pediment The gable end of a roof in classical architecture.

Penetrometer A device for testing the resistance of a material to penetration, usually used to make a quick, approximate determination of its compressive strength.

Periodic kiln A kiln that is loaded and fired in batches, as differentiated from a tunnel kiln, which is operated continuously.

Perlite Expanded volcanic rock, used as a lightweight aggregate in concrete and plaster, and as an insulating fill.

Perm A unit of vapor permeability.

Pier A caisson foundation unit.

Pilaster A vertical, integral stiffening rib in a masonry or concrete wall.

Pile A long, slender piece of material driven into the ground to act as a foundation.

Pile cap A thick slab of reinforced concrete poured across the top of a pile cluster to cause the cluster to act as a unit in supporting a column or grade beam.

Piledriver A machine for driving piles.

Pintle cap A metal device used to transmit compressive forces between superimposed columns in Mill construction.

Pitch The slope of a roof or other plane, often expressed as inches of rise per foot of run; a dark, viscous hydrocarbon distilled from coal tar; a viscous resin found in wood.

Pitched roof A sloping roof.

Plainsawing Sawing a log into dimension lumber without regard to the direction of the annual rings.

Plain slicing Cutting a flitch into veneers without regard to the direction of the annual rings.

Planing Smoothing the surface of a piece of wood or stone.

Plaster A cementitious material, usually based on gypsum or portland cement, applied to lath or masonry in paste form, to harden into a finish surface.

Plasterboard *See* **gypsum board**.

Plaster of Paris Pure calcined gypsum.

Plaster screeds Intermittent spots or strips of plaster used to establish the level to which a large plaster surface will be finished.

Plasticity The ability to retain a shape attained by pressure deformation.

Plate A broad sheet of rolled steel $\frac{1}{4}$ inch (6.35 mm) or more in thickness; a two-way concrete slab; a horizontal top or bottom member in a platform frame wall structure.

Plate girder A large beam made up of steel plates, sometimes in combination with steel angles, welded, bolted, or riveted together.

Plate glass Glass of high optical quality produced by grinding and polishing both faces of a glass sheet.

Platform frame A wooden building frame composed of closely spaced members nominally 2 inches (51 mm) in thickness, in which the wall members do not run past the floor framing members.

Plumb Vertical.

Plumb cut A saw cut that produces a vertical (plumb) surface at the lower end of a sloping rafter.

Plumbing up The process of making a steel building frame vertical and square.

Ply A layer, as in a layer of felt in a built-up roof membrane or a layer of veneer in plywood.

Plywood A wood panel composed of an

odd number of layers of wood veneer bonded together under pressure.

Pointing The process of applying mortar to the surface of a mortar joint after the masonry has been laid, either as a means of finishing the joint or to repair a defective joint.

Polybutene A sticky, masticlike tape used to seal nonworking joints, especially between glass and mullions.

Polycarbonate An extremely tough, strong, usually transparent plastic used for window and skylight glazing, light fixture globes, and other applications.

Polyethylene A thermoplastic widely used in sheet form for vapor retarders, moisture barriers, and temporary construction coverings.

Polymer A compound consisting of repeated chemical units.

Polymercaptans Compounds used in high-range gunnable sealants.

Polystyrene foam A thermoplastic foam with thermal insulating properties.

Polysulfide A high-range gunnable sealant.

Polyurethane Any of a large group of resins and synthetic rubber compounds used in sealants, varnishes, insulating foams, and roof membranes.

Polyurethane foam A thermosetting foam with thermal insulating properties.

Polyvinyl chloride (PVC) A thermoplastic material widely used in construction products, including plumbing pipes, floor tiles, wallcoverings, and roofing membranes.

Portal frame A rigid frame; two columns and a beam attached with moment connections.

Portland cement The gray powder used as the binder in concrete, mortar, and stucco.

Posttensioning The compressing of the concrete in a structural member by means of tensioning high-strength steel tendons against it after the concrete has cured.

Pour To cast concrete; an increment of concrete casting carried out without interruption.

Powder-driven Inserted by a gunlike tool using energy provided by an exploding charge of gunpowder.

Precast concrete Concrete cast and cured in a position other than its final position in the structure.

Predecorated gypsum board Gypsum board finished at the factory with a decorative layer of paint, paper, or plastic.

Preformed cellular tape sealant A sealant inserted into a joint in the form of a compressed sponge impregnated with compounds that cure to form a watertight seal.

Preformed solid tape sealant A sealant inserted into a joint in the form of a flexible strip of solid material.

Prehung door A door that is hinged to its frame in a factory or shop.

Pressure-equalized wall design Curtain wall design that utilizes the *rain-screen principle*.

Pressure-treated lumber Lumber that has been impregnated with chemicals under pressure, for the purpose of retarding either decay or fire.

Prestressed concrete Concrete that has been pretensioned or posttensioned.

Prestressing Applying an initial compressive stress to a concrete structural member, either by pretensioning or posttensioning.

Pretensioning The compressing of the concrete in a structural member by pouring the concrete for the member around stretched high-strength steel strands, curing the concrete, and releasing the external tensioning force on the strands.

Priming Covering a surface with a coating that prepares it to accept another type of coating or sealant.

Protected membrane roof A membrane roof assembly in which the thermal insulation lies above the membrane.

Punty A metal rod used in working with hot glass.

Purlins Beams that span across the slope of a pitched roof to support the roof decking.

Putty A simple glazing compound used with small lights.

PVC *See* **polyvinyl chloride.**

Q

Quarry An excavation from which building stone is obtained; the act of taking stone from the ground.

Quarry bed A plane in a building stone that was horizontal before the stone was cut from the quarry.

Quarry sap Excess water found in rock at the time of its quarrying.

Quarry tile A large clay floor tile, usually unglazed.

Quartersawn Lumber sawn in such a way that the annual rings run roughly perpendicular to the face of each piece.

Quartersliced Veneer sliced in such a way that the annual rings run roughly perpendicular to the face of each veneer.

Quoin A corner reinforcing of cut stone or bricks in a masonry wall, usually done for decorative effect.

R

Rabbet A longitudinal groove cut at the edge of a member to receive another member; also called a *rebate*.

Raft A mat footing.

Rafter A framing member that runs up and down the slope of a pitched roof.

Rail A horizontal framing piece in a panel door; a handrail.

Rainscreen principle A theory by which wall cladding is made watertight by providing wind-pressurized air chambers behind joints to eliminate air pressure differentials between the outside and inside that might transport water through the joints.

Raised access flooring *See* **access flooring.**

Rake The sloping edge of a pitched roof.

Raker A sloping brace for supporting sheeting around an excavation.

Ratchet A mechanical device with sloping teeth that allow one piece to be advanced against another in small increments, but not to move in the reverse direction.

Ray A tubular cell that runs radially in a tree trunk.

RBM *See* **reinforced brick masonry.**

Reflective coated glass Glass onto which a thin layer of metal or metal oxide has been deposited to reflect light and/or heat.

Reinforced brick masonry (RBM) Brickwork into which steel bars have been embedded to impart tensile strength to the construction.

Reinforced concrete Concrete work into which steel bars have been embedded to impart tensile strength to the construction.

Relative humidity A decimal fraction

representing the ratio of the amount of water vapor contained in a mass of air to the amount of water it could contain under the existing conditions of temperature and pressure.

Removable glazing panel A framed sheet of glass that can be attached to a window sash to increase its thermal insulating properties.

Reshoring Inserting temporary supports under concrete beams and slabs after the formwork has been removed, to prevent overloading prior to full curing of the concrete.

Resilient flooring A manufactured sheet or tile flooring of asphalt, polyvinyl chloride, linoleum, rubber, or other resilient material.

Resin A natural or synthetic, solid or semisolid organic material of high molecular weight, used in the manufacture of paints, varnishes, and plastics.

Retaining wall A wall that resists horizontal soil pressures at an abrupt change in ground elevation.

Ridge board The board against which the tips of rafters are fastened.

Rigid frame Two columns and a beam or beams attached with moment connections; a moment-resisting building frame.

Rise A difference in elevation, such as the rise of a stair from one floor to the next, or a rise per foot of run in a sloping roof.

Riser A single vertical increment of a stair; the vertical face between two treads in a stair; a vertical run of plumbing, wiring, or ductwork.

Rivet A structural fastener on which a second head is formed after the fastener is in place.

Rock anchor A posttensioned rod or cable inserted into a rock formation for the purpose of tying it together.

Rock wool An insulating material manufactured by forming fibers from molten rock.

Roof deck The structural surface that supports the waterproof layer of a roof.

Roofer One who installs roof coverings.

Roof window An openable window designed to be installed in the sloping surface of a roof.

Rotary-sliced veneer A thin sheet of wood cut by rotating a log against a long, sharp knife blade in a lathe.

Rough arch An arch made from masonry

units that are rectangular rather than wedge-shaped.

Rough carpentry Framing carpentry, as distinguished from finish carpentry.

Roughing in The installation of mechanical, electrical, and plumbing components that will not be exposed to view in the finished building.

Rough opening The clear dimensions of the opening that must be provided in a wall frame to accept a given door or window unit.

Rowlock A brick laid on its long edge, with its end exposed in the face of the wall.

Rubble Unsquared stones.

Run Horizontal dimension in a stair or sloping roof.

Runner channel A steel member from which furring channels and lath are supported in a suspended plaster ceiling.

Running bond Brickwork consisting entirely of stretchers.

Runoff bar A small rectangular steel bar attached temporarily at the end of a prepared groove for the purpose of permitting the groove to be filled to its very end with weld metal.

R-value A numerical measure of resistance to the flow of heat.

S

Safing Fire-resistant material inserted into a space between a curtain wall and a spandrel beam or column, to retard the passage of fire through the space.

Sand cushion terrazzo Terrazzo with an underbed that is separated from the structural floor deck by a layer of sand.

Sand-mold brick, sand-struck brick A brick made in a mold that was wetted and then dusted with sand before the clay was placed in it.

Sandstone A sedimentary rock formed from sand.

Sandwich panel A panel consisting of two outer faces of wood, metal, or concrete bonded to a core of insulating foam.

Sapwood The living wood in the outer region of a tree trunk or branch.

Sash A frame that holds glass.

Scarf joint A glued end connection between two pieces of wood, using a sloping cut to create a large surface for the glue bond, to allow it to develop the full tensile strength of the wood it connects.

Scratch coat The first coat in a three-coat application of plaster.

Screed A strip of wood, metal, or plaster that establishes the level to which concrete or plaster will be placed.

Screw port A three-quarter circular profile in an aluminum extrusion, made to accept a screw driven parallel to the long axis of the extrusion.

Screw slot A serrated slot profile in an aluminum extrusion, made to accept screws driven at right angles to the long axis of the extrusion.

Scupper An opening through which water can drain from the edge of a flat roof.

Sealant gun A tool for injecting sealant into a joint.

Seated connection A connection in which a steel beam rests on top of a steel angle fastened to a column or girder.

Security glass A glazing sheet with multiple laminations of glass and plastic, designed to stop bullets.

Sedimentary rock Rock formed from materials deposited as sediments, such as sand or sea shells, which form sandstone and limestone, respectively.

Segregation Separation of the constituents of wet concrete caused by excessive handling or vibration.

Seismic Relating to earthquakes.

Seismic load A load on a structure caused by movement of the earth relative to the structure during an earthquake.

Self-drilling Drills its own hole.

Self-furring lath Metal lath with dimples that space the lath away from the sheathing behind to allow plaster to penetrate the lath.

Self-tapping Creates its own screw threads on the inside of a hole.

Self-weight The weight of a beam or slab.

Set To cure; to install; to recess the heads of nails; a punch for recessing the heads of nails.

Setting block A small block of synthetic rubber or lead used to support the weight of a sheet of glass at its lower edge.

Shading coefficient The ratio of total solar heat passing through a given sheet of glass to that passing through a sheet of clear double-strength glass.

Shaft An unbroken vertical passage through a multistory building, used for elevators, wiring, plumbing, ductwork, and so on.

Shaft wall   A wall surrounding a shaft.

Shake   A shingle split from a block of wood.

Shale   A rock formed from the consolidation of clay or silt.

Shear   A deformation in which planes of material slide with respect to one another.

Shear connection   A connection designed to resist only the tendency of one member to slide past the other, and not to resist any tendency of the members to rotate with respect to one another.

Shear panel   A wall, floor, or roof surface that acts as a deep beam to help stabilize a building against deformation by lateral forces.

Shear stud   A piece of steel welded to the top of a steel beam or girder so as to become embedded in the concrete fill over the beam and cause the beam and the concrete to act as a single structural unit.

Sheathing   The rough covering applied to the outside of the roof, wall, or floor framing of a light frame structure.

Shed   A building or dormer with a single-pitched roof.

Sheeting   A stiff material used to retain the soil around an excavation; a material such as polyethylene in the form of very thin, flexible sheets.

Sheet metal   Flat rolled metal less than $\frac{1}{4}$ inch (6.35 mm) in thickness.

Shelf angle   A steel angle attached to the spandrel of a building to support a masonry facing.

Shim   A thin piece of material placed between two components of a building to adjust their relative positions as they are assembled; to insert shims.

Shingle   A small unit of water-resistant material nailed in overlapping fashion with many other such units to render a sloping roof watertight; to apply shingles.

Shiplap   A board with edges rabbeted so as to overlap flush from one board to the next.

Shop drawings   Detailed plans prepared by a fabricator to guide the shop production of such building components as cut stonework, steel or precast concrete framing, curtain wall panels, and cabinetwork.

Shoring   Temporary vertical or sloping supports of steel or timber.

Shotcrete   A low-slump concrete mixture deposited by being blown from a nozzle at high speed with a stream of compressed air.

Shrinkage-temperature steel   Reinforcing bars laid at right angles to the principal bars in a one-way slab, for the purpose of preventing excessive cracking caused by drying shrinkage or temperature stresses in the concrete.

Sidelight   A tall, narrow window alongside a door.

Siding   The exterior wall finish material applied to a light frame wood structure.

Siding nail   A nail with a small head, used to fasten siding to a building.

Silicone   A polymer used for high-range sealants, roof membranes, and masonry water repellants.

Sill   The strip of wood that lies immediately on top of a concrete or masonry foundation in wood frame construction; the horizontal bottom portion of a window or door; the exterior surface, usually sloping to shed water, below the bottom of a window or door.

Sill sealer   A resilient, fibrous material placed between a foundation and a sill to reduce air infiltration between the outdoors and indoors.

Single-hung window   A window with two overlapping sashes, the lower of which can slide vertically in tracks, and the upper of which is fixed.

Single-strength glass   Glass approximately $\frac{3}{32}$ inch (2.5 mm) thick.

Single tee   A precast slab element whose profile resembles the letter *T*.

Sitecast   Concrete poured and cured in its final position in a building.

Skylight   A fixed window installed in a roof.

Slab on grade   A concrete surface lying upon, and supported directly by, the ground beneath.

Slag   The mineral waste that rises to the top of molten iron or steel, or to the top of a weld.

Slaked lime   Calcium hydroxide.

Slate   A metamorphic form of clay, easily split into thin sheets.

Sliding window   A window with one fixed sash and another that moves horizontally in tracks.

Slip-critical connection   A steel connection in which high-strength bolts clamp the members together with sufficient force that the load is transferred between them by friction.

Slip forming   Building multistory sitecast concrete walls with forms that rise up the wall as construction progresses.

Slump test   A test in which wet concrete or plaster is placed in a cone-shaped metal mold of specified dimensions and allowed to slump under its own weight after the cone is removed. The vertical distance between the height of the mold and the height of the slumped mixture is an index of its working consistency.

Slurry   A watery mixture of insoluble materials.

Smoke developed rating   An index of the toxic fumes generated by a material as it burns.

Smoke shelf   The horizontal area behind the damper of a fireplace.

Soffit   The undersurface of a horizontal element of a building, especially the underside of a stair or a roof overhang.

Soffit vent   An opening under the eave of a roof used to allow air to flow into the attic or the space below the roof sheathing.

Soft mud process   Making bricks by pressing wet clay into molds.

Soldier   A brick laid on its end, with its narrow face toward the outside of the wall.

Sole plate   The horizontal piece of dimension lumber at the bottom of the studs in a wall in a light frame building.

Solid-core door   A flush door with no internal cavities.

Solid slab   A concrete slab, without ribs or voids, that spans between beams or bearing walls.

Solid tape sealant   *See* **preformed solid tape sealant.**

Solvent   A liquid that dissolves another material.

Sound Transmission Class (STC)   An index of the resistance of a partition to the passage of sound.

Space frame, space truss   A truss that spans with two-way action.

Span   The distance between supports for a beam, girder, truss, vault, arch, or other horizontal structural device; to carry a load between supports.

Spandrel   The wall area between the head of a window on one story and the sill of a window on the floor above; the area of a wall between adjacent arches.

Spandrel beam   A beam that runs along the outside edge of a floor or roof.

Spandrel glass   Opaque glass manufactured especially for use in spandrel panels.

Spandrel panel   A curtain wall panel used in a spandrel.

Span rating   The number stamped on a sheet of plywood or other wood building panel to indicate how far in inches it may span between supports.

Specifications   The written instructions from an architect or engineer concerning the quality of materials and workmanship required for a building.

Spirit level   A tool in which a bubble in an upwardly curving, cylindrical glass vial indicates whether a building element is level or not level, plumb or not plumb.

Splash block   A small precast block of concrete or plastic used to divert water at the bottom of a downspout.

Spline   A thin strip inserted into grooves in two mating pieces of material to hold them in alignment; a ridge of material intended to lock to a mating groove.

Split jamb   A door frame fabricated in two interlocking halves, to be installed from the opposite sides of an opening.

Staggered truss system   A steel framing system in which story-high trusses, staggered one-half bay from one story to the next, support floor decks with both their top and bottom chords.

Standing seam   A sheet metal roofing seam that projects at right angles to the plane of the roof.

Stay   A cable used to stabilize a structure against lateral deflection or wind uplift.

STC   *See* **Sound Transmission Class.**

Steam curing   Aiding and accelerating the setting reaction of concrete by the application of steam.

Steel   Iron with a controlled amount of carbon, generally less than 1.7 percent.

Stick system   A metal curtain wall system that is largely assembled in place.

Stiffener plate   A steel plate attached to a structural member to support it against heavy localized loadings or stresses.

Stiff mud process   A method of molding bricks in which a column of damp clay is extruded from a rectangular die and cut into bricks by fine wires.

Stile   A vertical framing member in a panel door.

Stirrup   A vertical loop of steel bar used to reinforce a concrete beam against diagonal tension forces.

Stirrup-tie   A stirrup that forms a complete loop, as differentiated from a U-stirrup, which has an open top.

Stool   The interior horizontal plane at the sill of a window.

Storm window   A sash added to the outside of a window in winter to increase its thermal resistance and decrease air infiltration.

Story pole   A strip of wood marked with the exact course heights of masonry for a particular building, used to make sure that all the leads are identical in height and coursing.

Straightedge   To strike off the surface of a concrete slab using screeds and a straight piece of lumber or metal.

Strain   Deformation under stress.

Stress   Force per unit area.

Stressed-skin panel   A panel consisting of two face sheets of wood, metal, or concrete bonded to perpendicular spacer strips.

Stretcher   A brick or masonry unit laid in its most usual position, with the broadest surface of the unit horizontal and the length of the unit parallel to the surface of the wall.

Striated   Textured with parallel scratches or grooves.

Stringer   The sloping wood or steel member supporting the treads of a stair.

Strip flooring   Wood finish flooring in the form of long, narrow tongue-and-groove boards.

Stripping   Removing formwork from concrete; sealing around a roof flashing with layers of felt and bitumen.

Structural bond   The interlocking pattern of masonry units used to tie two or more wythes together in a wall.

Structural glazed tile   A hollow clay masonry unit with glazed faces.

Structural mill   The portion of a steel mill that rolls structural shapes.

Structural silicone flush glazing   Glass secured to the face of a building with strong, highly adhesive silicone sealant so as to eliminate the need for any metal to appear outside the inner face of the glass.

Structural terra-cotta   Molded components, often highly ornamental, made of fired clay, designed to be used in the facades of buildings.

Structural tubing   Hollow steel cylindrical or rectangular shapes made to be used as structural members.

Stucco   Portland cement plaster used as an exterior cladding or siding material.

Stud   One of a series of small, closely spaced wall framing members; a heavy steel pin.

Subcontractor   A contractor who specializes in one area of construction activity, and who usually works under a general contractor.

Subfloor   The loadbearing surface beneath a finish floor.

Subpurlin   A very small roof framing member that spans between joists or purlins.

Substructure   The occupied, below-ground portion of a building.

Superplasticizer   A concrete admixture that makes wet concrete extremely fluid without additional water.

Superstructure   The above-ground portion of a building.

Supply pipe   A pipe that brings clean water to a plumbing fixture.

Supporting stud   A wall framing member that extends from the sole plate to the underside of a header and that supports the header.

Surface bonding   Bonding a concrete masonry wall together by apply a layer of glass-fiber-reinforced stucco to both its faces.

Suspended ceiling   A finish ceiling that is hung on wires from the structure above.

Suspended glazing   Large sheets of glass hung from clamps at their top edges so as to eliminate the need for metal mullions.

T

Tackless strip   A wood strip with projecting points used to fasten a carpet around the edge of a room.

Tagline   A rope attached to a building component to help guide it as it is lifted by a crane or derrick.

Tapered edge The longitudinal edge of a sheet of gypsum board, which is recessed to allow room for reinforcing tape and joint compound.

Tee A metal or precast concrete member with a cross section resembling the letter *T*.

Tempered glass Glass that has been heat-treated to increase its toughness and its resistance to breakage.

Tendon A steel strand used for prestressing a concrete member.

Tensile strength The ability of a structural material to withstand stretching forces.

Tensile stress A stress caused by stretching of a material.

Tension A stretching force; to stretch.

Tension control bolt A bolt driven by means of a splined end that breaks off when the bolt has reached the required tension.

Terne An alloy of lead and tin, used to coat sheets of carbon steel or stainless steel for use as metal roofing sheet.

Terrazzo A finish floor material consisting of concrete with an aggregate of marble chips selected for size and color, which is ground and polished smooth after curing.

Thatch A thick roof covering of reeds, straw, grasses, or leaves.

Thermal break A section of material with a low thermal conductivity, installed between metal components to retard the passage of heat through a wall or window assembly.

Thermal bridge A component of higher thermal conductivity that conducts heat more rapidly through an insulated building assembly, such as a steel stud in an insulated stud wall.

Thermal conductivity The rate at which a material conducts heat.

Thermal insulation A material that greatly retards the passage of heat.

Thermal resistance The resistance of a material or assembly to the conduction of heat.

Thermoplastic Having the property of softening when heated and rehardening when cooled.

Thermosetting Not having the property of softening when heated.

Thrust A lateral or inclined force resulting from the structural action of an arch, vault, dome, suspension structure, or rigid frame.

Thrust block A wooden block running perpendicular to the stringers at the bottom of a stair, whose function is to hold the stringers in place.

Tie A device for holding two parts of a construction together; a structural device that acts in tension.

Tieback A tie, one end of which is anchored in the ground, with the other end used to support sheeting around an excavation.

Tie beam A reinforced concrete beam cast as part of a masonry wall, whose primary purpose is to hold the wall together, especially against seismic loads, or cast between a number of isolated foundation elements to maintain their relative positions.

Tier The portion of a multistory steel building frame supported by one set of fabricated column pieces, commonly two stories in height.

Tie rod A steel rod that acts in tension.

Tile A fired clay product that is thin in cross section as compared to a brick, either a thin, flat element (ceramic tile or quarry tile), a thin, curved element (roofing tile), or a hollow element with thin walls (flue tile, tile pipe, structural clay tile); also a thin, flat element of another material, such as an acoustical ceiling unit or a resilient floor unit.

Tilt-up construction A method of constructing concrete walls in which panels are cast and cured flat on the floor slab, then tilted up into their final positions.

Timber Standing trees; a large piece of dimension lumber.

Tinted glass Glass that is colored with pigments, dyes, or other admixtures.

Toe nailing Fastening with nails driven at an angle.

Tongue-and-groove An interlocking edge detail for joining planks or panels.

Tooling The finishing of a mortar joint by pressing and compacting it to create a particular profile.

Topping A thin layer of concrete cast over the top of a floor deck.

Topping-out Placing the last member in a building frame.

Top plate The horizontal member at the top of a stud wall.

Topside vent A water-protected opening through a roof membrane to relieve pressure from water vapor that may accumulate beneath the membrane.

Torque Twisting action; moment.

Torsional stress Stress resulting from the twisting of a structural member.

Tracheids The longitudinal cells in a softwood.

Traffic deck A walking surface placed on top of a roof membrane.

Transit-mixed concrete Concrete mixed in a drum on the back of a truck as it is transported to the building site.

Travertine A richly patterned, marble-like form of limestone.

Tread One of the horizontal planes that make up a stair.

Tremie A large funnel with a tube attached, used to deposit concrete in deep forms or beneath water or slurry.

Trim accessories Casing beads, corner beads, expansion joints, and other devices used to finish edges and corners of a plaster wall or ceiling.

Trimmer A beam that supports a header around an opening in a floor or roof frame.

Trowel A thin, flat steel tool, either pointed or rectangular, provided with a handle and held in the hand, used to manipulate mortar, plaster, or concrete. Also, a machine whose rotating steel blades are used to finish concrete slabs; to use a trowel.

Truss A triangular arrangement of structural members that reduces nonaxial forces on the truss to a set of axial forces in the members. *See also* **Vierendeel truss.**

Tuck pointing The process of removing deteriorated mortar from the zone near the surface of a brick wall, and inserting fresh mortar.

Tunnel kiln A kiln through which clay products are passed continuously on railroad cars.

Turn-of-nut method A method of achieving the correct tightness in a high-strength bolt by first tightening the nut snugly, then turning it a specified additional fraction of a turn.

Two-way action Bending of a slab or deck in which bending stresses are approximately equal in the two principal directions of the structure.

Two-way concrete joist system A reinforced concrete framing system in which columns directly support an orthogonal grid of intersecting joists.

Two-way flat plate A reinforced concrete framing system in which columns directly support a two-way slab that is planar on both its surfaces.

Two-way flat slab A reinforced concrete framing system in which columns with mushroom capitals and/or drop panels directly support a two-way slab that is planar on both its surfaces.

Type X gypsum board A fiber-reinforced gypsum board used where greater fire resistance is required.

U

Unbonded construction Posttensioned concrete construction in which the tendons are not grouted to the surrounding concrete.

Uncoursed Laid without continuous horizontal joints.

Undercarpet wiring system Flat, insulated electrical conductors that are run under carpeting, and their associated outlet boxes and fixtures.

Undercourse A course of shingles laid beneath an exposed course of shingles at the lower edge of a wall or roof, in order to provide a waterproof layer behind the joints in the exposed course.

Underfire The floor of a fireplace.

Underlayment A panel laid over a subfloor to create a smooth, stiff surface for application of finish flooring.

Underpinning The process of placing new foundations beneath an existing structure.

Unfinished bolt An ordinary carbon steel bolt.

Unit-and-mullion system A curtain wall system consisting of prefabricated panel units secured with site-applied mullions.

Unit system A curtain wall system consisting entirely of prefabricated panel units.

Unreinforced Constructed without steel reinforcing bars or welded wire fabric.

Upside-down roof A membrane roof assembly in which the thermal insulation lies above the membrane.

Urea formaldehyde A water-based foam used as thermal insulation.

U-stirrup An open-top, U-shaped loop of steel bar used as reinforcing against diagonal tension in a beam.

V

Valley A trough formed by the intersection of two roof slopes.

Valley rafter A diagonal rafter that supports a valley.

Vapor barrier *See vapor retarder.*

Vapor retarder A layer of material intended to obstruct the passage of water vapor through a building assembly. Also called, less accurately, *vapor barrier*.

Vault An arched surface.

Vee joint A joint whose profile resembles the letter *V*.

Veneer A thin layer, sheet, or facing.

Veneer plaster A wall finish system in which a thin layer of plaster is applied over a special gypsum board base.

Veneer plaster base The special gypsum board over which veneer plaster is applied.

Vent spacer A device used to maintain a free air passage above the thermal insulation in an attic or roof.

Vermiculite Expanded mica, used as an insulating fill or a lightweight aggregate.

Vertical bar An upright reinforcing bar in a concrete column.

Vertical grain lumber Dimension lumber sawed in such a way that the annual rings run more or less perpendicular to the faces of each piece.

Vierendeel truss A truss with rectangular panels and rigid joints. The members of a Vierendeel truss are subjected to strong nonaxial forces.

Vitrification The process of transforming a material into a glassy substance by means of heat.

W

Waferboard A building panel made by bonding together large, flat flakes of wood.

Waffle slab A two-way concrete joist system.

Wainscoting A wall facing, usually of wood, cut stone, or ceramic tile, that is carried only partway up a wall.

Waler A horizontal beam used to support sheeting or concrete formwork.

Wane An irregular rounding of a long edge of a piece of dimension lumber caused by cutting the lumber from too near the outside surface of the log.

Washer A steel disk with a hole in the middle, used to spread the load from a bolt, screw, or nail across a wider area of material.

Water-cement ratio A numerical index of the relative proportions of water and cement in a concrete mixture.

Water-resistant gypsum board A gypsum board designed for use in locations where it may be exposed to occasional dampness.

Water-smoking The process of driving off the last water from clay products before they are fired.

Waterstop A synthetic rubber strip used to seal joints in concrete foundation walls.

Water-struck brick A brick made in a mold that was wetted before the clay was placed in it.

Water table The level at which the pressure of water in the soil is equal to atmospheric pressure; effectively, the level to which ground water will fill an excavation.

Water vapor Water in its gaseous phase.

Wattle and daub Mud plaster (daub) applied to a primitive lath of woven twigs or reeds (wattle).

Weathered joint A mortar joint finished in a sloping, planar profile that tends to shed water to the outside of the wall.

Weathering steel A steel alloy that forms a tenacious, self-protecting rust layer when exposed to the atmosphere.

Weatherstrip A ribbon of resilient material used to reduce air infiltration through the crack around a sash or door.

Weep hole A small opening whose purpose is to permit drainage of water that accumulates inside a building component.

Weld A joint between two pieces of metal formed by fusing the pieces together, usually with the aid of additional metal melted from a rod or electrode.

Welding The process of making a weld.

Weld plate A steel plate anchored into the surface of concrete, to which another steel element can be welded.

"Wet" systems Construction systems that utilize considerable quantities of water on the construction site, such as masonry, plaster, and terrazzo.

Wide-flange section Any of a wide range

of steel sections rolled in the shape of a letter *I* or *H*.

Wind brace A diagonal structural member whose function is to stabilize a frame against lateral forces.

Winder A stair tread that is wider at one end than at the other.

Wind load A load on a building caused by wind pressure and/or suction.

Wind uplift Upward forces on a structure caused by negative aerodynamic pressures that result from certain wind conditions.

Wired glass Glass in which a wire mesh was embedded during manufacture.

Wracking Forcing out of plumb.

Wrought iron A form of iron that is soft, tough, and fibrous in structure, containing about 0.1 percent carbon and 1 to 2 percent slag.

Wythe (Rhymes with "scythe" and "tithe.") A vertical layer of masonry one masonry unit thick.

Y

Yield strength The stress at which a material ceases to deform in a fully elastic manner.

Z

Z-brace door A door made of vertical planks held together and braced on the back by three pieces of wood whose configuration resembles the letter Z.

Zero-slump concrete A concrete mixed with so little water that it has a slump of zero.

Zoning ordinance A law that specifies in detail how land may be used in a municipality.

INDEX

—

LUMBER SIZES

25.41mm = 1 inch

NOMINAL	ACTUAL	METRIC
2 x 4	1½" x 3½	38 x 89
2 x 6	1½" x 5½"	38 x 140
2 x 8	1½" x 7¼	38 x 184
2 x 10	1½" x 9¼"	38 x 235
2 x 12	1½" x 11¼"	38 x 286

25.41mm = 1 inch